Product Design and Testing for Automotive Engineering

Volume II

Product Design and Testing for Automotive Engineering

Volume II

YOUNG J. CHIANG

SAE INTERNATIONAL®

Warrendale, Pennsylvania, USA

400 Commonwealth Drive
Warrendale, PA 15096-0001 USA
E-mail: CustomerService@sae.org
Phone: 877-606-7323 (inside USA and Canada)
724-776-4970 (outside USA)
Fax: 724-776-0790

Library of Congress Catalog Number 2024938706
http://dx.doi.org/10.4271/9781468607697

ISBN-Print 978-1-4686-0770-3
ISBN-PDF 978-1-4686-0769-7
ISBN-epub 978-1-4686-0772-7

To purchase bulk quantities, please contact: SAE Customer Service

E-mail: CustomerService@sae.org
Phone: 877-606-7323 (inside USA and Canada)
724-776-4970 (outside USA)
Fax: 724-776-0790

Visit the SAE International Bookstore at books.sae.org

Publisher
Sherry Dickinson Nigam

Product Manager
Amanda Zeidan

Production and Manufacturing Associate
Michelle Silberman

From the Publisher

I am thrilled to introduce our groundbreaking series *Design of Experiments for Product Reliability Growth*, which is divided into four distinct volumes. *Product Design and Testing for Automotive Engineering* marks the second installment of our exploration of this critical subject.

We are publishing each volume separately to ensure in-depth coverage of each topic, allowing readers to focus on specific aspects of the subject matter for a deeper understanding. After careful consideration of how our readers will be using the content, we opted to continue to the page numbering system throughout the volumes for continuity.

Volume I	Fundamentals of Design of Experiments for Automotive Engineering
Volume II	Product Design and Testing for Automotive Engineering
Volume III	Manufacturing Reliability Growth for Automotive Engineering
Volume IV	Business Reliability Growth for Automotive Engineering

Thank you for joining us on this journey. We look forward to your continued support and engagement with the upcoming volumes in the series.

Sincerely,

Sherry Nigam

Sherry Nigam
Publisher, SAE Books

Contents

Preface

This is a book about product design for reliability and sustainability via the design of experiments (DOEs). It is an application-oriented text with an emphasis on deploying the DOEs for product reliability growth throughout an enterprise as a "statistical thinking" tool, instead of focusing on a specific statistical approach to designing experiments. Emphasis is laid on how to grow product reliability using the DOEs. Product reliability growth via process improvements and system optimizations is demonstrated using real-world examples, mostly extracted from common practices in automotive engineering. Theoretical concepts that are important to the understanding of the DOEs are rephrased in practical exercises.

Product development efforts are concurrently focused on a product configuration at the lowest feasible cost by means of a thorough understanding of its functionality, reliability, and sustainability, while today's central theme for proactive improvement calls for building high reliability into products. It is imperative to avoid field failures during a product's estimated lifetime in service, which consequently leads to delighted customers and elimination of the high expense of repairing failed goods and fixing underlying causes. Henry Ford said, "Failure is simply the opportunity to begin again, this time more intelligently." Product reliability growth is in a perpetual motion, but intellectual vitality rolls with us in light of the DOEs.

Much of the challenge and opportunity in engineering design comes from stochastic material properties and manufacturing processes in nature. "A large safety factor does not necessarily translate into a reliable product. Instead, it often leads to an overdesigned product with reliability problems," as addressed in the article "Failure Analysis Beats Murphy's Law," Mechanical Engineering (September 1993). On problem-solving, a thinker is encouraged to poise oneself on the understanding of physics of failure that is related to exploiting distinguished product properties and characterizing specific identifiable variables and attributes. Drawing on his extensive product development experience, the author presents a comprehensive process for ensuring product performance and reliability over the entire life cycle.

We hope this will be a comprehensive textbook of applied DOEs for product reliability growth and a reference manual for practicing engineers in various functional areas of product-oriented business, including research, design, testing, manufacturing, operation, supplier quality assurance, marketing, and warranty analysis. While lessons hereby learned from the book can be appropriated for product development in the real world, comments and suggestions are motivated for further improvements.

Young J. Chiang, PhD
会而必议 议而必决 决而必行 行而必果

Acronyms

2_R^{k-p} - 2-level fractional factorial design having k variables with resolution R

3_R^{k-p} - 3-level fractional factorial design having k variables with resolution R

360° Evaluation - Evaluating performance with inputs from supervisors, peers, and other employees

3 Spheres - Quality management, assurance, and control

3Ts - Task, treatment, and tangibles in service design

4P - People, place, procedure, and policies

4S - Surroundings, suppliers, systems, and skills

5M - Manpower, machine, method, material, and measurement

5R - Responsiveness, reliability, rhythm, responsibility, and relevance

5S - Select/sort, straighten/set in place, shine, standardize, and sustain

5w2h - Who, what, when, where, why, how, and how much

6M - Manpower, machine, method, material, measurement, and mother nature

6S - Select, set in place, shine, standardize, sustain, and safe

7 Factors - Team, provider, patience, task, work environment, equipment and technology, and organization and management

7P - Proper prior planning prevents pitifully poor performance

7 Wastes - Over-production, transport, waiting, inventory, defects, over-processing, and unnecessary movement

8D - 8 disciplines for structured step-by-step problem-solving

80/20 Rule - 80% of the problems resulting from 20% of the causes

AAI - Appearance Approval Inspection

AALA - American Association for Laboratory Accreditation

AAR - Appearance Approval Report

ADV - Analysis/Development/Validation (GM Specific)

AEB - Automatic Emergency Braking (AEB)

Ah (mAh) - Amp-Hours (10^{-3} Amp-Hours)

AI - Artificial Intelligence

AIAG - Automotive Industry Action Group

AIQ - Average Incoming Quality; Average quality level before inspection

A2LA - American Association of Laboratory Accreditation

ALT - Accelerated Life Testing

ANCOVA - Analysis of Covariance

ANN - Artificial Neural Network

ANOM - Analysis of Means

ANOVA - Analysis of Variance

AOQ - Average Outgoing Quality (Average quality level leaving the inspection point after rejection and acceptance of a number of lots)

AOQL - Average Outgoing Quality Limit (Maximum value of the AOQ)

AOZ - Accept on Zero

AP - Advanced Product

APQP - Advanced Product Quality Planning

AQC - Attribute Quality Characteristic

AQL - Acceptable Quality Level

AQP - Advanced Quality Planning

ARIMA - Autoregressive Integrated Moving Average

ARL - Averaged Run Length

ARMA - Autoregressive Moving Average

ASIC - Application-Specific Integrated Circuit

ASIL - Automotive Safety Integrity Level

ASME - American Society of Mechanical Engineers

ASQ - American Society of Quality

ASTM - American Society for Testing and Materials

AVL - Approved Vendor List

B7 (Q7) - Basic 7 QC Tools, i.e., Fishbone Diagram, Histogram, Pareto Analysis, Flowchart, Scatter Plot, Run Chart, and Control Chart

BCWP - Budgeted Cost of Work Performed

BCWS - Budgeted Cost of Work Scheduled

BDD - Binary Decision Diagram

BEV - Battery Electric Vehicles

BIST - Built-in Self-Test

BIT - Built-in Test

BJT - Bipolar Junction Transistors

BLDC - Brushless Direct Current

BN - Bayesian Network

BOM - Bill of Material

BPI - Business Performance Improvement

CA - Corrective Actions

CAAM - China Association of Automobile Manufacturers

CAD - Computer-aided Design

CAE - Computer-aided Engineering

CAM - Computer-aided Manufacturing

CAN - Controller Area Network

CAPA - Corrective and Preventive Action
CBP - Customer Benefits Package
CC - Critical Characteristic
CCD - Central Composite Design
c-Chart - Count Chart (Attribute Quality Control Chart for Number of Defects)
CCPM - Critical Chain Project Management
CDF - Cumulative Distribution Function
CF - Carbon Fiber
CFRP - Carbon Fiber Reinforced Polymer or Carbon Fiber Reinforced Plastics
CFT - Cross Functional Team
CI - Confidence Interval
CIP - Continuous Improvement Process
CLT - Central Limit Theorem
CMM - Coordinate Measuring Machine
CMM - Capability Maturity Model
CONC - Cost of Non-conformance
COP - Customer-Oriented Process
COQ - Cost of Quality
COR - Cost of Reliability
CPCV - Constant-Power-Constant-Voltage
CPM - Critical Path Method, either PERT or Gantt Chart
CPP - Critical Process Parameter
CPPD - Collaborative Product Process Design
CPV - Cost per Vehicle
CQI - Continuous Quality Improvement
CQM - ASQ Certified Quality Manager
CRD - Completely Random Design
CRM - Customer Relationship Management
CRS - Constant Returns to Scale
CS1 - Control Shipping 1
CS2 - Control Shipping 2
CTQ - Critical-to-Quality
CUSUM - Cumulative Sum Charting
CV - Coefficient of Variation
DCO - Document Change Order
DCP - Dimensional Control Plan
DCP - Dynamic Control Plan (Ford)
DDDM - Degradation-Data-Driven Method
DEA - Data Envelopment Analysis

DETMAX - Determinant Maximizing Algorithm (for D-Optimality)
DFA - Design for Assembly
DFM - Design for Manufacture
DFMA - Design for Manufacture and Assembly
DOC - Depth of Charge
DOD - Depth of Discharge
DOE - Design of Experiments
DOE-E - Design of Experiments Based on Exponential Distribution
DOE-F - Design of Experiments Based on F-Distribution
DOE-ln - Design of Experiments Based on Lognormal Distribution
DOE+ML - Design of Experiments and Machine Learning, Joint Application of
DOE-t - Design of Experiments Based on t-Distribution
DOE-W - Design of Experiments Based on Weibull Distribution
DOF or DF - Degree of Freedom
DFMEA - Design Failure Mode and Effects Analysis
DFT - Design for Testability
DLM - Direct Linearization Method
DMADV - Define, Measure, Analyze, Design, Verification
DMAIC - Define, Measure, Analyze, Improve, and Control (Kaizen—5 Steps)
DMU - Design Making Unit
DRACAS - Data Reporting Analysis and Corrective Action System
DRF - Datum Reference Frame
DSC - Dynamic Stability Control
DSD - Definitive Screening Designs
DVP&R - Design Verification Plan and Report
ECFA - Event and Causal Factor Analysis
ECN - Engineering Change Notice
ECR - Engineering Change Request
EMI - Electromagnetic Interference
EMOO - Evolutionary Multi-Objective Optimization
EMS - Expected Means Squares
ERP - Enterprise Resource Planning
ES - Engineering Specification
ES - Earned Schedule
ESC - Electronic Stability Control
ESD - Electrochemical Storage Devices
ESS - Environmental Stress Screening
EV - Electric Vehicle
EVM - Earned Value Management

EVOP - Evolutionary Operation Process
EWMA - Exponentially Weighted Moving Average
EWMV - Exponentially Weighted Moving Variance
EWS - Early Warning System
Exp, exp, e - Exponential Function
FAI - First Article Inspection or Final Acceptance Inspection
FAST - Function analysis System Technique Diagram
FDI - Fault Detection and Isolation
FDM - Fused Deposition Modeling
FEA - Finite Element Analysis
FEM - Finite Element Methods
FFA - Functional Failure Analysis
FFF - Fused Filament Fabrication
FHA - Functional Hazard Analysis
FIFO - First In First Out
FIM - Fisher Information Matrix
FIR - Fast Initial Response
FMEA - Failure Mode and Effects Analysis
FMECA - Failure Modes, Effects, and Criticality Analysis
FMEDA - Failure Modes, Effects, and Diagnostic Analysis
FMVT - Failure Mode Verification Testing
FOZ - Feature-of-Size
FPGA - Field Programmable Gate Array
FQPR - Field Quality Problem Report
FRACAS - Failure Reporting, Analysis, and Corrective Action System
FRB - Failure Review Board
FSLT - Full System Life Test
FSR - Functional Safety Requirement
FTA - Fault Tree Analysis
FTC - First Time Capability
GA - Genetic Algorithm
Gage R&R - Gage Repeatability and Reproducibility
GaN - Gallium Nitride
GD3 - Good Design, Good Discussion, Good Dissection
GD&T - Geometric Dimensioning and Tolerancing
GERT - Graphic Evaluation and Review Technique
GF - Glass Fiber
GFRP - Glass Fiber Reinforced Polymer or Glass Fiber Reinforced Plastics
GK - General Knowledge

GNSS - Global Navigation Satellite System (e.g., GPS + IMU)
GPa - Gega Pascal
GPC - Gage Performance Curve
GPS - Global Positioning System
GPU - Graphics Processing Unit
GR - Gage Repeatability
GR&R (GRR) - Gage Repeatability and Reproducibility
GTD - Getting Things Done
GTO - Gate Turn-off Thyristor
GVDP - Global Vehicle Development Process (GM Practice)
H_0 - Null hypothesis
H_1 - Alternative hypothesis
HACCP - Hazard Analysis and Critical Control Points
HALT - Highly Accelerated Life Test
HARA - Hazard Analysis and Risk Assessment
HASS - Highly Accelerated Stress Screening
HAZOP - Hazard and Operability
HCF - High Cycle Fatigue
HEV - Hybrid Electric Vehicle
HFCV - Hydrogen Fuel Cell Vehicle
HIC - Head injury criterion
IATF - International Automotive Task Force
IC - Integrated Circuit
IDOV - Identify, Design, Optimize, Validate
IEEE - Institute of Electrical and Electronics Engineers
IGBT - Insulated-Gate Bipolar Transistor
IGCT - Integrated Gate Commutated Thyristor
IID - Independent and Identically Distributed
IM - Induction Motor
IMDS - International Material Data System
IMP - Integrated Master Program
IMU - Inertial Measurement Unit
IOT or IoT - Internet of Things
IPMA - International Project Management Association
ISIR - Initial Sample Inspection Report
ISO - International Organization for Standardization
IsoPlot - 3D Iso-surface Plot
ITPV - Incidents per Thousand Vehicles
JIT - Just in Time

KAIZEN - Change for the Better
KBF - Key Business Factors
KCC - Key Control Characteristic
KISS - Keep it Simple, Stupid
KLT - Key Life Tests
KPC - Key Product Characteristic
KPI - Key Product (Process) Indicator or Key Performance Index
LCF - Low Cycle Fatigue
LCL - Lower Control Limit
LFP - LiFePO4 Battery
Li-Air - Lithium–Air (Battery)
Lidars - Laser Light Detection and Ranging Systems
Li-ion - Lithium–Ion (Battery)
LIN - Local Internet Network
LKA - Lane Keeping Assistance
LMC - Least Material Condition
LN, Ln, or ln - Lognormal
LRP - Logistics Requirement Plan
LRU - Line Replaceable Unit
LSL - Lower Specified Limit
LSM - Linear Scheduling Method
LTBF - Longest Time between Failures
LTPD - Lot Tolerance Percentage Defective Acceptable in Production Lots
MBD - Management by Decree
MCO - Manufacturing Change Order
MDT - Mean Down Time
MEOST - Multiple Environment Over Stress Tests
MILTFP-41 - Make it Like the Finished Print- For Once
ML - Machine Learning
MLE - Maximum-Likelihood Estimation
MMC - Maximum Material Condition
MME - Method of Moments Estimator
MOPSO - Multi-Objective Particle Swarm Optimization
MOSFET - Metal-Oxide Semiconductor Field-Effect Transistor
MORT - Management Oversight and Risk Tree
MPa - Mega Pascal
MRR - Median Rank Regression
MRS - Marginal Rate of Substitution
MSE - Mean Squared Error

M&TE - Measurement and Test Equipment

MTBF - Mean Time between Failures

MTBF-SR - Mean Time between Failures with Scheduled Replacements)

MTBM - Mean Time between Maintenance

MTBOF - Mean Time between Operational Failures

MTBR - Mean Time between Replacements

MTS - Make to Stock

MTTF - Mean Time to Failure

MTTR - Mean Time to Repair

MVBS - Manufacturing Validation Build Saleable (GM-specific)

N_2 - An Interface Matrix, Representing Interfaces between System Elements

N7 - New 7 QC Tools (Affinity Diagram/Brainstorming, Relation Diagram, Tree Diagram, Matrix Diagram/QFD, Arrow Diagram/PERT, Process Decision Chart, and Matrix Data Analysis/Principal Component Analysis)

NCP - Non-Conforming Products

NCR - Non-Conformance Report

NDT - Non-Destructive Test

NHPP - Non-Homogeneous Poisson Process

NHTSA - National Highway Traffic Safety Administration

NIST - National Institute of Standards and Technology, Dept. of Commerce, USA

NPI - New Product Introduction

np-chart - Attribute Quality Control Chart for Number of Defective Products

NTF - No Trouble Found

NTSA - National Transportation Safety Administration, Dept. of Transportation, USA

OACD - Orthogonal-Array Composite Designs

OC - Operating Characteristic

OC Curve - Operating Characteristic Curve

OCV - Open Circuit Voltage

OEE - Overall Equipment Effectiveness

OEM - Original Equipment Manufacturer

O&SHA - Operational and Support Hazard Analysis

OSHA - Occupational Safety and Health Administration

OVAT - One Variable at a Time

PAA - Product Application Agreement

PAT - Process Analytic Technology

PBS - Product Breakdown Structure

PCA - Permanent Corrective Actions

PCB - Printed Circuit Board

PCBA - Printed Circuit Board Assembly

P-chart - Attribute Quality Control Chart for Proportion of Nonconforming Units

PCI - Process Capability Index
PCP - Production Control Plan
PDAS - Plan–Do–Study–Action
PDCA - Plan–Do–Check–Act cycle
PDF - Probability Density Function
P-Diagram - Parameter Diagram
PDM - Product Data Management
PDPC - Process Decision Program Chart
PERT - Program Evaluation and Review Technique
PEV - Plug-in Electric Vehicles
PFD - Process Flow Diagram or Probability of Failure on Demand
PFMEA - Process Failure Mode and Effects Analysis
PHEV - Plug-in-Hybrid Electric Vehicle
PHR - Part Handling Review
PL - Product Line
PLM - Product Life Management
PMA - Program Management Administration
PMBoK - Project Management Body of Knowledge
PMHF - Probabilistic Metric for Hardware Failure
PMI - Project Management Institute
PMO - Preventive (or Planned) Maintenance Optimization
PMSM - Permanent Magnet Synchronized Motor
Poka Yoke - Mistake- Proofing; Fool Proofing
POSEC - Prioritize by Organizing, Streamlining, Economizing, and Contributing
PP100 - Problems Reported per 100 Vehicles
PPAC - Product Performance Agreement Center
PPAP - Production Part Approval Process
PPC - Production Planning and Control
PPHR - Parts per Hundred Rubber Parts
PPM - Parts per Million
PPV - Product and Process Validation
PQP - Product Quality Planning
PQRR - Program Quality Readiness Review (GM)
PRR - Problem Reporting and Resolution
PSO - Particle Swarm Optimization
PSS - Passive Safety Systems
PSW - Part Submission Warrant
PV - Part-to-Part Variation
PWM - Pulse-Width Modulation

Q1 - Ford Q1, considered worldwide as an indication of exceptional quality
QA - Quality Assurance
QALT - Quantitative Accelerated Life Test
QbD - Quality by Design
QC - Quality Control
QC - Quality Circles
QCC - Quality Control Circle
QCCAR - Quality Control Corrective Action Request
QFD - Quality Function Deployment
QIP - Quality Improvement Program
QIT - Quality Improvement Team
QL - Quality Lever
QLS - Quality Loss Function or Quadratic Loss Function
QMD - Quality Measurement Data
QMS - Quality Management System
QS 9000 - Quality System Requirements 9000
QSA - Quality System Assessment
QSB - Quality System Basics
QSR - Quality System Requirements or Quality Service Reviews
QVR - Quality Verification Report
R, R_{adj}, R_{pred} - Regression, adjusted, and predictive correlations, respectively
R&C - Reliability and Confidence Level (e.g., a combination of R ≥ 95% and C = 75% means that someone is 75% Confident that Reliability ≥ 95%)
RBD - Reliability Block Diagram
RCA - Root Cause Analysis
RCDQ - Reactive Customer-Driven Quality
R-chart - Chart Used to Monitor the Dispersion of a Process
RCL - Robustness Checklist
RCM - Reliability-Centered Maintenance
RDM - Reliability Demonstration Matrix
RDOE - Reliability Design of Experiments
RDT - Reliability Demonstration Test
Red X - Primary Cause
REML - Residual Maximum Likelihood
RFD - Reliability Function Deployment
RFQ - Request for Quotation
RFS - Regardless of Feature Size
RFT - Right First Time
RG - Reliability Growth

RGCA - Reliability Growth Cause Analysis

RMA - Return Material Analysis

RMS (rms) - Root Mean Squared (i.e., Root Sum Squares Method)

ROCOF - Rate of Occurrence of Failures

RoHS - Restriction of Hazardous Substances

ROI - Return on Investment

RPN - Risk Priority Number

RQL - Rejectable Quality Level

RQT - Reliability Qualification Tests

R@R - Run at Rate (GM-specific)

RSM - Response Surface Method

RSS - Root Sum Squares Method (i.e., Root Mean Squared Method)

SAE - Society of Automotive Engineers

SASIG - Strategic Automotive product data Standards Integration Group

SBU - Strategic Business Unit

S Chart - Control Chart for Standard Deviations

SCM - Supply Chain Management

SEV - Smallest Extreme Value

SFDC - Shop Floor Data Collection

SFF (S_{FF}) - Safe Failure Fraction

SFMEA - System Failure Mode and Effects Analysis

SiC - Silicon Carbide

SIF - Stress Intensity Factor

SLA - Supplier Launch Audit (GM-specific)

SLAM - Simultaneous Localization and Mapping

SLM - Ship Logistic Management

SLS - Selective Laser Sintering

SMED - Single-Minute Exchange of Die for flexible manufacturing

SOC or SoC - State of Charge

SOH - State of Health

SOP - Start of Production

SOP - Standard Operating Procedure

SORP - Start of Regular Production

SOTIF - Safety of the Intended Functionality

SPC - Statistical Process Control

SPOF - Single Point of Failure

SQA - Supplier Quality Assurance

SQE - Supplier Quality Engineer or Supplier Quality Engineering

SQIP - Supplier Quality Improvement Program

SS (6-Sigma) - Six Sigma

STBF - Shortest Time between Failures

Supplier CR - Supplier Customer Readiness

SVHC - Substances of Very High Concern

SWIP^{3}E - Standard, Workpiece, Instrument, Person/Procedure/Policy, and Environment (Six essential elements of a generalized measuring system)

SWOT - Strengths, Weaknesses, Opportunities, and Threats

TAAF - Test, Analyze, And Fix

TBD - To Be Decided or To Be Determined

TCO - Total Cost of Ownership

TE - Tooling and Equipment Certification

TGR - Things Gone Right

TGW - Things Gone Wrong

TNI - Trouble Not Identified

TOP - Time Optimized Process

TOPS - Team Oriented Problem Solving (Ford Motor Company-specific)

TPM - Total Productive Maintenance

TPMS - Triply Periodic Minimal Surfaces

TQM - Total Quality Management

TRIZ - Russian, i.e., Theory of Inventive Problem Solving (TIPS)

TV - Total Variation

TWI - Training within Industry

UAV - Unmanned Aerial Vehicle

u-chart - Attribute Quality Control Chart for Frequency of Defects

UCL - Upper Control Limit

UCS - Ultimate Compressive Strength

USL - Upper Specified Limit

UTS - Ultimate Tensile Strength

VDA6.3 - Verband der Automobilindustrie (German); Automotive Quality Management System Process Audit

VEVA - Value Engineering and Value Analysis

VIF - Variance Inflation Factor

VMEA - Variation Mode and Effect Analysis

VOC - Voice of Customer

VRS - Variable Returns to Scale

VSM - Value Stream Mapping

V2V - Vehicle to Vehicle

WBS - Work Breakdown Structure

Nomenclature

A, B, C, …	Design factors
a, b, c, …	Coded variables
C	Confidence level
C_b	Battery capacity
C_p	Process capability index, usually for symmetric distribution; two-sided
C_{pk}	Process capability index, usually for skewed distribution; one-sided
C_{Pm}	Process capability index, a departure of mean μ away from process target
[C]	Conference matrix
[D]	Design matrix
D_c	Depth of charge
E (GPa)	Modulus of elasticity or Young's modulus
E_{11}, E_{22}, and E_{33} (GPa)	Moduli of elasticity, namely Young's moduli, for orthotropic materials
E_T (GPa)	Tensile modulus of elasticity
E[ ]	Expected value
F	F-distribution
F (N)	Force (Newton)
f( )	Function of ( )
G (GPa)	Shear modulus of elasticity
G_{23}, G_{31}, and G_{12} (GPa)	Shear moduli of elasticity for orthotropic materials
H, h, or hr	Hour
H_V (kgf/mm^2, GPa)	Vickers hardness: rectangular pyramidal indenter
I_{zod} (kJ/m^2 or J/m^2)	Izod notched impact strength at 23°C (ISO 180/1A)
[I]	Information matrix or Fisher information matrix
L( )	Likelihood
L_{ln}( )	Likelihood of natural-log-transformed data
Ln( ) or ln( )	Natural logarithmic transformation of ( )
Log( ) or log( )	Logarithmic transformation of ( )
LR	Likelihood ratio
P_p	Process performance
P_{pk}	Process performance index for a stable process
R	Reliability

R	Resistance, electric
R	Model correlation, e.g., regression model
R_{adj}	Adjusted model correlation
R_{pred}	Predictive correlation, based on predicted values
R_a **(μm)**	Surface roughness—average
R_{rms} **(μm)**	Surface roughness—root mean squared (rms)
R_z **(μm)**	Surface roughness—maximum
S_h	State of health
SS	Sum of squares
T (°C), Temp (°C)	Temperature in °C
T_k **(°K)**	Temperature in °K
T (ton)	Mass for mm-sec-ton system
t	t-distribution (Student's t-distribution)
t (sec)	Time
S	Sample standard deviation or sample error
S_c	State of charge
S^2	Sample variance
u, v, and w (m; mm)	Displacements in x-, y-, and z-directions, respectively
V	Volts
(X, Y, Z), (x, y, z)	Cartesian coordinate system
Y	Response or objective function
Y_p	Predicted value for Y
α	p-value
α **(μm/m/°C)**	Coefficient of linear thermal expansion; ⊥ and // to mold or casting flow
α_x, α_y, **and** α_z **(μm/m/°C)**	Coefficients of linear thermal expansion in x-, y-, and z-directions
α_1, α_2, **and** α_3 **(μm/m/°C)**	Coefficients of linear thermal expansion in 1-, 2-, and 3-directions
β	Coefficients of linear moisture expansion
β_x, β_y, **and** β_z	Coefficients of linear moisture expansion in x-, y-, and z-directions
β_1, β_2, **and** β_3	Coefficients of linear moisture expansion in 1-, 2-, and 3-directions
ε	Residual
ε	Strain
ε_{Creep}	Creep rupture strain
ε^e **and** ε_e	Elastic strain
ε^p **and** ε_p	Plastic strain
ε_{eq}	Equivalent strain
$\varepsilon_{eq}{}^p$	Equivalent plastic strain
ε_{xx}, ε_{yy}, **and** ε_{zz}	Normal strains
ε_{yz}, ε_{zx}, **and** ε_{xy}	Shear strains
ε_{ucs}	Ultimate compressive strain

ε_{uts}	Ultimate tensile strain
ε_1, ε_2, and ε_3	Principal strains
ε_{11}, ε_{22}, and ε_{33}	Normal strains defined in the (1, 2, 3) coordinate system
ε_{23}, ε_{31}, and ε_{12}	Shear strains (tensor) defined in the (1, 2, 3) coordinate system
ε_{11c}, ε_{22c}, and ε_{33c}	Ultimate compressive strains along the primary orthotropic material axes
ε_{11t}, ε_{22t}, and ε_{33t}	Ultimate tensile strains along the primary orthotropic material axes
ε_{23u}, ε_{31u}, and ε_{12u}	Ultimate shear strains in primary orthotropic material coordinates (1, 2, 3)
μ	Mean
ρ, ρ_f, and ρ_m (g/cm$_3$)	Density (overall), density of fiber, and density of matrix, respectively
σ	Standard deviation
σ (MPa)	Stress
σ_{eq} (MPa)	Equivalent stress, e.g., von Mises stress
σ_f (MPa)	Fatigue limit, also called endurance limit
$\sigma_{f'}$ (MPa)	Fatigue strength coefficient
σ_{uts} (MPa)	Ultimate tensile strength
σ_{ucs} (MPa)	Ultimate compressive strength
σ_{xx}, σ_{yy}, and σ_{zz} (MPa)	Normal stresses
σ_{xy}, σ_{yz}, and σ_{zx} (MPa)	Shear stresses
σ_y, $\sigma_{0.2\%}$ (MPa)	Yield strength, i.e., stress at 0.2% (tensile) or −0.2% (compressive) strain σ_{YC} (MPa) yield strengths in compression
σ_{YT} (MPa)	Yield strengths in tension
σ_1, σ_2, and σ_3 (MPa)	Principal stresses
σ_2	Variance
τ, τ_{yz}, τ_{zx}, and τ_{xy} (MPa)	Shear stresses
ν	Degrees of freedom
χ^2	Chi-square distribution

Introduction

Product reliability growth through statistical tolerance design and testing based on the design of experiments (DOEs) is the central theme, as it is utilized to accommodate the stochastic nature of part fabrications and assembly variations. The automotive industry is at a crossroad of electrification and product design, and testing the new path that each manufacturer chooses is paramount for its future. Given that product design and testing are accomplished using real-world tests (e.g., by strain gages) or validated computer simulations (e.g., by finite element methods), the reliability DOEs based on different statistical distributions shall be elaborated perceptively.

Chapter 7: Statistical tolerance design (analysis and synthesis) plays an important role in the manufacturing process and material selection. Assembly of parts by default constitutes another statistical game of DOEs. As a design parameter, each dimension or feature, which affects the product assembly and cost, may have a "non-normal" probability distribution. Consequently, there exists a nonlinear relationship between the response (assembly or cost) and contributing factors (part dimensions, features, and assembly sequences). The statistical distribution of each dimension or feature is first identified using the maximum-likelihood estimator, the method of moments estimator, or other estimators. Then, the statistic tolerance analysis and synthesis of the assembly can be carried out by applying Monte Carlo simulations, root sum squared method, finite element tolerance analysis with skin model shapes, homogenous transformations for tolerance analysis, perturbation analysis and direct linearization method, and T-map algorithm or other methods.

Chapter 8: Life cycles of mechanical and electronic components in automobiles are often not normally distributed with respect to the related design parameters in statistics. The reliability DOEs provides a system-level estimate of product lifetime that often obeys a log-location-scale distribution in statistics, especially in accelerated life testing. Failure modes attributed to the lognormal distribution are worked in detail, and various transformations are introduced for parts and system design.

Chapter 9: Reliability DOEs based on Weibull distribution (DOE-W) using the maximum-likelihood method is often applied to characterizing product lifetime, because of its capability in modeling skewed data and versatility in fitting various mechanical failure modes and material degradation. Reliability demonstration testing based on Weibull distribution has its way around the intricate maze of product life prediction. How to assess Weibull parameters by the maximum-likelihood method, method of moments, and median rank regression is hence illustrated. The application of DOE-W is demonstrated using real-world problems. As a subsidiary of DOE-W, the reliability DOE based on exponential distribution (DOE-E) is also addressed.

Chapter 10: It is essential to do product life prediction and grow its reliability against a specific potential failure mode in a finite time domain at the product design stage. Various designs for reliability tests such as qualitative accelerated life test and stepwise accelerated life test are applicable. Stepwise accelerated tests conducted using the cumulative exposure model, Weibull statistics, and lognormal approximation are explained in detail. The credibility of analysis relies on the coupling of physics of failure and its inherent statistical distribution. The physics of failure can be as simple as a linear model or as complicated as material fatigue subjected to dynamic load steps.

Chapter 11: Product life prediction based on the reliability DOEs is legitimate and realistic when built upon proper inferential mechanisms based on natural laws of conversion (e.g., physics of failure, material conversion, and structural dynamics) in combination with stochastic models (e.g., Weibull statistics). Inferential mechanisms corresponding to frequently used natural laws, including the Arrhenius model, the Eyring model, oxidation, corrosion, rusting, crack propagation, creep, photovoltaics, and automobile ride, are demonstrated using real-world problems. Note that fatigue life-cycle prediction is a counting method, not an inferential mechanism. For example, the fatigue life of a bolt can be accounted for in terms of one or more of the following inferential mechanisms: corrosion, creep, mechanical stresses/strains, hydrogen embrittlement, material softening, material phase transformation, etc.

7

Tolerancing for Assembly and Manufacturing

A part is a material shaped up to the specified dimensions on the drawing with proper tolerances. Each part dimension with the generated tolerance range has its own individual statistical function inherently. A dimension is a nominal size value of a part feature, while tolerancing a dimension is to designate the range of size values, within which the part can function properly in its expected lifetime. Tolerancing a dimension may vary from part to part with respect to the manufacturing procedure, working environment, material usage and degradation, and certain stochastic factors. Assembly of parts by default constitutes another statistical game of DOEs. Monte Carlo simulations and finite element tolerance analysis with skin model shapes turn out to be promising methods for general applications. A review of Geometric Dimensioning and Tolerancing (GD&T) is presented in the last section of this chapter for reference.

7.1. Tolerance Design

Analytical modeling of assemblies provides a quantitative basis for the evaluation of performance and reliability because of dimensional variations. The desired applicable range of an unknown dimension shall be derived from assembling the required parts together, as obtained retroactively from its constituent dimensions of individual contributing members in a stochastic manner. When dealing with tolerance design problems, two types of problem resolution occur:

(a) Tolerance analysis

(b) Tolerance synthesis

They are defined reciprocally as shown in Figure 7.1.1. In tolerance analysis, the component tolerances are specified, and the resulting assembly variation and yield are calculated. Tolerance synthesis involves the allocation of the specified assembly tolerance among the component dimensions of an assembly to ensure a specified yield. Thus, it is also called tolerance allocation.

When analytical modeling of mechanical assemblies in tolerance analysis is conducted, mating between parts is generally classified into two categories—clearance and interference. For example, when a shaft is inserted into a hole, if the shaft is smaller than the hole the mating creates a clearance; on the other hand, the mating creates interference. The minimum stated clearance or maximum stated interference for mating parts is called allowance. Averages, namely sample means, and sample standard deviations have to be dealt with carefully because they are usually utilized for tolerance design instead of population means and standard deviations.

FIGURE 7.1.1 Tolerance analysis and tolerance synthesis (allocation).

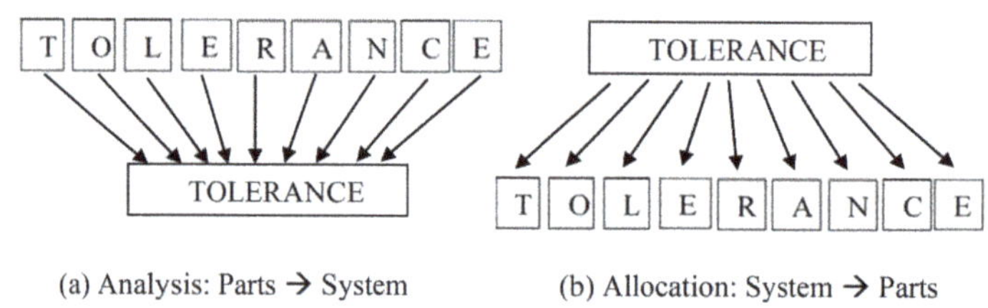

In addition to the variability as designed, it is important to distinguish the latent variabilities because of the manufacturing process, operations after production, and customer usage as revealed when the product is in use. In the tolerance specification, it is desired that all the potential product variabilities in various stages be taken into account. Integration of tolerance design (analysis and synthesis) and statistical production control into the product life-cycle management can be eventually made possible for shortening the product development cycle [Barai and Pop-Iliev 2009].

7.1.1. Analytic Methods

The advantages of applying analytic methods to resolving tolerance complexity are to improve product reliability and to reduce cost through controlled variation and robust design as opposed to the more traditional tolerance assignment. The choice of a proper way to predict assembly tolerances, which is inherent with the limiting deviations of a dimensional chain in the assembly, will affect the manufacturing accuracy and assembly interchangeability of components [Cox 2006, Craig 1995, Fischer 2011]. Therefore, the economy of production and operation depends on it. To resolve the tolerance relation in a dimensional chain, eight widely applied modeling algorithms are as follows:

1. Worst-case (worst-limit) method [Tynes 2012].
2. Closed-form equations based on the root sum squares (RSS) method [Greenwood and Chase 1990].
3. Method of Monte Carlo simulations [Wu et al. 2009].
4. Maximum-likelihood method.
5. 3-Dimensional (3D) tolerance analysis—homogeneous transformations/small displacement torsor (SDT).
6. 3D tolerance analysis—finite element analysis (FEA)/skin model shapes.
7. T-Map (tolerance map) method [Davidson et al. 2002].
8. Neural network-based approach [Chih et al. 2007].

The worst-case method is a quick examination of the potential assembly problem (e.g., interference or not). Both the RSS method and Monte Carlo simulations are based on the statistical tolerance design, which addresses a tolerance interval that is in a range likely to contain a defined proportion of a dimensional population. To calculate tolerance intervals, one must stipulate the proportion of the dimensional population and the desired confidence level, which is the probability that the named proportion is actually included in the interval. The method of moments and maximum-likelihood method are utilized to identify the statistical distribution of an individual dimension, as a preparatory step for Monte Carlo simulations, when the measured data of the dimension of concern are available [Evans 1975a, 1975b, Meadows 2009].

The RSS method provides the statistical tolerance analysis in a closed-form manner. The analytics of 3D coordinate transformations for tolerance design, as extended tool based on the RSS method, is getting more robust than ever, but still not as reliable as Monte Carlo simulations. In order to have more confidence in the calculated results, one is encouraged to use the method of Monte Carlo simulations no matter whether it is a 1-, 2-, or 3D tolerance design.

In general, there are five main sources of dimensional changes that ought to be taken into consideration when doing tolerance analysis of a mechanical assembly [Green 2005]:

1. Dimensional variations in size.
2. Geometric variations in form, orientation, location, runout, profile, and symmetry.
3. Flexible deformations, such as beam deflection and creep as loaded in an assembly.
4. Thermal expansions, which subjected to temperature change and phase transformation.
5. Kinematic variations, e.g., rigid body motions rendering an adjustment between mating parts [Joskowicz et al. 1997].

Starting with a standard-compliant technical drawing to complete tolerance specification from a statistical point of view, one is able to know the facets of dimensional management including all the potential variations listed above. Physical meanings of geometric and kinematic variations are demonstrated in Figure 7.1.2(a) and 7.1.2(b), respectively, while flexible deformations and thermal expansions/contractions can be applied as add-on variations to the geometric variations whenever needed. Typical case studies in the automotive engineering can be found in [Becker et al. 2018, Bruyère et al. 2007, Chiang and Barber 2002, Choi et al. 1998, Kroczak and Dudziak 2011, Musia et al. 2004, Wang et al. 2006].

FIGURE 7.1.2 Schematics of geometric (red) and kinematic (green) variations.

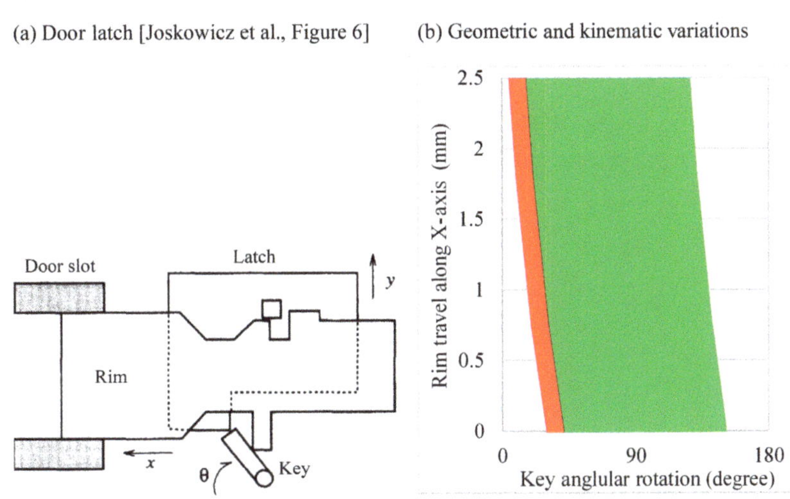

Gear whining emanating from an electric powertrain is a typical phenomenon resulting from kinematic variations in operation. The fundamental cause of gear whine comes from gear transmission errors, which refer to the deviation of the actual position of the output gear from the position it would occupy if the gear drive were perfectly conjugated. There are three different potential transmission errors in a loaded electric powertrain in operation:

1. Rigid Body Motion (Geometric Variation): It comes from a gear mandrel displacement that is taken into consideration to account for the cam-setting error because of tooth backlash and shaft clearance in rigid body motion with respect to the gear reference axis.
2. Flexible Bearing and Shaft (Flexible Deformation): This accounts for the oscillating behavior, which originates from the variation in the limited number of balls that come in contact with the flexible bearing outer ring.
3. Flexible Gear Teeth (Flexible Deformation): Another comes from the kinematic error of flexible gear teeth as engaged in operation.

Each transmission error would result in a time-varying number of teeth in contact and tooth contact condition (e.g., area and shape). Gear whine is a structure-born noise due to flexible gear teeth that generate fluctuating vibrations through high sound tonality in resonance.

7.1.2. Linear Dimensional Chain

To begin with tolerance analysis or tolerance allocation, one has to identify the critical tolerance to be studied and relate it to factors such as dimensions, angle, and surface texture. The size, tolerance, and limit deviations of the resulting dimension may rely on the size and tolerance of part dimensions. Two ways to define the relationship between the parts and the assembly need (e.g., examining interference) are as follows: One is the linear dimensional chain for simple geometry, and the other is vector-loop model for more complex geometry [Shen et al. 2005].

A linear dimensional chain is a set of independent parallel dimensions, which are connected to others to create a geometrically closed circuit. They can be dimensions specifying the mutual position of components on one part as shown in Figure 7.1.3(a) or dimensions of several parts in an assembled unit as shown in Figure 7.1.3(b). The assembly equation can be written in terms of the given geometric dimensions (without considering geometric features and other factors) as

$$Z = A - B - C - D \quad (7.1.1)$$

or

$$A - B - C - D - Z = 0 \quad (7.1.2)$$

FIGURE 7.1.3 An example of linear dimensional chains.

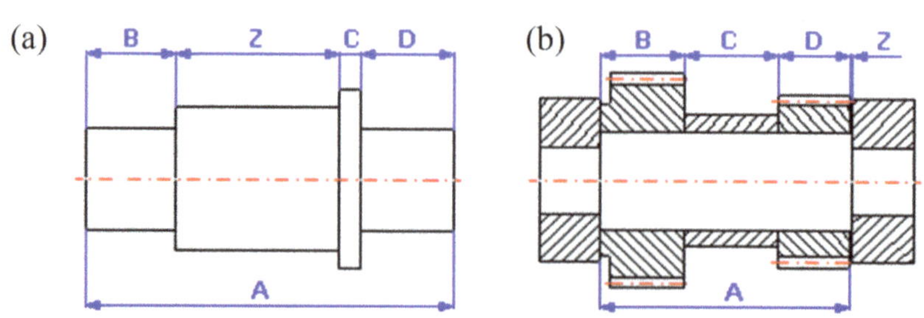

This statistical variation evaluated using a linear chain of dimensions represented by the above equation is called one-dimensional (1D) analysis. A 1D chain is set up by creating a cross section of a model and adding the tolerance values for each feature in a straight line. The dimensional stack-up consists of separate partial components (input dimensions) and ends with a closed component (resulting dimension) as Equation (7.1.1) exhibits. Partial components (A, B, C, and D) are dimensions either directly dimensioned in the drawing, derived from manufacturing, or obtained from possibly assembly operations. The closed component (Z) in the given chain represents the resulting manufacturing or assembly dimension, which is the result of combining partial dimensions as a scaled manufacturing dimension, possibly assembly clearance or interference of a component. Depending on how the change in a partial component affects the change in the closed component, two types of components are distinguished in dimensional chains:

(a) Increasing Components: Partial components, the increase in which results in an increase in the closed component, e.g., dimension A.

(b) Decreasing Components: Partial components, the increase in which results in a decrease in the closed component, e.g., B.

7.1.3. Vector-Loop Model

Vector loops present the assembly constraints that locate the parts of the assembly relative to each other in order to relate the critical tolerance to the product geometry. Although a vector-loop model is applicable to 3D parts, it lends itself to a "quasi-1D" path that enters a part through a joint, gets connected to the datum reference frame (DRF) via the datum path, further gets connected to another joint via a second datum path, and then exits to the next adjacent part in the assembly. DRFs are typical of a 3D view. A typical DRF is made up of three planes, such as "face side" and two "datum edges." For example, these three planes are marked A, B, and C, where A is the face side, B is the first datum edge, and C is the second datum edge. In this case, the DRF is A/B/C. A/B/C is shown at the end of the feature control frame to show from where the measurement is taken.

Typically, the functional datums addressed above are not used to manufacture the part. The manufacturing datums are typically different from the functional datums to save cost and improve process speed and production repeatability. A tolerance analysis may be needed in many cases to convert between the functional datums and the manufacturing datums [Kovar 2016].

Vector loops are created by simply linking together the datum paths, such that all the dimensions, forms, and even kinematics will be datum-referenced as illustrated in Figure 7.1.4. The vector loop identified is then joined tip to tail, forming a closed chain, passing through each part in the assembly in succession. An apparent dimensional tolerance is dealt with by default. Nevertheless, the geometric tolerance associated with an individual feature is to be converted into a "± tolerance," from which both upper limit (UL) (also called resultant condition) and lower limit (LL) (also called virtual condition) are derived by taking position tolerance, material condition, etc., into consideration [Li and Roy 2001, Lin and Chen 1994]. However, this method based on vector-loop model is applicable to assembling components that are consistent with normal distributions only, because the formulated nonlinear equations for most practical applications are linearized in the problem-solving process.

FIGURE 7.1.4 Stack-up diagram and its corresponding vector-loop diagrams with regard to Figure 7.1.3(b).

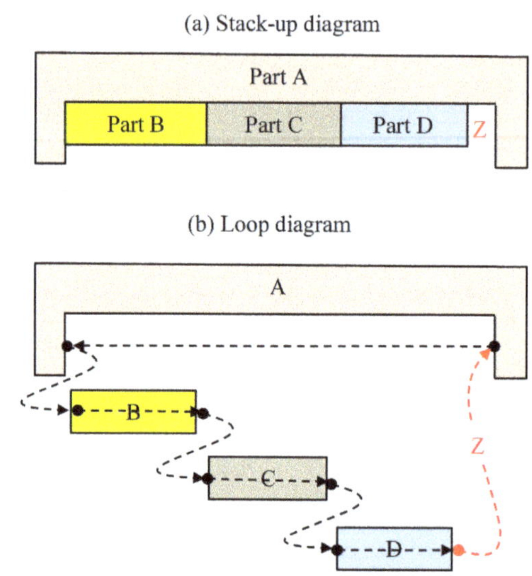

7.1.4. Tolerance Analysis Example: One-Way Clutch Assembly

For example, a simple automotive one-way clutch assembly consists of four kinds of mechanical parts: an inner hub, an outer ring, four rollers, and four springs, as shown in Figure 7.1.5. This kind of mechanical mechanism can be found in a lawn mower. During the operation in the clockwise direction, the springs push the rollers into the wedge-shaped space between the ring and the hub. When a counterclockwise torque is applied to the hub, rollers bind and force the ring to turn with the hub. No clockwise torque can be transferred from the hub to the ring, because rollers slip. Given that dimensions a (hub height), c (roller diameter), and e (inner diameter of race ring) may vary as manufactured, the following two questions are raised to quantify the "binding behavior of the rollers":

(a) How much is the spring gap?

(b) Where is the roller center to the rotating center stochastically?

FIGURE 7.1.5 One-way clutch assembly [Carr 1993, Chase 1999, Creveling 1997, Greenwood and Chase 1990].

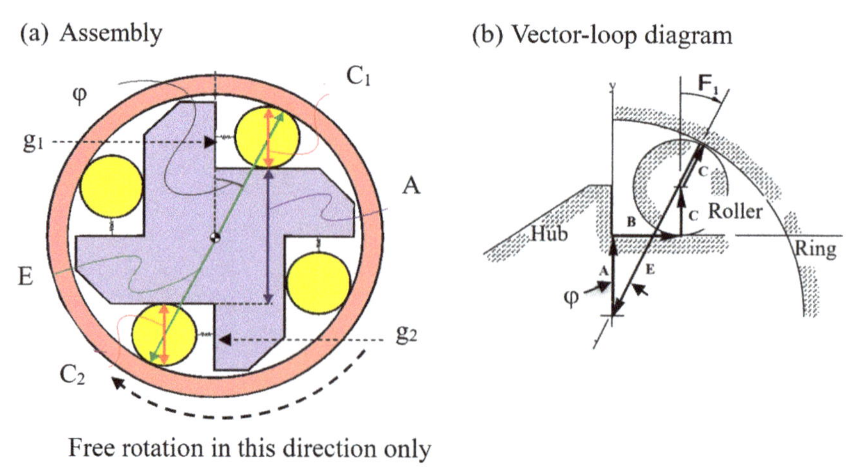

Locating the dimensional components in both the x-direction and y-direction of the roller center and the spring gap, one is able to locate the spring gaps and the angular position of the roller center in terms of the three known dimensions A, $C = C_1 = C_2$, and E as

$$g_1 = \tfrac{1}{2}\,[(E - \tfrac{1}{2}C_1 - \tfrac{1}{2}C_2)^2 - (A + \tfrac{1}{2}C_1 + \tfrac{1}{2}C_2)^2]^{1/2} - \tfrac{1}{2}C_1 \quad (7.1.3)$$

$$g_2 = \tfrac{1}{2}\,[(E - \tfrac{1}{2}C_1 - \tfrac{1}{2}C_2)^2 - (A + \tfrac{1}{2}C_1 + \tfrac{1}{2}C_2)^2]^{1/2} - \tfrac{1}{2}C_2 \quad (7.1.4)$$

and

$$\varphi = \cos^{-1}[(A + \tfrac{1}{2}C_1 + \tfrac{1}{2}C_2) / (E - \tfrac{1}{2}C_1 - \tfrac{1}{2}C_2)] \quad (7.1.5)$$

where

A is the hub height
C_1 and C_2 are the roller diameters for roller 1 and 2, respectively; nominally, $c_1 = c_2$
E is the inner diameter of the rim
g_1 and g_2 are the gaps for roller 1 and 2, respectively
φ is the angular location of the roller center, as measured from the vertical axis

The four dimensions, i.e., A, C_1, C_2, and E, considered here are assumed to be normally distributed. The above three simultaneous equations provide the assembly constraints and they are called assembly equations, which determine the effectiveness of the performance of the one-way clutch. Note that Equation (7.1.3) resembles Equation (7.1.4). After the assembly equations are derived, one can do the tolerance analysis using the worst-case method, the RSS method, Monte Carlo simulation, or others.

7.1.5. Tolerance Allocation

When an assembly tolerance is given as required according to design requirements or engineering specifications, e.g., clearance between piston crown and disc in an automotive caliper-disc brake, as shown in Figure 7.1.6 (system design), while its related component tolerances are to-be-decided or unknown, it is then imperative to allocate component tolerances in order to meet the need. This is called tolerance synthesis or tolerance allocation; i.e., limit deviations of individual components are to be designed using the known limit deviation of a closed-loop component assembly mandated by the functional demand. Statistical tolerance synthesis is a tool for intelligently allocating individual component tolerances, stringent or not, as well as assessing the likelihood of unacceptable tolerance combinations. Indirect component tasks are resolved when designing a functional group within an assembly. This allows an engineering organization to create more manufacturable and less expensive products. An example of tolerance synthesis of mechanisms by a robust design approach can be also found in [Caro et al. 2005].

FIGURE 7.1.6 Tolerance allocation for clearances between the disc and piston crowns in an automotive disc brake system [Hallmann et al. 2020].

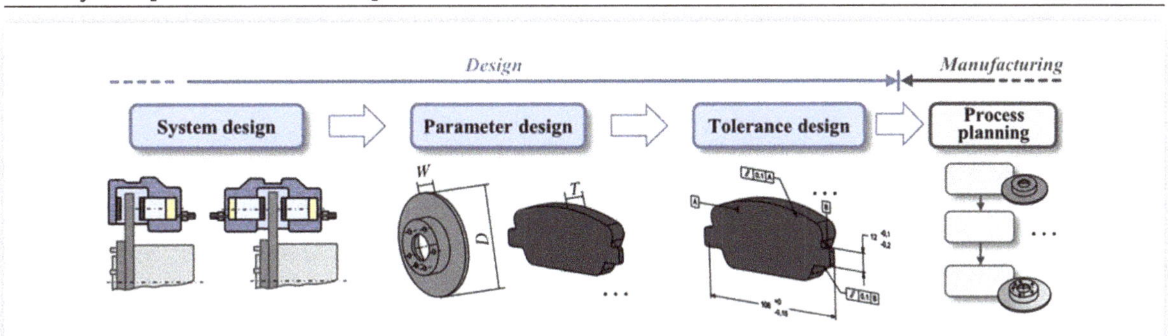

7.1.6. Average, Variance, Skewness, and Kurtosis

As each dimension is a random variable subjected to manufacturing capability [Zhu and Ting 2001], assembly of parts by default constitutes another statistical game of DOEs [Albrecht et al. 2013]. A design parameter such as a dimension or feature, which affects the response, either performance or failure mode, may have a "non-normal" probability distribution, and consequently, there exists a nonlinear relationship between the response and the design parameter [Liu et al. 2008]. When the number of simulations is large enough, e.g., 5000 and above, the simulation will result in a statistical set of data, including average, variance, skewness, and kurtosis, respectively, as

$$\mu = [\sum_{n=1}^{N} d_n] / N \qquad \text{(Average measuring location)} \tag{7.1.6}$$

$$\sigma^2 = [\sum_{n=1}^{N} (d_n - \mu)^2] / N \qquad \text{(Variance measuring spread)} \tag{7.1.7}$$

$$\beta_3 = \frac{[\sum_{n=1}^{N} (d_n - \mu)^3] / N}{\sigma^3} \qquad \text{(Skew measuring symmetry)} \tag{7.1.8}$$

and

$$\beta_4 = \frac{[\sum_{n=1}^{N} (d_n - \mu)^4] / N}{\sigma^4} \qquad \text{(Kurtosis measuring peakedness)} \tag{7.1.9}$$

where

- μ is the mean
- σ is the standard deviation
- β_3 is the coefficient of skewness
- β_4 is the coefficient of kurtosis

The statistical density function of a given set of data is supposedly characterized using the above four moments in Figure 7.1.6. Many design parameters are usually involved in a tolerance design, as the nature of a typical assembly may not lend itself to an analysis by simple techniques subject to no realistic complication. A technique for statistical tolerance analysis is strongly encouraged to cover all four moments: mean, standard deviation, skewness, and kurtosis. Statistical tolerance synthesis relies on distribution function zones [Zhang et al. 1999].

7.2. Dimensional Management

Dimensional management, for either product development or solving assembly problems for existing products, allows the effective management of deviations induced by different manufacturing processes to obtain a product that meets its design intent but is less sensitive to variations in partial components. Furthermore, the impact of form defects in particular on conducting the assembly simulation should be also considered [Homri et al. 2016]. It is the key to improving productivity by enhancing the manufacturing process capability, reducing the product cycle time, and growing the product reliability. Comprehensive dimensional management is therefore a prerequisite for economic success in competition. In other words, its implementation would go from tolerance allocation to tolerance-cost optimization, as illustrated in Figure 7.2.1.

FIGURE 7.2.1 Classification of tolerance specification, allocation, analysis, synthesis, and cost optimization via an automotive disc brake system [Hallmann et al. 2020].

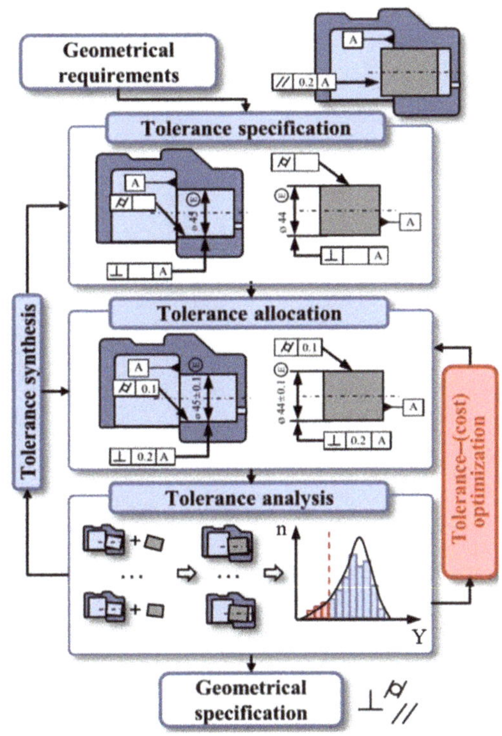

7.2.1. Cost Function for Part Fabrication

The cost function equation consists of two components. One is the fixed cost, and the other is the variable cost. A more generalized cost function can be expressed as

$$C(X) = C_{fixed} + C_V(X)$$

where

$C(X)$ is the total production cost at production volume X
C_{fixed} is the total fixed costs, such as tooling, setup, and other prior operations
C_V is the variable cost, i.e., cost of producing parts, as a function of number of parts
X is the number of production units

Tight tolerances will increase price, as shown in Figure 7.2.2. When tighter tolerances are requested, either expensive high-precision machinery or extra processes with more tools and operations are required. Furthermore, tight tolerances, additional inspection, and testing are required to verify that the part is in spec.

When a tolerance allocation method is employed to determine how to adjust component tolerances for the purpose of achieving a desired final tolerance, information on costs can be incorporated to find out the most cost-effective approach [Dong 1997]. Therefore, tolerance allocation definitely incorporates the tolerance analysis approach to component GD&T into process capability, cost, and mean shift [Hallmann et al. 2020]. As guided by tolerance allocation demonstrated by [Ullman 2003], the relative manufacturing cost versus tolerance is plotted in Figure 7.2.2. It can be seen that special operations such as honing and grinding are relatively expensive.

FIGURE 7.2.2 Variations of cost with respect to tolerance of machined steel parts.

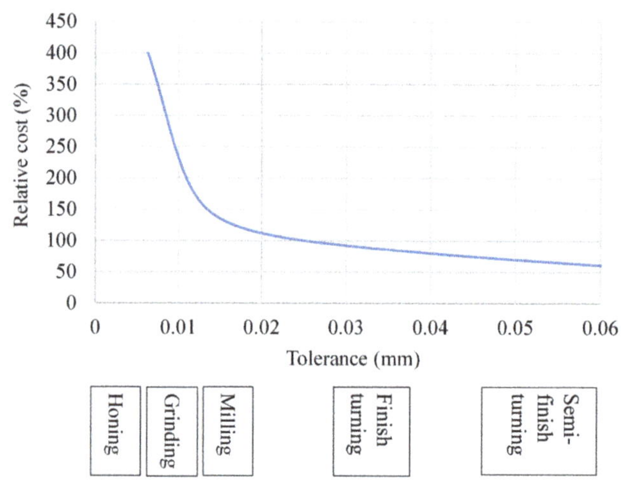

A universally applied formula for the cost function of fabricating mechanical parts has been debatable. A reciprocal equation for a generalized cost function related to the tolerance is given as follows [Chase and Greenwood 1988, Spotts 1973]:

$$\text{Cost} = p + q / T + r / T^2 \tag{7.2.1}$$

where

p is the constant, i.e., fixed cost or setup cost

q is the coefficient relating the cost variation linearly to the inverse of tolerance

r is the coefficient relating the cost variation quadratically to the inverse of tolerance value

T is the original tolerance

Nevertheless, the relative cost data for machining of general steel are given as follows.

Tolerance	Operation	Relative cost (normalized)
0.381 mm (0.015″)	Rough turning	25%
0.0762 mm (0.003″)	Semi-finish turning	60%
0.0254 mm (0.001″)	Finish turning	100% (Base)
0.0127 mm (0.0005″)	Grinding	200%
0.00635 mm (0.00025″)	Honing	380%

Regression models fitting the relative cost-versus-tolerance data given above for fabricating a generic steel part using Minitab lead to the following two equations:

$$Cost_{steel} = 0.20565 + 0.02273 / T_{steel} \quad (R^2 = 99.8\% \ \& \ R_{adj}^2 = 99.7\%) \tag{7.2.2}$$

and

$$Cost_{steel} = 0.23183 + 0.02160 / T_{steel} + 0.00000907 / T_{steel}^2 \quad (R^2 = 99.8\% \ \& \ R_{adj}^2 = 99.6\%) \tag{7.2.3}$$

where

R is the correlation

R_{adj} is the adjusted correlation

In this case, it can be seen that the model without the reciprocal "quadratic term" performs no worse than the one with the reciprocal quadratic term.

The following reciprocal equation [Chase and Greenwood 1988] is sometimes adopted herein for simplicity for the purpose of demonstrating the optimization procedure based on tolerance allocation for part fabrication,

$$Cost = p + q / T \tag{7.2.4}$$

This may not be true for some precision parts or other materials.

7.2.2. Fabrication Capability

The tolerance of a part is a function of its characteristic length or size, as presented using the following equation [Fortini 1967]

$$T = P\,L^{1/3} \tag{7.2.5}$$

where

L is the size or characteristic length

P is the proportionality

7.2.3. Control Plan Methodology

A control plan is a quality tool to identify and monitor the activity required to control the critical inputs or key outputs for a process so the process will continually meet customer requirements over time. Control plans are usually monitored by the quality assurance team and its related functional teams. Without control plans, most processes will gradually drift into a state of disrepair because of employee turnover, knowledge loss, and short term production priorities.

Manufacturing tolerances serve to satisfy not only the functional requirements given in the product definition model but also the manufacturing constraints, such as machine accuracy and minimum extra machining thickness. Control plans are written descriptions of the systems for controlling parts and processes. Separate control plans applied to the manufacturing tolerances can be classified into the following three distinct phases:

1. *Prototype*: A description of the dimensional measurements, and material and performance tests that will occur during prototype build.
2. *Pre-Launch*: A description of the dimensional measurements, and material and performance tests that will occur after prototype and before full production.
3. *Production*: A comprehensive documentation of product and process characteristics, process controls, tests, and measurement systems that will occur during mass production.

A control plan lists all product and process inspection points required to deliver a defect-free outcome and is essential for maintaining process control over the long run. The control plan methodology reduces waste and improves the quality of products by identifying process characteristics and their sources of variation. Changes in the product/process characteristics, characteristic measurement, and control method are addressed in the control plan as a living document. What makes an effective control plan? Available design and manufacturing information to gain a better understanding of the process can be sought utilizing the following:

(a) Clear product specifications—clearly defined and linked to customer requirements.
(b) Process flow diagram.
(c) Failure mode and effects analysis (design failure mode and effect analysis (DFMEA) and process failure mode effects analysis (PFMEA)).
(d) Design reviews.
(e) Program assumption and lessons learned from similar products.
(f) Analytic skills such as metallurgy, DOE, and FEA.
(g) Benchmarking and quality function deployment (QFD).
(h) Team knowledge including internal audit.

In practice, control plans are typically a document (in Excel) that lists the product and process characteristics that must be monitored during the design and manufacturing process, including measurement methods and necessary reaction plans for deviant conditions.

Item (column)	Description
Part/process number	This is typically the process step number that matches both the process flow diagram and PFMEA.
Process name/description	A clear description of each process step, i.e., "Drive screws 1 and 2."
Machine, device jig, and tools for manufacturing	Devices and reference numbers (also noted on the devices themselves) for the major items needed to complete the task, i.e., "Electric screwdriver 987-01A1."
Characteristics—number	This is a reference number for each characteristic being monitored. Control plans typically monitor product and process characteristics. For example, when manufacturing a disposable coffee cup, a product characteristic might be the overall height of the cup, and a process characteristic might be the curing temperature for the adhesive joining the top to the bottom of the cup.
Characteristics—product	If the characteristic being monitored is related to the product itself, fill in a brief description of the characteristic, i.e., "Screw torque."
Characteristics—process	If the characteristic being monitored is related to the process, fill in a brief description of the characteristic, i.e., "Glue Temperature."
Critical to quality (CTQ)	Note "Yes" if the characteristic is CTQ.
Product/process specification and tolerance	Note the specific product or process specification and tolerance for the characteristic being monitored, i.e., "Back-out screw torque between 26 and 29 in.-lb."
Evaluation and measurement techniques	Note the method used for measuring the characteristic, i.e., "Digital Calipers."
Sample size	Note the size of the sample, when a random sample is drawn from production at a certain frequency. For example, a sample size of 5 is commonly used with some control charts. This is also referred to as the subgroup size in control charting.
Sample frequency	Note how many times per shift or hour a sample group is to be measured.
Control method	Note the method by which process control is maintained, i.e., "Control Chart."
Reaction plan	Note how the associate is to react if a deviant condition is found.

Control plans for dimensions must be constantly updated as changes are made; should be audited as part of the total quality management (TQM) system that is designated by advanced quality product planning (APQP).

7.3. Deterministic Tolerance Design

When tolerance design (analysis and synthesis) is accomplished with worst-limit models, also called worst-case models, it is referred to as deterministic tolerance design (analysis and synthesis). Despite the limitations of deterministic tolerance design, it is a viable design practice for the following two reasons: (a) ease of use and (b) providing a basis for statistical tolerance design.

7.3.1. Worst-Limit Method

The basis of tolerance analysis in design is an analytical model for the accumulation of tolerances in a mechanical assembly of component parts. The purpose of the worst-limit method is to find the minimum and maximum values that can be achieved by the resulting dimension for a random combination of real input dimensions. In such an analysis, the assembly tolerance is determined by summing the component tolerances linearly. It is also called worst-case method obeying the following arithmetic law. A worst-case study assumes that all the dimensional tolerances occur at their worst-limits at the same time—the statistical distribution of each dimension is statistically uniform-distributed. Each component dimension tolerance T_i is assumed to be at its extrema (extreme limits) that result in the worst possible assembly limits.

One-dimensional assemblies: $T = \sum T_i$ (7.3.1)

or Multi-dimensional assemblies: $T = \sum |\partial H(x_i) / \partial x_i| T_i$ (7.3.2)

where

T_i is the nominal dimension tolerance of component i

$H(x_i)$ is the assembly function, describing the resulting dimension of the assembly, e.g., a clearance or interference

The partial derivative given in the above equation represents the sensitivity of the assembly tolerance to the variation in individual dimension characteristics, such as the directional cosine for an alignment of the direction. Nevertheless, each dimension as fabricated may have a distinct statistical distribution other than the uniform distribution. The low occurring probability of extreme limits renders calculation futile, as for but not limited to a normally distributed function. Because the worst-limit method will result in a futile tolerance control that may incur a high cost, it should not be used for design purpose but for the first approximation and for reference only.

7.3.2. One-Way Clutch Example by DOEs

The hub (Figure 7.1.5) is attached to a shaft, which in turn is attached to a driving mechanism. This assembly is designed to allow rotation in only one direction. The example of a one-way clutch example formulated using the vector-loop model is to be analyzed here using the method of worst-case. The nominal values and tolerances for the three dimensions (i.e., a, c, and e) are given as follows:

Variable (mm)	−1: Low (mm)	1: High (mm)
A = 46.74 ± 0.156	46.584	46.896
C_1 = 22.86 ± 0.013	22.847	22.873
C_2 = 22.86 ± 0.013	22.847	22.873
E = 101.6 ± 0.156	101.444	101.756

Since $C_1 = C_2$, let $C = C_1 = C_2$. There are eight and only eight possible runs based on the extrema of variates A, C, and E. They are listed as follows:

- A = 46.584, C = 22.847, and E = 101.444
- A = 46.896, C = 22.847, and E = 101.444
- A = 46.584, C = 22.873, and E = 101.444
- A = 46.896, C = 22.873, and E = 101.444
- A = 46.584, C = 22.847, and E = 101.756
- A = 46.896, C = 22.847, and E = 101.756
- A = 46.584, C = 22.873, and E = 101.756
- A = 46.896, C = 22.873, and E = 101.756

After letting

- $-1 \le a = \{A - [½(46.896 + 46.584)]\}/[½(46.896 - 46.584)] \le 1$
- $-1 \le b = \{B - [½(22.873 + 22.847)]\}/[½(22.873 - 22.847)] \le 1$
- $-1 \le c = \{C - [½(101.756 + 101.444)]\}/[½(101.756 - 101.444)] \le 1$

After the changes in variables are performed as above, the eight run conditions can be rewritten into a full factorial design (2^3) for the eight combinations of the extrema for the spring gap. The eight spring gaps are calculated using Equation (7.2.3) and registered in Table 7.3.1. Similarly, the angular locations of these eight runs are calculated using Equation (7.2.5) and registered in Table 7.3.2.

TABLE 7.3.1 Full factorial design 2^3 for combinations of extreme limits for spring gap.

Run	a	c	e	ac	ae	ce	ace	Mean	Gap
1	−1	−1	−1	1	1	1	−1	1	6.994
2	1	−1	−1	−1	−1	1	1	1	6.697
3	−1	1	−1	−1	1	−1	1	1	6.929
4	1	1	−1	1	−1	−1	−1	1	6.631
5	−1	−1	1	1	−1	−1	1	1	7.325
6	1	−1	1	−1	1	−1	−1	1	7.033
7	−1	1	1	−1	−1	1	−1	1	7.260
8	1	1	1	1	1	1	1	1	6.967
Contrast	−0.295	−0.065	0.334	−0.000	0.0027	0.0004	1.1E−05		
Effect	−0.148	−0.033	0.1668	−0.000	0.0013	0.0002	5.7E−06	6.979 (average)	

TABLE 7.3.2 2^3 Full factorial design for combinations of extreme limits for stop angle.

Run	a	c	e	ac	ae	ce	ace	Mean	Angle
1	−1	−1	−1	1	1	1	−1	1	27.947
2	1	−1	−1	−1	−1	1	1	1	27.458
3	−1	1	−1	−1	1	−1	1	1	27.871
4	1	1	−1	1	−1	−1	−1	1	27.380
5	−1	−1	1	1	−1	−1	1	1	28.371
6	1	−1	1	−1	1	−1	−1	1	27.891
7	−1	1	1	−1	−1	1	−1	1	28.296
8	1	1	1	1	1	1	1	1	27.815
Contrast	−0.486	−0.076	0.429	−0.001	0.0044	0.0008	1.8E−05		
Effect	−0.243	−0.038	0.215	−0.000	0.0022	0.0004	9.0E−06	27.879	

The extreme limits given above are denoted using the (−1, 1) coding system as shown in Tables 7.3.1 and 7.3.2, on the basis of normalized dimensionless variables, where −1 for low limit and +1 for the high limit. The contrast of "factor a" is calculated using "column a multiplied by column Spring" as follows:

$$a_{contrast} = (-1)(6.994) + 1\,(6.697) + (-1)(6.929) + 1\,(6.631)$$

$$+ (-1)(7.325) + 1\,(7.033) + (-1)(7.260) + 1\,(6.967)$$

$$= -0.2949$$

Thus, $a_{effect} = \frac{1}{2}\,a_{contrast} = -0.1475$

Each effect, main or interactive, is half of its corresponding contrast value, since it is a two-level factorial design. The design contrasts are calculated based on the design matrix of 2^3 as described in Volume 1, Chapter 2, and they are displayed in Tables 7.3.1 and 7.3.2 without further detailed derivations. Since all the interactions (ac, ae, ce, and ace) are small, the predictive equations for the spring gap and stop angle are obtained as follows:

$$g(a, c, e) \approx 6.979 - 0.1475\ a - 0.0327\ c + 0.1668\ e \qquad (g = g_1 = g_2)$$

and $$\varphi(a, c, e) \approx 27.879 - 0.486\ a - 0.076\ b + 0.429\ e$$

Finally, the maximum and minimum values for the spring gap and stop angle can be calculated using the predictive equations given above, respectively, as

$$g_{min}(a, c, e) = g_{min}(1, 1, -1) = 6.632\ \text{mm}$$

$$g_{max}(a, c, e) = g_{max}(-1, -1, 1) = 7.326\ \text{mm}$$

$$\varphi_{min}(a, c, e) \approx \varphi_{min}(1, 1, -1) = 26.89^{\circ}$$

and $$\varphi_{max}(a, c, e) \approx \varphi_{min}(-1, -1, 1) = 28.87^{\circ}$$

7.3.3. Deterministic Tolerance Synthesis

When the factorial DOEs are applied to do the tolerance synthesis, the multifactorial interactions will not be considered as the processes used to generate the tolerances of each component are basically assumed to be independent. A designer may consider the dimensions with tolerances as factors, engineer the manufacturing cost as a response, and select a proper design for the experiment. In addition to the traditional DOEs given in Volume 1, Chapters 2–5, one may incorporate the "constraint" column to represent the tolerance stack-up constraints. After the tolerance stack-up is calculated and the response to each treatment (row; run conditions) is computed, the experimenter can select the row with the smallest response value to fit the tolerance stack-up constraint as the most feasible solution.

7.3.4. Loss Function by Deterministic Tolerance Design

For activities engaged in reliability improvement of products and processes, the proposed concept of loss function [Taguchi 1987] is used mainly to quantify and compare the merits of design improvements in terms of monetary values. The expected saving from an improvement in reliability, i.e., reduced variation in performance around the target, can be translated into cost competitiveness.

Basic ideas covered in this section are to be demonstrated using the following example: to design for display control by a single-sided circuit board. Product robustness is measured on the basis of electric potential (voltage) between two critical points; the desired output value (Y) is 250 ± 20 mV, while each rejected sample will cost the company $40 for the rework.

The five factors related to the output performance in terms of mV ($=10^{-3}$ V) are listed in Table 7.3.3. Each factor has two extreme design levels, which are perceived to be the market demands from the supplier's service department. Factorial design 2_{III}^{5-2} constructed for eight runs, as displayed in Table 7.3.4, is employed to explore the main effects of individual factors. Note that the design matrix is constructed utilizing the following two alias equations:

$$D = -A\ C$$

and $$E = A\ B\ C$$

TABLE 7.3.3 Factors and design levels for experimental tests of a circuit board.

Variables (nominal ± tolerance)	Level (−)	Level (+)	Variate (−1, 1)
A (160 ± 8): transistor	152 hFE	168 hFE	a = (C − 160)/8
B (120 ± 6): resistor B (right side)	114 kΩ	126 kΩ	b = (B − 120)/6
C (40 ± 4): resistor A (left side)	36 kΩ	44 kΩ	c = (A − 40)/4
D (552 ± 25): oscillator, rated	527	577	d = (D − 552)/25
E (0.8 ± 0.04): jumper pin size—OD	0.76 mm	0.84 mm	e = (E − 0.8)/0.04

Significance levels can be also identified using t-ratios based on the combination of the three small effects from factor E, interaction AB, and interaction BC. It is shown that factors A, B, C, and D are statistically significant. The predictive equation is thus

$$Y_p = -3.75\,a - 6.75\,b + 9.25\,c + 9.75\,d$$

Another way to check the significance of each effect is to follow the engineering sense. When the absolute value of the design contrast because of a factor (or interaction) is less than 2% of the average outcome, i.e., 5.085 (254.25 × 2%), it is assumed to be insignificant. The findings are coincident with those based on the statistical t-test.

The predicted values (Y_p) corresponding to the eight treatments are also listed in Table 7.3.4. The correlation between the predicted values (Y_p) and the physical test data (Y) is 98%, which is high enough to conclude that the model is definitely adequate.

TABLE 7.3.4 Factorial design 2_{III}^{5-2} for evaluating voltage output of a circuit board.

Treatment	A	B	C	D	E	AB	BC	Y	Y_p
1	−1	−1	−1	−1	−1	1	1	246	245.75
2	1	−1	−1	1	1	−1	1	259	257.75
3	−1	1	−1	−1	1	−1	−1	232	232.25
4	1	1	−1	1	−1	1	−1	243	244.25
5	−1	−1	1	1	1	1	−1	279	283.75
6	1	−1	1	−1	−1	−1	−1	260	256.75
7	−1	1	1	1	−1	−1	1	275	270.25
8	1	1	1	−1	1	1	1	240	243.25
Contrast	−7.5	−13.5	18.5	19.5	−3.5	−4.5	1.5		
Effect	−3.75	−6.75	9.25	9.75	−1.75	−2.25	0.75	*254.25* (average)	
Error	1.702	1.702	1.702	1.702	1.702	1.702	1.702		
t-ratio	2.204	3.967	5.436	5.73	1.028	1.322	0.441		
$t_{3,1\text{-}90\%}$	1.638	1.638	1.638	1.638	1.638	1.638	1.638		
Significant?	Yes	Yes	Yes	Yes	No	No	No		

Notes:
Y (mV) is the electric potential (10^{-3} V)
Y_p (mV) is the predicted electric potential
D = −AC and E = ABC
R(Y, Y_p) = 98.24% (correlation)

The next step is to deploy the relationship between tolerances (variances) and fabrication cost. Factorial effects resulting from the DOEs show that no interaction is influential. Thus, the analysis of variance (ANOVA) is then applied in order to seek out the relative contributions from individual main effects in light of sums of squares. The percentage of contribution of each factor is attributed to its individual contribution to the subtotal sum of squares (SS_T) without considering the grand average (SS_{ave}). Note that the sum of squares of an individual factor stands for how much it has contributed to the subtotal sum of squares because of the variation of the factor of concern, when moving from the low level to the high level.

One loss function for dealing with the value of tolerancing is identified as the cost competition between fabrication and design for assembly (DFA), i.e., the case of "the closer to the desired nominal value the better," as [Taguchi 1987]

$$\text{Loss} = K\,(Y - Y_0)^2 = K\,(\text{Tolerance})^2 = K\,T^2 \qquad \text{(without mean shift)}$$

or

$$\text{Loss} = K\,[T^2 + (Y_{ave} - Y_0)^2] \qquad \text{(with mean shift)}$$

where

K is the coefficient

T^2 is the variability due to the given tolerance

Y_{ave} is the average

Y_0 is the target or nominal value

The loss function given above consists mainly of two curves, depicted in Figure 7.3.1. One is the cost of fabrication precision as demonstrated in Figure 7.2.1, and the other is the assembly unreliability [Joseph 2004]. Manufacturing cost increases with high-precision fabrication with small tolerances, but it reduces with easy assembly of parts with small tolerances. With regard to rejecting parts that are outside of the required tolerance, DFA requires an inspection of each dimension with very strict guidelines.

FIGURE 7.3.1 Loss function of manufacturing in terms of fabrication precision and DFA.

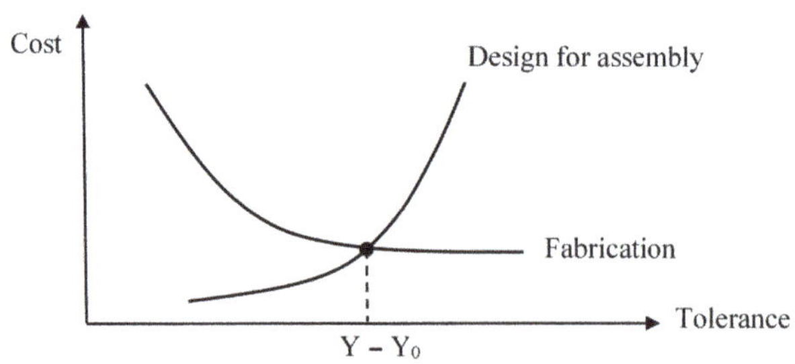

Since no interactions are effective, it is therefore assumed that the cost is proportional to the corresponding size of variation, i.e., the mean square in Table 7.3.5. To the maximum but still staying within the allowable tolerance of ±20 mV, the cost equation given on the basis of $40 rework for being out of the tolerance range is

$$\$40 = K\,(20)^2$$

$$\text{Thus,}\quad K = 0.1\ \$/(\text{mV})^2$$

TABLE 7.3.5 ANOVA of factorial design 2^4 with no interactions.

Variance	SS	DOF	MS	$F_{u,v}$	P-value	Contribution by SS
Regression	$SS_R = 1922$	4	480.5	$F_{4,3} = 20.74$	1.6%	96.5% (SS_R/SS_T)
A	$SS_A = 112.5$	1	112.5	$F_{1,3} = 4.86$	11.5%	5.6% (SS_A/SS_T)
B	$SS_B = 364.5$	1	364.5	$F_{1,3} = 15.73$	2.9%	18.3% (SS_B/SS_T)
C	$SS_C = 684.5$	1	684.5	$F_{1,3} = 29.55$	1.2%	34.4% (SS_C/SS_T)
D	$SS_D = 760.5$	1	760.5	$F_{1,3} = 32.33$	1.1%	38.2% (SS_D/SS_T)
Error	$SS_{error} = 69.5$	3	23.17	—	—	3.5% (SS_{error}/SS_T)
Subtotal	$SS_T = 1991.5$	7	284.5	—	—	100%
Grand ave.	SS_{ave}	1	—	—	—	—
Grand total	SS	8	—	—	—	—

of which mV is mini volts (unit of voltage). By the current design tolerance, one expects to have a loss because of the current practice of design tolerance.

$$\text{Loss} = 0.1\ T^2 = 0.1\ T^2 = 0.1\ (284.5) = \$28.45$$

Because the mean square is the estimate of the variance for that corresponding source of variability, the gage repeatability and producibility are calculated as, respectively,

$$\sigma^2_{error} = MS_E = SS_{error} / 3 = 69.5 / 3 = 23.7$$

The response of each treatment can be assumed to be a normal population with mean variances contributed by individual dimensions. Since the sum of square because of the random error is part of SS_A, SS_B SS_C, or SS_D, the relative contributions to the overall variance contributed by the variances because of these four factors are then

$$\sigma^2_A : \sigma^2_B : \sigma^2_C : \sigma^2_D = (5.6\% - 3.5\%) : (18.3\% - 3.5\%) : (34.4\% - 3.5\%) : (38.2\% - 3.5\%)$$

$$= 2.1\% : 14.8\% : 30.9\% : 34.7\%$$

$$= 1.47\% : 18.1\% : 37.9\% : 42.5\% \quad \text{(Adjusted on the basis of 100\%)}$$

Note that the contribution because of the error is excluded from the contributions of individual factors, as shown above. The loss models abide by individual components are, respectively,

$$\text{Loss-A} = \$28.45\ (1.47\%) = \$0.42$$
$$\text{Loss-B} = \$28.45\ (18.1\%) = \$5.16$$
$$\text{Loss-C} = \$28.45\ (37.9\%) = \$10.77$$
$$\text{Loss-D} = \$28.45\ (42.5\%) = \$12.10$$

According to the calculated potential losses of individual factors, the first cost reduction target is to reduce the oscillator tolerance (component D).

Assume that it will cost \$1.50 to have a new oscillator that has a tolerance of ±3% instead of the original ±4.53%, as derived from ±[½(577 − 527)]/[½(577 + 527)]. The cost-saving with the new oscillator is then

$$\text{Cost reduction-D} = \text{Loss-D}\,[1 - (3\% / 4.53\%)^2] - \$1.50$$
$$= \$12.10\,(56.14\%) - \$1.50$$
$$= \$5.30 \text{ per oscillator}$$

If the annual sales volume of the circuit board is 100,000 units, the cost reduction per year by using the new oscillator will be \$530,000.

7.4. Statistical Tolerance Design by RSS Method

In mechanical engineering, a product design is composed of multiple features with stochastic characteristics and dimensions, of which each comes with tolerance values that control the statistical aspects of those features. A statistical tolerance design is used to understand how these stochastic tolerances contribute to the various performance characteristics of the design. A statistical tolerance analysis is to take the variation of a set of inputs to calculate the expected variation of an output of interest. It should be employed to detect potential problems with mechanical assemblies [Bender 1968] instead of the worst-limit method.

A standard procedure for statistical tolerance analysis is to determine the first four moments of the associated probability density function (PDF) and use these to choose a distribution that describes the system variability. However, in some cases, the actual individual part dimensions fall near the center of the tolerance range and the part dimensions are tightly grouped and within the tolerance range. Assembly of dimensions with these types of tolerance distributions can be solved using only the first two moments, i.e., the mean and standard deviation, excluding higher moments such as skewness and kurtosis. The RSS method, using the mean and standard deviation, is used to do the stack-up analysis of an assembly tolerance chain with large numbers of components in it.

7.4.1. Tolerance Analysis Using RSS Formulae

The RSS method, which is a widely applied statistical tolerance analysis procedure, is based on such an assumption that each component variation follows a normal (i.e., Gaussian) distribution, or almost, and all performance characteristics have a linear relationship with the dimension. This simulation is a technique of setting up a mathematical or logical model of a real situation and then performing sampling experiments on the model. Solid models are more precise and convenient than two-dimensional (2D) engineering drawings as used for deploying the statistical tolerance analysis [Hura et al. 1988]. For simple fit problems, a 1D stack-up is generally sufficient. Assume that outcome y is a function of variates x_1, x_2, x_3, …, and x_N as

$$y = H(x_1, x_2, x_3, \ldots, x_N) \tag{7.4.1a}$$

or approximated by the following quadratic model

$$y \approx \beta_0 + \sum_{n=1}^{N} \beta_n x_n + \sum_{i=1}^{N} \sum_{j=1}^{N} \beta_{ij} x_i x_j + \varepsilon \tag{7.4.1b}$$

of which β_0, β_n, x_n, and β_{ij} are constant and coefficients, respectively. The last term (ε) is the residual and it is assumed to be normally distributed, if the model is proven to be adequate. Taking the differential of variable y, one has an extended Taylor series expansion as follows:

$$dy = \left(\frac{\partial y}{\partial x_1}\right) dx_1 + \left(\frac{\partial y}{\partial x_2}\right) dx_2 + \ldots + \left(\frac{\partial y}{\partial x_N}\right) dx_N$$

$$+\ 1/2 \sum_{i=1}^{N} \sum_{j=1} x_i\, x_j + \text{higher-order terms}$$

$$\text{Then,}\quad \Delta y \approx (\partial y / \partial x_1)\, \Delta x_1 + (\partial y / \partial x_2)\, \Delta x_2 + \ldots + (\partial y / \partial x_N)\, \Delta x_N \tag{7.4.2a}$$

$$\text{or}\quad \sigma_y \approx (\partial y / \partial x_1)\, \sigma_{x1} + (\partial y / \partial x_2)\, \sigma_{x2} + \ldots + (\partial y / \partial x_N)\, \sigma_{xN} \tag{7.4.2b}$$

The corresponding variance of outcome y that can be sourced from the variances of individual independent variables up to the second order, without interactive terms, can be obtained by taking squares on both sides of the above equation.

$$\sigma_y^2 = (\partial y / \partial x_1)^2\, \sigma_{x1}^2 + (\partial y / \partial x_2)^2\, \sigma_{x2}^2 + \ldots + (\partial y / \partial x_N)^2\, \sigma_{xN}^2 \tag{7.4.2c}$$

Similarly, letting $T_y = 3\sigma_y$ (natural tolerance), the related output tolerance is then the variation of y with respect to individual dimensional variations. It can be written as

$$T_y^2 = \left[\ \left(\frac{\partial y}{\partial x_1}\right)^2 T_1^2 + \left(\frac{\partial y}{\partial x_2}\right)^2 T_2^2 + \ldots + \left(\frac{\partial y}{\partial x_N}\right)^2 T_N^2 \tag{7.4.3}$$

where

T_y is the tolerance on fit of the goal

T_n is the tolerance corresponding to design parameter x_n, n = 1, 2, 3, …, and N

$\partial y/\partial x_n$ is the slope, i.e., variation of outcome y with respect to variate x_N

N is the total number of design parameters

The above analytics is based on the second-order tolerance expansion, and it is called the RSS method. Given that the linear relationship as shown in Figure 7.4.1 between response y and five design variables x_1, x_2, x_3, x_4, and x_5, for example, the outcome y is

$$y = x_1 + x_2 + x_3 + x_4 + x_5 \tag{7.4.4}$$

then each partial differentiation in Equation (7.4.3) turns out to be the geometric alignment between vector y and vector x_k, (k = 1, 2, 3, 4, or 5) in the space, i.e., $\cos(\theta_{ky}) = \cos(0°) = 1$ for the assembly configured in Figure 7.4.1. Thus, Equation (7.4.3) reduces to

$$T_y^2 = T_1^2 + T_2^2 + T_3^2 + T_4^2 + T_5^2 \tag{7.4.5}$$

$$\text{or}\quad T_y = [T_1^2 + T_2^2 + T_3^2 + T_4^2 + T_5^2]^{1/2} \tag{7.4.6}$$

The above equation shows that if each dimension varies according to a normal (Gaussian) distribution statistically, the tolerances add as the RSS. Mathematical operations of statistical variables in normal distributions are listed in Table 7.4.1, in which population means and standard deviations are demonstrated. Statistical analyses based on these algorithms are called the explicit probability method.

TABLE 7.4.1 RSS formulae for tolerance analysis based on normal distributions.

Operation	Mean	Standard deviation*
$a = K$	$\mu_a = \mu_K = K$	$\sigma_a = 0$
$b = K x$	$\mu_b = K \mu_x$	$\sigma_b = K^{1/2} \sigma_x$
$c = x + y$	$\mu_c = \mu_x + \mu_y$	$\sigma_c = (\sigma_x^2 + \sigma_y^2)^{1/2}$
$d = x - y$	$\mu_d = \mu_x - \mu_y$	$\sigma_d = (\sigma_x^2 + \sigma_y^2)^{1/2}$
$e = x y$	$\mu_e = \mu_x \mu_y$	$\sigma_e = \mu_x \mu_y [(\sigma_x/\mu_x)^2 + (\sigma_y/\mu_y)^2]^{1/2}$
$f = x/y$	$\mu_f = \mu_x/\mu_y$	$\sigma_f = (\mu_x/\mu_y) [(\sigma_x/\mu_x)^2 + (\sigma_y/\mu_y)^2]^{1/2}$
$g = x^n$	$\mu_g = \mu_x^n$	$\sigma_g = n (\sigma_x/\mu_x) \mu_x^n$
$h = x^{-n}$	$\mu_h = \mu_x^{-n}$	$\sigma_h = n (\sigma_x/\mu_x) \mu_x^{-n}$

where
- K is the constant
- x and y are variables
- n is the real number
- μ is the mean
- σ is the population standard deviation; sample standard deviation S may be used instead, when the number of samples is significantly large

* Some formulae are approximations only

As dealing with a unidirectional assembly that has N dimensions aligned in one and only one direction, the total tolerance can be assessed using the RSS method as [Chase and Greenwood 1988]

$$T = [\sum_{n=1}^{N} T_n^2]^{1/2} \tag{7.4.7}$$

Assume that an entire tolerance T (including ± regions) is divided into K standard deviations. According to Equation (7.4.7), one may relate the total sample standard deviation to the sample standard deviations of the N constituents as

$$K\sigma_T = [\sum_{n=1}^{N} (K\sigma_n)^2]^{1/2} = K[\sum_{n-1}^{N} \sigma_n^2]^{1/2}$$

Thus, $\sigma_T = [\sum_{n=1}^{N} \sigma_n^2]^{1/2}$

and $\sigma_T^2 = \sum_{n-1}^{N} \sigma_n^2$

As long as the operating relationship is linear, the additivity of standard deviations of individual dimensions is valid. In practice, an entire tolerance (±) is commonly assumed to correspond to six sample standard deviations, 6S, i.e., 3S on each side, as a so-called naturalized tolerance. Here, S is used for sample standard deviation and σ is used for population standard deviation. It means that the UL and LL of the resulting dimension are at 3S distance from the mean value. The area of the normal probability density curve between both limits then equals 99.73% of the total area and represents the portion of products meeting the specification requirements. The area outside these limits equals 0.27% and represents off-size products. However, the minimum requirement coming from an automotive manufacturer to a tier-one supplier may request a "±4S" capability, i.e., 99.994% for the process yield [AIAG 2008], with 63 rejects per million components produced; a "±5S" capability with 0.6 rejects per million components produced is preferred.

Equation (7.4.7) can be modified for a multidirectional assembly as

$$T = \{ \sum_{n=1}^{N} [|\partial H(x_1, x_2, \ldots, x_N) / \partial x_n|^2 \, T_n^2]\}^{1/2} \tag{7.4.8}$$

Based on the central limit theorem, if one draws samples from a population that is greater than or equal to 30 the sample mean is approximately a normally distributed random variable. It means that the variability contributions of all terms in the linear combination become negligible when the data originate from many samples, even if the original distribution function of dimension x_n is not really normal-distributed. In spite of the relaxed distributional assumptions for the part dimensions, it has to be emphasized that the assembly criterion is again approximately normally distributed and its assembly mean μ_y coincides with its desired corresponding nominal value.

7.4.2. Modified Root-Mean-Squared Method

When skewness and kurtosis are taken into consideration, the RSS method for a general case with non-normal-distributed dimensions may not be a proper approach. However, this can be corrected using the PDF-oriented correction factors; it is called the modified root-mean-squared method. Equation (7.4.8) is then rewritten as

$$T = \{ \sum_{n=1}^{N} [\, C_n \, |\partial H(x_1, x_2, \ldots, x_N) / \partial x_n|^2 \, (T_n)^2]\}^{1/2} \tag{7.4.9}$$

where

T is the assembly tolerance

T_n is the individual tolerance of dimension n, where $1 \le n \le N$

$H(x_N)$ is the assembly function

C_n is the correction factor accounting for abnormality of dimension n, usually $1 \le C_n \le 2$

The above equation yields tolerance stacking formulae that essentially agree with Equation (7.4.8), except that an inflation factor, c, has been added as a correction factor [Greenwood and Chase 1990]. Typically, a correction factor for applying the RSS method falls in the range of 1.48 to 1.8 [Drake 1999] as exhibited in Figure 7.4.1, depending on the PDF.

FIGURE 7.4.1 Correction factors for applying the RSS method to tolerance analysis in different statistical distributions [Scholz 1995].

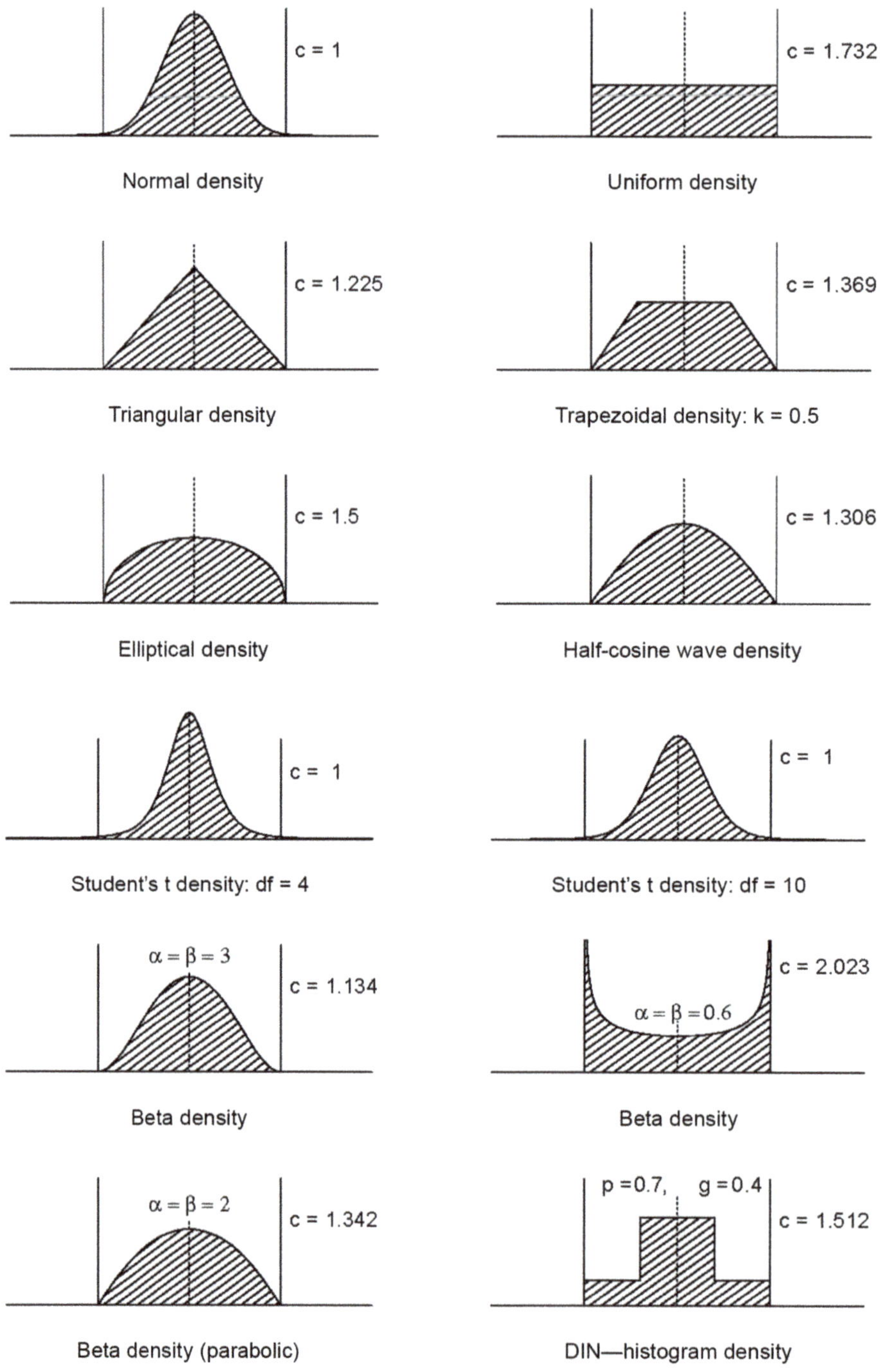

Example 7.4.1 is given to illustrate how to apply statistical tolerance analysis using mathematical formulae in Table 7.4.1. In general, the geometry considered for a linear case study is simplified to be a 1D problem without any complication of vector operations for the purpose of illustration.

EXAMPLE 7.4.1

A simple gear set is assembled as shown in Figure 7.4.2. One of the functional requirements is to assure the axial clearance ($F > 0$). Given that $a = 140 \pm 0.03$, $b = 20 \pm 0.01$, $c = 120 \pm 0.02$, and $d = 300.10 \pm 0.04$, how much is the axial clearance if dimensions a, b, c, and d are the only dimensions of concern?

Solution:

The loop chain of dimensions to predict the nominal, upper control limit (UCL), and lower control limit (LCL) can be formulated as

$$F = -b + d - c - b - a$$

Two different approaches to predicting the clearance (gap) are taken as follows:

- Worst-Limit Study:
  - Nominal: $F = -20 + 300.10 - 120 - 20 - 140 = 0.10$
  - UL: $F_U = 0.01 + 0.04 + 0.02 + 0.01 + 0.03 = 0.11$
  - LL : $F_L = -0.01 - 0.04 - 0.02 - 0.01 - 0.03 = -0.11$
  - Clearance/interference: $-0.01 \leq F \leq 0.21$
- RSS:
  - Nominal: $F = -20 + 300.10 - 120 - 20 - 140 = 0.10$
  - UL: $F_U = -[0.01^2 + 0.04^2 + 0.02^2 + 0.01^2 + 0.03^2]^{1/2} = 0.0557$
  - LL : $F_L = -[0.01^2 + 0.04^2 + 0.02^2 + 0.01^2 + 0.03^2]^{1/2} = -0.0557$
  - Clearance: $0.0443 \leq F \leq 0.1557$

In general, results from RSS method are more truthful.

FIGURE 7.4.2 Dimension chain for linear tolerance stack-up analysis of a gear set [Navas 2015].

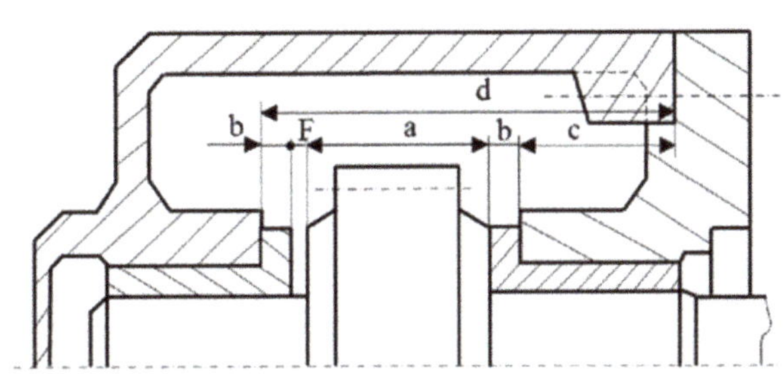

The statistical tolerance analysis based on a root-mean-squared approach using Table 7.4.1 is effective. However, all parts exist in the space in three-dimensional forms. It would be very hard to apply the closed-form equations to space vectors. Advanced statistical methods are sometimes used for the tolerance analysis because they permit non-normal distributions as models of component variation. These methods can give much better estimates of the number of rejects than simple statistical analysis, when the component distributions are well-known non-normal functions. Three other general statistical methods that have been applied to tolerance analysis are Monte Carlo simulation, method of moments, and maximum-likelihood method.

7.4.3. Tolerance Allocation by Machining Capability

A rotating shaft is assembled in its housing with bearings, as shown in Figure 7.4.3. A positive clearance with no possibility of interference is mandated for such a design. A high cost that could incur because of tight tolerances is not desired. How the tolerances are put on allocation to reduce the cost. Assume that the clearance between the bearing and the retaining ring is expected to be 0.508 ± 0.381 mm. Based on the linear dimensional chain, one has the following assembly equation for clearance (H):

$$H = -A + B - C + D - E + F - G$$

where

$A = 1.283 \pm 0.0381$
$B = 203.200 \pm 0.2032$
$C = 12.936 \pm 0.0635$
$D = 10.160 \pm 0.0508$
$E = 185.860 \pm 0.1524$
$F = 10.160 \pm 0.0508$
$G = 12.934 \pm 0.0635$

FIGURE 7.4.3 Rotating shaft assembled in the housing with bearings [Chase and Greenwood 1988].

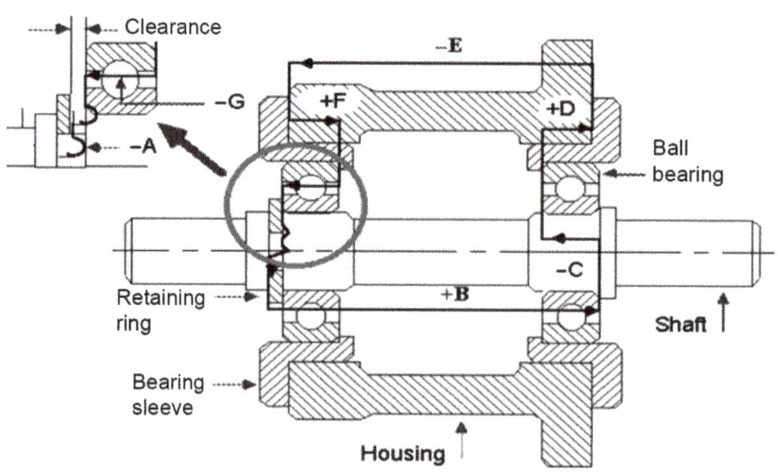

Tolerance data given above are the initial tolerances of geometric dimensions and features for the round-one design. The nominal value and tolerance based on the RSS method are derived, respectively, as follows:

$$H_{nominal} = -A + B - C + D - E + F - G$$
$$= -1.283 + 203.200 - 12.936 + 10.160 - 195.860 + 10.160 - 12.934$$
$$= 0.508 \text{ mm}$$
$$T_H = (T_A^2 + T_B^2 + T_C^2 + T_D^2 + T_E^2 + T_F^2 + T_G^2)^{1/2}$$
$$= (0.0381^2 + 0.2032^2 + 0.0635^2 + 0.0508^2 + 0.1524^2 + 0.0508^2 + 0.0635^2)^{1/2}$$
$$= 0.2814 \text{ mm}$$

Thus, H = 0.508 ± 0.2814, of which the tolerance is tighter than the allowed value (i.e., 0.381 mm).

Although the tolerance calculated above (i.e., 0.2814 mm) meets the expected tolerances (i.e., 0.381 mm), the head of the financial department requests a cost reduction, and therefore, a tolerance allocation scheme is to be applied for that purpose. Assume that the machining capability is proportional to $L^{1/3}$, where L is the size, as given by Equation (7.2.5); i.e., $T = P\,L^{1/3}$.

The characteristics that are going to take tolerance modifications have been chosen, and they are B, D, E, and F. Having the same proportionality for all these four parts, one has the new assembly equation for tolerances

$$T_H^2 = T_A^2 + T_{B*}^2 + T_C^2 + T_{D*}^2 + T_{E*}^2 + T_{F*}^2 + T_G^2$$
$$= T_A^2 + T_C^2 + T_G^2 + P^2\,(T_{B*}^2 + T_{D*}^2 + T_{E*}^2 + T_{F*}^2) \tag{7.4.10}$$

or

$$T_H^2 = T_A^2 + T_C^2 + T_G^2 + P^2\,(L_B^{2/3} + L_D^{2/3} + L_E^{2/3} + L_F^{2/3}) \tag{7.4.11}$$

where

* denotes the new tolerances that are to be used for characteristics B, D, E, and F

L_B, L_D, L_E, L_F are the sizes of characteristics B, D, E, and F, respectively

Substituting the corresponding tolerances of parts A, C, and G, as well as dimensions of parts B, D, E, and F, into the above equation, one has

i.e., $$0.381^2 = 0.0381^2 + 0.0635^2 + 0.0635^2 + P^2\,(203.2^{2/3} + 10.16^{2/3} + 195.859^{2/3} + 10.16^{2/3})$$

Thus, $P = 0.04179$

As allocated, the new tolerances for parts D, E, F, and G are, respectively,

$$T_{B*} = P\,L_B^{1/3} = 0.04179\,(203.2)^{1/3} = 0.2457 \text{ mm}$$
$$T_{D*} = P\,L_D^{1/3} = 0.04179\,(10.16)^{1/3} = 0.0905 \text{ mm}$$
$$T_{E*} = P\,L_B^{1/3} = 0.04179\,(185.86)^{1/3} = 0.2427 \text{ mm}$$

and

$$T_{F*} = P\,L_B^{1/3} = 0.004179\,(10.16)^{1/3} = 0.0905 \text{ mm}$$

Each new tolerance listed above is scaled from the originally given tolerance according to its corresponding part size.

7.4.4. Tolerance Allocation by Cost–Benefit Equation

Assume that the associated cost function is proportional to the tolerance constraint of the target value as shown in Figure 7.2.1. The cost function is proportional to the statistical RSS model for tolerance accumulation constraint:

$$(\text{Cost function}) \propto \Sigma T_i^2 \tag{7.4.12}$$

or

$$(\text{Cost function}) = -\psi\,\Sigma T_i^2 \tag{7.4.13}$$

of which ψ is the cost sensitivity (proportionality) and is expected to be constant. This is called the Lagrange multiplier method. It is an efficient tolerance design tool that yields a closed-form solution for the optimum tolerances. It applies merely to continuous cost–tolerance relationships, because it requires a continuous first derivative.

The "negative sign" in Equation (7.4.13) means that an increasing tolerance (denominator) results in a decreasing cost (numerator). A partial differentiation of the cost function with respect to each individual tolerance, T_n, where n = 1, 2, …, and N, will lead to the minimum cost as

$$\partial(\text{Cost function}) / \partial T_n = -\psi\,[\partial(\Sigma T_i^2) / \partial T_n] \tag{7.4.14}$$

i.e.,

$$\partial(\text{Cost function}) / \partial T_n + \psi\,[\partial(\Sigma T_i^2) / \partial T_n] = 0$$

It means that the variation of the cost function with respect to the tolerance change in a dimensional characteristic is proportional to the variation of the tolerance constraint of the target value with respect to such a tolerance change.

Specifically, the cost function as the reciprocal of tolerance, i.e., Equation (7.2.4), i.e., $\text{cost}_i = p_i + q_i/T_i$, is taken here for the given example, and coefficient q_i relates the cost to the tolerance of dimensional characteristic i, where i = A, B, C, D, E, F, and G for the components addressed in the given example. Assume that the following data are provided by the component suppliers for both setup and variable costs due to the original tolerances:

$$\text{Cost}_A = p_A + q_A / T_A = \$0.15 + q_A / T_A \tag{7.4.15}$$

$$\text{Cost}_B = p_B + q_B / T_B = \$4.75 + \$1.20\,(0.2032) / T_B \tag{7.4.16}$$

$$\text{Cost}_C = p_C + q_C / T_C = \$2.50 + q_C / T_C \tag{7.4.17}$$

$$\text{Cost}_D = p_D + q_D / T_D = \$2.80 + \$2.75\,(0.0508) / T_D \tag{7.4.18}$$

$$\text{Cost}_E = p_E + q_E / T_E = \$4.50 + \$0.88\,(0.1524) / T_E \tag{7.4.19}$$

$$\text{Cost}_F = p_F + q_F / T_F = \$2.80 + \$2.75\,(0.0508) / T_F \tag{7.4.20}$$

and

$$\text{Cost}_G = p_G + q_G / T_G = \$2.50 + q_G / T_G \tag{7.4.21}$$

where

p_i is a constant, i = A, B, C, D, E, F, and G
q_i is the proportionality, i = A, B, C, D, E, F, and G

The total cost function is

$$\text{Cost function} = \text{Cost}_A + \text{Cost}_B + \text{Cost}_C + \text{Cost}_D + \text{Cost}_E + \text{Cost}_F + \text{Cost}_G \tag{7.4.22}$$

All the data on p_i and q_i given above are estimated on the basis of original tolerances and the related costs. For example, when the given tolerance for dimensional characteristic B is 0.2032, its cost is \$1.20. Thus, its $q_B = \$1.20 * 0.2032$. Note that the tolerances of dimensional characteristics A, C, and G are fixed, and thus, there is no need to estimate their respective cost sensitivities.

Let us consider dimension A first. Substituting Equation (7.4.22) into Equation (7.4.14) yields the cost sensitivities for dimension A,

$$\text{LHS:} \quad \partial(\text{Cost function}) / \partial T_A = \partial(\text{Cost}_A) / \partial T_A = -q_A / T_A^2$$

$$\text{RHS:} \quad -\psi\left[\partial\left(\sum \text{Ti}^2\right) / \partial T_A\right] = -\psi\,(2T_A)$$

Since LHS = RHS, one has $-q_A/T_A^2 = -\psi\,(2T_A)$, which yields

$$\psi = \tfrac{1}{2}\, q_A / T_A^3 \tag{7.4.23}$$

By the same token, one can have $\psi = \frac{1}{2}\, q_B/T_B^3 = \frac{1}{2}\, q_C/T_C^3 = \frac{1}{2}\, q_D/T_D^3 = \frac{1}{2}\, q_E/T_E^3 = \frac{1}{2}\, q_F/T_F^3 = \frac{1}{2}\, q_G/T_E^3$. Thus,

$$\psi = \tfrac{1}{2}\, q_n / T_n^3 \tag{7.4.24}$$

where n = 1, 2, 3, …, N for a general case of N dimensions. Since parameter ψ is the global proportionality, which is assumed to be a constant, one can therefore refer each cost–benefit value to the first tolerance as

$$q_n / T_n^3 = q_1 / T_1^3 \tag{7.4.25}$$

Plugging the above equation into Equation (7.4.5) for N-dimensional characteristics yields

$$\begin{aligned} T^2 &= \sum T_n^2 \\ &= T_1^2 + (q_2 / q_1)^{2/3}\, T_1^2 + (q_3 / q_1)^{2/3}\, T_1^2 + \ldots + (q_N / q_1)^{2/3}\, T_1^2 \\ &= T_1^2\, [1 + (q_2 / q_1)^{2/3} + (q_3 / q_1)^{2/3} + \ldots + (q_N / q_1)^{2/3}] \end{aligned}$$

$$\text{or} \quad T_1 = T / [1 + (q_2 / q_1)^{2/3} + (q_3 / q_1)^{2/3} + \ldots + (q_N / q_1)^{2/3}]^{1/2} \tag{7.4.26}$$

Substituting the above equation into Equation (7.4.25) to eliminate T_1 leads to the minimum cost resulting from varying the dimensional characteristics.

$$q_n = T_n^3\, q_1 / T_1^3$$

$$= \frac{q_1\, T_n^3}{T^3 / [1 + (q_2 / q_1)^{2/3} + (q_3 / q_1)^{2/3} + \ldots + (q_N / q_1)^{2/3}]^{3/2}}$$

$$= \{q_1\, [1 + (q_2 / q_1)^{2/3} + (q_3 / q_1)^{2/3} + \ldots + (q_N / q_1)^{2/3}]^{3/2}\}\, (T_n / T)^3$$

$$= (q_1^{2/3} + q_2^{2/3} + q_3^{2/3} + \ldots + q_N^{2/3})^{3/2}\, (T_n / T)^3 \qquad (7.4.27)$$

or

$$T_n = \frac{q_n^{1/3}}{(q_1^{2/3} + q_2^{2/3} + q_3^{2/3} + \ldots + q_N^{2/3})^{1/2}}\, T \qquad (7.4.28)$$

For the given example,

$$T_n = \frac{q_n^{1/3}}{(q_A^{2/3} + q_{B*}^{2/3} + q_C^{2/3} + q_{D*}^{2/3} + q_{E*}^{2/3} + q_{F*}^{2/3} + q_G^{2/3})^{1/2}}\, T$$

Here, the above equation is rewritten individually for the components that get involved with the cost reduction only.

$$T_B = \frac{q_{B*}^{1/3}}{(q_A^{2/3} + q_{B*}^{2/3} + q_C^{2/3} + q_{D*}^{2/3} + q_{E*}^{2/3} + q_{F*}^{2/3} + q_G^{2/3})^{1/2}}\, T \qquad (7.4.29a)$$

$$T_D = \frac{q_{D*}^{1/3}}{(q_A^{2/3} + q_{B*}^{2/3} + q_C^{2/3} + q_{D*}^{2/3} + q_{E*}^{2/3} + q_{F*}^{2/3} + q_G^{2/3})^{1/2}}\, T \qquad (7.4.29b)$$

$$T_E = \frac{q_{E*}^{1/3}}{(q_A^{2/3} + q_{B*}^{2/3} + q_C^{2/3} + q_{D*}^{2/3} + q_{E*}^{2/3} + q_{F*}^{2/3} + q_G^{2/3})^{1/2}}\, T \qquad (7.4.29c)$$

and

$$T_F = \frac{q_{F*}^{1/3}}{(q_A^{2/3} + q_{B*}^{2/3} + q_C^{2/3} + q_{D*}^{2/3} + q_{E*}^{2/3} + q_{F*}^{2/3} + q_G^{2/3})^{1/2}}\, T \qquad (7.4.29d)$$

Note that "*" is used to denote the cost sensitivities of the corresponding dimensional characteristics B, D, E, and F that are to be altered to reduce the cost, regardless of dimensional characteristics A, C, and G. Because their corresponding dimensional characteristics remain the same, cost sensitivities q_A, q_C, and q_G are kept constant. Equations (7.4.29a)–(7.4.29d) exhibit the relationships between T_B, T_D, T_E, and T_F and any given assembly tolerance T. With the allowable, i.e., T = 0.381 mm, desired for cost reduction specifically for this problem, one can calculate the four unknowns, T_B, T_D, T_E, and T_F directly.

$T_B = 0.2004$ mm (Originally $T_B = 0.2032$ mm)

$T_D = 0.1806$ mm (Originally $T_D = 0.0508$ mm)

$T_E = 0.1735$ mm (Originally $T_E = 0.1524$ mm)

and $T_F = 0.1806$ mm (Originally $T_F = 0.0508$ mm)

Substituting the above equations into Equations (7.4.15)–(7.4.21) and summing up the cost differences between the new tolerances and the old tolerances, one can calculate how much the cost reduction in total, denoted by ΔCost, as

$$\begin{aligned}\Delta Cost &= \Delta Cost_B + \Delta Cost_D + \Delta Cost_E + \Delta Cost_F \\ &= \{[\$4.75 + (\$1.20)(0.2032/0.2004)] - (\$4.75 + \$1.20)\} \\ &\quad + \{[\$2.80 + (\$2.75)(0.0508/0.1806)] - (\$2.80 + \$2.75)\} \\ &\quad + \{[\$4.50 + (\$0.88)(0.1524/0.1735)] - (\$4.50 + \$0.88)\} \\ &\quad + \{[\$2.80 + (\$2.75)(0.0508/0.1806)] - (\$2.80 + \$2.75)\} \\ &= -\$4.48\end{aligned}$$

7.4.5. Nonlinear Case: Capability Study on One-Way Clutch

On the one-way clutch depicted in Figure 7.1.3, Equations (7.1.5) and (7.1.6) are identified to locate the center of the ball bearing. The next step is to figure out the spring gap and the angular position of the roller center when subjected to dimensional variations of hub thickness a, roller diameter c, and inner radius of the race ring e, whose dimensions and tolerances are given as follows:

$$A \pm T_a = 46.74 \pm 0.156$$
$$C_1 \pm T_c = 22.86 \pm 0.013$$
$$C_2 \pm T_c = 22.86 \pm 0.013$$
and
$$E \pm T_e = 101.6 \pm 0.156$$

The UCL and LCLs for the spring gap are, respectively, 6.85 and 7.15, i.e., 5% variation of the nominal size of the spring length (7 mm as given), as demanded.

7.4.5.1. Capability Study on Gap

How can the process capability of the one-way clutch with respect to such a packaging requirement for the spring operate properly? First, let us calculate the nominal value of spring gap g_1. By Equation (7.1.3),

$$\begin{aligned}g_1 &= \tfrac{1}{2}[(E - \tfrac{1}{2}C_1 - \tfrac{1}{2}C_2)^2 - (A + \tfrac{1}{2}C_1 + \tfrac{1}{2}C_2)^2]^{1/2} - \tfrac{1}{2}C_1 \\ &= \tfrac{1}{2}[(101.6 - 11.43 - 11.43)^2 - (46.74 + 11.43 + 11.43)^2]^{1/2} - 11.43 \\ &= 6.9808 \end{aligned} \quad \text{(a)}$$

Taking partial differentiations of all terms on both sides of Equation (7.1.3) with respect to the three design parameters (i.e., a, c, and e) yields, respectively,

$$\begin{aligned}\partial g_1/\partial A &= -\tfrac{1}{2}(A + \tfrac{1}{2}C_1 + \tfrac{1}{2}C_2)[(E - \tfrac{1}{2}C_1 - \tfrac{1}{2}C_2)^2 - (A + \tfrac{1}{2}C_1 + \tfrac{1}{2}C_2)^2]^{-1/2} \\ &= -\tfrac{1}{2}(46.74 + 22.86)[(101.6 - 22.86)^2 - (46.74 + 22.86)^2]^{-1/2} \\ &= -0.9451\end{aligned} \quad \text{(b)}$$

$$\begin{aligned}\partial g_1/\partial C_1 &= -\tfrac{1}{2}\{(A + \tfrac{1}{2}C_1 + \tfrac{1}{2}C_2)[(E - \tfrac{1}{2}C_1 - \tfrac{1}{2}C_2)^2 - (A + \tfrac{1}{2}C_1 + \tfrac{1}{2}C_2)^2]^{-1/2} - 0.5 \\ &= -\tfrac{1}{2}(46.74 + 22.86)[(101.6 - 22.86)^2 - (46.74 + 22.86)^2]^{-1/2} - 0.5 \\ &= -1.5451\end{aligned} \quad \text{(c)}$$

and

$$\begin{aligned}\partial g_1/\partial E &= \tfrac{1}{2}(E - \tfrac{1}{2}C_1 - \tfrac{1}{2}C_2)[(E - \tfrac{1}{2}C_1 - \tfrac{1}{2}C_2)^2 - (A + \tfrac{1}{2}C_1 + \tfrac{1}{2}C_2)^2]^{-1/2} \\ &= \tfrac{1}{2}(101.6 - 22.86)[(101.6 - 22.86)^2 - (46.74 + 22.86)^2]^{-1/2} \\ &= 1.0692\end{aligned} \quad \text{(d)}$$

The tolerance for the spring gap is hence calculated using Equation (7.4.8):

$$T_{g1} = [(\partial g_1 / \partial A)^2 (T_A)^2 + (\partial g_1 / \partial C)^2 (T_C)^2 + (\partial g_1 / \partial E)^2 (T_E)^2]^{1/2}$$

$$= [(-0.9451)^2 (0.156)^2 + (-1.4451)^2 (0.013)^2 + (1.0692)^2 (0.156)^2]^{1/2}$$

$$= [0.02174 + 0.000353 + 0.002782]^{1/2} \quad (e)_1$$

$$= 0.2234 \qquad (3S_g \text{ is taken as the natural tolerance}) \quad (e)_2$$

Thus, $g_1 = 6.9808 \pm 0.2234$ (f)

and $S_{g1} = T_{g1} / 3 = 0.07447$ (Sample standard deviation of spring gap) (g)

Process capability index C_{pk} for g_1 can then be obtained using Equation (4.6.7) as

$$C_{pk,g1} = \min(|6.9808 - 6.85| / 0.2234, |7.15 - 6.9808| / 0.2234)$$

$$= 0.56 \quad (h)$$

which is the process capability of the spring gap as designed. Since process capability $C_{pk,g1} < 1.67$, the design is not robust and there is room for improvement. The relative sensitivities of the spring gap with respect to variations of the three independent dimensions can be obtained from Equation $(e)_1$ as

$$A: C: E = 0.02174 : 0.000353 : 0.02782$$

$$= 43.55\% : 0.71\% : 55.74\% \quad (i)$$

Tolerances of dimensions A and B can be improved for enhancing the required C_{pk}.

7.4.5.2. Capability Study on Angular Position

Next, let us calculate the nominal angular position of the roller center using Equation (7.1.5).

$$\varphi = \cos^{-1}[(A + ½ C_1 + ½ C_2) / (E - ½ C_1 - ½ C_2)]$$

$$= \cos^{-1}[(46.74 + 11.43 + 11.43) / (101.6 - 11.43 - 11.43)]$$

$$= 27.881° \quad (j)$$

Taking partial differentiations of the terms on both sides of Equation (7.1.5) with respect to the three given design parameters a, c, and e, one has, respectively,

$$\partial\varphi / \partial A = -\{1 - [(A + ½ C_1 + ½ C_2) / (E - ½ C_1 - ½ C_2)]^2\}^{-1/2} / (E - ½ C_1 - ½ C_2)$$

$$= -\{1 - [(46.74 + 22.86) / (101.6 - 22.86)]^2\}^{-1/2} / (101.6 - 22.86)$$

$$= -0.02716 \qquad \text{(k)}$$

$$\partial\varphi / \partial C = -\{1 - [(A + ½ C_1 + ½ C_2) / (E - ½ C_1 - ½ C_2)]^2\}^{-1/2}$$

$$[(A + E) / (E - ½ C_1 - ½ C_2)^2]$$

$$= -\{1 - [(46.74 + 22.86) / (101.6 - 22.86)]^2\}^{-1/2}$$

$$[(46.74 + 101.6) / (101.6 - 22.86)^2]$$

$$= -0.05116 \qquad \text{(l)}$$

and

$$\partial\varphi / \partial E = -\{1 - [(A + ½ C_1 + ½ C_2) / (E - C)]^2\}^{-1/2}$$

$$[(A + ½ C_1 + ½ C_2) / (E - ½ C_1 - ½ C_2)^2]$$

$$= -\{1 - [(46.74 + 22.86) / (101.6 - 22.86)]^2\}^{-1/2}$$

$$[(46.74 + 22.86) / (101.6 - 22.86)^2]$$

$$= -0.02401 \qquad \text{(m)}$$

The tolerance for the spring gap is hence calculated using Equation (7.4.8).

$$T_\varphi = [(\partial\varphi / \partial A)^2 (T_a)^2 + (\partial\varphi / \partial C)^2 (T_C)^2 + (\partial\varphi / \partial E)^2 (T_E)^2]^{1/2}$$

$$= [(-0.02716)^2 (0.156)^2 + (-0.05116)^2 (0.013)^2 + (-0.02401)^2 (0.156)^2]^{1/2}$$

$$= [0.00001795 + 0.000000442 + 0.00001402]^{1/2} \qquad (n)_1$$

$$= 0.00569 \qquad (3S_\varphi \text{ is taken as the natural tolerance}) \qquad (n)_2$$

Thus, $S_\varphi = T_{g1} / 3$

$$= 0.001898 \text{ (Sample standard deviation of angular position)} \qquad \text{(o)}$$

Process capability index C_{pk} for φ can then be obtained using Equation (3.4.7) as

$$C_{pk,\varphi} = \min(|27.8809 - 27.8| / 0.00569, |27.9 - 27.8809| / 0.00569)$$

$$= 3.36 \qquad \text{(p)}$$

which is the process capability of angular position. Since process capability $C_{pk,\varphi} > 1.67$, it fits a robust design. The relative sensitivities of the angular position with respect to variations of the three independent dimensions can be obtained from Equation $(n)_1$ as

$$A : C : E = 0.00001795 : 0.000000442 : 0.00001402 = 55.37\% : 1.36\% : 43.26\%$$

7.5. Statistical Tolerance Design by Monte Carlo Simulations

As a more generalized problem-solving process, Monte Carlo simulations are conducted using the identifiable PDFs of individual dimension or feature and then applying Equation (7.4.1b) directly for analyzing the statistical tolerance stack-up. It is based on the use of a random number generator to simulate the effects of manufacturing variations on assemblies. The method of Monte Carlo simulations is such a powerful tool for tolerance design of mechanical assemblies that it can be used for a general case study involving both nonlinear assembly functions and non-normal distributions. Advantages of Monte Carlo simulations over the explicit probability method (e.g., RSS method) are listed as follows:

(a) The probability of occurrence is taken into consideration, when assembly tolerances are established.
(b) The flexibility to handle any situation that can be mathematically modeled using nominal data and tolerances simultaneously. It is the most realistic method for tolerance analysis.
(c) It can also be used for nondimensional studies.
(d) The methodology is strongly related to GD&T.

The method of Monte Carlo simulations is a standard procedure for statistical tolerance analysis that implicitly accommodates the first four moments of the associated PDF. It is an implicit probability method that can operatively analyze the variability of the process before the life production cycle has occurred. A statistical distribution function (i.e., Weibull distributed data) can be assessed using Monte Carlo simulations with the given parameters. The effects can then be estimated and used to develop a reliability model [Kremer et al. 2020].

7.5.1. Concerns of Dimensional Control over Normality Assumption

Quality control methods are usually formulated under the assumption of normal distribution. These techniques do not utilize information on the third and fourth moments, and are valid only if the large sample size is warranted. Even if the distribution shape is well-known by design, and this information is used to refine component tolerances, the existing quality control techniques may not be sufficiently robust to predict out-of-control conditions if higher moments vary. The normality assumption is true only if the central limit theorem is proven to be applicable. The following open issues remain:

(a) Tolerance zones are modeled as normal distributions. Tolerance zones $\pm 3\sigma$ used in the RSS analysis represent six-sigma distribution; i.e., the zone size given in a GD&T feature control frame represents 99.73% of the modeled tolerance zone with the remaining 0.27% being simulated outside the tolerance zone. This is called the natural tolerance zone.
(b) All components in the analysis are modeled as rigid bodies, which need to be refined for flexible bodies. Thermal effects and mechanical loads that render deformations are not represented in the model.

In fact, the functional equation is usually nonlinear and input variables are not normally distributed for realistic applications. Because of its generality, and applicability to most skewed PDFs, Weibull distributions may be employed to model performance behaviors or failure modes of a mechanical system. Utilizations of Weibull statistics are recommended as a versatile approach to tolerance analysis. Approximations of lognormal and normal distributions by Weibull distribution functions are feasible.

The formulae in Table 7.4.1 are mainly used to illustrate the concept of statistical tolerance analysis; nevertheless, it is hard to fulfill the calculations based on RSS formulae in solving some realistic tolerance analysis problems encountered in a 3D mechanical assembly with complex geometry. The RSS method is a quick but effective approach to resolving the statistical tolerance stack-up of some scenarios where the inputs are normal and the assembly relationships are linear, but a second-order tolerance analysis is required to address the real world of manufacturing.

The method of Monte Carlo simulations is a valid alternative course. The method of Monte Carlo simulations is the most popular and simplest method for nonlinear statistical tolerance analysis in practice [Yan et al. 2015]. Nevertheless, if all involved part dimensions can be described by normal distributions, what results from the method of Monte Carlo simulations is expected to equal that obtained from the RSS method.

7.5.2. Monte Carlo Simulations

The "Monte Carlo" methodology is a simulation algorithm, also called probability simulation, based on the random generation of input dimensions within the selected tolerance ranges according to the rated distribution functions of dimensional characteristics. It simulates fabrication of each component, an assembly, and a subsequent checking of functional conditions such as clearances or interferences. While simulations with many dimensional tolerances can be cumbersome, doing complex Monte Carlo simulations may boil down to the following steps:

1. Identify a set of constraint equations, also called transfer functions, in light of the measurable objective function of the activity or process, e.g., the clearance between two identified planes. A constraint equation (transfer function) relates the objective function to the given variates (dimensions).
2. Decide the most fit statistical model, such as normal, lognormal, Weibull, and other distributions, using the measured data derived from production tools if possible, as described in Volume 1, Section 2.8. If no data are available, one may assume a normal distribution.
3. Define the statistical parameters, e.g., the mean and standard deviation for a normal distribution or the shape factor, scale factor, and guaranteed value for a Weibull distribution.
4. Create random data using random generator in computer software such as Excel, Minitab, and Reliasoft.
5. Map the first set of random data of all dimensions of concern back to its corresponding statistical model. Calculate the objective function value using all the constraint equations for this first set of data. Note that these constraint equations could be quite nonlinear.
6. Repeat steps (4) and (5) for the user-defined number of simulations that is usually more than 5000 times, while a number of 30,000 times or more makes the better fit. In other words, there will be more than 5000 objective function value calculated from more than 5000 virtual assemblies.
7. Plot the histogram using the 5000 data or more to evaluate the process capability.
8. Conduct the tolerance allocation scheme to improve the product performance and/or financial impact.

With companion by computer software, e.g., Excel and Minitab, engineers can easily perform a Monte Carlo analysis based on the procedure outline above in order to:

(a) Simulate product results while accounting for the variability in the inputs.
(b) Optimize process settings.
(c) Identify critical-to-quality factors.
(d) Find a solution to reduce defects.

Fortunately, each dimension tolerance is stochastic by nature and can be assigned a value randomly using Monte Carlo simulations. Each geometric tolerance (e.g., flatness) can be treated as a dimensional tolerance whose nominal value is zero, where Monte Carlo simulations apply.

7.5.3. Characterizing Dimensions Using Statistical Distributions

Each dimension is assumed to have a certain statistical distribution function, e.g., normal, lognormal, or Weibull distribution. First, a random number generator has to be identified. The results obtained from the random number generator are used to describe the likelihood of reaching various results in the statistical model. After a random number is generated, it is scaled down to a number between 0 and 1, which is mapped onto the accumulative statistical distribution function of the dimension. The apparent dimension is then the dimensional size producing the accumulative probability following the statistical distribution function. Monte Carlo is an efficient tool to simulate complex physical and mathematical systems, but large sample sizes are usually required to obtain acceptable accuracy.

7.5.3.1. Random Number Generation

For Monte Carlo simulations, one isolates a number of key variables that control and describe the outcome of the experiment and then assigns a probability distribution after many random samples are performed. The mixed congruent method is a popular choice for generating a sequence of random numbers. It starts with calculating the next random number from the last one obtained, such as

$$r_{n+1} = (a\, r_n + c)\ (\text{modulo } m) \tag{7.5.1}$$

where

r_n is the known random number, which begins with an initial seed corresponding to n = 0
r_{n+1} is the next random number to be generated
a is a coefficient
c is a constant
m is the total data to be generated

For example, the following parameters can be used for a reasonable size of data for 8 (m = 8): a = 5, c = 7, and seed r_0 = 4. This is one of the simple equations to generate random numbers. After a random number is generated from Equation (7.5.1) or other generators, it is scaled down to a number between 0 and 1 such that a uniform distribution function for the generated random number is presumed to be true.

In Excel, RAND() is the formula utilized to generate a set of random numbers, falling between 0 and 1, which can be interpreted as the cumulative distribution function (CDF) value selected randomly. Assume that one is dealing with dimension 48.31 ± 2.67, which is assumed to be normally distributed. When the formula for generating random numbers is initiated in Excel, i.e., putting "=RAND()" in Excel cell, one gets a number (namely x) that is equally likely to assume any value between 0 and 1. Variate x = RAND(), which shows up in the cell is the probability corresponding to a CDF, ranges from 0 to 1 as obtained from random generation.

Assume that variate x can be rescaled to the normal distribution function using NORM.INV(RAND(), mean, standard_dev), where mean = 48.31 and standard_dev = 2.67/3 = 0.8888. The CDF of N(48.31, 0.8888) is the smooth curve demonstrated in Figure 7.5.1, while the staggered line depicts the original data as given. The Excel NORM.INV function calculates the inverse of the CDF for a supplied value of normal-distributed x, according to the supplied distribution mean and standard deviation. For example, when RAND() = 99%, NORM.INV(99%, 48.31, 0.8888) → 50.38. This means that scores greater than or equal to 50.38 comprise the top 1% of the population.

The number of generations greater than 30,000 is recommended for each case study in order to have an accurate and truthful replication of the population mean and standard deviation, especially when taking skewness and kurtosis into consideration.

FIGURE 7.5.1 CDF of N(48.31, 0.8888).

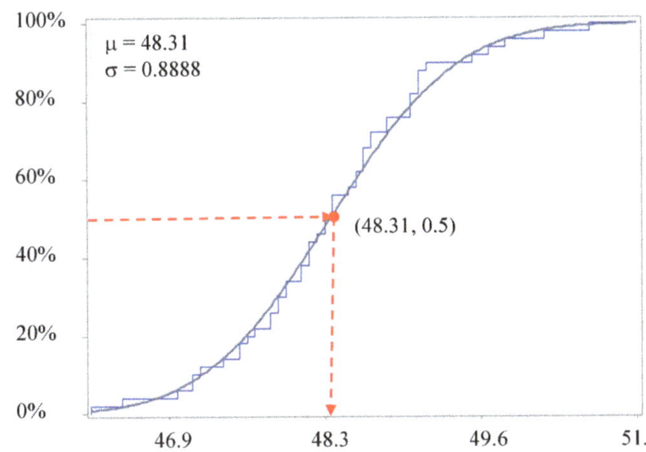

7.5.3.2. Dimensional Characterization with No Tolerance Data

One serious limitation to the application of statistical tolerance analysis as a design tool is that little information is available on the distribution type of the given dimension in the early design stages. Even for standard items, large variations between different batches are frequently encountered, while quality control records of previous production batches just disclose the percent yield or mean and standard deviation, but not the statistical distribution. For such a case, one may perform the tolerance analysis and allocation techniques based on a normal distribution for each tolerance with many Monte Carlo simulations, as being guarded by the central limit theorem in statistics. The mean (μ) and standard deviation (σ) of such an assumed normal distribution are calculated, respectively, as

$$\mu = \tfrac{1}{2}\,(\text{Upper limit} + \text{Lower limit}) \tag{7.5.2}$$

and

$$\sigma = (\text{Upper limit} - \text{Lower limit})\,/\,6 \tag{7.5.3}$$

of which the natural tolerance (±3σ) is assumed to be true for the tolerance range. Generic random number generation can be obtained using either Excel or Minitab, described as follows:

(a) Excel: Go to Formulae ➔ Statistics, then

I. The Excel function to use is NORM.INV(RAND(), mean, standard). This maps the random number generated by RAND() to the normal distribution to search for a random dimension corresponding to the specified tolerance range with the given mean and standard deviation.

II. Then, one may run the computer 30 × 10^3 times or so by copying "NORM.INV (probability, mean, standard)" to the next cell (downward) so many times. A bar chart will exhibit the resulting "bell-shaped" curve with the overall average and standard deviation.

(b) Minitab: Go to Calc > Random Data; then,

I. Select the desired distribution with its corresponding parameters, e.g., mean and standard deviation for normal distribution, and shape factor and scale factor for a two-parameter Weibull distribution.

II. Specify the number of random variables to generate.

III. Identify the column to store the generated data.

7.5.3.3. Dimensional Characterization with a Known Statistical Distribution

After a significant number of sample data are taken from prototypes that are manufactured using the production tools, one is able to figure out the statistical distribution associated with each individual dimensional tolerance at this moment. The statistical distribution function of each individual dimension can be sought out using two different methods (a) method of moments and (b) maximum-likelihood method, given that the measured data are available.

When the statistical distribution is known, it is straightforward to utilize the inverse transformation method to obtain the corresponding cumulative probability from the CDF for the selected random observation p_r, which comes from a random generator, as

$$F(x_q) = P\{x \leq x_q\} = p_q, \text{ where } q = 1, 2, 3, \ldots, Q \tag{7.5.4}$$

which is the cumulative probability of the desirable random observation x_q that is associated with the statistical distribution function being sought. The inverse distribution function for continuous variables $F^{-1}(p_q)$ is then

$$x_q = F^{-1}(p_q) \tag{7.5.5}$$

After a complete set of dimensions with tolerances is characterized in each round of simulation, the measurable object function of an assembly, e.g., the distance between two points, the distance of a point to a plane, or an angle between two planes, is then calculated using constraint equations formulated using vectoral relationships based on the apparent dimensions associated with the specified GD&T for the parts of the assembly.

EXAMPLE 7.5.1

An O-ring is installed onto the vessel gland for sealing the coolant. When the end plate is fastened on the top to compress the O-ring vertically, what is the tolerance as the amount of compression imposed on the basis of (a) worst-limit study, (b) RSS method, and (c) Monte Carlo simulations, respectively? The gland height is 4.3 ± 0.13 mm, and the O-ring diameter is 5.33 ± 0.13 mm.

Solution:

Considering the direction of compression, one may obtain the squeeze of O-rings from the following three different methods.

(a) Worst-Limit Study: $\delta_W = (4.32 - 5.33) \pm (0.13 + 0.13) = -2.01 \pm 0.26$ mm

(b) RSS Method: $\delta_R = -2.01 \pm [(0.13)^2 + (0.13)^2]^{1/2} = -2.01 \pm 0.18385$ mm

(c) Monte Carlo Simulations After Taking 5000 Sample Assemblies: $\delta_M = -2.01 \pm 0.1987$ mm

The tolerance resulting from the RSS method (closed-form solution) is 7.5% less than that of the Monte Carlo simulations. The method based on Monte Carlo simulations is usually considered more accurate than the RSS method, while both are generally considered acceptable. The tolerance resulting from the worst-limit study is not accepted.

7.5.4. Example: Actuating Mechanism

The procedure for conducting Monte Carlo simulations is presented here using an example for studying the amount of protrusion (r_3) of an actuating pin, as depicted in Figure 7.5.2 [Yan et al. 2015]. Based on the linear chain of dimensional study of a closed-loop system, the assembly equation can be written in terms of their geometric dimensions (A, B, C, D, E, Z, and θ) and features (F_1, F_2, F_3, and F_4) as follows:

$$-Z - A - B + C\cos(\theta) + D + E + F_1 + F_2 + F_3 + F_4\cos(\theta) = 0$$

$$\text{i.e.,}\quad Z = -A - B + (C + F_4)\cos(\theta) + D + E + F_1 + F_2 + F_3 \qquad (7.5.6)$$

where

$A = 118.00 \pm 0.14$ mm
$B = 30.00 \pm 0.05$ mm
$C = 40.00 \pm 0.15$ mm
$D = 5.00 \pm 0.02$ mm
$E = 155.00 \pm 0.10$ mm
$\theta = 55.00° \pm 0.17°$
$F_1 = 0.000 \pm 0.040$: profile at the pin head
$F_2 = 0.000 \pm 0.030$: flatness at the pin bottom
$F_3 = 0.000 \pm 0.024$: circularity at the contact end of the arm
$F_4 = 0.000 \pm 0.050$: concentricity at the pinned hole

FIGURE 7.5.2 Tolerance analysis of an actuating mechanism using Monte Carlo simulations [Yan et al. 2015].

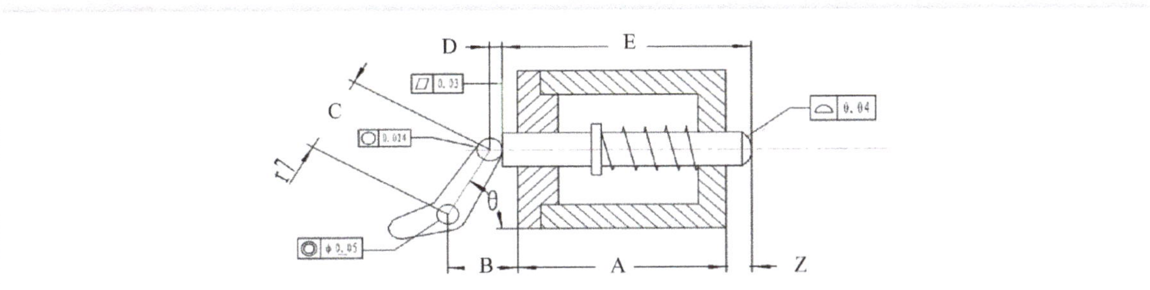

The procedure for conducting Monte Carlo simulations is given as follows:

1. Set the sample size, N.
2. If there are measured manufacturing data available for a certain dimensional characteristic or feature (to be called parameter here for convenience), one should identify its proper statistical distribution function such as a Weibull, normal (Gaussian), or lognormal distribution, or any other. For convenience, each parameter is assumed to follow a normal distribution N(μ, σ); mean is usually its nominal dimension, and sigma is one-sixth of its total tolerance range.
3. Generate N sets of dimensions and features involved in Equation (7.5.6). Each set is to be obtained from the same random number generator. An individual random number is scaled down to a number u_n (between 0 and 1) such that a uniform distribution function for the random number is presumably true. The scaled random number for each parameter (u_n) will be taken as the input to the inverse of its corresponding normal PDF. In summary, this can be implemented using Excel's NORM.INV(Rand(), Mean, Standard).

4. After a complete set of parameters is obtained, one can sample the assembly function using Equation (7.5.6) to obtain gap Z (outcome).
5. After completing all the N sets of sampling, one can estimate the assembly performance checking, i.e., sample mean, sample standard deviation, coefficient of skewness, and coefficient of kurtosis using Equations (7.1.6)–(7.1.9).
6. Change the sample size N, and repeat step (3) to step (6) until the sample size N is large enough to guarantee the analysis accuracy.
7. Plot sample mean, sample standard deviation, coefficient of skewness, and coefficient of kurtosis as functions of the number of simulations, as depicted in Figure 7.5.3. Each curve oscillates periodically but is usually centered at a value.

FIGURE 7.5.3 Sample mean, standard deviation, coefficient of skewness, and coefficient of kurtosis as functions of the number of simulations [Yan et al. 2015].

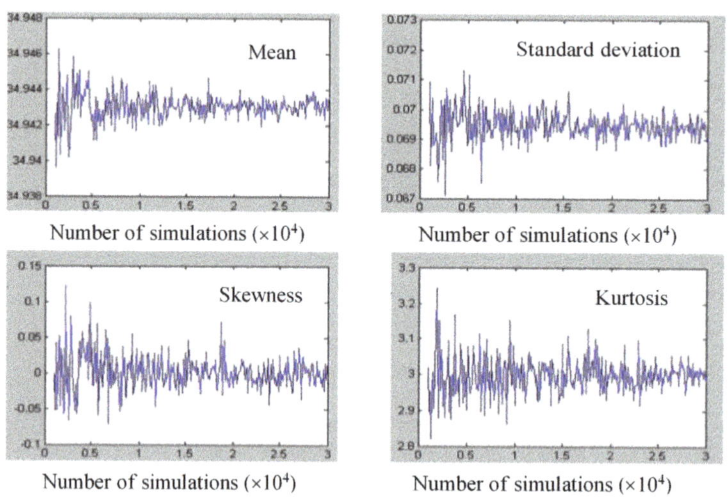

After 30,000 simulations, the results are given as follows: $\mu = 34.943$ mm, $\sigma = 0.0694$ mm, $\beta_3 = 0.0017$, and $\beta_4 = 3.0017$. The coefficient of skewness (β_3) is undesired, because the smaller the value, the more statistically meaningful. The coefficient of kurtosis (β_4) of the sum of uniform distributions is so close to 3.0, which is the value for a normal distribution as expected on the basis of statistical theory.

7.5.5. Example: Electric Circuits

A typical electrical circuit comprises components, of which the statistical distribution functions of their key characteristics can be identified (empirically, experimentally, theoretically) and their output (i.e., assembly equation) can be described mathematically. Then, the Monte Carlo method can be employed to predict the statistical significance of its output performance. For example, a network of resistors is arranged as shown in Figure 7.5.4. The output resistance is expected to be 428.00 ± 8.56 Ω. The assembly equation is derived as follows:

$$R_{network} = R_1 + R_2 + [R_3 R_4 / (R_3 + R_4)] + [R_5 R_6 / (R_5 + R_6) + R_7$$

Supplier A provides inexpensive resistors with variations within 5% (±5%), i.e., R_1 = 10.0 ± 0.5 Ω, R_2 = 300.0 ± 15.0 Ω, R_3 = 20.0 ± 1.0 Ω, R_4 = 20.0 ± 1.0 Ω, R_5 = 40.0 ± 2.0 Ω, R_6 = 10.0 ± 0.5 Ω, and R_7 = 10.0 ± 0.5 Ω. Assume that the resistance variation of each resistor has a normal distribution statistically. Is the supplier a good choice? Based on the Monte Carlo simulations, one obtains the following: $\mu = 427.92$ and $\sigma = 5.29$. Thus, the process capability of supplier A is

$$C_{pk} = \min[(427.92 - 419.44) / (3 \times 5.29), (436.56 - 427.92) / (3 \times 5.29)] = 0.534$$

The requirement for a supplier in the automotive industry is to have a $C_{pk} \geq 1.33$ (but $C_{pk} \geq 1.67$ is preferred), and thus, it is not a qualified supplier.

FIGURE 7.5.4 Resistor network.

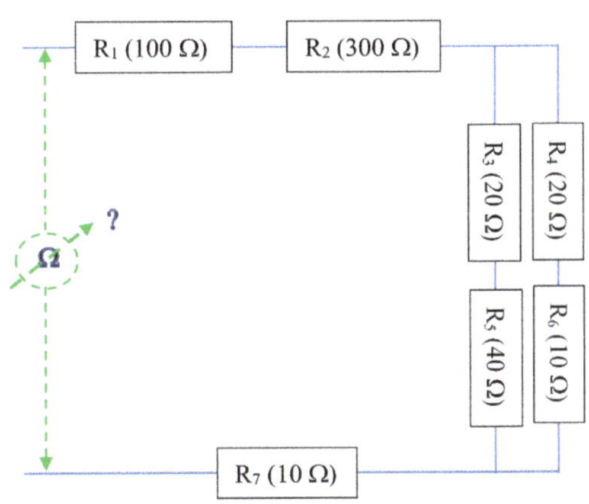

7.5.6. Histogram

The simulation repeats until a given number of assembled units are finished. For the evaluation purposes of the "statistical distribution profile" of the resulting dimension occurrence, this statistical set is further processed into the form of a histogram, which is a "rough" PDF of the statistical distribution of the target. When all the input distributions are normal distributions, the resulting distribution will also be normally distributed.

Consider the example of resistor network given in Figure 7.5.4. The histogram out of 10,000 simulations obtained from Monte Carlo simulations is depicted in Figure 7.5.5. The UCL and LCL are usually defined on the basis of market needs or the voice of the customer.

FIGURE 7.5.5 Histogram resulting from Monte Carlo simulations for the example of resistor network given in Figure 7.5.4.

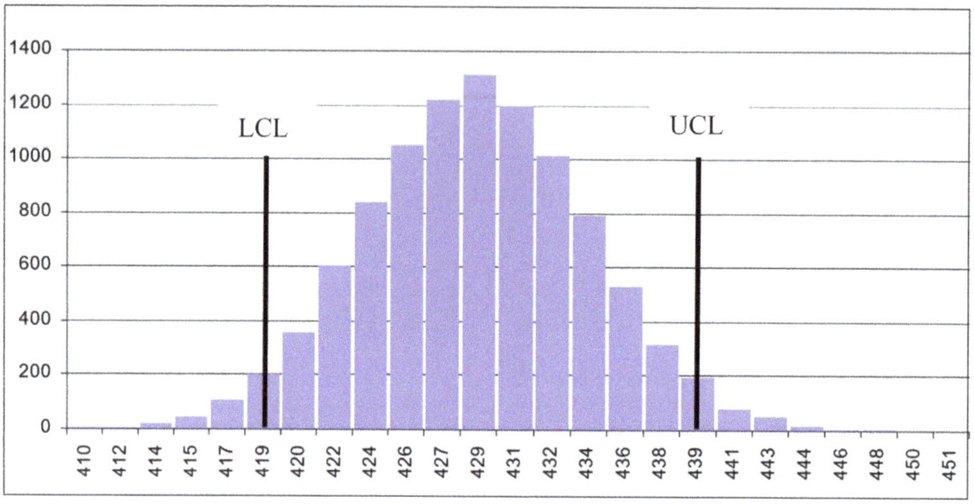

The process capability study can be implemented using Minitab, as Stat ➔ Quality Tools ➔ Capability Analysis. Rejection of the manufacturing process gives the expected number of unsuitable products, i.e., products with the resulting dimension of a "closed" component outside the interval given by the required limit sizes. The common engineering practice suggests that the manufacturing process is considered satisfactorily reliable at the level of $\pm 3\sigma$; i.e., the process comes with the maximum number of 2700 rejected pieces per million (0.27%) produced. However, the level of $\pm 4\sigma$ is recommended as the minimum requirement for key components in the automotive industry because of a possible mean shift; $C_{pk} = 1.33$ is regarded as the minimum requirement, while $C_{pk} = 1.67$ is expected for a robust production process.

The iterations may also presuppose the allocation of component tolerances according to, for example, the clearance deviation from the desired value. If the obtained clearance is too tight, the component tolerances are increased. If the clearance exceeds the specified value, the component tolerances are reduced. This procedure can have several iterations by trial and error or based on the methodology of DOEs until the desired clearance specification is fulfilled.

7.6. Mean Shift

In real practice, the assumption that the variation of each component is symmetrically distributed about the mean may not be valid. Mean shifts and drifts are very common in assembly processes [Guzmán and Hammett 2003]. Tolerance adjustments involve a shift of the mean and/or an expansion of the tolerance width, while a mean shift moves the nominal value of the checkpoint to a different value while maintaining the same tolerance width. An expansion of the tolerance width can be done only on one side (unilateral) or on both sides (bilateral) to accommodate a larger part-to-part variation.

In most production processes encountered in practice, a process mean may shift or drift in the long run although the process is in control. Evaluation of long-term characteristic tolerances, which describe the acceptable uncertainty of product characteristics in its lifetime of service, plays an important role in the design process. It is well known that reduction in tolerance values provides an improvement in product quality but also an improvement in manufacturing costs that increase steeply as the tolerances become tighter.

Systematic variability factors are because of a variety of reasons, such as material degradation, tool wear, human error, and uncontrollable working environment. The tolerance of a dimensional characteristic, T_i, and its related standard deviation without a mean shift is defined as the average of the difference between the UL and LL:

$$T_i = ½ (UL - LL) \tag{7.6.1}$$

$$\text{and} \quad C_{pi} = T_i / (3\sigma_i) = (UL - LL) / (6\sigma_i) \tag{7.6.2}$$

With mean shift,

$$T_k \equiv ½ (UL - LL)(1 - m_k) = (1 - m_k) T_i \tag{7.6.3}$$

$$\text{and} \quad C_{pk} \equiv T_k / (3\sigma_i) = (UL - LL)(1 - m_k) / (6\sigma_i) = (1 - m_k) C_{pi} \tag{7.6.4}$$

of which m_k is the mean shift factor. When $m_k = 0$, there is no shift and Equation (7.6.2) reduces to Equation (7.6.1). When $0 < m_k < 100\%$, there is an expected mean shift and the sample standard deviation is defined as

$$\sigma_i = \frac{T_i}{3 (1 - m_k) C_{pi}} \tag{7.6.5}$$

7.6.1. Tolerance Due to Mean Shift

Once the range of mean shift has been estimated for each component, the assembly tolerance is calculated using the following model instead of Equation (7.4.8):

$$T = \Sigma\, m_k\, |\partial f(x_k) / \partial x_k|\, T_k + [\Sigma (1 - m_k)^2 |\partial f(x_k) / \partial x_k|^2 T_k^2]^{1/2} \tag{7.6.6}$$

Parameter m_k is the mean shift factor for the kth dimensional characteristic, which is mainly affected by the mean shift of this single component. If there is no shift, $m_k = 0$. The assembly tolerance in the above equation has been split into two summations. The first term is the sum of the mean shifts added as a worst-limit. The second term is the sum of the component tolerances added statistically. Thus, the contributions of both mean shift or bias and part tolerance or variance on the resulting assembly tolerance are available [Stefano 2003]. Note that the midpoint tolerance zone is expressed as a fraction of the specified tolerance zone for the part dimension, ranging from 0 to 1.

A tolerance range (Figure 7.6.1) of $\pm 4.5\sigma$, evaluated over an extended time period, is sometimes used as the cut-off point for "Six-Sigma Quality" [Cavanagh et al. 2005]. It has less than 3.4 defects per million opportunities (DPMO; if normally distributed) and is suggested for adjusting for drift in product mean. However, in order to meet APQP requirement that $C_{pk} = 1.67$ for a "good" production process, $\pm 5\sigma$ with 0.573 DPMO (if normally distributed) is strongly encouraged in the automotive industry.

FIGURE 7.6.1 Dimensions characterized with distinct PDFs.

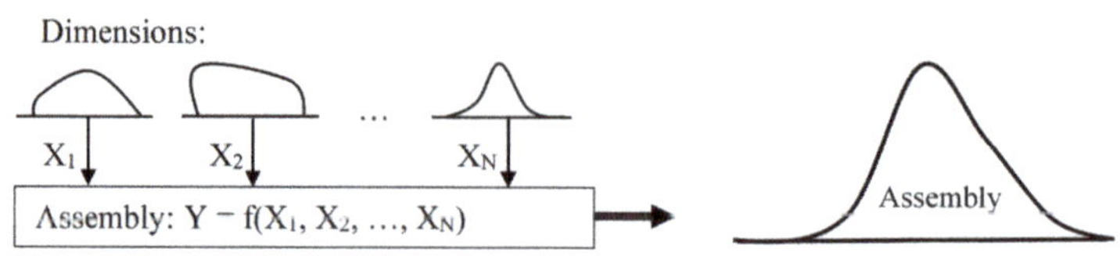

7.6.2. Mean Shift Due to Tool Wear Using Monte Carlo Simulation

A case study on the variability of product characteristics in the life of the manufacturing process with functional and assembly considerations was done by [Stefano 2003]. A key contribution to the process variability is because of continuous shift of the mean from the value at tool setup to the last value before another tool setup. The severity of shift is measured using the long-term standard deviation and defined as

$$\beta = \delta / \sigma_{LT} \quad (7.6.7a)$$

or

$$\delta = \beta\, \sigma_{LT} \quad (7.6.7b)$$

where

β is the dimensionless shift, also called mean shift ratio
δ is the amount of shift, usually one-sided
σ_{LT} is the long-term standard deviation, which is assumed to be a unitary error

The long-term standard deviation may comprise three components from different identifiable independent sources of variation, and it can be calculated from the following equation:

$$\sigma_{LT} = (\sigma_{comp}^2 + \sigma_{setup}^2 + \sigma_{mean\text{-}shift}^2)^{1/2} \quad (7.6.8)$$

where

σ_{comp} is the standard deviation inherent with the machining process
σ_{setup} is the standard deviation because of variability with tool setup
$\sigma_{mean\text{-}shift}$ is the standard deviation because of mean shift as a function of time

Standard deviation $\sigma_{mean\text{-}shift}$ represents the process capability of a tool in its life cycle, and its related amount of mean shift can be calculated as

$$\delta_{mean\text{-}shift} = \mu_{final} - \mu_{initial} \quad (7.6.9)$$

The weighting factor is the unknown in Equation (7.6.7a), and it is to be solved for how much the mean deviation is. Factors that affect weighting factor m_k are the confidence level $(1 - \alpha)$, the number of assembly dimensions (N), and the assortment of tolerance values that are taken into account for the equivalent number of dimensions $(n_{E,k})$,

$$n_{E,k} = \frac{\sum |\partial f(x_k) / \partial x_k|\, T_k}{|\partial f(x_k) / \partial x_k|\, T_k} \quad (7.6.10)$$

Let α be the probability that indicates the confidence interval (CI) (1 – α), in which statistically the α of the instances of the Monte Carlo simulations is contained. Assume that all component dimensions have the same mean shift ratio and a tolerance T_α at the same CI that is dependent on the value of the β factor, i.e.,

$$T_\alpha = T_\alpha(\beta) \tag{7.6.11}$$

It is also assumed that T_α contributes to the assembly tolerance (T_{mc}) that is to be derived from Monte Carlo simulation and has a unitary sensitivity factor.

$$|\partial f(x_k) / \partial x_k| = 1 \tag{7.6.12}$$

In order to perform Monte Carlo simulations, Equation (7.6.6), when incorporated with Equations (7.6.11) and (7.6.12), is so normalized that the assembly tolerance with N assembly dimensions and "dimension equivalents" can be expressed as

$$T_{mc} = N\ m\ T_\alpha(\beta) + N^{1/2}(1 - m)\ T_\alpha(\beta) \tag{7.6.13}$$

$$\text{Thus,}\quad m = \frac{T_{mc} - N^{1/2}\,T_\alpha(\beta)}{(N - N^{1/2})\ T_\alpha(\beta)} \tag{7.6.14}$$

7.7. Hole and Shaft Effects

Hole making is a major class of machining operations that are specifically used to make a hole into a work-piece, including lapping, honing, boring, counter-boring, counter-sinking, grinding, turning, reaming, milling, drilling, planning, and shaping. The standard of tolerance on fit is divided into two categories as explained in Table 7.7.1, one for holes and the other for shafts:

1. Hole Basis System: The minimum size of the hole is taken as the datum for the fit. The hole basic deviation class is designated as H, and the minimum tolerance of the H grade coincides with the basic size. One may add or subtract the allowance value from the hole diameter to get the shaft maximum diameter for interference and clearance fits, respectively.
2. Shaft Basis System: The maximum size of the shaft is taken as the datum for the fit. The hole basic deviation class is designated as h, and the minimum tolerance of the h grade coincides with the basic size. One may add or subtract the allowance value from the shaft diameter to get the hole minimum diameter, for interference and clearance fits, respectively.

There are three basic types of fits, i.e., clearance fit, transition fit, and interference fit, as shown in Figure 7.7.1.

TABLE 7.7.1 Hole basis system and shaft basis system.

Basis	Datum	Tolerance grade	Extreme diameter of interest
Hole	Minimum hole size	H6, H7, ..., H11	Maximum shaft diameter
Shaft	Maximum shaft size	h6, h7, ..., h11	Minimum hole diameter

FIGURE 7.7.1 Hole basis system and shaft basis system [ANSI B4.2 2020].

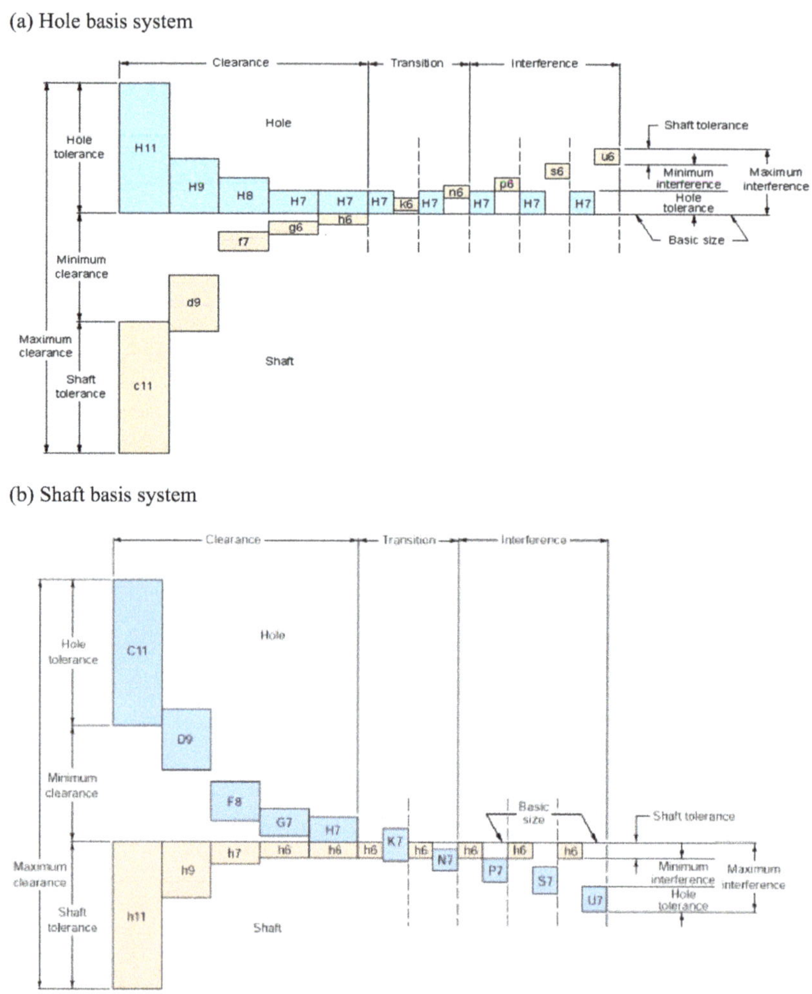

7.7.1. International Standards for Holes and Shafts

The hole is labeled with a capital letter, while the shaft is labeled with a lowercase letter. Each comes with a number that stands for the machining precision, such as H7 (hole or nut) and h7 (shaft or bolt) demonstrated in Figure 7.7.1. The tolerance grades corresponding to different kinds of mechanical operations are given in Figure 7.7.2.

Note that H7/h6 is a very common standard tolerance in practice, which gives a rather tight fit, but not so tight that one cannot put the shaft in the hole or turn the nut on the bolt by hand. The tolerances work in such a way that the hole should be made slightly larger than the base dimension. The actual amount bigger/smaller depends on the base dimension. For example, a hole of H7 means for an International Organization for Standardization (ISO) fit 10 (+0.015–0): up to 0.015 mm larger and 0.000 mm smaller than the base dimension of the shaft. For a shaft of the same size, h6 would mean 10 (+0–0.009), which is the opposite of H7: up to 0.000 mm larger and 0.009 mm smaller than the base dimension of the hole. This method of standard tolerances is also known as "limits and fits" and can be found in ISO 286-1:2010.

Tolerances generated by different machining tools are given in Figure 7.7.2. Depending on the mutual position of tolerance zones of the coupled parts, three types of fit can be provided for the mating:

1. Clearance Fit: The LL size of the hole is greater or at least equal to the UL size of the shaft, enabling a clearance between the hole and shaft in the coupling.
2. Transition Fit: The dimensions of the shaft and hole are "the same" in theory. It is a fit where both clearance and interference may occur in the coupling.
3. Interference Fit: The UL size of the hole is smaller or at least equal to the LL size of the shaft, ensuring some interference between the hole and shaft in the coupling.

FIGURE 7.7.2 General tolerances and operations (international tolerance grades 0–16).

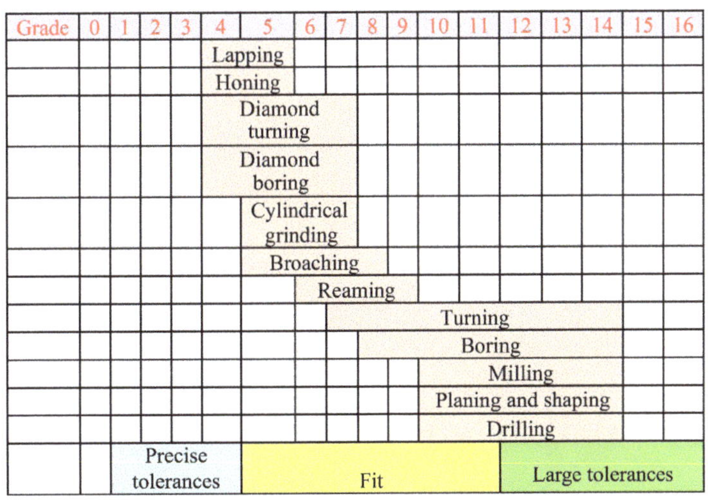

7.7.2. Virtual Condition

Two special dimensional management algorithms are adopted for two distinct fastening processes. One is for floating fasteners and the other is for fixed fasteners, as illustrated in Figure 7.7.3. A constant boundary is expected on the basis of collective effects of the maximum material condition (MMC) limit of size of a feature and the specified geometric tolerance. This is also called the virtual condition of a feature at MMC. The constant boundary can be further divided into the following two cases:

1. Floating Fastener Algorithm (Hole Basis System): This is referenced to the hole [Figure 7.7.4(a)]. The geometric tolerance will be taken by holes. The desired clearances and interferences in the fit are achieved by combining various shaft tolerance zones with the hole tolerance zone "H." In this system of tolerances and fits, the lower deviation of the hole is always equal to zero, and

$$\text{MMC hole} - \text{MMC shaft} = \text{Tolerance for all holes} \tag{7.7.1}$$

The hole clearance (tolerance) based on Figure 7.7.4(a) is 0.270 – 0.250 = 0.020, of which 0.250 is the shaft size at MMC.

2. Fixed Fastener Algorithm (Shaft Basis System): This is referenced to the shaft [Figure 7.7.4(b)]. The desired clearances and interferences in the fit are achieved by combining various tolerance zones of the hole with the shaft tolerance zone "h." In this system of tolerances and fits, the upper deviation of the hole is always equal to zero. It means that the geometric tolerance is to be shared by the mating features, i.e.,

$$\text{MMC hole} - \text{MMC shaft} = \text{Tolerance divided between mating features} \tag{7.7.2}$$

The hole clearance (tolerance) based on Figure 7.7.4(b) is 0.274 – 0.250 = 0.024 + 0.000, of which 0.250 is the shaft size at MMC, 0.024 is the threaded hole location tolerance at MMC, and 0 is the clearance hole location tolerance at MMC. In other words, the MMC hole size is calculated as 0.274 = 0.250 + 0.024 + 0.000.

FIGURE 7.7.3 Floating fasteners: bolts pass through one part as treaded into another and fixed fasteners: bolts pass through both parts as bound with nuts [McKinney 2000].

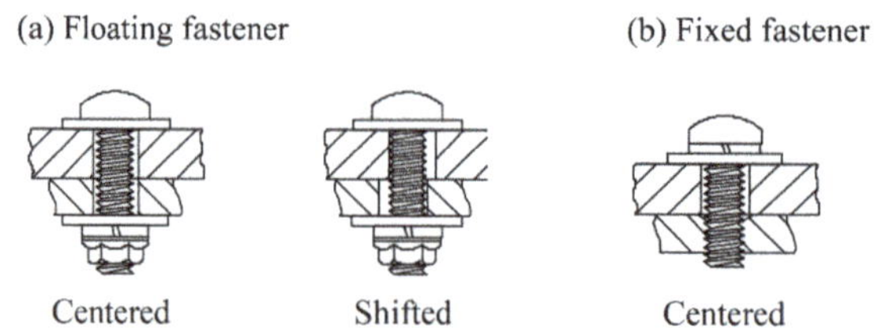

FIGURE 7.7.4 Assembly of fasteners—floating or fixed algorithms [Congorno 2010].

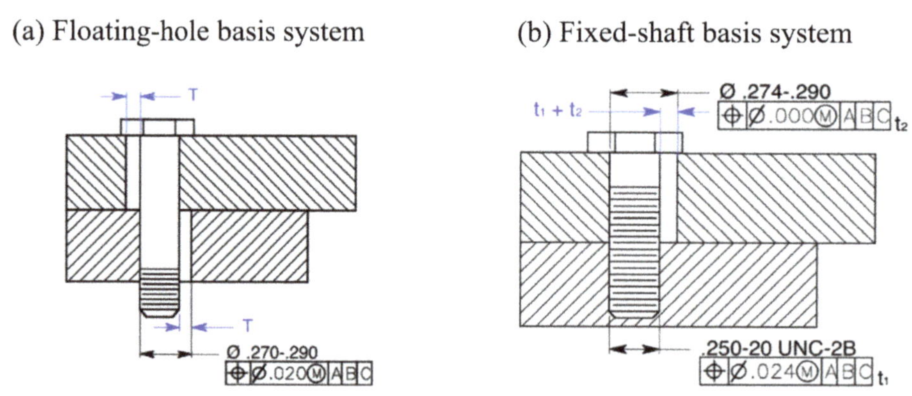

7.7.3. Resultant Condition

The resultant condition of a feature specified at MMC is a variable boundary generated by the collective effects of the least material condition (LMC) limit of size of a feature, the specified geometric tolerance, and any applicable bonus tolerance.

1. Hole: Hole dimension at LMC + geometric tolerance at MMC + bonus tolerance
2. Pin: Pin dimension at LMC – geometric tolerance at MMC – bonus tolerance

7.7.4. Equal Bilateral Tolerance of Pin and Hole

The worst-case boundaries of a hole and its mating pin, given in Figure 7.7.5, can be calculated as follows:

(A) Hole Size: Virtual condition ≤ hole size ≤ resultant condition

i.e., MMC − GT ≤ hole size ≤ LMT + GT + bonus

$(50 - 1) - 1 \leq$ hole size $\leq (50 + 1) + 1 + 2$

$48 \leq$ hole size ≤ 54

Thus, hole size = 51 ± 3

Note that the bonus tolerance comes from [1 − (−1)], originating from Φ50 ± 1.

(B) Pin Size: Virtual condition ≥ pin size ≥ resultant condition

i.e., MMC + GT ≥ pin size ≥ LMT − GT − bonus

$(46 + 1) + 1 \geq$ pin size $\geq (46 - 1) - 1 - 2$

$48 \geq$ pin size ≥ 42

Thus, pin size = 45 ± 3.

Note that the amount of bonus tolerance comes from Φ45 to 47.

FIGURE 7.7.5 Boundaries of a hole and its mating pin [Congorno 2010].

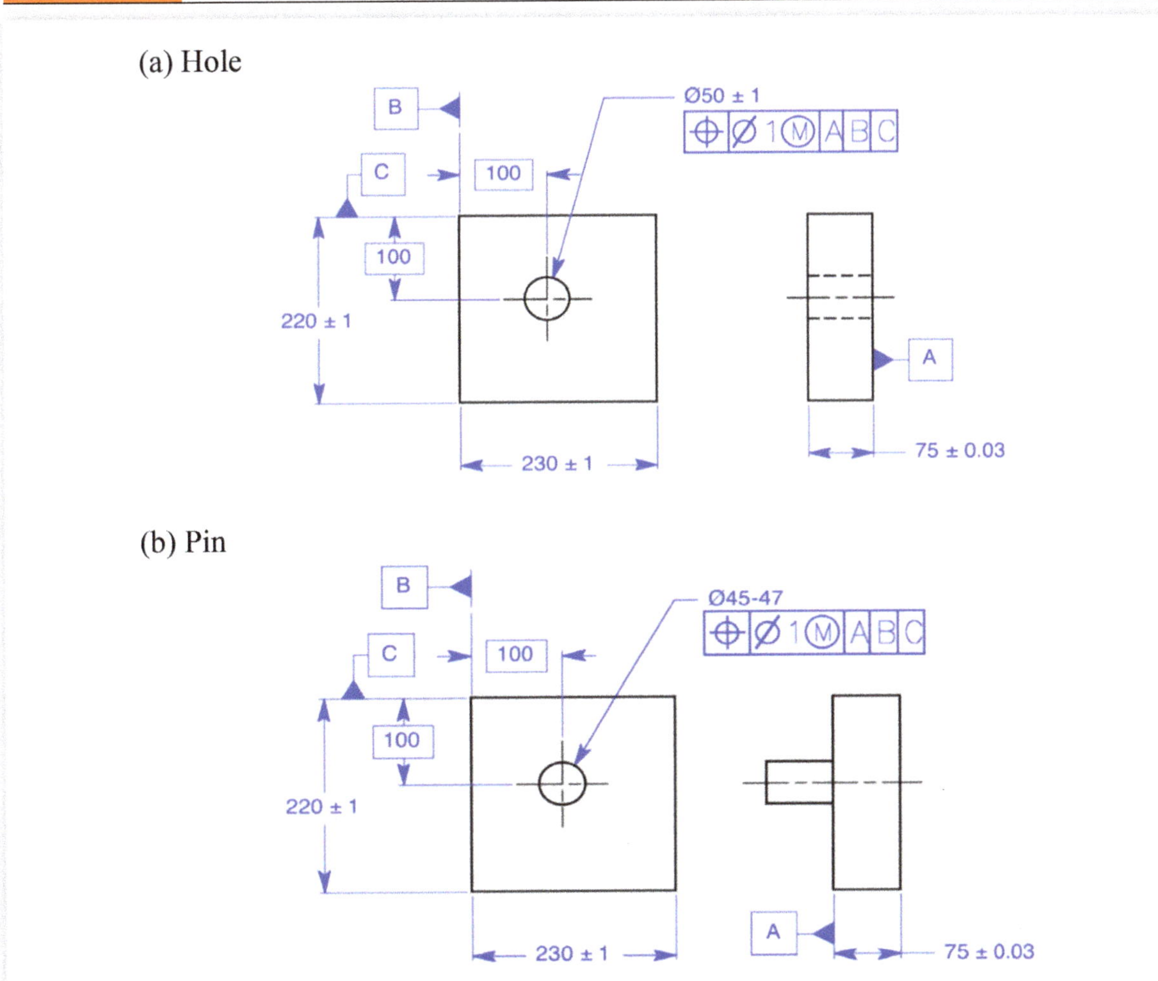

Courtesy of James D. Meadows.

7.7.5. Tolerance Stack-Up with Material Conditions

An assembly is shown in Figure 7.7.6(a). The objective is to find the gap between the upper two flanges with one circular hole each, put on top of a plate with two locating circular holes. Three tolerances are zeroed and no material condition is specified for any dowel pin for convenience, as shown in the figure. Note that a dowel pin is a type of fastener used to hold an object in place.

As it is located by the two dowel pins, how much is the resulting gap tolerance between the two flanges? Which dimensional variation has the greatest impact on the gap tolerance?

(A) The first step is to figure out the "± tolerance" inherent with the upper holes (holes #1 and #2) in light of the hole-pin match. Consider holes #1 and #2. The UL (resultant condition) can be calculated as

  (a) UL = 6.0 (hole size) + 0.4 (tolerance) + 1.0 (material condition) = 7.4 mm

  (b) LL = 5.0 (hole size) – 0.4 (tolerance) = 4.6 mm

  (c) The hole diameter with virtual tolerancing is 6.0 ± 1.4 mm

  So, the hole radius is 3.0 ± 0.7 mm

(B) The next step is to figure out the "± tolerance" inherent with the lower holes (holes #3 and #4) in light of the hole-pin match. Consider holes #3 and #4. The UL (resultant condition) can be calculated as

  (d) UL = 6.0 (hole size) + 0.0 (tolerance) + 1.0 (material condition) = 7.0 mm

  (e) LL = 5.0 (hole size) – 0.0 (tolerance) = 5.0 mm

  (f) The hole diameter with virtual tolerancing is 6.0 ± 1.0 mm

  So, the hole radius is 3.0 ± 0.5 mm

(C) Draw the vector-loop diagram, which is shown in Figure 7.7.6(b). This is a key step and probably the hardest one to make such a tolerance study. One may start with one side of the unknown gap.

(D) Dimensioning and ± tolerancing of all the parts in the vector-loop diagram, which is shown in Figure 7.7.6(c).

FIGURE 7.7.6 Tolerance stack-up with material conditions [Congorno 2010, Srinivasmurthy and Herle 2019].

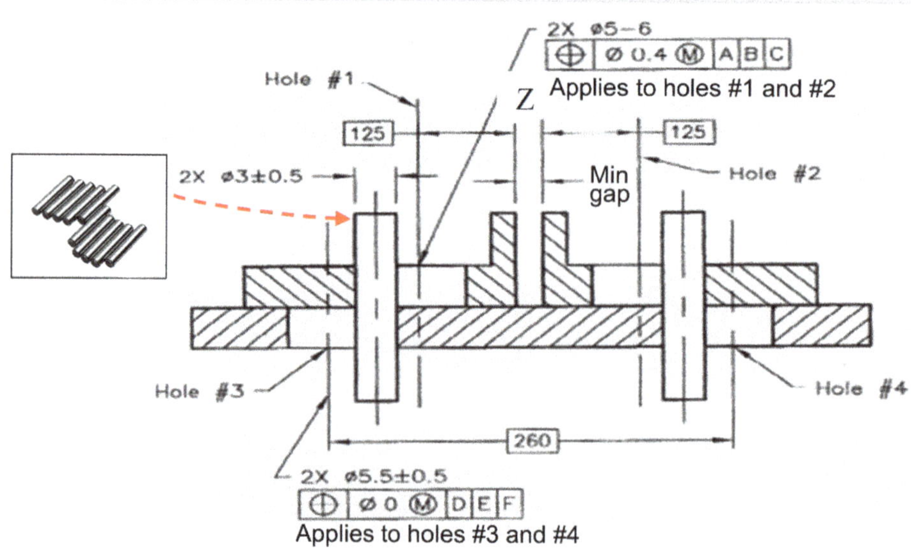

(a) Assembly drawing

Courtesy of James D. Meadows.

(Continued)

FIGURE 7.7.6 (Continued) Tolerance stack-up with material conditions [Congorno 2010, Srinivasmurthy and Herle 2019].

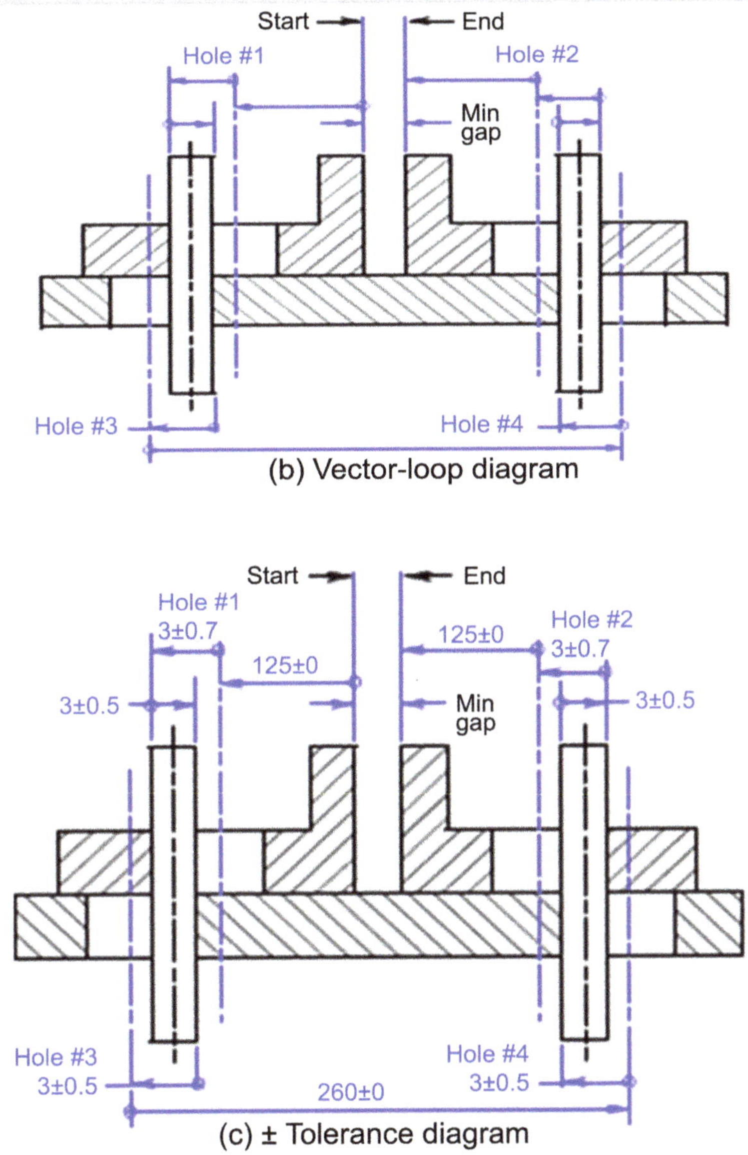

Courtesy of James D. Meadows.

7.8. Identification of Dimensions by Method of Moments Estimator (MME)

The method of moments, also called MME, is a method of estimation of population parameters of a statistical distribution. It utilizes statistical moments of the component distribution and the first and second partial derivatives of the assembly function to find the first two moments of the assembly distribution, Equations (7.5.1) and (7.5.2). These two moments are used to find the parameters of a statistical distribution such as the Weibull distribution and lognormal distribution. After the parameters of a statistical distribution are determined, the fraction outside of the assembly limits is found from numerical integrations, tabulated data, or in some cases by algebraic equations. The computation time is moderate compared to Monte Carlo simulation.

The MME can be the first approach to estimating the two parameters as the first approximation, although the MME is superseded by the maximum-likelihood estimation (MLE) based on the Fisher information matrix concerning asymptotic unbiasedness and minimal variance in many cases.

7.8.1. Moment-Generating Functions

Assume that a dimension has a two-parameter Weibull distribution. Given a collection of measured data that fits the Weibull distribution, one would like to estimate the Weibull parameters which best fit the data. Only the first two moments are applied when using the MME to assess the two-parameter Weibull model because there are only two unknowns to be resolved. Therefore, the MME begins with the first two moments:

$$\mu_1 = \frac{1}{N} \sum_{n=1}^{N} t_n \tag{7.8.1}$$

and

$$\mu_2 = \frac{1}{N} \sum_{n=1}^{N} t_n^2 \tag{7.8.2}$$

where

μ_k is the moments, where k = 1 or 2

N is the number of data available for each dimensional characteristic

The MME can be derived by hand calculation. The number of such equations is the same as the number of parameters to be estimated. One may set these two moments equal to the corresponding theoretical moments from the moment-generating function, $m_k(t)$, k = 1, and 2.

The moment-generating function derived from the two-parameter Weibull distribution is given as follows [Casella and Berger 2002]:

$$m_K(t) = \eta^k \, \Gamma(1 + k / \beta) = \eta^k \int_0^{\infty} x^{(k/\eta)} \exp(-k / \beta)\, dx \tag{7.8.3}$$

where

$m_K(t)$ is the kth theoretical moment, k = 1, 2, 3, or 4 that stands for the order of moment

Γ() is the gamma function

β is the shape factor

η is scale

The relationship between the estimated moments and the theoretical Weibull distribution-oriented moments can be formulated as follows [Murthy et al. 2004]:

$$\frac{\mu_2}{\mu_1^2} \approx \frac{m_2(t)}{[m_1(t)]^2} \tag{7.8.4}$$

Both $m_1(t)$ and $m_2(t)$ are defined by Equation (7.8.3). Substituting the theoretical moments based on Equation (7.8.3) into the above equation leads to

$$\frac{\mu_2}{\mu_1^2} \approx \frac{\eta^2 \, \Gamma(1+2/\beta)}{\eta^2 \, [\Gamma(1+1/\beta)]^2} = \frac{\Gamma(1+2/\beta)}{[\Gamma(1+1/\beta)]^2} \tag{7.8.5}$$

$$\text{i.e.,} \quad \frac{\Gamma(1+2/\beta)}{[\Gamma(1+1/\beta)]^2} - \frac{\mu_2}{\mu_1^2} = 0 \tag{7.8.6}$$

Since shape factor β is the only unknown in the above nonlinear equation, it can be solved iteratively by numerical analysis, such as the least squares method or Newton–Raphson's method. Once shape factor β is derived, the following equation can be used for estimating scale factor η based on Equation (7.8.3) as

$$\mu_1 \approx m_1(t) = \eta \, \Gamma(1+1/\beta)$$

$$\text{i.e.,} \quad \eta = \mu_1 / \Gamma(1+1/\beta) \tag{7.8.7}$$

7.8.2. Blend of Method of Moments and Monte Carlo Simulation

An alternative scheme which requires a moderately complicated program with moderate computation time is a blend of the method of moments and Monte Carlo simulation. This is the hybrid method based on Monte Carlo simulation to generate a smaller number of assembly values. The sample size is usually on the order of 1000 to 5000. The resulting assembly dimensions are used to calculate the statistical moments of the assembly distribution and then estimate the fraction of rejects. Most of the complexity of the method of moments is eliminated since numerical derivatives and the series summations are not needed to find the assembly moments from the component moments. As the sample size can be on the order of 10^3, the computation is greatly reduced from the simple Monte Carlo simulation.

7.9. Identification of Dimensions by MLE

It is essential to identify the statistical distributions of part dimensions and features in an assembly before the tolerance analysis is conducted. The maximum-likelihood method is a versatile approach besides Monte Carlo simulations. Assume that a measured dimension is denoted by t_q, where q = 1, 2, …, and Q. It means that the measurement is replicated Q times. Two statistical distributions are examined here for their potential fits to dimensional tolerances.

7.9.1. Log-Likelihood Function and Likelihood Equations for Normal Distribution

In sampling from a normal distribution with mean μ and variance σ^2, the likelihood function based on all the given data can be written as [Freeman 2010]

$$L(\mu, \sigma^2) = f(t_1)\, f(t_2) \ldots f(t_Q) = \prod_{q=1}^{Q} f(t_q) \tag{7.9.1}$$

By statistical theory, maximizing $L(\mu, \sigma^2)$ is equivalent to maximizing $\ln[L(\mu, \sigma^2)]$. The log-likelihood function is derived by taking a natural log of the above equation, of which f(t) is the PDF of a normal distribution, i.e., Equation (2.8.1).

$$\ln[L(\mu, \sigma^2)] = -\tfrac{1}{2} Q \ln(2\pi) - \tfrac{1}{2} Q \ln(\sigma^2) - \tfrac{1}{2} \left[\sum_{q=1}^{Q} \left(\frac{t_q - \mu}{\sigma}\right)^2 \right] \tag{7.9.2}$$

The likelihood equations for μ and σ^2 can be obtained by setting the first derivatives of the above equation with respect to these two parameters to zeros:

$$\frac{\partial\{\ln[L(\mu, \sigma^2)]\}}{\partial \mu} = \frac{1}{\sigma^2} \left[\sum_{t=1}^{Q} (t_q - \mu) \right] = 0 \tag{7.9.3}$$

and

$$\frac{\partial\{\ln[L(\mu, \sigma^2)]\}}{\partial (\sigma^2)} = \frac{-Q}{2\sigma^2} + \frac{1}{2\sigma^4} \left[\sum_{q=1}^{Q} (t_q - \mu)^2 \right] = 0 \tag{7.9.4}$$

The above two equations can be solved for two unknowns:

$$\mu_Q = \frac{1}{Q} \left[\sum_{q=1}^{Q} t_q \right] \tag{7.9.5}$$

and

$$\sigma_Q^2 = \frac{1}{Q} \left[\sum_{q=1}^{Q} (t_q - \mu)^2 \right] \tag{7.9.6}$$

where

μ_Q is the mean estimated from Q samples
σ_Q^2 is the variance estimated from Q samples

Note that the MLE of σ_Q^2 for the normal model is not the sample variance S^2.

7.9.2. Log-Likelihood Function and Likelihood Equations for Weibull Distribution

Weibull distribution is flexible enough to model a variety of data sets. Got right-skewed data? Weibull can model that. Left-skewed data? Sure, that is cool with Weibull. Symmetric data? Weibull is up for it. That flexibility is why reliability engineers use the Weibull distribution to evaluate the material strengths of almost everything from electric capacitors to mechanical bearings. It is also flexible to characterize dimensional tolerances. Here, it is given that the dimension follows a two-parameter Weibull distribution, of which the PDF is

$$f(t_q) = \frac{\beta}{\eta}\left(\frac{t_q}{\eta}\right)^{(\beta-1)} \exp\left[-\left(\frac{t_q}{\eta}\right)^{\beta}\right] \tag{7.9.7}$$

Then, the likelihood function based on the Q given data can be written as [Freeman 2010] follows:

$$L_w(\beta, \eta) = f(t_1)\, f(t_2) \ldots f(t_Q) = \prod_{q=1}^{Q} f(t_q) \tag{7.9.8}$$

The two Weibull parameters can be estimated by maximizing the likelihood given above. By statistical theory, maximizing $L_w(\beta, \eta)$ is equivalent to maximizing $\ln[L_w(\beta, \eta)]$, which is

$$\ln[L_w(\beta, \eta)] = \sum_{q=1}^{Q} \ln[f(t_q)]$$

$$= \sum_{q=1}^{Q} [\ln(\beta) - \ln(\eta) + (\beta - 1)\ln(t_q) - (\beta - 1)\ln(\beta) - (\frac{t_q}{\eta})^{\beta}]$$

$$= Q\,[\ln(\beta) - \beta\ln(\eta)] + (\beta - 1)\sum_{q=1}^{Q} \ln(t_q) - \sum_{q=1}^{Q} \left(\frac{t_q}{\eta}\right)^{\beta} \tag{7.9.9}$$

where

$L_w(\beta, \eta)$ is the likelihood function with two Weibull parameters
$\ln[L_w(\beta, \eta)]$ is the logarithmic transformation of $L_w(\beta, \eta)$
$f(t_q)$ is the Weibull PDF evaluated at t_q
q is the data number
Q is the total number of measured data

The likelihood equations for β and η can be obtained by setting the first derivatives of the above equation with respect to these two parameters to zeros.

$$\frac{\partial\{\ln[L_w(\beta, \eta)]\}}{\partial\beta} = \frac{-Q\,\eta}{\beta} + \eta \sum_{t=1}^{Q} \left[\frac{t_q^{\eta}}{\beta^{(\eta+1)}}\right] = 0 \tag{7.9.10}$$

and

$$\frac{\partial\{\ln[L_w(\beta, \eta)]\}}{\partial\eta} = \frac{Q}{\eta} - Q\ln(\beta) - \sum_{q=1}^{Q} \ln\left(\frac{t_q}{\beta}\right)\exp\left[\eta \ln\left(\frac{t_q}{\beta}\right)\right] + \sum_{q=1}^{Q} \ln(t_q) = 0 \tag{7.9.11}$$

Equation (7.9.10) can be rearranged to yield

$$\beta = \left[\frac{1}{Q}\left(\sum_{q=1}^{Q} t_q^{\eta}\right)\right]^{-\eta} \tag{7.9.12}$$

The above equation is then substituted into Equation (7.9.12). The resulting equation is arranged into the following form:

$$\eta = \left\{\frac{-1}{Q}\left[\sum_{q=1}^{Q} \ln(t_q)\right] + \left[\sum_{q=1}^{Q} t_q^{\eta} \ln(t_q)\right] / \left[\sum_{q=1}^{Q} t_q^{\eta}\right]\right\}^{-1} \tag{7.9.13}$$

from which η_Q is to be estimated from Q samples. After then, β_Q for Q samples is then obtained by substituting the obtained η_Q into Equation (7.9.12).

7.9.3. Identification of Dimensions Using Minitab

In order to identify the proper statistical distribution of a given set of data such as the measured value corresponding to a certain dimension in Minitab, one may go to

Stat → Quality Tools → Individual Distribution Identification

There are three measures that will tell which one of the 16 different statistical distributions generated in session windows and figures is the best fit, as described in Volume 1, Section 2.8.

7.10. Finite Element Tolerance Analysis with Skin Model Shapes

The concept of FEA is to utilize skin model shapes derived from solid models to present part-shape variabilities, which can be used for 3D statistical tolerance design of a stack-up, as shown in Figure 7.10.1. It employs discrete geometry schemes, such as surface meshes and points, just like finite element models [Semere et al. 2018]. When finite element meshes of parts with tolerance variations are assembled, they are called skin model shapes that function as a virtual representation of overall variational geometry of an assembly [ISO 2017, Schleich et al. 2014]. Note that coordinate transformations of shape deviations are carried out when assembling the element matrices formed in the local coordinate systems in the global finite element matrices (such as stiffness, damping, and mass) in the FEA.

FIGURE 7.10.1 Assembly for the tolerance analysis based on skin model shapes with mesh and magnified form deviations [Schleich and Wartzack 2016].

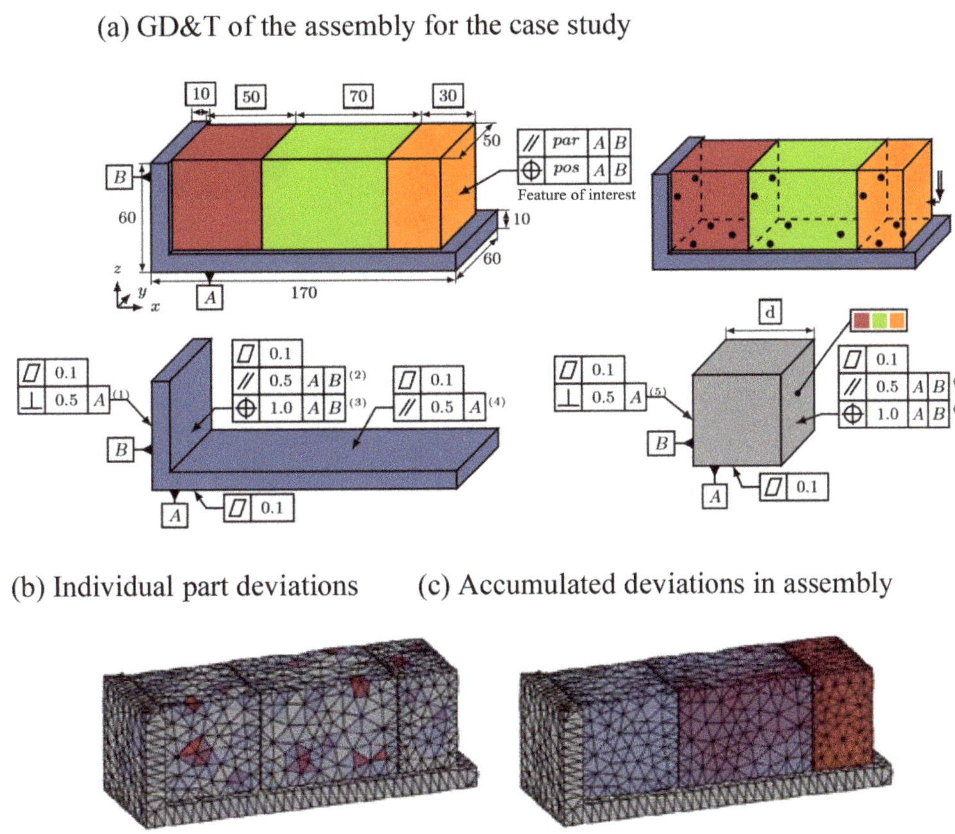

The idea of incorporating skin model shapes into finite element meshes in search of structural performance (e.g., deformations and stresses) is first provoked by the thermal loading under the special working environment of a gas turbine blade [Garaizar et al. 2016]. The application of this method in the context of tolerance analysis for a thermomechanical system also allows the evaluation of the effects of tolerances, deformations, and mobility on the motion of mechanisms with flexible members. For example, a combination of skin model shapes into the finite element model of a bushing for analyzing the structural deformations based on eight-node solid elements was carried out. The superposition of the nominal geometry and a skin model shape is shown in Figure 7.10.2. It demonstrates the realistic impact of small deviations because of manufacturing errors on the final geometry shape and stress level of assemblies [Anwer et al. 2014]. Different skin model shapes of the external cylindrical surface of a cylindrical bushing are modeled through the generation (by commercial code autoregressive-moving average (ARMAX) model), and the stacking and the random rotation of 16 profiles are used for each FEA.

FIGURE 7.10.2 Finite element-tolerant analysis [Corrado and Polini 2018].

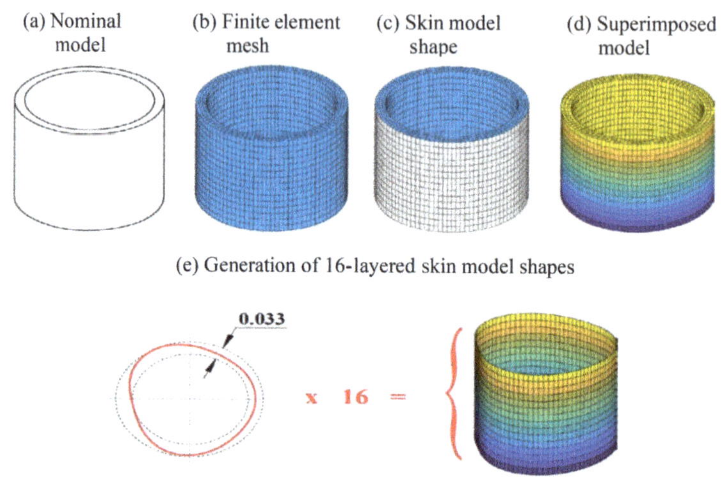

A finite element tolerance analysis procedure is suggested by [Corrado and Polini 2018]. How to automate reforming individual finite elements to accommodate the features with geometric variations in the preprocessor is a challenge, while how to handle the interference fits in finite element calculations utilizing proper contact algorithms is the key to success as observed using the Abaqus codes [Chiang et al. 2000].

One robust approach suggested here is to utilize the methodology of DOEs based on the calculated deformation or other performance indices resulting from Monte Carlo simulations that facilitate the deviations assigned to all discretized features. For example, when an engineer would like to know how four dimensions (factors) of interest affect the assembly tolerance, design matrices 2^4, 2^{4-1}, 2^{4-2}, 3^4, 3^{4-1}, and 3^{4-2}, composite design, or optimal design can be applied. The run conditions required for the FEA for each treatment of the design matrix can be obtained from Monte Carlo simulations.

7.11. Homogenous Transformations for Tolerance Analysis

A global design system with 3D solid models relates all feature control frames to each other and models DRFs and their precedence. It provides a basis for geometric validation of a GD&T scheme and downstream 3D tolerance design [Bernstein 1989]. Each geometric dimension has 3D tolerance zones. Therefore, it is logical that the effect of a specified tolerance may have an impact in 3D mode [Roy and Liu 1988]. Transformations in the space represent both the nominal relations between parts and the deviations caused by geometric variations allowed by the tolerances. A statistical estimate of the location of the nth part in an assembly, usually starting from a part or fixture, can be carried out. Representative assemblies are compatible with tolerance analysis based on a closed-form algorithm, except for form tolerances, as used in robotic applications [Whitney et al. 1994].

7.11.1. Homogeneous Transformations in 3D Space

Consider the tolerance propagation as a stack-up of kinematic constraints at mating nodes that appear in an assembly loop, as demonstrated in Figure 7.11.1. The constraint with vector-loop-based assembly models may be expressed as a concatenation of homogeneous transformation matrices [Rogers and Adams 1990], such as

$$[U] = [R_1][T_1]\ [R_2][T_2] \ldots [R_n][T_n] \ldots [R_N][T_N]\ [R_f] \tag{7.11.1}$$

where

[U] is the resultant matrix

$[R_n]$ is the rotational transformation matrix at node n, n = 1, 2, …, and N

$[T_n]$ is the translational matrix at node n, n = 1, 2, …, and N

$[R_f]$ is the final closure rotation

When the assembly is simulated by means of closed-loop constraints, i.e., the starting point and ending point of the vector loop for the assembly are the same, [U] = [I], which is the identity matrix. On the other hand, when there is an "open node" [U] represents the final clearance or interference and its associated orientation. It is assumed that the propagation of tolerances in three-dimensional space can be treated as the propagation of errors during transformations of Cartesian frames.

FIGURE 7.11.1 Kinematic constraints for mating between parts [Chase et al. 1997].

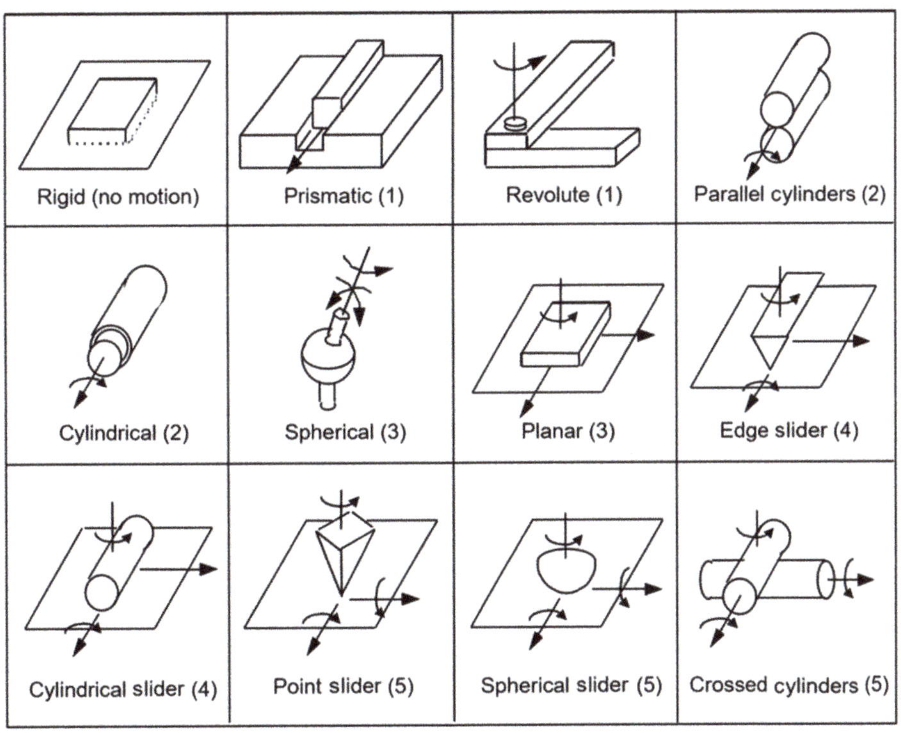

7.11.2. Translational and Rotational Matrices for Both Kinematic and Tolerance Variations

Rotations and translations are operations on free vectors and they preserve length of vectors and angles between vectors, in 3D space. There are three independent basic rotations and three independent basic translations. For a 3D assembly, homogeneous transformation matrices that account for both rotations and translations are listed here without derivations as follows [Whitney et al. 1994]:

$$[R_x] = \begin{bmatrix} 1 & 0 & 0 & 0 \\ 0 & \cos(\alpha) & -\sin(\alpha) & 0 \\ 0 & \sin(\alpha) & \cos(\alpha) & 0 \\ 0 & 0 & 0 & 1 \end{bmatrix}_{4x4} \quad (7.11.2)$$

$$[R_y] = \begin{bmatrix} \cos(\beta) & 0 & \sin(\beta) & 0 \\ 0 & 1 & 0 & 0 \\ -\sin(\beta) & 0 & \cos(\beta) & 0 \\ 0 & 0 & 0 & 1 \end{bmatrix}_{4x4} \quad (7.11.3)$$

$$[R_z] = \begin{bmatrix} \cos(\gamma) & -\sin(\gamma) & 0 & 0 \\ \sin(\gamma) & \cos(\gamma) & 0 & 0 \\ 0 & 0 & 1 & 0 \\ 0 & 0 & 0 & 1 \end{bmatrix}_{4x4} \quad (7.11.4)$$

and

$$[T_d] = \begin{bmatrix} 1 & 0 & 0 & u \\ 0 & 1 & 0 & v \\ 0 & 0 & 1 & w \\ 0 & 0 & 0 & 1 \end{bmatrix}_{4x4} \quad (7.11.5)$$

where

α, β, and γ are the angles of rotation around x-, y-, and z-axes, respectively

$[R_x]$, $[R_y]$, and $[R_z]$ are the rotation matrices around x-, y-, and z-axes, respectively

$[T_d]$ is the translation matrix, where u, v, and w are x-, y-, and z-displacements, respectively

The 4 × 4 matrix transformations given above can be regarded as a construction scheme used to cast necessary rotations or translations into one single matrix. All matrices have the same last row, i.e., [0 0 0 1], and they are called affine matrices. A transformation encoded by a series of affine matrices is called an affine transformation.

The rotation matrices, given by Equations (7.11.2)–(7.11.4), operate along the primary axes centered at the origin, namely (x, y, z) = (0, 0, 0). The order of rotation is important, and consistency must be maintained through the process of all transformations [Gilbert 1992]. The use of consistent 4 × 4 homogenous transformation matrices for the shape and location that are consistent with the way computer-aided design (CAD) systems record locations of items in a geometric model. Their derivatives can be also expressed in terms of multiplications of simply transformed matrices [Skowronski 1998, Yan et al. 2016].

The translation matrix given by Equation (7.11.5) means to move by u, v, and w along the x-, y-, and z-axes, respectively, where

$$[T_d] \text{ reduces to } [T_x], \quad \text{if } v = w = 0 \quad (7.11.6)$$

$$[T_d] \text{ reduces to } [T_y], \quad \text{if } u = w = 0 \quad (7.11.7)$$

and

$$[T_d] \text{ reduces to } [T_z], \quad \text{if } u = v = 0 \quad (7.11.8)$$

Equations (7.11.2)–(7.11.8) are applied to finite displacement (i.e., u, v, and w) and finite angular rotations (i.e., α, β, and γ) in coordinate transformations, such as from a feature coordinate system to its associated part coordinate system and then from the part coordinate system to the global base coordinate system.

Equations (7.11.2)–(7.11.8) are also applicable to small random variables such as tolerances. The ranges of random variables $\delta\alpha$, $\delta\beta$, $\delta\gamma$, δu, δv, and δw are restricted by a series of inequalities, which can effectively represent the variation of two mating features, such as their surfaces and the fit between them in the 3D space.

The idea of establishing a matrix model is to confirm the local DRFs and the related displacement matrix with respect to the local DRF for each feature to which a tolerance is applied. The general functional mating matrix, denoted as [F], for a feature can be expressed in a series of cascaded transformations as

$$[F] = [T_z]\,[T_y]\,[T_x]\,[R_z]\,[R_y]\,[R_x] \tag{7.11.9}$$

Substituting Equations (7.11.2)–(7.11.4) and (7.11.6)–(7.11.8) into the above equation leads to

$$[F] = \begin{bmatrix} c(\delta\gamma)\,c(\delta\beta) & -s(\delta\gamma)\,c(\delta\alpha) - c(\delta\gamma)\,s(\delta\beta)\,s(\delta\alpha) & s(\delta\gamma)\,s(\delta\alpha) - c(\delta\gamma)\,s(\delta\beta)\,c(\delta\alpha) & \delta u \\ s(\delta\gamma)\,c(\delta\beta) & c(\delta\gamma)\,c(\delta\alpha) + s(\delta\gamma)\,s(\delta\beta)\,s(\delta\alpha) & -c(\delta\gamma)\,s(\delta\alpha) + s(\delta\gamma)\,s(\delta\beta)\,s(\delta\alpha) & \delta v \\ -s(\delta\beta) & s(\delta\alpha)\,c(\delta\beta) & c(\delta\alpha)\,c(\delta\beta) & \delta w \\ 0 & 0 & 0 & 1 \end{bmatrix} \tag{7.11.10}$$

of which $c(\delta\alpha) \equiv \cos(\delta\alpha)$, $c(\delta\beta) \equiv \cos(\delta\beta)$, $c(\gamma) \equiv \cos(\delta\gamma)$, $s(\delta\alpha) \equiv \sin(\delta\alpha)$, $s(\delta\beta) \equiv \sin(\delta\beta)$, and $s(\delta\gamma) \equiv \sin(\delta\gamma)$.

For each feature, different tolerance types may result in the same matrix equation; however, the variation of each degree of freedom (DOF) (i.e., variable $\delta\alpha$, $\delta\beta$, $\delta\gamma$, δu, δv, or δw) is unique as manufactured. When the variations of all six variables are as small as tolerances errors (denoted especially as tolerances $\delta\alpha$, $\delta\beta$, $\delta\gamma$, δu, δv, and δw), $c(\delta\alpha) \approx 1$, $c(\delta\beta) \approx 1$, $c(\delta\gamma) \approx 1$, $s(\delta\alpha) \approx \delta\alpha$, $s(\delta\beta) \approx \delta\beta$, and $s(\delta\gamma) \approx \delta\gamma$. Thus, Equation (7.11.10) becomes

$$[F] = \begin{bmatrix} 1 & -\delta\gamma - \delta\beta\,\delta\alpha & \delta\beta & \delta u \\ \delta\gamma & 1 + \delta\gamma\,\delta\beta\,\delta\alpha & -\delta\alpha + \delta\gamma\,\delta\beta\,\delta\alpha & \delta v \\ -\delta\beta & \delta\alpha & 1 & \delta w \\ 0 & 0 & 0 & 1 \end{bmatrix} \tag{7.11.11}$$

After linearization, the above equation reduces to

$$[F] \approx \begin{bmatrix} 1 & -\delta\gamma & \delta\beta & \delta u \\ \delta\gamma & 1 & -\delta\alpha & \delta v \\ -\delta\beta & \delta\alpha & 1 & \delta w \\ 0 & 0 & 0 & 1 \end{bmatrix} \tag{7.11.12}$$

The approach to tolerance analysis is based on the concept of SDT, which describes the displacement of a geometric element by a translation vector and a linearized rotation vector. It is also called the Jacobian–Torsor matrix. How to conduct such a study is illustrated using the case study on tolerance analysis of the gap between two pins as fastened to a plate (Example 7.11.1).

EXAMPLE 7.11.1

Consider an assembly shown in Figure 7.11.2, where two pins are fastened to a plate. With the given relevant dimensions and their related tolerances as shown in Figure 7.11.3, what would be the gap (clearance) between these two pins? This is a study on how important contact errors are. Without the loss of generality of demonstrating the calculating procedure, it is assumed that all the tolerances are normally distributed and structural deformations (kinematic and elastoplastic) are neglected.

Solution:

The first step is to do the system identification of part mating mechanisms for cascaded transformations, as exhibited in Figure 7.11.3. Sequentially,

$$[T]_{system} = [F]_1[T_d]_{1-2}[F]_2[T_d]_{2-3}[T_d]_{3-4}[F]_{3-4}[T_d]_{4-5}[F]_5[T_d]_{5-6}[F]_{5-6}[T_d]_{6-7}[F]_7[T_d]_{7-8}[F]_8 \quad \text{(a)}$$

Translational transformations required for the eight related nominal dimensions can be obtained using Equation (7.11.5) as

$$[T_d] = \begin{bmatrix} 1 & 0 & 0 & u \\ 0 & 1 & 0 & v \\ 0 & 0 & 1 & w \\ 0 & 0 & 0 & 1 \end{bmatrix}_{4\times4} \quad \text{(b)}$$

where

$[T_d]_{1-2}$: u = 0, v = −20/2, w = −60/2
$[T_d]_{2-3}$: u = 0, v = 0, w = −60/2
$[T_d]_{3-4}$: u = 0, v = 0, w = 0
$[T_d]_{4-5}$: u = 0, v = 21, w = 0
$[T_d]_{5-6}$: u = 0, v = 0, w = 0
$[T_d]_{6-7}$: u = 0, v = 0, w = 60/2
$[T_d]_{7-8}$: u = 0, v = −20/2, w = 60/2

Summing up all v values (along y-axis as shown in Figure 7.11.3) leads to the nominal gap size of 1 mm, as −20/2 + 21 + 20/2 = 1. Note that plane 3 is datum A for pin 1, plane 4 is datum A for the plate, and plane 6 is datum A for pin 2.

Given that $\delta\gamma = 0$ due to the cylindrical symmetry, transformations required to account for tolerances (dimensional deviations) can be obtained using Equation (7.11.12) as

$$[F] \approx \begin{bmatrix} 1 & 0 & \delta\beta & \delta u \\ 0 & 1 & -\delta\alpha & \delta v \\ -\delta\beta & \delta\alpha & 1 & \delta w \\ 0 & 0 & 0 & 1 \end{bmatrix} \quad \text{(c)}$$

The distributions of the random variables given in the above equation are supposedly to be estimated by repeating the procedure for a large number of instances of the two profiles. For simplicity, it is assumed that

1. δw: It may take a negative value because the upper part (profile) moves downward upon the lower one in the assembly. Based on the fractal geometry of the contact zone, it is assumed that δw is proportional to the randomness of the cross section as [Armillotta 2016] follows:

$$\delta w = H\left[(\delta u)^2 + (\delta v)^2\right]^{1/2} \quad \text{(d)}$$

   where H represents the autocorrelation of fabrication (e.g., machining and forming) and $0 \le H \le$ 100%) as

   (a) H = 0: White noise, no correlation
   (b) H = 50%: Random movement of small deviations, also called Brownian motion
   (c) H = 100%: Analytic surface, whose motion is governed by $[(\delta u)^2 + (\delta v)^2]^{1/2}$

   Furthermore, δw varies with respect to the fastening force and related material properties applied in the joint. For simplicity, it is here assumed that $\delta w \approx 0$ for this case study.

2. δu and δv: It is assumed that "zone of error $\approx 6\sigma$ (or $\pm 3\sigma$)" for each of these two linear tolerances, given that all the tolerances are hypothetically assumed to be normally distributed.
3. $\delta\alpha$ and $\delta\beta$: They are expected to have zero mean and the same tolerance due to the symmetry cylindrically.

   Thus,

   $[F]_1$: $\delta v = 0.025$

   $[F]_2$: $\delta u = 0.025$, $\delta v = 0.025$, $\delta\alpha = 0.025/60$, and $\delta\beta = 0.025/60$

   $[F]_{3\text{-}4}$: $\delta u = 0.05 + 0.05$ and $\delta v = 0.05 + 0.05$

   $[F]_5$: $\delta v = 0.05$

   $[F]_{5\text{-}6}$: $\delta u = 0.05 + 0.05$ and $\delta v = 0.05 + 0.05$

   $[F]_7$: $\delta u = 0.025$, $\delta v = 0.025$, $\delta\alpha = 0.025/60$, and $\delta\beta = 0.025/60$

and $[F]_8$: $\delta v = 0.025$

   where

   $[F]_1$, $[F]_2$, $[F]_5$, $[F]_7$, and $[F]_8$ are the tolerances of forms and features

   $[F]_{3\text{-}4}$ and $[F]_{5\text{-}6}$ are contact errors

Substituting the above data into Equations (b) and (c), and successively into Equation (a), leads to the solution. The gap between the two pins that are fastened to a plate is 1 ± 0.08 (having standard deviation $\sigma = 0.0267$) without considering contact errors, but gap = 1 ± 0.13 (having standard deviation $\sigma = 0.0433$) when contact errors are included.

Nevertheless, it is recommended that the statistical distribution of every individual tolerance should be assessed using experimental data for applications in the real world. The statistical distributions of the random variables subjected to contact errors, i.e., δu and δv in matrices $[F]_{3\text{-}4}$ and $[F]_{5\text{-}6}$, can be estimated by repeating the procedure of measurement for a large number of instances of the profiles of the two mating planes.

FIGURE 7.11.2 Tolerance analysis of gap between two pins as fastened to a plate [Armillotta 2016].

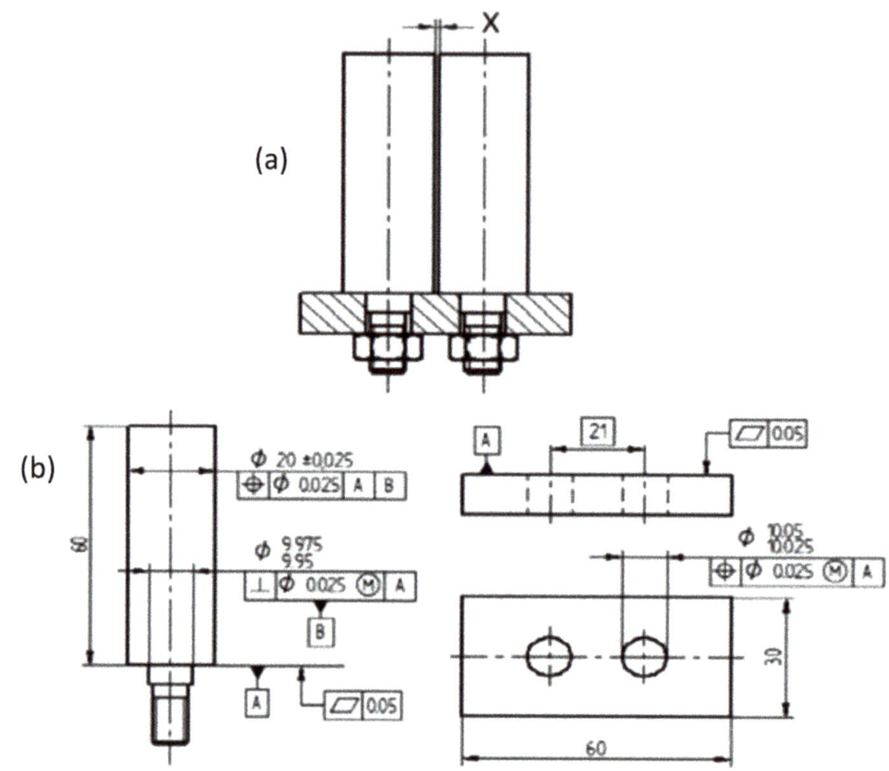

FIGURE 7.11.3 Mating sequence for the case study on two pins fastened to a plate [Armillotta 2016].

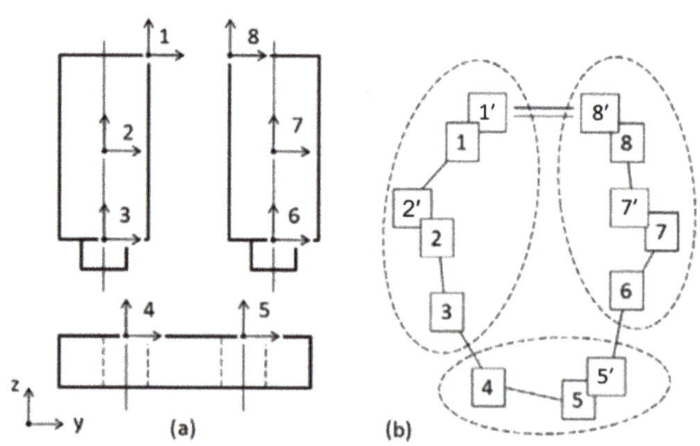

7.11.3. Stochastic Contact Variations

It is enabled to improve the precision of the 3D deviation (including dimensional and form/feature tolerances) analysis with contact errors based on homogeneous transformations as shown by Example 7.11.1. How stochastic contact variations between two contact surfaces gear up toward the application of Equation (7.11.12) is demonstrated using three distinct case studies, i.e., two datum planes, two hollow cylindrical shafts, and two-bearing supporting system, as exhibited in Figure 7.11.4.

FIGURE 7.11.4 Stochastic contact variations—planes, cylindrical shafts, and bearings.

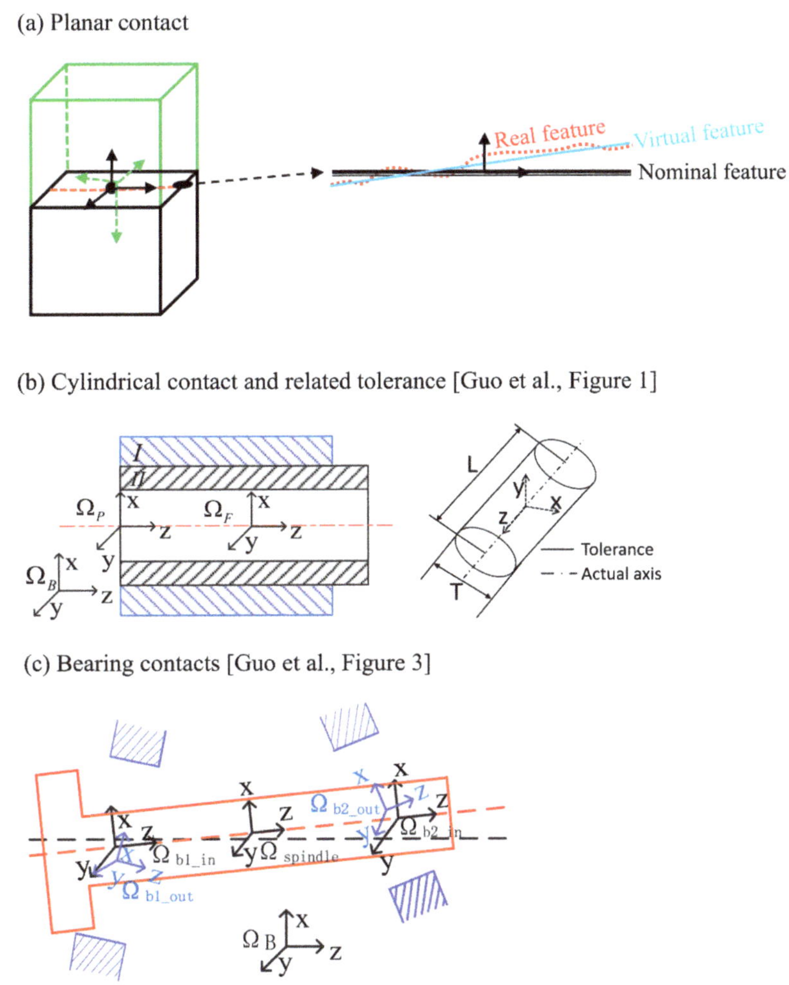

EXAMPLE 7.11.2

Consider the contact between two nominally planar profiles $z_1(x, y)$ and $z_2(x, y)$ with the same contact area of L × L, as shown in Figure 7.11.4(a). Assume that both are datum features for the respective parts. What is the contact error (tolerance) between these two datum planes?

Solution:

Given that the manufacturing errors associated with both datum features are small, the mating between two datum planes leads to $c(\delta\alpha) \approx 1$, $c(\delta\beta) \approx 1$, $c(\delta\gamma) = c(0) = 1$, $s(\delta\alpha) \approx \delta\alpha$, $s(\delta\beta) \approx \delta\beta$, and $s(\delta\gamma) = s(0) = 0$. The functional mating matrix between these two datum features can be assessed using Equation (7.11.12) as

$$[F]_m = \begin{bmatrix} 1 & 0 & \delta\beta & 0 \\ 0 & 1 & -\delta\alpha & 0 \\ -\delta\beta & \delta\alpha & 1 & \delta w \\ 0 & 0 & 0 & 1 \end{bmatrix}$$

The above tolerance mating function is formulated using the same coordinate system virtually attached to the common plane [Armillotta 2016].

Nevertheless, the coordinate systems for the two individual mating parts are exactly opposite to each other in directions as shown in Figure 7.11.4(a). Thus, the feasible functional mating matrix has to be modified with the coordinate mating matrix, namely $[F]_c$, which accounts for the coordinate transformation between two different reference coordinate systems accordingly, as

$$[F] = [F]_c[F]_m = \begin{bmatrix} -1 & 0 & 0 & 0 \\ 0 & -1 & 0 & 0 \\ 0 & 0 & -1 & 0 \\ 0 & 0 & 0 & 1 \end{bmatrix} \begin{bmatrix} 1 & 0 & \delta\beta & 0 \\ 0 & 1 & -\delta\alpha & 0 \\ -\delta\beta & \delta\alpha & 1 & \delta w \\ 0 & 0 & 0 & 1 \end{bmatrix}$$
$$= \begin{bmatrix} -1 & 0 & \delta\beta & 0 \\ 0 & -1 & \delta\alpha & 0 \\ \delta\beta & \delta\alpha & -1 & -\delta w \\ 0 & 0 & 0 & 1 \end{bmatrix}$$

where the two coordinate systems attached to both datums are virtually opposite to each other in all three directions.

Different algorithms have been proposed on how to obtain the realistic values of random variables $\delta\alpha$, $\delta\beta$, and δw [Armillotta 2016, Serrano-Mira et al., Guo et al. 2013] that would account for the true contact errors. A simplified answer is given in Example 7.11.1.

Nevertheless, according to [Armillotta 2016], random variables $\delta\alpha$, $\delta\beta$, and δw can be related to datum profile variables T_1 (flatness tolerance 1 along 1-axis) and T_2 (flatness tolerance 2 along 2-axis), which may be figured out with variable L (size), materials, and machining process, based on repeated measurements of production parts for a large number of samples. The computational effort is simplified if 1- and 2-axes are virtually coincident with x- and y-axes, respectively.

EXAMPLE 7.11.3

Consider the contact between two hollow circular shafts in a rotating machine, as shown in Figure 7.11.4(b). Circular profiles $z_1(x, y)$ and $z_2(x, y)$ have a surface of length L and diametric tolerance T, which are known as measured. What is the contact error (tolerance) between these two hollow shafts?

Solution:

The variation of the nominal contact surface can be expressed in terms of random variables α and β, as shown in Figure 7.11.4(b). Assume that the manufacturing errors are small. Thus, the diametric tolerance T can be expressed in terms of displacements u and v, which are of small values with the following constraints [Guo et al. 2013]:

$$-T/L \le \alpha \le T/L$$

$$-T/L \le \beta \le T/L$$

$$-T/2 \le u \le T/2$$

$$u + \beta(L/2) \le T/2$$

$$v + \alpha(L/2) \le T/2$$

$$\text{and} \quad u^2 + v^2 \le (T/2)^2$$

With regard to the angular rotations, $c(\delta\alpha) \approx 1$, $c(\delta\beta) \approx 1$, $c(\delta\gamma) = c(0) = 1$, $s(\delta\alpha) \approx \delta\alpha$, $s(\delta\beta) \approx \delta\beta$, and $s(\delta\gamma) = s(0) = 0$. The functional mating matrix between these two datum features can be assessed using Equation (7.11.12) as

$$[F]_m = \begin{bmatrix} 1 & 0 & \delta\beta & \delta u \\ 0 & 1 & -\delta\alpha & \delta v \\ -\delta\beta & \delta\alpha & 1 & 0 \\ 0 & 0 & 0 & 1 \end{bmatrix}$$

The tolerance mating function given above is formulated using the common coordinate system that exists virtually.

It is necessary to modify the tolerance mating matrix with the coordinate mating matrix, namely $[F]_c$, that accounts for the two different reference coordinate systems attached to the two cylindrical shafts, respectively, as

$$[F] = [F]_c [F]_m = \begin{bmatrix} -1 & 0 & 0 & 0 \\ 0 & -1 & 0 & 0 \\ 0 & 0 & -1 & 0 \\ 0 & 0 & 0 & 1 \end{bmatrix} \begin{bmatrix} 1 & 0 & \delta\beta & 0 \\ 0 & 1 & -\delta\alpha & 0 \\ -\delta\beta & \delta\alpha & 1 & \delta w \\ 0 & 0 & 0 & 1 \end{bmatrix}$$

$$= \begin{bmatrix} -1 & 0 & \delta\beta & 0 \\ 0 & -1 & \delta\alpha & 0 \\ \delta\beta & \delta\alpha & -1 & -\delta w \\ 0 & 0 & 0 & 1 \end{bmatrix}$$

Random variables $\delta\alpha$, $\delta\beta$, and δw can be related to datum profile variable T (cylindrical surface), which must accommodate contact length variable L (size), materials, machining process, and kinematic and plastic deformations after assembly (e.g., press fit) based on repeated measurements of production parts for a large number of samples.

EXAMPLE 7.11.4

What is the contact error (tolerance) induced by a bearing system with two supporting ball-bearing sets, as shown in Figure 7.11.4(c)?

Solution:

The 3D non-repetitive runout of a bearing system directly influences the precision of shaft rotation. The variation of a nominal contact surface can be expressed in terms of random angular variables $\delta\alpha$ and $\delta\beta$, while the assembly tolerance T consists of displacements δu, δv, and δw [Yang et al. 2011].

With regard to the angular rotation for such a bearing system, $c(\delta\alpha) \approx 1$, $c(\delta\beta) \approx 1$, $c(\delta\gamma) = c(0) = 1$, $s(\delta\alpha) \approx \delta\alpha$, $s(\delta\beta) \approx \delta\beta$, and $s(\delta\gamma) = s(0) = 0$. The functional mating matrix between these two features in contact with each other can be assessed using Equation (7.11.12) as [Guo et al. 2013] follows:

$$[F]_m = \begin{bmatrix} 1 & 0 & \delta\beta & \delta u \\ 0 & 1 & -\delta\alpha & \delta v \\ -\delta\beta & \delta\alpha & 1 & \delta w \\ 0 & 0 & 0 & 1 \end{bmatrix}$$

The tolerance mating function given above is formulated using the common coordinate system that exists virtually.

These two supporting bearings are often adjusted by an arbitrary angle γ in the assembly process. One may have a certain number (e.g., 60) discrete points around a circle to find in which assembly angle Γ_1 and Γ_2 leads to the minimum rotation error $\delta\gamma$ that is associated with angle γ. Thus, it is necessary to modify the tolerance mating matrix with the coordinate mating matrix accordingly, namely $[F]_c$, as

$$[F] = [F]_c [F]_m = \begin{bmatrix} 1 & -\delta\gamma & 0 & 0 \\ \delta\gamma & 1 & 0 & 0 \\ 0 & 0 & 1 & 0 \\ 0 & 0 & 0 & 1 \end{bmatrix} \begin{bmatrix} 1 & 0 & \delta\beta & \delta u \\ 0 & 1 & -\delta\alpha & \delta v \\ -\delta\beta & \delta\alpha & 1 & \delta w \\ 0 & 0 & 0 & 1 \end{bmatrix}$$

$$= \begin{bmatrix} 1 & -\delta\gamma & \delta\beta + \delta\gamma\,\delta\alpha & \delta u - \delta\gamma\delta v \\ \delta\gamma & 1 & \delta\beta\,\delta\gamma - \delta\alpha & \delta v + \delta\gamma\delta u \\ -\delta\beta & \delta\alpha & 1 & \delta w \\ 0 & 0 & 0 & 1 \end{bmatrix} \approx \begin{bmatrix} 1 & -\delta\gamma & \delta\beta & \delta u \\ \delta\gamma & 1 & -\delta\alpha & \delta v \\ -\delta\beta & \delta\alpha & 1 & \delta w \\ 0 & 0 & 0 & 1 \end{bmatrix}$$

Random variables $\delta\alpha$, $\delta\beta$, $\delta\gamma$, δu, δv, and δw can be related to datum profile variable T (cylindrical surfaces of bearings), which must accommodate contact length variable of individual bearing (size), materials, machining process, and kinematic and plastic deformations after assembly based on repeated measurements of production parts for a large number of samples.

7.12. Perturbation Analysis and Direct Linearization Method (DLM)

Assume that the variations of all the constraint functions are relatively small. The DLM based on Taylor series expansion is hence applied [Gao et al. 1998]. As an extended version of the vector-loop model, the direct linearization method is constructed using the following common component elements: vector chains, kinematic joints, assembly datums, dimensional tolerances, geometric feature tolerances, assembly tolerance limits, etc. [Chase et al. 1995, 1996, 1998]. Key features include the following:

(a) A to-be-decided assembly dimension of interest (e.g., gap and interference) is represented by one or several vector loops, which are then translated into a set of algebraic equations, mostly nonlinear.

(b) These equations are manipulated algebraically until an explicit expression of the to-be-decided assembly dimension is sought in terms of the contributing component elements such as dimensional tolerances, geometric feature tolerances, kinematic joints, and assembly datums.

(c) The resulting expression is used to derive tolerance sensitivities, by either partial differentiation via perturbation or numerical methods.

The process for solving tolerance analysis process based on the DLM is illustrated here using Example 7.12.1 after [Leishman and Chase 2010], without considering features, positions, joints, and flatness, for simplicity without the loss of generality of application. Besides Monte Carlo simulations, one common approach to resolving nonlinear problems encountered in a 3D tolerance analysis is the perturbation analysis that often requires direct linearizations of nonlinear terms. It is called the DLM, which bears the assumption that all the dimensions of concern have normal distributions just like the RSS method.

EXAMPLE 7.12.1

As depicted in Figure 7.12.1, a four-bar linkage comes with the following dimensions and operating parameters:

1. Lengths: $r_1 = 50 \pm 0.1$ mm, $r_2 = 20 \pm 0.05$ mm, $r_3 = 50 \pm 0.10$ mm, and $r_4 = 45 \pm 0.075$ mm
2. Operating parameters (i.e., inputs for the angular rotations): θ_1 (grounded) and $0 < \theta_2 \pm 0.0974° < 360°$

How do the position deviations of θ_3 and θ_4 (namely, $\delta\theta_3$ and $\delta\theta_4$) vary with respect to angle θ_2?

Solution:

Since it is a closed loop of vectors, the vector chain of position variations in a 2D mode, namely (x, y), can be expressed in terms of complex variables as [Tinoco and Durango 2019] follows:

$$0 = r_1 \exp(i\theta_1) + r_2 \exp(i\theta_2) + r_3 \exp(i\theta_3) - r_4 \exp(i\theta_4) \quad \text{(a)}$$

The above equation can be resolved in a 2D format along x- and y-axes, respectively, as

$$\text{x-axis:} \quad r_1 \cos(\theta_1) + r_2 \cos(\theta_2) + r_3 \cos(\theta_3) - r_4 \cos(\theta_4) = 0 \quad \text{(b)}$$

and

$$y-\text{axis}:\quad r_1 \sin(\theta_1)+r_2 \sin(\theta_2)+r_3 \sin(\theta_3)-r_4 \sin(\theta_4)=0 \tag{c}$$

Taking differentiations of Equations (b) and (c) with respect to all the given variables leads to [Tinoco and Durango 2019]:

$$dH_x=\cos(\theta_1)dr_1-[r_1\sin(\theta_1)]d\theta_1+\cos(\theta_2)dr_2-[r_2\sin(\theta_2)]d\theta_2$$

$$+\cos(\theta_3)dr_3-[r_3\sin(\theta_3)]d\theta_3-\cos(\theta_4)dr_4-[r_4\sin(\theta_4)]d\theta_4=0 \tag{d}$$

$$dH_y=\sin(\theta_1)dr_1+[r_1\cos(\theta_1)]d\theta_1+\sin(\theta_2)dr_2+[r_2\cos(\theta_2)]d\theta_2$$

$$+\sin(\theta_3)dr_3+[r_3\cos(\theta_3)]d\theta_3-\sin(\theta_4)dr_4-[r_4\cos(\theta_4)]d\theta_4=0 \tag{e}$$

Given that r_1, r_2, r_3, r_4, θ_1, and θ_2 are known, the above two equations can be rearranged as

$$\begin{Bmatrix} dH_x \\ dH_y \end{Bmatrix}=[A]\{dr_1,dr_2,dr_3,dr_4,d\theta_1,d\theta_2\}^T+[B]\begin{Bmatrix} d\theta_3 \\ d\theta_4 \end{Bmatrix}=\begin{Bmatrix} 0 \\ 0 \end{Bmatrix} \tag{f}$$

where

$$[A]=\begin{bmatrix} \cos(\theta_1) & \cos(\theta_2) & \cos(\theta_3) & -\cos(\theta_4) & -r_1\sin(\theta_1) & -r_2\sin(\theta_2) \\ \sin(\theta_1) & \sin(\theta_2) & \sin(\theta_3) & -\sin(\theta_4) & r_1\cos(\theta_1) & r_2\cos(\theta_2) \end{bmatrix} \tag{g}$$

and

$$[B]=\begin{bmatrix} -r_3\sin(\theta_3) & r_4\sin(\theta_4) \\ r_3\cos(\theta_3) & -r_4\cos(\theta_4) \end{bmatrix} \tag{h}$$

Note that r_1 is grounded to the x-axis as a fixed boundary condition. Assume that the deviation ranges of parameters r_1, r_2, r_3, r_4, θ_1, θ_2, θ_3, and θ_4 are small, and thus, the above equation is approximated by

$$\begin{Bmatrix} \delta H_x \\ \delta H_y \end{Bmatrix}=[A]\{\delta r_1,\delta r_2,\delta r_3,\delta r_4,\delta\theta_1,\delta\theta_2\}^T+[B]\begin{Bmatrix} \delta\theta_3 \\ \delta\theta_4 \end{Bmatrix}\approx\begin{Bmatrix} 0 \\ 0 \end{Bmatrix} \tag{i}$$

where $\delta r_1 = 0.2$ mm, $\delta r_2 = 0.1$ mm, $\delta r_3 = 0.2$ mm, $\delta r_4 = 0.15$ mm, $\delta\theta_1 = 0$ (datum), and $\delta\theta_2 = 0.1948°$. The above equation can be solved for unknowns $\delta\theta_3$ and $\delta\theta_4$ simultaneously for any input (θ_1 and θ_2) as

$$\begin{Bmatrix} \delta\theta_3 \\ \delta\theta_4 \end{Bmatrix} = -[B]^{-1}[A]\{\delta r_1, \delta r_2, \delta r_3, \delta r_4, \delta\theta_1, \delta\theta_2\}^T \quad \text{(j)}$$

The above equation can be rewritten as

$$\begin{Bmatrix} \delta\theta_3 \\ \delta\theta_4 \end{Bmatrix} = [S]\{\delta r_1, \delta r_2, \delta r_3, \delta r_4, \delta\theta_1, \delta\theta_2\}^T \quad \text{(k)}$$

where $[S] = -[B]^{-1}[A]$ is called the sensitivity matrix that relates the output vector to the input vector. A plot of variations of $\delta\theta_3$ and $\delta\theta_4$ with respect to θ_2 is exhibited in Figure 7.12.2.

This example is given only for the demonstration of DLM for position deviations. For a realistic application, more design tolerances, such as pin locations (position deviations) and diametric deviations of involute joints, are required. The application of the DLM was extended to describe steady-state dynamic motions (speeds and accelerations) of the four-bar linkage [Leishman and Chase 2010] with pin position deviations and diametric deviations of involute joints. Nevertheless, the directional preference of dynamic effects (e.g., eccentricity of pin positions and tilting of involute joints) cannot be accounted for. The finite element tolerance analysis with skin model shapes could be the better method for applications in the real world, but it is much more time-consuming.

FIGURE 7.12.1 An open four-bar linkage.

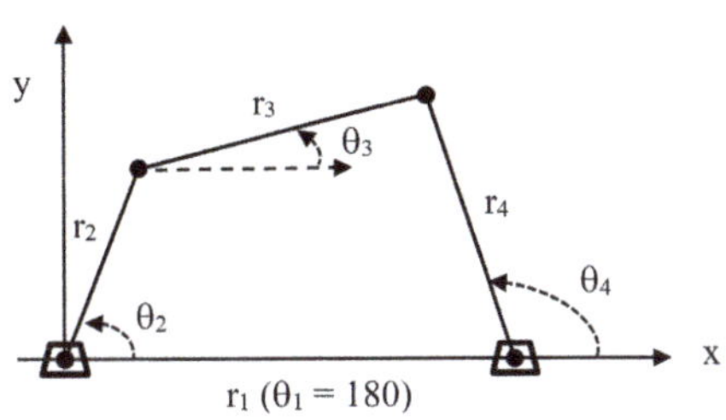

FIGURE 7.12.2 Variations of angular position deviations $\delta\theta_3$ and $\delta\theta_4$ with respect to angle θ_2 for the four-bar linkage (Figure 7.12.1).

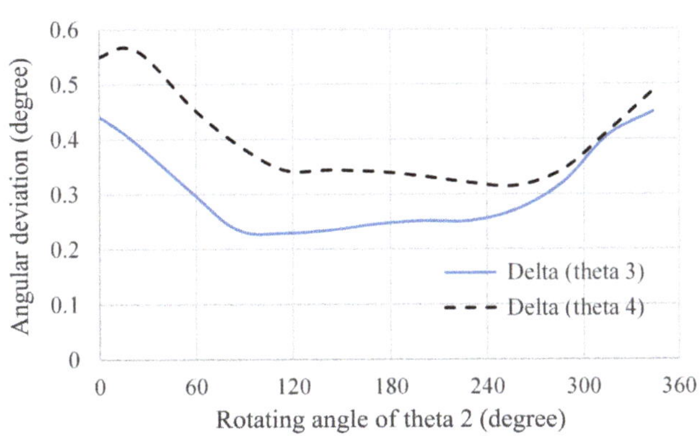

7.13. T-Map Algorithm

A tolerance map, also called T-map, prompts a curve-fitting scheme to describe a hypothetical Euclidean point space, of which the size and shape reflect all variational possibilities for a targeted tolerance feature [Arroyave-Tobón et al. 2017, Marziale and Polini 2011]. Steps toward the T-map tolerance analysis are as follows:

1. To extract the stack loop (similar to vector loop) of interest.
2. To form T-map for each feature in the stack loop in its respective local coordinate system: The tolerance of interest is the range of points resulting from a one-to-one mapping based on the variational possibilities of a feature, within its tolerance zone, to the Euclidean point space [Davidson et al. 2002]. As determined to summarize a combination of tolerance variations on top of a feature, such as size, position, and orientation, it is constructed from a basis simplex and described in a local coordinate system [Ameta et al. 2007, 2010, 2011].
3. To transform all T-maps to fit the global coordinate frame in the assembly.
4. To synthesize the overall functional tolerance contributed by each feature.

The implementation of T-map algorithm can be automated for 3D analysis in a general-purpose tolerance analysis software in conjunction with a CAD system, just like Monte Carlo simulations. Nevertheless, some obstacles remain for practical applications. One challenge is the consolidation of different statistical distributions that are associated with individual dimensions and features of interest. Another challenge is on how to accommodate contact errors in the assembly. The global T-map may have dependent small displacements, and thus, it rises to the third challenge of having a robust computing process [Chitale et al. 2018].

7.14. Tolerance Design Procedure and Software

Several commercial codes are available for doing statistical tolerance analysis and synthesis such as 3DCS, ANATOLEFLEX, CETOL 6σ, Mech Master, PolitoCAT, RD&T, Sigmund, TAA, and VisVSA. Each of them is designed to be integrated into primary 3D CAD environments, in which an automotive manufacturer designs parts and assemblies. Component tolerances are all known or specified, and the resulting assembly tolerance can be calculated using these packages. It is feasible to do statistical tolerance analysis and result visualization for systems in motion [Stuppy et al. 2010]. Direct tasks are unambiguous in calculation and are usually used for checking components and assembly units manufactured according to the specific drawing, following known limit deviations of all partial components, while the limit deviation of the closed component is set.

Each commercial software for statistical tolerance design is generally enabled to utilize GD&T reports to communicate sufficient tolerance design for the product under development. Thus, a manufacturer is able to effectively create a live model for statistical tolerance analysis during conceptual design stage for improving the product reliability and eventually followed by a tolerance allocation study prior to production assembly for improving the profit. Some software packages in use for major CAD systems are listed as follows:

Software	Provider	Major CAD system	Model	Methodology
VSA	Tecnomatix	I-DEAS/CATIA/UG	Variation	MC
3DCS	DCS	CATIA/NX/3DEXPERIENCE	Variation	MC/RSS/WC
eM-TolMate	Tecnomatix	CATIA/NX//UG/I-DEAS	Variation	MC
CETOL 6σ	Sigmetrix	SOLIDWORKS/CATIA/NX	Vector loop	MC/RSS/WC
Sigmund	Varatech	SOLIDWORKS/Solid Edge/Creo	Vector loop	MC/RSS/WC

Note that MC, RSS, and WC stand for Monte Carlo simulation, root-mean-squared method, and worst-case study, respectively. The tolerance analysis models are explained as follows:

(a) Variation Model: The statistical variation analysis model takes advantage of the principles of statistics to relax the component tolerances for an objective such as cost reduction or better quality, as presented in Section 7.4.4. Each component's variation is modeled as a statistical distribution, and these distributions are summed to predict the distribution of the assembly measurement.

(b) Vector Loop: A vector-loop tolerance model presents how the manufactured dimensions (e.g., lengths and angles) are combined sequentially for a proper assembly, as illustrated in Figure 7.1.4. The vector loops gear up for dimensional, form, and kinematic variations.

Although advanced 3D analysis methods for tolerance design such as homogeneous transformations, skin shape models, and T-map algorithms are available, Monte Carlo and RSS based on vector loops and variational methods are still more welcome for general-purpose statistical tolerance analysis software packages.

7.14.1. Variation Simulation Analysis (VSA)

The work procedure following VSA, mainly based on Monte Carlo simulations, is considered as the industry standard [Siemens n.d.]. Commercial software such as VSA 3D (VSA-3D) makes an abstraction of solid models from CAD and their associated GD&T information and proceeds with the dimensional management process described as follows:

1. Define objectives (define dimensional requirements)
2. Identify components
3. Create an assembly sequence plan
4. Create all the parts needed for constructing the assembly
5. Create tolerances—GD&T for these parts
6. Create functional feature models
7. Check GD&T
8. Define assembly operations
9. Create measurement points
10. Define assembly measurements
11. Generate tolerance files
12. Generate an assembly
13. Generate a VSA-3D model
14. Run simulations
15. Generate and evaluate reports
16. Verify the assembly
17. Perform "DOEs" or "What if?" analysis
18. Establish a measurement plan based on the analysis results
19. Determine in situ actual process capability and compare it to design intent
20. Feedback as a product-perfecting procedure
21. Take corrective and preventive actions

VSA facilitates statistical analysis of dimensional variations in parts and assemblies, mostly based on Monte Carlo simulations. Tolerance-specific entities and attributes are interactively extracted from CAD models.

Feature attributes are varied within the specified tolerance range, and user-defined statistical distributions are used in simulation runs to determine the contributors, the extent of contributions, sensitivities, and statistical distribution of the analyzed part dimension or assembly clearance/interference. VisVSA classifies tolerances specification into four groups, i.e., size, location, orientation, and form [Ramnath et al. 2018]:

1. Size Tolerances—equal bilateral, unequal bilateral, and unilateral. Size tolerance can be applied to features-of-size only.
2. Location Tolerances—position, surface profile, total runout, and ±.
3. Orientation Tolerances—perpendicularity, parallelism, angularity, and total runout.
4. Form Tolerances—flatness, surface profile, straightness, and cylindricity.

Although CAD solid models can be imported into VSA-3D as an assembly, the assembly sequence in solid models will be altered when applying VSA-3D. The user interactively creates an abstracted "feature model" for each dimension to be included in the analysis. Abstract objects can be created independently or from actual CAD geometry. Abstraction of the geometry and selected dimensions is done in terms of points, lines, planes and distances, and angles between them. For example,

(a) A cylindrical feature is represented by "point + vector (axis) + radius + length."
(b) A planar feature is represented by "point + vector (normal) + width + length."
(c) The dimension of interest is defined in terms of point coordinate, point-to-point, point-to-line, point-to-plane, gap/flush [Wickman and Söderberg 2007], angle, and maximum or minimum virtual clearance.

It is necessary to specify the datums before assigning them to a particular tolerance feature frame. When a feature of size (FOZ) is located with a position tolerance, the effect of position callout has to be taken into consideration. If tolerances and datums are specified with material modifiers, such as MMC and LMC, there exist bonus and shift tolerances and these material modifiers can be specified to a FOZ for a straightness, positional, or orientation tolerance only. VisVSA uses a point-based analysis method, which does not conform to true 3D tolerance zones and geometric variations. The results from VisVSA are prone to the knowledge and expertise of the user.

VisVSA uses Monte Carlo simulations to predict the amounts of variation [Siemens n.d.]. For the amounts of variation, all the dimensions with tolerances in the model are assigned a statistical distribution (Gaussian, by default). It has been integrated into most solid modeling software, such as CATIA, NX, Pro-E, and I-DEAS, for tolerance design, as being used by most automotive manufacturers. Nevertheless, VisVSA requires that CAD data be sent through one of their translators to bring the data into the Vis-Mockup (*.jt file), which works as a preprocessor for VisVSA. An application of VisVSA in automotive engineering is exhibited in [Ding et al. 2014, Pawar et al. 2011].

7.14.2. CETOL 6σ

CETOL 6σ is a tolerance design software from Sigmetrix, LLC (McKinney, TX) and allows users to tackle complex tolerance analysis challenges in large-scaled assemblies. In addition to being fully compatible with Pro/ENGINEER, CATIA V5, and SolidWorks, CETOL 6σ accommodates tolerance management needs through a dimensional analysis (DA) system that is engineered to support product development processes from conceptual design through the final manufacture.

The CETOL 6σ solution provides product development teams with early evaluations in the design process. The software integrates directly into CATIA allowing product development teams in a variety of industries to work directly in Dassault Systèmes CATIA user's modeling environment.

Chang'an Automobile Co., Ltd. (Chongqing, China) uses Sigmetrix CETOL 6σ tolerance analysis software as part of its research and dimensional engineering and manufacturing program.

7.14.3. 3DCS

3DCS is a DA software for performing variation analysis simulations. Variation analysis of tolerances is the process of analyzing and predicting assembly variation based on component tolerances, piece part geometry, and locating schemes, in a "REAL Virtual Prototype [Dimensional Control Systems Inc., Troy, MI, US]." It works by replicating the build process with deviated parts in a virtual environment, mainly through Monte Carlo simulations, by which the deviations from tolerances are randomly varied within specified ranges based on corresponding statistical distributions. The 3DCS offers both a CAD-independent solution and a CATIA V5 integrated solution. Similarly, not only is 3DCS for NX integrated into NX CAD, but it is also further integrated with Teamcenter. The 3DCS analysis data are stored in the NX CAD model, meaning that any place the model is stored or managed takes the 3DCS data along with it.

3DCS-FEA, a compliant modeler, adds FEA to Monte Carlo simulations, letting the user accurately predict the behavior of flexible parts and the effects of manufacturing processes like welding, clamping, bolting, and riveting. It also incorporates thermal loadings and gravity force effects to see how part weights and thermal loadings affect the dimensional tolerances and structural performance of the assembly.

7.14.4. ANATOLEFLEX

ANATOLEFLEX is a specialty software that provides solutions to tolerance analysis problems [Falgaronea et al. 2016] of flexible components that require complex and difficult modeling tasks. It provides the convenience for users to take into account assembly sequence, joint with defaults, composite or metallic material properties, form default from manufacturing processes, and contact modeling in order to build a full and realistic simulation of loaded assembly.

7.15. Overview of GD&T

This is an overview of GD&T for engineers and designers without a technical background in mechanical drawing. Regarded as geometrical product specifications by ISO, GD&T provides communication among major engineering communities, including research, design, testing and manufacturing, and program management teams. GD&T is the engineering product standard, which describes the geometric features of a product and their operational relationships, expressed as tolerances [Nassef and El Maraghy 1995], to each other and their functional interfaces with mating parts, assemblies, etc.

The current American standard is evolved from American Society of Mechanical Engineers *(ASME): Dimensioning and Tolerancing.* Symbols related to GD&T can be classified into two major categories: geometric characteristic symbols and material conditions, as given in Figures 7.15.1 and 7.15.2, respectively. By providing uniformity in drawing specifications and interpretation, GD&T reduces guesswork throughout the manufacturing process, improving quality, lowering costs, and shortening deliveries.

	Maximum material condition	Least material condition	Regardless of feature size
Symbol	Ⓜ	Ⓛ	None
Status	FOZ has the greatest amount of material within the tolerance zone	FOZ has the least amount of material within the tolerance zone	FOZ is free
Example	Largest dowel size and smallest hole size	Smallest dowel size and largest hole size	—

FIGURE 7.15.1 Basic geometric characteristic symbols [ASME Y14.5M 2014].

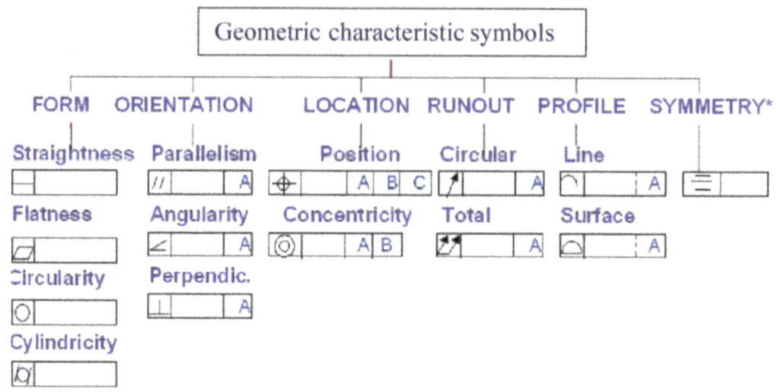

FIGURE 7.15.2 Material conditions.

	Maximum material condition	Least material condition	Regardless of feature size
Symbol	M (circled)	L (circled)	None
Status	Feature of size has the greatest amount of material within the tolerance zone	Feature of size has the least amount of material within the tolerance zone	Feature of size is free
Example	Largest dowel size and smallest hole size	Smallest dowel size and largest hole size	—

A GD&T note consists of a feature constraint frame, which describes the type of constraint to apply to the given feature. Constraints of a feature include the geometric dimensions (e.g., the radius of a circle) and the location of the feature (e.g., location of the center of a circle). A basic dimension is a numerical value used to describe the theoretically exact size, profile, orientation, or location of a feature, feature size, or datum target [Henzold 2020].

Although 2D engineering drawing [You and Yang 1998] has been in practice in the last century in the automotive industry, it would be more representative, straightforward, and thus less time-consuming to do dimensioning and tolerancing on solid models directly. Solid modeling provides orthographic views for product definition, can be used to check the path of assembly, and permits automatic interference or clearance checking [Gossard et al. 1988]. Feature-based assembly of parts provides convenience for statistical tolerance analysis. These features are parametric through their dimensions and tolerancing in a way such that they can be reconstructed when the related dimensions vary. DA is frequently used by engineers and physicists to reduce the complexity of fundamental equations describing the behavior of a system to the simplest and most economical form.

7.15.1. Datum

A typical DRF is made up of three planes. There are typically six DOFs that need to be considered by a designer before choosing which feature is A, B, or C. For example, when A is the primary datum, B is the secondary, and C is the tertiary datum, the primary datum controls the most DOFs. The tertiary datum controls the least DOFs. For the example drawing given in Figure 7.7.5(a), datum A controls three DOFs, datum B controls two DOFs, and datum C controls one DOF. As 3 + 2 + 1 = 6, all six DOFs are constrained.

7.15.2. Dimensions

GD&T is the first step to document design intent and contents. A dimension is defined as a numerical value expressed in appropriate units of measure and used to define the size, location, geometric characteristics, or surface texture of a part of a feature [ASME 2019, International Bureau of Weights and Measures 2006]. A feature is applied to a physical portion of a part, such as a surface, pin, tab, hole, or slot. However, the measurement of dimensions must refer to certain points, axes, and/or planes, which are called datums [McCale 1999]. A dimension may include the following six elements:

1. A dimension line, with its arrowheads, shows the direction and extent of a dimension.
2. Extension lines are used to indicate the extension of a surface or point to a location preferably outside the part outline.
3. A limited length (or area indication) is used to identify a limited length or area, which is to receive additional treatment or consideration within limits specified on the drawing.
4. A leader is used to direct a dimension, note, or symbol to the intended place or the drawing.
5. A reading direction provides the orientation to read a note or dimension.
6. Symmetry.

Terms in product design used for dimensioning mechanical parts are explained as follows:

1. Nominal Dimension: The desired dimension, annotated on 3D models and 2D drawings.
2. Basic Size: The exact theoretical size, limiting dimensions in either the plus or minus direction beginning from the nominal dimension.
3. Limits: The specified maximum and minimum dimensions. They can be bilateral such as 12.0 ± 0.1, or unilateral such as 12.0 – 0.2, 12.0 + 0.0.
4. Tolerance: The differences in the two limits stated, e.g., 0.1 – (–0.1) = 0.2 for bilateral and 0 – (–0.2) = 0.2 for unilateral for the two examples given above.
5. Natural Tolerance (or Naturalized Tolerance): It is a hypothetical tolerance based on a statistical distribution. If the data set fits normal distribution, i.e., equals to $\pm 3\sigma$, where σ is the standard deviation from the mean. It warrants that 99.73% of the production dimension falls within the limits. In practice, the standard deviation and mean are often replaced by sample standard deviation and average.

A dimension may have modifying symbols to characterize the type of constraint sometimes, e.g., R10.0, where R is a modifying symbol for "Radius," meaning a radius of 10.0 units. As presented in Figure 7.15.1, modifying symbols used to constrain a design feature include:

(a) Tangent Plane: to specify the tangency of a plane to another.
(b) Diameter: to denote the diameter of a circle.
(c) Spherical Diameter: to denote the diameter of a sphere.
(d) Radius: to denote the radius of a circle.
(e) Spherical Radius: to denote the radius of a sphere.
(f) Controlled Radius: Radii taken at all points on the part contour shall neither be smaller than the specified minimum limit nor larger than the maximum limit.
(g) Reference: to denote a redundant dimension with respect to the constraint of a part or feature, which usually comes without any tolerance as used for information purpose.
(h) Arc Length: to present a linear dimension for a length measured on a curved outline.

7.15.3. Tolerancing

The total amount of a specific dimension is permitted to vary according to the designated tolerance limits or conditions. A tolerance is defined as the difference between the maximum and minimum limits. When a tolerance permits the dimension to vary in both directions, it is called bilateral tolerance, while one direction for a unilateral tolerance. As illustrated in Figure 7.15.1, geometric tolerancing can be classified into the following five groups:

(a) Form: ***Straightness***, ***flatness***, ***circularity*** (roundness), and cylindricity are used for deciding how far an actual surface or feature is permitted to vary from the specified form designated on the drawing.

(b) Profile: ***Profile of a line*** and ***profile of a surface*** are used for determining how far an actual surface or feature is permitted to vary from the desired form designated in the drawing and/or vary relative to a datum or datums.

(c) Orientation: ***Angularity***, ***perpendicularity***, and ***parallelism*** are used for measuring how far an actual surface or feature is permitted to vary in orientation, relative to a datum or datums.

(d) Location: ***Position***, ***concentricity***, and ***symmetry*** are used for locating how far an actual size feature is permitted to vary from the perfect location implied by the drawing as related to a datum, or datums, or other features.

(e) Runout: ***Circular runout*** or ***total runout*** is used for identifying how far an actual surface or feature is permitted to vary from the desired form implied by the drawing during full rotation of the part around a datum axis.

It would be encouraged to keep dimensions loose at the early product development stage, as many automotive hardware parts are driven by arbitrary size constraints. Tolerances can have modifying symbols to clarify any potentially ambiguous design intent. Frequently applied additional modifying symbols for tolerances include the following [ASME Y14.5M 2014]:

(a) MMC: It tells that the feature contains the maximum amount of material permitted by the tolerance size dimension for that feature, such as smallest hole size and largest shaft size.

(b) LMC: It tells that the feature contains the minimum amount of material permitted by the tolerance size dimension for that feature, such as the largest hole size and smallest shaft size.

(c) Projected Tolerance Zone: It is employed to predict the final dimensions and locations of features on a component or assembly subject to tolerance stack-up, as applied to a hole in which a pin, stud, screw, or bolt is located.

(d) Free State: It means that the geometric tolerance applies in its free state. A free-state variation is used to describe the distortion of a part after the removal of applied forces during manufacture.

(e) Statistical Tolerance: It means that a tolerance is decided on statistical tolerancing.

(f) Between: It shows that the tolerance applies to a limited segment of a surface between designated extremities.

It is not economical to build actual parts to evaluate the performance response or detect the failure modes as being subjected to dimensional variations. Nowadays, engineers can build up a computer model to simulate the response before the first prototype is constructed. It would unveil permissible variations that are established by tolerances on other dimensions, in notes or feature control frames. Most commercial codes in the area of tolerance analysis and synthesis usually can meet the following functional needs:

(a) Receiving solid models built by a CAD system

(b) Generating GD&T variations

(c) Providing GD&T checking
(d) Specifying assembly sequences and methods
(e) Doing 3D measurements
(f) Doing 3D analyses
(g) Identifying relative importance of each contributing factor

The traditional method of intuitively assigning relative importance in GD&T is valid only when a linear relationship exists between the design variables and the final response. In other words, multifactorial interactions and higher-order effects are generally not considered. However, statistical tolerance design, including analysis and allocation, using Monte Carlo simulations, moment generation, and maximum-likelihood method will provide the insight into nonlinear relationships.

7.15.4. Material Conditions

A material condition mandates an "offset zone" by further offsetting the nominal boundary of a part by an amount equal to the tolerance on one side of the nominal after the tolerance zone is specified. Offsets are obtained for the MMC or for the LMC. The difference between these two zones (e.g., LMC for a hole and MMC for its mating dowel) comprises the additional material-originated tolerance zone, an envelope within which the boundary of the part must lie [Shah and Miller 1990]. These callouts also remove one GD&T rule that states that all geometry tolerances are controlled independently of the feature size. MMC can be also applied to the following GD&T symbols:

(a) Straightness (axis)
(b) Parallelism
(c) Perpendicularity
(d) Angularity
(e) True position

When it is desired to assure that two mating parts, such as a shaft in a hole, never interfere with each other, or limit the amount of interference between these two parts in case they are at their worst tolerances, MMC at the hole can be called out. The objective is to take a shaft to go through a hole with the clearance between these two parts. When geometric control and material control are used together, they form the true maximum envelope or virtual condition with a bonus tolerance that the part can be in and still be to specification. If MMC is not specified, the GD&T tolerance may include a given "bonus tolerance" that enlarges the position tolerance of the holes.

On the other hand, if a shaft is made smaller using LMC, the design configuration gains even more bonus tolerance. This extra bonus can be added to the GD&T tolerance and would widen the conform zone that would accommodate potential impacts of straightness, parallelism perpendicularity, angularity, and/or true position tolerances.

7.15.5. Surface Textures

Roughness of surface texture is a measure of surface irregularities, on top of the specified surface form and waviness. It is usually denoted in micrometer (μm). Surface roughness is usually associated with the manufacturing process. Assume that there are N measured deviations of the surface valleys and peaks. The surface roughness can be characterized using one of the following methods:

(a) Surface Roughness Average (R_a): It is defined as the arithmetic average deviation of the surface valleys and peaks; for the given example,

$$R_a = (z_1 + z_2 + \ldots + z_N) / N \quad (7.15.3)$$

(b) Root Mean Square (R_{rms}, R_s, or R_q): It is defined as the square root of squared deviations of the surface valleys and peaks; for the given example,

$$R_s = [(z_1^2 + z_2^2 + \ldots + z_N^2) / N]^{1/2} \quad (7.15.4)$$

(c) Peak Roughness (R_z): It is defined as the maximum deviation of the surface valleys and peaks; for the given example,

$$R_z = \mathrm{Max}(|z_1|, |z_2|, \ldots, |z_N|) \quad (7.15.5)$$

The position of the specifications of the surface texture in a complete symbol is demonstrated in Figure 7.15.3, while the grades of surface roughness according to ISO 1032 are listed in Table 7.15.1.

FIGURE 7.15.3 Specification of surface texture.

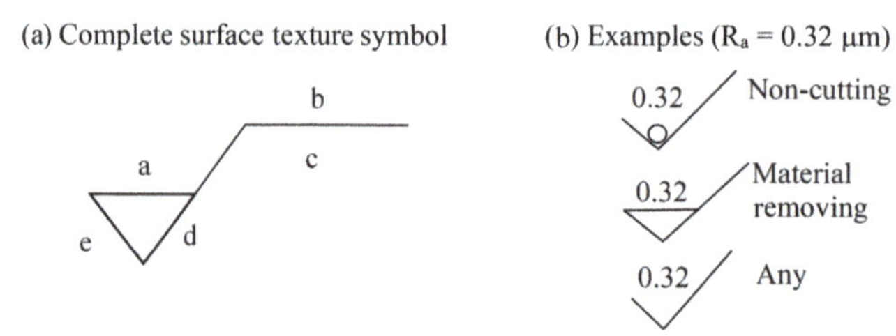

TABLE 7.15.1 Surface roughness (μm) grades.

Grade	N1	N2	N3	N4	N5	N6	N7	N8	N9	N10	N11	N12
R_a	0.025	0.05	0.1	0.2	0.4	0.8	1.6	3.2	6.3	12.5	25	50
R_z	0.1	0.25	0.4	0.8	1.6	3.15	6.3	12.5	25	40	80	160

7.15.6. Units

The Système International (SI) units for dimensions are commonly in use in the automotive industry, although the US customary units were used. SI units based on meter for length are the valid choice in general practice. Nevertheless, SI units based on mm (mini-meter) for length are the popular choice in the automotive industry in accordance with the size and precision control required for on-ground vehicles.

References

AIAG (2008), *Advanced Product Quality Planning and Control Plan*, 2nd Edition, Southfield, MI.

Albrecht, M. C., Nachtsheim, C. J., Albrecht, A. A., and Cook, R. D. (2013), "Experimental Design for Engineering Dimensional Analysis," *Technometrics*, 55(3), pp. 257-270.

Ameta, G., Davidson, J., and Shah, J. (2007), "Using Tolerance-Maps to Generate Frequency Distributions of Clearance and Allocate Tolerances for Pinhole Assemblies," *Journal of Computing and Information Science in Engineering*, 7(4), pp. 347-359.

Ameta, G., Davidson, J., and Shah, J. (2010), "Influence of Form on Tolerance Map-Generated Frequency Distributions for 1D Clearance in Design," *Precision Engineering*, 34(1), pp. 22-27.

Ameta, G., Serge, S., and Giordano, M. (2011), "Comparison of Spatial Math Models for Tolerance Analysis: Tolerance-Maps, Deviation Domain, and TTRS," *Journal of Computing and Information Science in Engineering*, 11(2), pp. 021004.

ANSI (2020), "Preferred Metric Limits and Fits," ANSI B4.2, Reaffirmed 2020, Secretariat: ASME, 72 pages.

Anwer, N., Schleich, B., Mathieu, L., and Wartzack, S. (2014), "From Solid Modelling to Skin Model Shapes: Shifting Paradigms in Computer-Aided Tolerancing," *CIRP Annals*, 63(1), pp. 137-140.

Armillotta, A. (2016), "Tolerance Analysis Considering Form Errors in Planar Datum Features," *Procedia CIRP*, 43, pp. 64-69.

Arroyave-Tobón, S., Teissandier, D., and Delos, V. (2017), "Tolerance Analysis with Polytopes in HV-Description," *Journal of Computing and Information Science in Engineering*, 17(4), 041011.

ASME (2014), "Dimensioning and Tolerancing," ASME Y14.5M-2014, ASME, New York.

ASME Standard (2019), "ASME Y14.5 2018: Dimensioning and Tolerancing," American Society of Mechanical Engineers, New York, 344 pages.

Barai, A. and Pop-Iliev, R. (2009), "Integration of Design Tolerance Analysis and Statistical Production Control in the Product Lifecycle Management," *CIRP CAT Conference 2009*, Annecy, France.

Becker, J., Nemeth, T., Wegmann, R., and Sauer, D. (2018), "Dimensioning and Optimization of Hybrid Li-Ion Battery Systems for EVs," *World Electric Vehicle Journal*, 9(2), p. 19.

Bender, A. Jr. (1968), "Statistical Tolerancing as it Relates to Quality Control and the Designer," *SAE Technical Paper 680490*, https://doi.org/10.4271/680490.

Bernstein, N. X. (1989), "Representation and Analysis of Dimension and Tolerance Information in Solid Modeling," MSc thesis, Department of Mechanical Engineering, Ben-Gurion University of Negev, Beer Shava, Israel.

Bruyère, G., Dantan, J., Bigot, R., and Martin, P. (2007), "Statistical Tolerance Analysis of Bevel Gear by Tooth Contact Analysis and Monte Carlo Simulation," *Mechanism and Machine Theory*, 42(10), pp. 1326-1351.

Caro, S., Bennis, F., and Wenger, P. (2005), "Tolerance Synthesis of Mechanisms: A Robust Design Approach," *ASME Journal of Mechanical Design*, 27(1), pp. 86-94.

Carr, C. D. (1993), "A Comprehensive Method for Specifying Tolerance Requirements for Assemblies," ADCATS Report No. 93-1, Brigham Young University, Provo, UT.

Casella, G. and Berger, R. L. (2002), *Statistical Inference*, 2nd Edition, Duxbury Press, Belmont.

Cavanagh, R. R., Neuman, R. P., and Pande, P. S. (2005), *What Is Design for Six Sigma?*, McGraw-Hill, New York.

Chase, K. W. (1999), "Tolerance Analysis of 2-D and 3-D Assemblies," ADCATS Report No. 99-4, Mechanical Engineering Department, Brigham Young University, Provo, UT.

Chase, K. and Greenwood, W. (1988), "Design Issues in Mechanical Design," *Manufacturing Review*, 1(1), pp. 50-59.

Chase, K., Gao, J., and Magleby, S. (1995), "General 2-D Tolerance Analysis of Mechanical Assemblies with Small Kinematic Adjustments," *Journal of Design and Manufacturing*, 5(4), pp. 263-274.

Chase, K., Gao, J., Magleby, S., and Sorenson, C. (1996), "Including Geometric Feature Variations in Tolerance Analysis of Mechanical Assemblies," *IIE Transactions*, 28(10), pp. 795-807.

Chase, K. W., Magleby, S. P., and Glancy, C. G. (1997), "A Comprehensive System for Computer-Aided Tolerance Analysis of 2-D and 3-D Mechanical Assemblies," *Proceedings of the 5th International Seminar on Computer-Aided Tolerancing*, Toronto, Canada, 1997.

Chase, K. W., Gao, J., and Magleby, S. P. (1998), *Advanced Tolerancing Techniques*, John Wiley & Sons, New York, pp.103-137.

Chiang, Y. J. and Barber, G. C. (2002), "Operating Effort and Related Statistical Tolerance Analyses of Automotive Latching Mechanisms," *International Journal of Materials and Product Technology*, 17(5/6), pp. 353-367.

Chiang, Y. J., Shih, C. D., Lin, C. C., Lee, H. K., Tseng, Y. Y., Tsai, K. C., and Cheng, Y. H. (2000), "Multi-Variable Effects on Sealing Pressure between Tires and Rims," *International Journal of Vehicle Design*, 23(1/2), pp. 78-93.

Chih, W., Fu, J., and Wang, K. (2007), "A Neural Network Based Approach for the Tolerance Analysis," *Journal of Chinese Institute of Industrial Engineers*, 24(5), pp. 366-377.

Chitale, A., Davidson, J. K., and Shah, J. J. (2018), "Statistical Tolerance Analysis with Sensitivities Established with Tolerance-Maps," *ASME 2018 IDETC/CIE 2018*, Quebec, Canada, August 26–29, 2018.

Choi, J., Lee, S., and Choi, D. (1998), "Tolerance Optimization for Mechanisms with Lubricated Joints," *Multibody System Dynamics*, 2(2), pp. 145-168.

Congorno, G. (2010), "Tolerance Stack-Up Analysis," Technical Training Consultants, file:///H:/Books/Design%20of%20Experiments%20for%20Product%20Reliability%20Growth/7-Tolerancing%20with%20Design%20for%20Assembly%20and%20Manufacturing/Tolerance%20Stackup%20Analysis.pdf.

Corrado, A. and Polini, W. (2018), "FEA Integration in the Tolerance Analysis Using Skin Model Shapes," *Procedia CIRP*, 75, pp. 285-290.

Cox, N. D. (2006), *How to Perform Statistical Tolerance Analysis*, ASQC, Milwaukee, WI; ISBN: 978-0-87389-010-6.

Craig, M. (1995), "Using Dimensional Management," *Mechanical Engineering*, 117, pp. 86-88.

Creveling, C. M. (1997), *Tolerance Design- A Handbook for Developing Optimal Specifications*, Addison Wesley Longman, Inc., Reading, MA; ISBN: 0133052346.

Davidson, J. K., Mujezinovic, A., and Shah, J. J. (2002), "A New Mathematical Model for Geometric Tolerances as Applied to Round Faces," *Journal of Mechanical Design*, 124(4), pp. 609-622.

Ding, G. D., Chiang, Y. J., Xu, Q. Z., Song, Y. G., Wang, K. K., Luan, X. Q., and Zhao, H. C. (2014), "Statistical Tolerance Analysis of Clearance between Piston and Valve in Hot State Using VisVSA," *Internal Combustion Engine and Power Plant*, 31(1), pp. 11-13.

Dong, Z. (1997), "Tolerance Synthesis by Manufacturing Cost Modeling and Design Optimization," *Advanced Tolerancing Techniques*, Edited by Zhang, H.-C., John Wiley & Sons, New York, pp. 233-260.

Drake, P. Jr. (1999), *Dimensioning and Tolerancing Handbook*, McGraw-Hill, New York; ISBN: 0-07018131-4.

Evans, D. H. (1975a), "Statistical Tolerancing: The State of the Art, Part 1," *Journal of Quality Technology*, 7(1), pp. 1–12.

Evans, D. H. (1975b), "Statistical Tolerancing: The State of the Art, Part 2," *Journal of Quality Technology*, 7(1), pp. 1–12.

Falgaronea, H. et al. (2016), "Variation Simulation during Assembly of Non-Rigid Components. Realistic Assembly Simulation with ANATOLEFLEX Software," *Procedia CIRP*, 43, pp. 202-207.

Fischer, B. R. (2011), *Mechanical Tolerance Stackup and Analysis*, CRC Press, Boca Raton; ISBN: 1439815720.

Fortini, E. T. (1967), *Dimensioning for Exchangeable Manufacture*, Industrial Press, New York.

Freeman, L. J. (2010), "Statistical Methods for Reliability Data from Designed Experiments," PhD dissertation in Statistics, Virginia Polytechnic Institute and State University, Blacksburg, VA.

Gao, J., Chase, K., and Magleby, S. (1998), "Generalized 3-D Tolerance Analysis of Mechanical Assemblies with Small Kinematic Adjustments," *IIE Transactions*, 30(4), pp. 367-377.

Garaizar, O. R., Qiao, L., Anwer, N., and Mathieu, L. (2016), "Integration of Thermal Effects into Tolerancing Using Skin Model Shapes," *Procedia CIRP*, 43, pp. 196-201.

Gilbert, O. L. (1992), "Representation of Geometric Variations Using Matrix Transformations for Statistical Tolerance Analysis in Assemblies," MS thesis, Department of Mechanical Engineering, Massachusetts Institute of Technology, Cambridge, MA.

Gossard, D. C., Zeffante, R. P., and Sakurai, H. (1988), "Representing Dimensions, Tolerances, and Features in MCAE Systems," *Computer Graphics and Applications*, 8, pp. 51-59.

Green, P. (2005), *The Geometrical Tolerancing Desk Reference: Creating and Interpreting ISO Standard Technical Drawings*, Elsevier, Oxford; ISBN: 978-0-7506-6821-7.

Greenwood, W. and Chase, K. (1990), "Root Sum Squares Tolerance Analysis with Nonlinear Problems," *Journal of Engineering for Industry*, 112(4), pp. 382-384.

Guo, J. et al. (2013), "A Tolerance Analysis Method for Rotating Machinery," *Procedia CIRP*, 10, pp. 77-83.

Guzmán, L. and Hammett, P. (2003), "A Tolerance Adjustment Process for Dimensional Validation of Stamping Parts and Welded Assemblies," *SAE Technical Paper 2003-01-2871*, https://doi.org/10.4271/2003-01-2871.

Hallmann, M. et al. (2020), "From Tolerance Allocation to Tolerance-Cost Optimization: A Comprehensive Literature Review," *The International Journal of Advanced Manufacturing Technology*, 107, pp. 4859-4912.

Henzold, G. (2020), *Geometrical Dimensioning and Tolerancing for Design, Manufacturing and Inspection*, 3rd Edition, CRC Press, Boca Raton, 486 pages; ISBN: 978-0128240618.

Homri, L., Dantan, J., and Levasseur, G. (2016), "Comparison of Optimization Techniques in a Tolerance Analysis," *Procedia CIRP*, 43, pp. 184-189.

Hura, C. M., Caron, N. O., Kanik, J. A., and Hunt, W. A. (1988), "Statistical Tolerance Analysis Using Solid Modeling," *SAE Technical Paper 881245*, https://doi.org/10.4271/881245.

International Bureau of Weights and Measures (2006), *The International System of Units*, 8th Edition, BIPM, Sèvres, France; ISBN: 92-822-2213-6.

ISO (2017), "Preview Geometrical Product Specifications (GPS)—Geometrical Tolerancing—Tolerances of Form, Orientation, Location and Run-Out," ISO 1101.

Joseph, V. R. (2004), "Quality Loss Functions for Nonnegative Variables and Their Applications," *Journal of Quality Technology*, 36, pp. 129-138.

Joskowicz, L., Sacks, E., and Srinivasan, V. (1997), "Kinematic Tolerance Analysis," *Computer-Aided Design*, 29(2), pp. 147-157.

Kovar, S. (2016), "Datum Reference Frame Position and Shape Tolerances for Tolerance Analysis," www.tul.cz, Technical University of Liberec, Retrieved September 22, 2019, file:///H:/Books/Design%20of%20Experiments%20for%20Product%20Reliability%20Growth/7-Tolerancing%20with%20Design%20for%20Assembly%20and%20Manufacturing/Datum%20reference%20frame%20Position%20and%20shape%20tolerances%20Tolerance%20analysis.pdf.

Kremer, A. et al. (2020), "Reliability Prediction Using Design of Experiments," *Proceedings of the 30th European Safety and Reliability Conference*, Edited by Baraldi, P., Di Maio, F., and Zio, E., Research Publishing, Singapore.

Kroczak, J. and Dudziak, M. (2011), "Tolerance Analysis of Involute Splines," *Proceedings of the World Congress on Engineering 2011*, Vol. III, London, UK, July 6–8, 2011.

Leishman, R. C. and Chase, K. W. (2010), "Direct Linearization Method Kinematic Variational Analysis," *Journal of Mechanical Design*, 132, p. 071003.

Li, B. and Roy, U. (2001), "Relative Positioning of Toleranced Polyhedral Parts in an Assembly," *IIE Transactions*, 33(4), pp. 323-336.

Lin, P. D. and Chen, J. F. (1994), "Analysis of Errors in Precision for Closed Loop Mechanisms," *Journal of Mechanical Design*, 116, pp. 197-203.

Liu, S. et al. (2008), "Non-Normal Statistical Tolerance Analysis Using Analytic Convolution Method," *Journal of Advanced Manufacturing Systems*, 7(1), pp. 127-130.

Marziale, M. and Polini, W. (2011), "Review of Variational Models for Tolerance Analysis of an Assembly," *Journal of Engineering Manufacture*, 225(3), pp. 305-318.

McCale, M. R. (1999), "A Conceptual Data Model of Datum Systems," *Journal of Research of the National Institute of Standards and Technology*, 104(4), pp. 349-400.

McKinney, Tex (2000), "Fixed and Floating Fastener Variation," www-eng.lbl.gov, Chapter 23.

Meadows, J. D. (2009), *Geometric Dimensioning and Tolerancing Handbook: Applications, Analysis & Measurement*, ASME Press, New York; ISBN: 9780971440166.

Murthy, D. P., Xie, M., and Jiang, R. (2004), *Weibull Models*, John Wiley & Sons, Inc., Hoboken.

Musia, R. A. et al. (2004), "Simulation-Based Tolerance Stackup Analysis in Machining," *Transactions of NAMRI/SME*, 32, pp. 533-540.

Nassef, A. O. and El Maraghy, H. A. (1995), "Allocation of Geometric Tolerance," *CIRP Annual*, 46(1), pp. 101-106.

Navas, H. V. G. (2015), "Monte Carlo Model Applied to Tolerance Analysis of Mechanical Assembly Sets," *Engineering and Automation*, 2(5), pp. 55-58.

Pawar, S. Y. et al. (2011), "Tolerance Stackup Analysis and Simulation Using Visualization VSA," *International Journal of Advanced Engineering Technology*, 2(1), pp. 169-175.

Ramnath, S. et al. (2018), "Comparative Study of Tolerance Analysis Methods Applied to a Complex Assembly," *Procedia CIRP*, 75, pp. 208-213.

Rogers, D. F. and Adams, J. A. (1990), *Mathematical Elements for Computer Graphics*, 2nd Edition, McGraw-Hill, New York.

Roy, U. and Liu, C. R. (1988), "Feature Based Representational Scheme of Solid Modeler for Providing Dimensioning and Tolerancing Information," *Robotics and Computer-Integrated Manufacturing*, 3(4), pp. 335-345.

Schleich, B. and Wartzack, S. (2016), "A Quantitative Comparison of Tolerance Analysis Approaches for Rigid Mechanical Assemblies," *Procedia CIRP*, 43, pp. 172-177.

Schleich, B., Anwer, N., Mathieu, L., and Wartzack, S. (2014), "Skin Model Shapes: A New Paradigm Shift for Geometric Variations Modelling in Mechanical Engineering," *Computer-Aided Design*, 50, pp. 1-15.

Scholz, F. (1995), "Tolerance Stack Analysis Methods," Boeing Information & Support Services, Seattle, WA, Retrieved September 5, 2019.

Semere, D. et al. (2018), "Skin Solid Based Tolerance and Variation Analysis," *Procedia CIRP*, 72, pp. 726-731.

Shah, J. J. and Miller, D. (1990), "A Structure for Supporting Geometric Tolerances in Product Definition Systems for CIM," *Manufacturing Review*, 3(1), p. 2331.

Shen, S. et al. (2005), "A Comparative Study of Tolerance Analysis Methods," *Journal of Computing and Information Science in Engineering*, 5(3), 247-256.

Siemens (n.d.), VisVSA Online Help, Version 4.0.

Skowronski, V. (1998), "Calculating Derivatives in Statistical Tolerance Analysis," *Computer-Aided Design*, 30(5), pp. 367-375.

Spotts, M. F. (1973), "Allocation of Tolerances to Minimize the Cost of Assembly," *Journal of Engineering for Industry*, 95, pp. 762-764.

Srinivasmurthy, S. and Herle, M. (2019), "Tolerance Stack-up Analysis, Its Benefits and Steps Involved in Mechanical Design," ASM Technologies, https://www.asmltd.com/tolerance-stack-analysis-benefits-steps-involved-mechanical-design/#wrap.

Stefano, P. (2003), "Tolerance Analysis and Synthesis Using the Mean Shift Model," *Journal of Mechanical Engineering Science*, 217, pp. 149-159.

Stuppy, J. et al. (2010), "Statistical Tolerance Analysis and Result Visualization for Systems in Motion," *Proceedings of the 11th International Design Conference (DESIGN 2010)*, Dubrovnik, Croatia, 2010, pp. 1431-1440.

Taguchi, G. (1987), *System of Experimental Design*, UNIPUB, Kraus International Publications, New York.

Tinoco, H. A. and Durango, S. (2019), "Tolerance Analysis of Planar Mechanisms Based on a Residual Approach: A Complementary Method to DLM," *Mathematical Problems in Engineering*, 2019, pp. 1–13, Article ID 9067624.

Tynes, J. (2012), *Make It Fit: Introduction to Tolerance Analysis for Mechanical Engineers*, CreateSpace Independent Publishing Platform; ISBN: 978-1482350258.

Ullman, D. G. (2003), *The Mechanical Design Process*, 3rd Edition, McGraw-Hill, New York.

Wang, R. et al. (2006), "Statistical Tolerance Transfer for Complex Sheet Metal Part Forming," *5th International Conference on Mechanics and Materials in Design*, Porto, Portugal, July 24–26, 2006.

Whitney, D. E., Gilbert, O. L., and Jastrzebski, M. (1994), "Representation of Geometric Variations Using Matrix Transformations for Statistical Tolerance Analysis in Assemblies," *Research in Engineering Design*, 6, pp. 191-210.

Wickman, C. and Söderberg, R. (2007), "Perception of Gap and Flush in Virtual Environments," *Journal of Engineering Design*, 18, pp. 175-193.

Wu, F. et al. (2009), "Improved Algorithm for Tolerance Allocation Based on Monte Carlo Simulation and Discrete Optimization," *Computers and Industrial Engineering*, 56(4), pp. 1402-1413.

Yan, H. et al. (2015), "Application of Monte Carlo Method in Tolerance Analysis," *Procedia CIRP*, 27, pp. 281-285.

Yan, H., Cao, Y., and Yang, J. (2016), "Statistical Tolerance Analysis Based on Good Point Set and Homogeneous Transform Matrix," *Procedia CIRP*, 43, pp. 178-183.

Yang, Z., Hong, J., and Liu, J. (2011), "Theoretical Model to Analyze the Non-Repetitive Run-Out (NRRO) of Ball Bearings," *Advanced Science Letters*, 4, pp. 2522-2527.

You, C. F. and Yang, S. S. (1998), "Automatic Feature Recognition from Engineering Drawing," *International Journal of Advanced Manufacturing*, 14(7), pp. 495-507.

Zhang, C., Luo, J., and Wang, B. (1999), "Statistical Tolerance Synthesis Using Distribution Function Zones," *International Journal of Production Research*, 37, pp. 3995-4006.

Zhu, J. and Ting, K. L. (2001), "Performance Distribution Analysis and Robust Design," *Journal of Mechanical Design*, 123(1), pp. 11-17.

Problems

P7.1: For the component shown in Figure P7.1(a), the production tolerances of the given dimensions, i.e., A–G, are supposed to assure that the distance between the two orifice centers (i.e., Z) is maintained at least to have a CI of 99.5% within the range. Please conduct a tolerance analysis of the horizontal distance between two orifice holes by Monte Carlo simulations. The dimensional chain is given in Figure P7.1(b) for reference.

Hint: $X = B - D + (C - E)\cos(A) - G\sin(A)$ and $Y = (C - E)\sin(A) + G\cos(A) - F$.

FIGURE P7.1 Plate with two orifice holes [https://www.mitcalc.com/doc/tolanalysis3d/help/en/example2.htm].

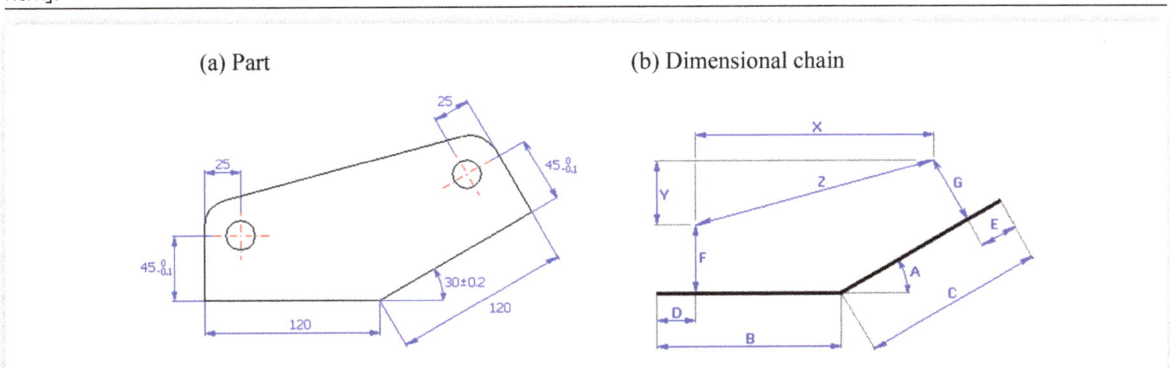

Courtesy of Miroslav Petele.

P7.2: For the engine slider-crank mechanism shown in Figure P7.2(a), the production tolerances (i.e., A–F) of the given dimensions are supposed to assure that the distance between the piston top and the cylinder head (i.e., Z) is properly maintained. Using the tolerance analysis of a 2D chain by Monte Carlo simulations to figure out the unknown clearance, i.e., Z. The dimensional chain is given in Figure P7.2(b) for reference.

Hint: $Z = E + F - D - X$ where $X = B\cos(\alpha) + C\{1 - [(B/C)\sin(\alpha) + A/C]^2\}$

FIGURE P7.2 2D tolerances of a slider-crank mechanism [www.mitcalc.com].

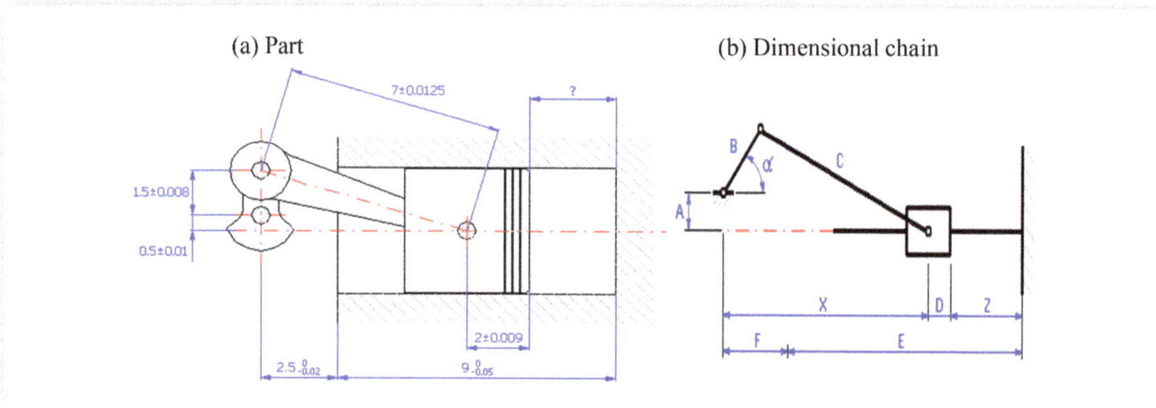

Courtesy of Miroslav Petele.

P7.3: On the resistors given in Figure 7.5.4, supplier B can provide resistors within ±2% in resistance variation, how much is the process capability (C_{pk}) of the supplier?

8

Lognormal Transformations in DOE

DOE methods combined with reliability engineering make a powerful combination for product development such as reliability DOEs, which are the DOEs involved with various reliability goals, e.g., maximizing product life cycles against fatigue. By statistics, life cycles of automotive components in operation are often not normally distributed with respect to the related design parameters. Reliability DOEs are built upon creating proper stochastic models based on reliability statistics in combination with physical laws and provide a system-level estimate for the output engineering parameters. In DOE-based accelerated life testing, the lifetime is often assumed to follow log-location-scale distributions, such as lognormal distribution (DOE-LN) and Weibull distribution (DOE-W). Lognormality-based reliability DOEs are an important statistical discipline as part of reliability DOEs.

8.1. Reliability DOEs

Reliability is the probability that a product or a system will perform its intended function without failure for a specified in-service lifetime under various working conditions. Reliability analysis refers to the use of data to evaluate a product's ability to perform over time. It is regarded as a tool to assess product failures, involving the characterization of failure modes of the products using statistical distributions such as Weibull (including exponential) and lognormal functions in combination of laws of physics. Based on the fitted statistical distribution function, failures are mitigated, warranty returns are predicted, maintenance actions are exploited, and preventive actions are taken. In design for reliability (DFR), reliability analysis can be comfortably carried out using experimental design techniques in conjunction with the physics of failure. By adopting the methodology with combined statistics and physics, reliability analysis is a powerful tool for developing robust products that operate with minimal defects. Based on DOEs, DFR utilizes predictive engineering based on the lifetime data to build reliability into future products, in addition to quantifying the existing product reliability [Hamada 1995]. An overview of designing experiments for reliability data can be found in [Vining et al. 2015].

The response is usually the product life in its respective units, such as kilometers for automotive systems, number of cycles for mechanical rotating components, and months (years) for electronic parts. These measures expressed in terms of life expectancy of such products are quite nonlinear and do not follow the normal distribution as assumed for the DOEs based on t-distribution (DOE-t) and DOEs based on F-distribution (DOE-F) (e.g., response surface method (RSM) based on F-distribution), even for normal-distributed design variables as shown in Figure 8.1.1. The typical lifetime data of automotive products, including mechanical and electronic components, are more adequately modeled by either the Weibull or lognormal distribution functions. Note that an exponential distribution is a subset of the Weibull distribution.

FIGURE 8.1.1 Normal-distributed design variable (e.g., strain) yielding non-normal-distributed life cycles.

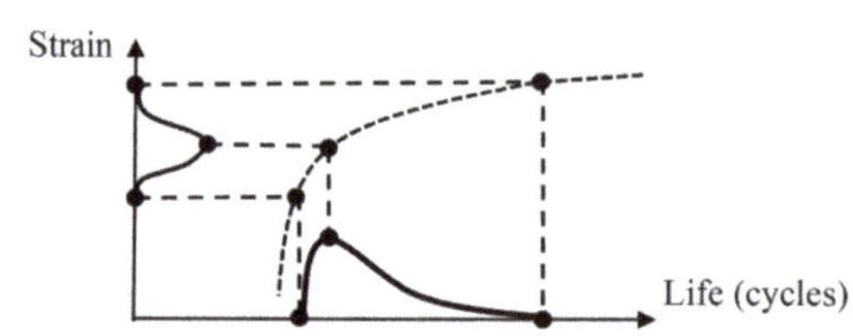

ANOVA for factorial designs is quite sensitive to the assumption that the data pertaining to the product characteristics for analysis have to be normally distributed. Methodologies applied for reliability DOEs follow the principles of DOEs, while the response is not limited to normality assumption.

As the first step toward reliability analysis, one can use commercial codes to test the fit of the collected data against various distributions, for example, by clicking the "Use all distributions" option while using Minitab. As reliability prediction is important to failure prevention, highly accelerated growth techniques based on reliability statistics, i.e., lognormal distribution and Weibull distribution, are frequently employed to examine the high degree of reliability growth while minimizing the effect of noises on the design cycle. Reliability DOEs based on lognormal distribution and Weibull distribution are called DOE-LN and DOE-W, respectively. As a unified methodology in practice, solution procedures based on Weibull and lognormal distributions look much alike, but their perspective reliability predictions may lead to a significant difference statistically [Meeker 1984].

8.1.1. Logarithmic Transformations

As inspired by the fact that the normal distribution has very desirable properties in practice, it is useful to transform a random variable into another variable that is normally distributed. For product life prediction, a logarithmic transformation is sometimes more suitable than an equidistant linear distribution of the design. A logarithmic transformation measures observations automatically according to a ratio scale and reduces problems associated with percentage changes from baseline. Variables such as fatigue of electronic packaging materials and biochemical measurements typically show a skewed distribution, which can often be made symmetric using a natural log transformation [Keene 1995].

Three frequently used natural log-location-scale distributions applied in reliability analysis, i.e., lognormal, Weibull, and exponential functions, are listed in Table 8.1.1. Note that the exponential distribution function is a special case of the Weibull distribution function.

TABLE 8.1.1 Frequently used natural log-location-scale distributions in reliability analysis.

Variable t	Statistical distribution of y	Location	Standardized variable
Normal	t: Normal	μ_t, i.e., mean of t	$z = (t - \mu_t)/\sigma_t$
Lognormal	y = ln(t): Normal	μ_y, i.e., mean of y	$z = [\ln(t) - \mu_y]/\sigma_y$
Weibull	y = ln(t): SEV	$\ln(\eta)$	$z = \beta\,[\ln(t) - \ln(\eta)]$
Exponential	y = ln(t): SEV	$\ln(\lambda^{-1})$, as $\lambda^{-1} = \eta$	$z = \ln(t) - \ln(\lambda^{-1})$

where

t is the observation ∈ test data acquired; no transformation, if t is normally distributed
μ_t is the sample mean, i.e., the expected value of variable t
μ_y is the sample mean of y, where y = ln(t)
σ_y is the standard deviation of y = ln(t)
z is the variate corresponding to the standardized distribution
SEV is the smallest extreme-value distribution (Chapter 9)
η is the scale factor of the Weibull distribution (Chapter 9)
β is the shape factor of the Weibull distribution (Chapter 9)
λ: $\lambda = 1/\eta$; Weibull distribution turns into an exponential distribution, when $\beta = 1$

The scale of the PDF in life cycles is assumedly compressed at a high load (strain), as demonstrated schematically in Figure 8.1.2. It means that the failure mode remains the same, but the time of occurrence decreases at the high level only. Statistically speaking, when the product life expectancy decreases as a factor (e.g., strain) changes from the low level to the high level, the PDF curves may still have the same shape, while the scale of the curve has changed. In such a case, a life characteristic can be chosen to represent the curve and utilized to investigate the effect of potential factors on the product life [Guo and Mettas 2010], instead of considering the entire scope of its PDF.

Note that the shape factor used in the Weibull distribution can be the same across all the treatments (runs) for some experimental designs, but varies for others. This is to be further addressed in Chapter 9.

FIGURE 8.1.2 Influence of design factor level (e.g., strain) on the scale factor.

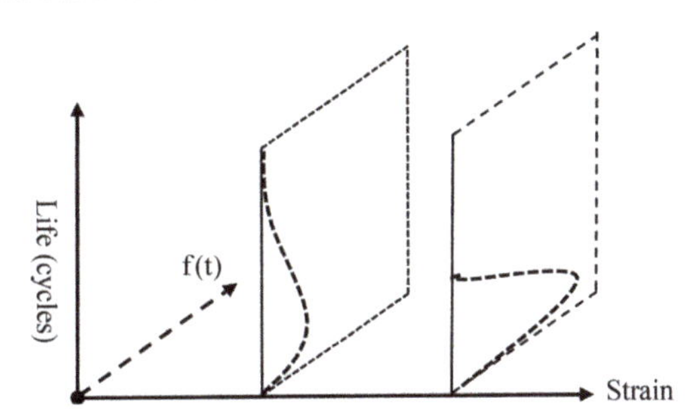

8.1.2. Data Types

Even in well-controlled DOEs, imperfect or missing data may occur. Imperfect or missing data may reduce the related statistical power and attain biased estimates, even leading to invalid conclusions. In practice, there are three different types of data commonly observed:

1. Complete Data: If a specimen is tested up to failure and the exact failure time (cycles) is known, the failure time (cycles) is called complete data. The data are recorded using a single number.
2. Interval Data: If a specimen is failed and the exact failure time is not known but the failure is known to occur within a time range, this time range is called interval data. For example, an electric fan is found to fail after it has been in use sometime between the 32nd month and the 34th month. Then, the data can be recorded as (32, 34).
3. Censored (Suspended) Data: If a specimen has not failed but the test does not continue anymore accidentally or on purpose, the last observation is called censored or suspended data.

These data types are often used for time responses, but they can be also used for other responses such as stress and chemical reaction.

8.1.3. Data Censoring

Data censoring has been in general practice for engineering tests for the sake of cost reduction, shortening delivery time, and/or limited number of test specimens. Obviously, either ignoring the censored data or treating them as failure lifetimes underestimates the system reliability. There are three ways to deal with data censoring, and they are described as follows [Zhang 2000]:

(A) Type I Censored Data (Time Censoring): Tests are run over an observed time period in such a way that an individual's lifetime will be known exactly only if it is less than some predetermined life value. If there is no unexpected accident, then the total number of data N consists of two different kinds of data, i.e., $N = N_f + N_s$, where N_f is the number of data for individuals that fail and N_s is the number of data for individuals that are suspended. Life test data overwhelmingly come from type 1 censored tests (time censoring) because a schedule dictates the time at which the test will end. As N items are placed on the test, a decision is made to suspend the test after time T_s. Then, the overall data are t_f ($n_f = 1, 2, \ldots, N_f$) and $t_s = T_s$ ($n_s = 1, 2, \ldots, N_s$), where N_f is the total number of specimens that fail and N_s is the total number of specimens suspended.

(B) Type II Censored Data (Failure Censoring): The experiment starts at a fixed time of zero, t = 0, and terminates when a fixed number of units have failed. For example, a total of N items are placed on the test at a fixed time of zero (t = 0), and the test is terminated at the time of the N_f^{th} item's failure. Thus, only a fixed number N_f (< N) observations in a sample size of N items are observed. It has no predetermined physical value in the specification. Some statistical theories and estimation methods have been developed for type II censored data, such as the MLE. Type II censoring is uncommon in practice. Nevertheless, most research on statistical methods for censored data has dealt with type II censoring because it is technically simpler.

(C) Random Multiply-Censored Data: It is censored progressively. There is neither a predetermined test time, nor a fixed number of test specimens to make the termination decision. It is the choice influenced by the results of the study up to termination time. Field data are overwhelmingly multiply-censored because of competing failure modes, random entry into a study, and variation in use rates.

If a response (observation) is censored, the corresponding residual is also to be censored in the residual plot when conducting a DOE.

Experimenters may tweak the censored data with the maximum-likelihood method or median rank regression (MRR) instead of using ANOVA including multifactorial linear regression, since the exact value for each observation is required for the use of ANOVA or multifactorial linear regression. Either MLE or MRR may be utilized to estimate model parameters by maximizing the probability of occurrence of the observed data.

8.1.4. Type I Censoring

For type I censoring, the total duration of the experiment is fixed, while the number of suspended units is random. Type I censoring can be further divided into three follow-up categories [Stovig 2014]:

(a) Right Censoring: Right censoring occurs if some units still function at the termination time. For example, it is determined to suspend the experiment at 10^8 cycles in a fatigue test. Therefore, the censored time will be smaller than the actual failure time. Right censoring is the most common type of censoring in the survival analysis of highly reliable products that will last for a long time and can result in the impracticality of conducting life testing at the designed normal usage conditions. Reliability function R(t) applies in this case. For example, $R(2) = 1 - F(2)$ if right-censored at $t = 2$, referring to Figure 8.1.3.

(b) Left Censoring: Event of interest occurs before the experimental observation begins. Left censoring happens when the test units are not checked until the moment after the study has begun. For example, the time for vehicle body parts to gather rust lends an example of left censoring. At the moment when the study begins, some test units may have gathered rust. Vehicle body parts in rust before the first observed time are thus left-censored. Hence, the censored time will be greater than the actual failure time. Failure function F(t) applies in this case; for example, F(0.5) refers to the accumulated area for $0 \le t \le 0.5$ in Figure 8.1.3, if left-censored occurs at 0.5.

(c) Interval Censoring: Both right- and left-censoring schemes are applied [Yang and Pan 2013]. The difference in failure functions, i.e., $F(t_n) - F(t_{n-1})$, can be used; for example, $F(2) - F(0.8)$ refers to the accumulated area for the interval of $0.8 \le t \le 2.0$ in Figure 8.1.3.

When type I censoring of a test specimen occurs, the experimenter has partial information about its survival time but is not privy to its exact survival time.

FIGURE 8.1.3 Example PDF used for examining type I censoring.

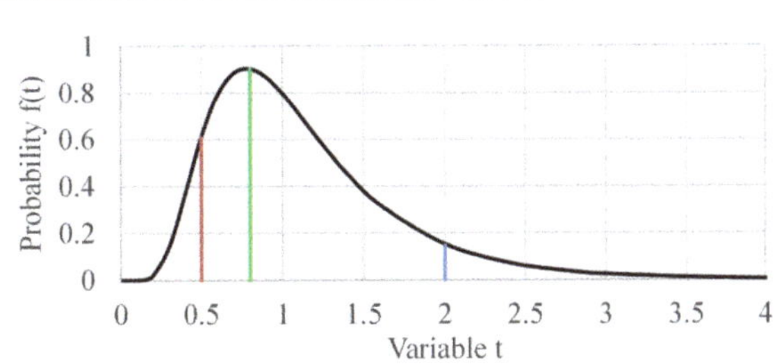

8.2. Lognormal Distribution

The continuous probability distribution of a random variable whose logarithm is normally distributed is called a lognormal distribution [Nelson and Kielpinski 1976]. For reliability study, a logarithmic scale is able to more uniformly present the structure of collected data when the independent variable of interest becomes large and has a wide range of values that vary by several orders of magnitudes. Recently, more and more product life prediction and failure analysis have led to the justification of lognormal distribution for some practical applications, such as

(A) Electronic Packaging: Insulated gate bipolar transistor (IGBT) modules based on accelerated lifetime test [Liu et al. 2013], conduction phenomena in the thin zinc-oxide nonlinear varistors, coating of electric motor windings, and the ball grid array connections appeal to follow lognormal distribution.
(B) Civil Engineering: The compressive strength of concrete also exhibits a lognormal distribution instead of the normal distribution [Silvestri et al. 2008].
(C) Mechanical Parts: Fatigue strength of brittle materials such as ceramics and spring steels follows the lognormal distribution better than the Weibull distribution.
(D) Chemical Analysis: Doping concentrations in chemical analysis also fall in lognormal distribution.
(E) Bioengineering: It is verified that the lognormal distribution is suitable for modeling blood pressure [Makuch et al. 1979]. This can be used for monitoring the driver's health situation.

8.2.1. Lognormal PDF

When the PDF of a random variable looks bell-shaped and its peak value is skewed toward the left with a long tail to the right as depicted in Figure 8.2.1, the corresponding random variable may fit lognormal distribution well. Given that t is an independent random variable, its natural logarithmic transformation is expressed as

$$y = \ln(t) \tag{8.2.1}$$

of which $0 < t < \infty$. It means that a random variable of lognormal distribution takes only positive real values. Assume that random variable y happens to fit a normal distribution $N(\mu_y, \sigma_y)$ well, i.e.,

$$y \sim N(\mu_y, \sigma_y) \tag{8.2.2}$$

where

μ_y is the mean value of variable y
σ_y is the standard deviation
σ_y^2 is the variance of variable y

The notation for random variable t that is lognormally distributed with mean μ_y and variance σ_y^2 is

$$t \sim \ln(\mu_y, \sigma_y) \equiv \text{lognormal}(\mu_y, \sigma_y) \tag{8.2.3}$$

If y has a normal distribution, then $t = \exp(y)$ has a lognormal distribution. A statistical result of the multiplicative product of some independent random variables, each of which is positive, is considered as a lognormal process [Holgate 1989].

FIGURE 8.2.1 Lognormal PDFs [median at t = 1.0, i.e., y = ln(t) = 0].

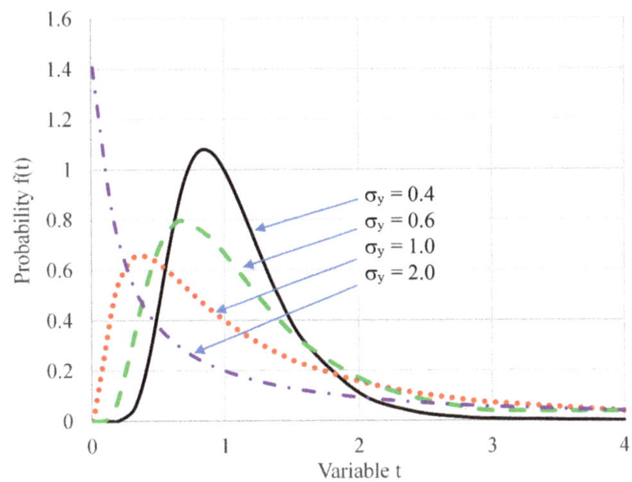

In light of the PDF of normally distributed variables, the PDF of lognormally distributed variable t can be written as

$$f(t) = \frac{1}{t\,\sigma_y (2\pi)^{1/2}} \exp\{-\tfrac{1}{2}[\frac{\ln(t) - \mu_y}{\sigma_y}]^2\} \quad \text{if } t > 0 \tag{8.2.4}$$

$$\text{and} \quad f(t) = 0 \quad \text{if } t \leq 0 \tag{8.2.5}$$

For example, variable t can be the time or the stress level when the failure occurs. A lognormal distribution is also called an anti-lognormal distribution, because it is not the distribution of the logarithm of a normal variable but is instead the anti-log of a normal variable as addressed by [Brezina 1963].

Let z be the standard normal variable for the lognormal distribution. The relationship between physical variable t and its standard normal variable can be expressed as

$$z = [\ln(t) - \mu_y] / \sigma_y \tag{8.2.6}$$

$$\text{or} \quad t = \exp(\mu_y + z\,\sigma_y) \tag{8.2.7}$$

8.2.2. Lognormal CDF

Data with all positive values, skewed distributions with low mean values, and large variance are suitable for fitting a lognormal PDF. The lognormal CDF is the integration of its PDF from zero to the value of interest.

$$F(t) = \frac{1}{\sigma_y (2\pi)^{1/2}} \int_0^t \frac{1}{u} \exp\{-\tfrac{1}{2}[\frac{\ln(u) - \mu_y}{\sigma_y}]^2\}\, du \tag{8.2.8a}$$

By changing the variable, on the basis of $\tau = \{[\ln(u) - \mu_y]/\sigma_y\}$ and $d\tau = [1/(\sigma_y u)]\, du$, one may deduce

$$F(t)] = \frac{1}{(2\pi)^{1/2}} \int_{-\infty}^{[\ln(t)-\mu_y]/\sigma_y} \exp(-½\, \tau^2)\, d\tau\} \tag{8.2.8b}$$

Solving Equation (8.2.8a) leads to

$$F(t) = ½\, \{1 + \mathrm{erf}[\frac{\ln(t)-\mu_y}{2^{1/2}\, \sigma_y}]\} \tag{8.2.8c}$$

of which erf is the error function. The reliability function of a failure mode that fits the lognormal distribution is then

$$R(t) \equiv 1 - F(t) = ½\, \{1 - \mathrm{erf}[\frac{\ln(t)-\mu_y}{2^{1/2}\, \sigma_y}]\} \tag{8.2.9}$$

8.2.3. Characteristics of Lognormal Distribution

The lognormal distribution of random t is a two-parameter distribution with parameters μ_y and σ_y. The mean, variance, median, and mode of sample variable t that has a lognormal distribution are, respectively, as follows:

$$\mathrm{Mean}[t] = E[t] = \mu_t = \exp(\mu_y + ½\, \sigma_y^2) \quad \text{(Average)} \tag{8.2.10}$$

$$\mathrm{Median}[t] = \exp(\mu_y) \quad \text{(Measure of central tendency)} \tag{8.2.11}$$

$$\mathrm{Mode}[t] = \exp(\mu_y - \sigma_y^2) \quad \text{(Highest probability of occurrence)} \tag{8.2.12}$$

and

$$\mathrm{Variance}[t] = \sigma_t^2 = \exp(2\mu_y + \sigma_y^2)\, [\exp(\sigma_y^2) - 1] \tag{8.2.13}$$

where random variable t ranges from 0 to ∞, so does its mean and variance. Given that $t \sim \ln(\mu_y, \sigma_y^2)$,

1. Standard deviation σ_y is the shape factor of the lognormal distribution function.
2. Median $\exp(\mu_y)$ is the scale factor that provides a robust estimate of the location parameter.

8.2.4. Multiplicative Property of Two Lognormal Distributions

An important property of the lognormal distribution is its multiplicative property. The product (and ratio) of two lognormal independent random variables is also a lognormally distributed random variable. This property states that if two independent random variables, s and t, are distributed, respectively, as $\ln(\mu_x, \sigma_x^2)$ and $\ln(\mu_y, \sigma_y^2)$, then the product of s and t is distributed as $\ln[\mu_x\, \mu_y, (\sigma_x^2 + \sigma_y^2)]$. In light of Equation (8.2.4), the joint PDF of variables s and t is

$$f(t) = \frac{1}{2\, \pi\, s\, t\, \sigma_x\, \sigma_y} \exp\{-½\, [\frac{\ln(s)-\mu_x}{\sigma_x}]^2 + [\frac{\ln(t)-\mu_y}{\sigma_y}]^2\} \tag{8.2.14}$$

8.2.5. Curve Fitting to the Lognormally Distributed Data

The lognormal distribution is popular for characterizing time-to-event data, especially for modeling the material strength and fatigue life of brittle materials such as ceramic parts and spring steels. The system identification can be carried out as follows: After the logarithmic transformations of the data set t_i (where subscript i denotes the number of data) are taken, the distribution of $\ln(t_i)$ tends to be normal-distributed. In other words, for a given set of data, one may make a chart of the probability of $1 - F(t_i)$ as the vertical axis versus ln(t) as the horizontal axis to see whether the plot seemingly follows a straight line, in order to decide the distributional fit of a lognormal function. When it fits normal distribution, the plot tends to follow a straight line whose slope is σ_y^{-1} [Das et al. 2017, Das and Park 2012].

After the system identification, one may estimate the PDF parameters using different algorithms, such as moment estimate and maximum-likelihood method, based on the given lognormally distributed data.

8.2.6. Standardized Residuals for Lognormal Distribution

Residual plots are widely used in linear regression analyses, in which residuals are assumed to be normally distributed. By examining the pattern of residual plots, one can identify whether there are additional variables that should be included in the regression model. In linear regression, each standard residual for the lognormal distribution follows the standard normal distribution that resembles Equation (8.2.4).

$$z_q = [\ln(t_q) - \mu_{yq}] / \sigma_{yq} \tag{8.2.15}$$

where

z_q is the standard residual
t_q is the qth data under the qth run condition (treatment)
μ_{yq} is the sample mean
σ_{yq} is the sample standard deviation

If μ_{yq} and σ_{yq} are independent of run conditions (treatments), then Equation (8.2.15) reduces to Equation (8.2.6) as

$$z_q = [\ln(t_q) - \mu_y] / \sigma_y \tag{8.2.16}$$

As long as an observation (t_q) is censored, the corresponding residual is also to be censored in the corresponding residual plot.

8.3. DOE-LN by Means of DOE-t or DOE-F

If there are N data that result from the qth treatment in the DOE being addressed, the average is expected to approach μ_{yq} when sample size N is large. The difference among all the data within treatment is then due to the stochastic property within treatment q, and it is governed by the lognormal distribution. In the meantime, the difference among the Q treatments is governed by the deterministic approximation of a polynomial in the DOE.

Assume that the stochastic lifetime of a component (t) can be described by a lognormal distribution. As for the purpose of demonstrating how the DOE-LN proceeds via DOE-t or DOE-F, a simplified model of three design factors is here considered as an example. An observation corresponding to treatment q can be rewritten as

$$t_q = \gamma_{q,0}' + \gamma_{q,1}' x_1 + \gamma_{q,2}' x_2 + \gamma_{q,3}' x_3 + \gamma_{q,12}' x_1 x_2 + \gamma_{q,13}' x_1 x_3 + \gamma_{q,23}' x_2 x_3 + \gamma_{q,123}' x_1 x_2 x_3 + \varepsilon_q' \quad (8.3.1)$$

where

t_q is the lifetime to failure with treatment q in the DOE
ε_q' is the residual or random error
x_1, x_2, and x_3 are the design factors 1, 2, and 3
$\gamma_{q,0}'$, $\gamma_{q,1}'$, $\gamma_{q,2}'$, $\gamma_{q,3}'$, $\gamma_{q,12}'$, $\gamma_{q,13}'$, $\gamma_{q,23}'$, and $\gamma_{q,123}'$ are the intercept and coefficients

8.3.1. Logarithmic Transformation

Assume that the response for treatment q is lognormally distributed, and then, random error term ε_q' in the DOE model given above within treatment q may not be normally distributed supposedly. Therefore, after the logarithmic transformation of t is taken, Equation (8.3.1) becomes

$$\ln(t_q) = \gamma_{q,0} + \gamma_{q,1} x_1 + \gamma_{q,2} x_2 + \gamma_{q,3} x_3 + \gamma_{q,12} x_1 x_2 + \gamma_{q,13} x_1 x_3 + \gamma_{q,23} x_2 x_3 + \gamma_{q,123} x_1 x_2 x_3 + \varepsilon_q \quad (8.3.2)$$

where

$\ln(t_q)$ is the transformed data t_q
ε_q is the residual or random error after transformation
$\gamma_{q,0}$, $\gamma_{q,1}$, $\gamma_{q,2}$, $\gamma_{q,3}$, $\gamma_{q,12}$, $\gamma_{q,13}$, $\gamma_{q,23}$, and $\gamma_{q,123}$ are intercept and coefficients after transformation

Random error term ε_q in the DOE model given above, as applied to within treatment q, is now normally distributed. As long as the variances associated with all treatments are consistent statistically, Equation (8.3.2) can be rewritten for the entire model as

$$\ln(t) = \gamma_0 + \gamma_1 x_1 + \gamma_2 x_2 + \gamma_3 x_3 + \gamma_{12} x_1 x_2 + \gamma_{13} x_1 x_3 + \gamma_{23} x_2 x_3 + \gamma_{123} x_1 x_2 x_3 + \varepsilon \quad (8.3.3)$$

Consequently, the solution procedures of DOE-t and DOE-F (using ANOVA) can be applied if it is a complete design matrix without censored data. However, the maximum-likelihood method has to be used to estimate the parameters in DOE-LN analyses if censored or interval data are present.

8.3.2. Example: Tool Life Subjected to Turning Operations

Turning is the simplest but most applied machining process, in which a single-point cutting tool removes material from the surface of a rotating cylindrical workpiece. The cutting tool is fed linearly in a direction parallel to the axis of rotation at a certain rate, called feed rate. The turning speed of the rotating workpiece is the cutting speed, and the amount of material to be removed from the workpiece in the radial direction is the depth of cut. These three operating parameters constitute the most important factors that have a great influence on the tool life. An experimental assessment of the tool life based on fractional factorial design 3^{3-1} was studied by [Yang and Tarng 1998]. Design levels of these three operating parameters are given in Table 8.3.1. The design matrix and test results are listed in Table 8.3.2, in which

$$y = \ln(t) \tag{8.3.4}$$

TABLE 8.3.1 Operating parameters for fractional factorial design 3^{3-1} of metal cutting [Yang and Tarng 1998].

Variables	Level (−1)	Level (0)	Level (+1)
A: Cutting speed (m/min)	135	210	285
B: Feed rate (mm/rev)	0.08	0.2	0.32
C: Depth of cut (mm)	0.6	1.1	1.6

TABLE 8.3.2 Design matrix of fractional factorial 3^{3-1} ($C = A^2B$) for metal cutting [Yang and Tarng 1998].

Run	A	B	C	A	B	C	t (sec)	y = ln(t)	$y_p = \ln(t_p)$	$t_{p,mean}$	$t_{p,median}$
1	−1	−1	−1	135	0.08	0.6	2645	7.880	7.899	2714	2695
2	−1	0	0	135	0.2	1.1	2060	7.630	7.657	2130	2115
3	−1	1	1	135	0.32	1.6	1733	7.458	7.415	1660	1672
4	0	−1	1	210	0.08	1.6	1310	7.178	7.179	1312	1321
5	0	0	−1	210	0.2	0.6	1198	7.088	6.965	1059	1066
6	0	1	0	210	0.31	1.1	734	6.599	6.745	850.1	856.1
7	1	−1	0	285	0.08	1.1	854	6.750	6.813	910.0	916.5
8	1	0	1	285	0.2	1.6	765	6.640	6.579	719.7	724.8
9	1	1	−1	285	0.32	0.6	216	5.375	5.377	216.3	217.9

where t (sec) is tool life, defined as the elapsed period of cutting time when the average flank wear land of the tool is less than 0.3 mm or the maximum flank wear land is greater than 0.6 mm.

Since sampling of the tool life of all treatments comes far apart, a logarithmic transformation of each piece of data is taken. The data analysis of tool life is done with Minitab (Stat → Regression → Regression → Fit Regression Model). The "most insignificant" factor or interaction that comes with the largest p-value is weeded out one by one until all p-values are less than 10% that is here defined as the parting significance level based on F-distribution. As exhibited by the ANOVA presented in Table 8.3.3, the final predictive equations for the tool life is

$$y_p = \ln(t_p) = 9.830 - 0.01177\,A + 5.70\,B - 1.866\,C - 0.03629\,A\,B + 0.00881\,A\,C \tag{8.3.5}$$

$$(R = 99.5\%,\ R_{adj} = 98.7\%,\ \text{and}\ R_{pred} = 94.5\%)$$

The correlation, adjusted correlation, and predictive correlation are strong enough to justify the effectiveness and adequacy of the above equation.

TABLE 8.3.3 ANOVA on tool life subjected to turning operations.

Source of variation	DOF	SS_{adj}	MS_{adj}	F-ratio	P-value
Regression	5	4.354	0.871	61.8	0.3%
A	1	0.325	0.325	23.1	1.7%
B	1	0.103	0.103	7.28	7.4%
C	1	0.191	0.191	13.6	3.5%
AB	1	0.224	0.224	15.9	3.8%
AC	1	0.229	0.229	16.3	2.7%
Error	3	0.0423	***0.0141***	—	—
Subtotal		8	4.397		
Average		1			
Total		9			

The life estimate can be predicted in two different ways:

1. Mean Life: Let the predicted tool life in the time domain be represented by its mean value. The tool life pertaining to each run (treatment) is then calculated using Equation (8.2.10).

$$t_{p,q} = E[t_q] = \exp(\mu_{yq} + ½\, \sigma_y^2), \quad q = 1, 2, 3, 4, 5, 6, 7, 8, \text{ and } 9$$

$$\text{where} \quad \mu_{yq} \approx \ln(t_{p,q})$$

$$\text{and} \quad \sigma_y^2 = 0.0141 \quad (MS_{adj} \text{ of Error in Table 8.3.3})$$

where subscript q is the number of treatments as disclosed in Table 8.3.2 and mean μ_{yq} is the mean value calculated in the logarithmic domain.

2. Median Life: Let the predicted tool life in the time domain be represented by its median value. The tool life pertaining to each run (treatment) is then calculated using Equation (8.2.11).

$$t_{p,q} = \text{Median}[t_q] = \exp(\mu_{yq}), \quad q = 1, 2, 3, 4, 5, 6, 7, 8, 9$$

As long as the random error is small, the lifetimes predicted using both mean life and median life should be close to each other.

Besides the logarithmic transformation, another transformation using $y^{1/2}$ is also tried out for the purpose of comparison. It can be proven that modeling using either ln(y) or $y^{1/2}$ data performs better than the original y data in the time domain.

8.4. Maximum-Likelihood Method for DOE-LN

When censored data are present in the DOEs, the methodology of implementing DOE-LN via DOE-t or DOE-F procedures is jeopardized and cannot be applied anymore. The maximum-likelihood method can be used instead to estimate lognormal parameters in DOE-LN analyses. A likelihood function for DOE-LN originates from regarding lognormal parameters as unknown values and then finding them via the joint probability density of the observations of all treatments by minimizing the variance. The observations are assumed to be independent and identically distributed (IID). This estimator for the parameters of a statistical function by the maximum-likelihood method is important in the stochastic world, since its asymptotic unbiasedness results from the minimal variance.

8.4.1. Coupling between Process Acceleration Model and Lognormal Distribution Model

Since $t \sim \ln(\mu_y, \sigma_y^2)$ as described by Equation (8.2.3), $y = \ln(t)$ has a mean μ_y. If there are Q data that result from the qth treatment in the DOE being addressed, the average is expected to approach μ_{yq} if sample size n is large. The difference among all the data within treatment is then due to the stochastic property within treatment q, and it is governed by the lognormal distribution. In the meantime, the difference among the Q treatments is governed by the deterministic approximation of a polynomial in the DOE, i.e., Equation (8.3.2) or Equation (8.3.3), as

$$\mu_{yq} \approx \ln(Y_q) \tag{8.4.1}$$

8.4.2. MLE

Maximum-likelihood method has been traditionally employed to estimate the parameters of a statistical PDF, based on the given observations. It provides estimates of the means and covariance matrix, which can be utilized to get consistent estimates of the parameters of interest by maximizing the likelihood function. This method is called MLE. One major merit of MLE is its applicability to an experimental design with censored data. When the probability of one or more censored data on a variable is unrelated to the value of the variable itself or to the values of other variables inherent with the data set, the censoring is called missing completely at random.

MLE has been in use with a wide range of statistical distributions, such as normal, t-student, lognormal, Weibull, exponential, gamma, and binomial. In order to apply MLE to a certain set of data of interest, one has to find out which statistical distribution is fit for the given set of data in advance.

MLE is invariant with respect to certain types of transformations of the original data. For example, variable t is logarithmic-transformed to variable y, as

$$y = \ln(t) \tag{8.4.2}$$

As long as the transformation is one-to-one mapping and the transformation itself is independent of the parameters of the PDFs to be estimated, the PDFs of variable y and variable t are related to each other by their integrated probabilities in its own differential step as

$$f(t)\,|dt| = f(y)\,|dy|$$

$$\text{Thus,}\quad f(t) = f(y)\,|dy / dt| = f(y)\,|d[\ln(t)] / dt| = f(y) / |t| \tag{8.4.3}$$

where

f(y) is the PDF of variable y, of which $y = g(t)$

f(t) is the PDF of variable t

d[ln(t)]/dt is the differentiation of the transformation function with respect to variable t

Equation (8.4.3) reduces to

$$f(y) = |t|\, f(t) \tag{8.4.4}$$

Therefore, there is no statistical parameter in the lognormal PDF explicitly involved in the relationship between f(y) and f(t) as the transformation proceeds. In other words, the parameters of lognormal distribution f(t) to be estimated are the same as those obtained for the normal distribution f(y) when the maximum-likelihood method is applied.

8.4.3. Null Hypothesis for Maximum-Likelihood Method

The maximum likelihood refers to the rule—if the data points are all highly likely under specific parametric values of a function, their product will be the "most likely" outcome. The maximum of that function can be found by taking partial differentiations with respect to the parameters, right after the likelihood function is formed. Assume that all the data are in due test-to-failure time. The likelihood function and logarithmic likelihood function based on a given set of data with Q treatments in DOEs can be written as, respectively,

$$L_{\ln}(\mu_y, \sigma_y, t_q) = \prod_{q=1}^{Q} f(t_q) \tag{8.4.5}$$

and

$$\ln[L_{\ln}(\mu_y, \sigma_y, t_q)] = \sum_{q=1}^{Q} \ln[f(t_q)] \tag{8.4.6}$$

where

t_q is the lifetime within treatment q
μ_y is the mean of variable y, given that y = ln(t)
σ_y is the standard deviation of variable y, given that y = ln(t)
$L_{\ln}(\mu_y, \sigma_y, t_q)$ is the likelihood function with two lognormal parameters, i.e., μ_y and σ_y
$\ln[L_{\ln}(\mu_y, \sigma_y, t_q)]$ is the logarithmic transformation of $L_{\ln}(\mu_y, \sigma_y, t_q)$
$\prod$ is the multiplication
$f(t_q)$ is the lognormal PDF evaluated at data t_q
q is the treatment number, which is "row q" in the DOE matrix, q = 1, 2, …, Q
Q is the total number of treatments, which is also the total number of test data if no replication

Consider a general curve-fitting case up to the third order with no replication, for simplicity but without losing its generality, the lognormal mean value of the observed data within treatment q can be approximated by

$$\mu_q' = \gamma_0 + \sum_{i=1}^{k} \gamma_i x_{iq} + \sum_{i=1}^{k} \sum_{j=1}^{k} \gamma_{ij} x_{iq} x_{jq} + \sum_{i=1}^{k} \sum_{j=1}^{k} \sum_{m=1}^{k} \gamma_{ijm} x_{iq} x_{jq} x_{mq} + \ldots + \varepsilon_q \tag{8.4.7}$$

where

μ_q' is the lognormal mean value, after the logarithmic transformation has been taken
γ_0, γ_i, γ_{ij}, and γ_{ijm} are the effect coefficients to be estimated using the maximum-likelihood method
x_{iq}, x_{jq}, and x_{mq} are the test variables, such as −1, 0, +1, or real values

In order to identify the statistical significance of each effect (intercept and coefficients) in the above equation, the following hypotheses come along:

$$H_o: \gamma_0 = 0 \qquad H_1: \gamma_0 \neq 0 \tag{8.4.8}$$

$$H_o: \gamma_i = 0 \qquad H_1: \gamma_i \neq 0 \qquad (i = 1, 2, 3, \ldots) \tag{8.4.9}$$

$$H_o: \gamma_{ij} = 0 \qquad H_1: \gamma_{ij} \neq 0 \qquad (i, j = 1, 2, 3, \ldots) \tag{8.4.10}$$

and

$$H_o: \gamma_{ijm} = 0 \qquad H_1: \gamma_{ijm} \neq 0 \qquad (i, j, m = 1, 2, 3, \ldots) \tag{8.4.11}$$

where H_o is the null hypothesis and H_1 is the alternative hypothesis. The likelihood function is evaluated against each observed data and time t_q; thus, the parameters of the model are to be obtained by maximizing $\ln[L_{ln}(\mu_y, \sigma_y, t_q)]$.

8.4.4. Maximum-Likelihood Method with Test-to-Failure, Suspended, and Interval Data

Failure-time censoring (i.e., type I censoring) is usually unavoidable in product life tests, and it affects the statistics of the associated likelihood function. Interval data are also observed in the real-time scheduled inspection of product facilities, e.g., milling machines and light bulbs. All in all, three different kinds of data are available for most engineering cases:

1. Test-to-failure data
2. Right-suspended data (type I censoring)
3. Interval data

Equation (8.4.5) can be expanded to accommodate all three types of data given above [Meeker et al. 1998] as follows:

$$L_{ln}(\sigma_y, \gamma_0, \gamma_i, \gamma_{ij}, \gamma_{ijm}) = L_{failure}\, L_{suspended}\, L_{interval}$$

$$= \{\prod_{f=1}^{Q_f} f(t_f)\}\, \{\prod_{s=1}^{Q_s} [R(t_s)]\}\, \{\prod_{n=1}^{Q_n} [F(t_{n,2}) - F(t_{n,1})]\} \tag{8.4.12}$$

where

Q_f is the total number of test-to-failure specimens; subscript f for number order
Q_s is the total number of suspended (censored) test specimens; subscript s for number order
Q_n is the total number of interval data; subscript n for number order
t_s is the life of specimen, up to the moment of suspension (censor)
t_f is the life of specimen, up to failure
$t_{n,2}$ and $t_{n,1}$ are the time interval between time "2" and time "1," where $t_{n,1} < t_{failure} < t_{n,2}$

Note that shape factor σ_y is assumedly a constant throughout the entire model, while μ_y varies with respect to the treatment (run condition). Plugging Equations (8.2.4), (8.2.8a), (8.2.9), and (8.2.10) into the three terms on the right side of Equation (8.4.12), one has the following three terms [Meeker et al. 1998, Wu and Hamada 2009]:

1. Test-to-Failure Data

$$L_{failure} = \prod_{f=1}^{Q_f} f(t_f) = \prod_{f=1}^{Q_f} \frac{1}{t_f \sigma_y (2\pi)^{1/2}} \exp\{-\tfrac{1}{2} [\frac{\ln(t_f) - \mu_{yf}}{\sigma_y}]^2\} \tag{8.4.13}$$

2. Right-Suspended Data (Type I Censoring)

$$L_{suspended} = \prod_{s=1}^{Q_s} [R(t)] = \prod_{s=1}^{Q_s} \{1 - \frac{1}{(2\pi)^{1/2}} \int_{-\infty}^{[\ln(t_s) - \mu_{ys}] / \sigma_y} \exp(-\tfrac{1}{2} \tau^2)\, d\tau\} \tag{8.4.14}$$

3. Interval Data

$$L_{interval} = \prod_{i=1}^{Q_i} \{F(t_{i,2}) - F(t_{i,1})\} = \prod_{i=1}^{Q_i} \{\frac{1}{(2\pi)^{1/2}} \int_{-\infty}^{[\ln(t_{i,2}) - \mu_{yi}] / \sigma_y} \exp(-\tfrac{1}{2} \tau^2)\, d\tau - \frac{1}{(2\pi)^{1/2}} \int_{-\infty}^{[\ln(t_{i,1}) - \mu_{yi}] / \sigma_y} \exp(-\tfrac{1}{2} \tau^2)\, d\tau\} \tag{8.4.15}$$

Note that τ is a dummy variable that is only used for time integration.

If the null hypothesis is true, the test data $\ln(t_q)$ resulting from treatment q should approximate μ_q' in Equation (8.4.7). Following the concept of Equation (8.4.1), one has

$$\ln(t_{yq}) \approx \mu_{yq} \approx \gamma_0 + \sum_{i=1}^{k} \gamma_i x_{iq} + \sum_{i=1}^{k} \sum_{j=1}^{k} \gamma_{ij} x_{iq} x_{jq} + \sum_{i=1}^{k} \sum_{j=1}^{k} \sum_{m=1}^{k} \gamma_{ijm} x_{iq} x_{jq} x_{mq} = \gamma_0 + \gamma_1 x_{1s} + \gamma_2 x_{2s} + \ldots \tag{8.4.16}$$

When accelerated life tests are applied to properly represent the product operating conditions with a shorter testing time, engineers usually terminate the accelerated life test upon successfully running to a multiple of a prespecified bogey without any failure. This testing philosophy is called "bogey test" or "extended test" in the automotive industry [Wang 1991] and is a subset of type I right-censored tests in standard reliability literature.

Next, consider only test-to-failure and suspended data without interval data. One may replace μ_{yq} by $\ln(t_{yq})$ and then substitute Equation (8.4.16) into Equation (8.4.13) for test-to-failure data and Equation (8.4.16) into Equation (8.4.14) for suspended data as

$$L_{ln}(\sigma_y, \gamma_0, \gamma_i, \gamma_{ij}, \gamma_{ijm}) = \{\prod_{f=1}^{Q_f} f(t_f)\}\ \{\prod_{s=1}^{Q_s} [R(t_s)]\} \quad (8.4.17)$$

where
$$\prod_{f=1}^{Q_f} f(t_f) = \prod_{f=1}^{Q_f} \frac{1}{t_f\,\sigma_y\,(2\pi)^{1/2}} \exp\{-\tfrac{1}{2}\,[\frac{\ln(t_f)-(\gamma_0+\gamma_1 x_{1s}+\gamma_2 x_{2s}+\ldots)}{\sigma_y}]^2\} \quad (8.4.18)$$

and
$$\prod_{s=1}^{Q_s} [R(t_s)] = \prod_{s=1}^{Q_s} [1 - \frac{1}{(2\pi)^{1/2}}\{\int_{-\infty}^{[\ln(t_s)-(\gamma_0+\gamma_1 x_{1s}+\gamma_2 x_{2s}+\ldots)]/\sigma_y} \exp(-\tfrac{1}{2}\tau^2)\,d\tau\} \quad (8.4.19)$$

Taking a logarithmic transformation of Equation (8.4.17) leads to the lognormal likelihood function as

$$\ln[L_{ln}(\sigma_y, \gamma_0, \gamma_i, \gamma_{ij}, \gamma_{ijm})] = \{\sum_{f=1}^{Q_f} \ln[f(t_f)]\}\ \{\sum_{s=1}^{Q_s} \ln[R(t_s)]\} \quad (8.4.20a)$$

8.4.5. Maximum-Likelihood Regression with Test-to-Failure and Right-Suspended Data

Most cases encountered in automotive engineering have both test-to-failure and right-suspended data available. The idea for the MLE is to find the value of the parameter(s), for which the data have the highest probability. As the maximum-likelihood method applies when carrying out the design optimization, unknown parameters σ_y and unknown constants γ_0, γ_i, γ_{ij}, and γ_{ijm} are to be obtained by solving the following simultaneous equations [Schmee and Hahn 1979]:

$$\frac{\partial\,[L_{ln}(\sigma_y, \gamma_0, \gamma_i, \gamma_{ij}, \gamma_{ijm})]}{\partial\,\sigma_y} = 0 \quad (8.4.20b)$$

$$\frac{\partial\,[L_{ln}(\sigma_y, \gamma_0, \gamma_i, \gamma_{ij}, \gamma_{ijm})]}{\partial\,\gamma_0} = 0 \quad (8.4.21)$$

$$\frac{\partial\,[L_{ln}(\sigma_y, \gamma_0, \gamma_i, \gamma_{ij}, \gamma_{ijm})]}{\partial\,\gamma_i} = 0 \quad (8.4.22)$$

$$\frac{\partial\,[L_{ln}(\sigma_y, \gamma_0, \gamma_i, \gamma_{ij}, \gamma_{ijm})]}{\partial\,\gamma_{ij}} = 0 \quad (8.4.23)$$

and
$$\frac{\partial\,[L_{ln}(\sigma_y, \gamma_0, \gamma_i, \gamma_{ij}, \gamma_{ijm})]}{\partial\,\gamma_{ijm}} = 0 \quad (8.4.24)$$

Here are the conditions, under which the MLEs of model standard variation σ_y and regression constants γ_0, γ_i, γ_{ij}, and γ_{ijm} of Equation (8.4.16) exist and the calculated results are unique [Scholz 1981]. The formulations given above are based on the assumption that the test data are consistent in their common variance, i.e., σ_y^2. The MLE is asymptotically unbiased and has asymptotically minimal variance.

When dealing with various statistical distributions, there are two commonly applied methods for calculating reliability bounds, i.e., confidence intervals. They are the likelihood ratio (LR) bound (LRB) method and the Fisher information matrix (FIM) method. Both methods are applicable after the parameters are obtained from the MLE. The LRB method is based on the chi-squared distribution, and it is generally more accurate than FIM bounds when the sample size is small, but they differ very little if the sample size is large enough [Vander Wiel and Meeker 1990].

8.4.6. Standardized Residuals for Lognormal Distribution

The Weibull distribution is used to model the lifetime, leading to a smallest extreme value (SEV) distribution in the log-lifetime domain. Residual plots are widely used in regression analyses for checking the model, in which residuals are assumed to be normally distributed. By examining the pattern of residual plots, one can identify whether there are additional variables that should be included in the regression model.

Let z_{qn} stand for the standardized variable for the statistical distribution of y = ln(t), which constitutes a smallest extreme distribution as

$$z_{qn} = (y_{qn} - \mu_q) / \sigma_y = [\ln(t_{qn}) - \mu_q] / \sigma_y \tag{8.4.25}$$

where n is the number of data replicated within treatment q. Consider n = 1 for simplicity, but without the loss of generality. If the data follow a lognormal distribution, the standardized residuals can be obtained from the following equation:

$$\begin{aligned} \varepsilon_q &= [\ln(t_q) - \mu_q] / \sigma_y \\ &= [\ln(t_q) - (\gamma_0 + \gamma_1 x_{1q} + \gamma_2 x_{2q} + \gamma_3 x_{3q} + \ldots)] / \sigma_y \end{aligned} \tag{8.4.26}$$

The probability plot based on the above equation is used to check whether the residuals follow a standard normal distribution.

8.5. Statistical Inference Using LR by χ^2 Distribution

The LR test (LRT) is a test of hypothesis in which two different MLEs of a parameter are compared in order to decide whether to reject or not to reject a restriction on the parameter. For the DOEs, it is used for testing the significance of an effect by removing its corresponding coefficient (parameter) from the fitting equation, i.e., Equation (8.4.16). In order to justify the validity of each individual effect, namely coefficient γ_{test}, which is the generalized variate used to represent anyone of γ_o, γ_i, γ_{ij}, and γ_{ijm}, the following LRT applies [Guo and Mettas 2010]:

$$LR(\eta_{test}) \equiv -\ln\left\{\left[\frac{L_{ln}(All - \gamma_{test})}{L_{ln}(All)}\right]^2\right\} = -2\ln\left\{\frac{L_{ln}(All - \gamma_{test})}{L_{ln}(All)}\right\} \tag{8.5.1}$$

where

LR is the LR for effect η_{test} that accounts for the effectiveness of coefficient γ_{test}
γ_{test} is the coefficient γ, one of γ_o, γ_i, γ_{ij}, and γ_{ijm}, whose effectiveness is under the test
η_{test} is the effect represented by coefficient γ_{test}, which is under test
$L_{ln}(All)$ is the likelihood value with all coefficients γ_o, γ_i, γ_{ij}, and γ_{ijm}
$L_{ln}(All - \gamma_{test})$ is the likelihood value with all coefficients γ_o, γ_i, γ_{ij}, and γ_{ijm}, except γ_{test}

The negative sign on the right side in the above equation is to assure that there will be a positive LR since $L_{ln}(All - \gamma_{test})$ is generally smaller than $L_{ln}(All)$. The statistical inference can be made from the following hypotheses:

- H_0 (*Null Hypothesis*): Effect η_{test} that is represented by γ_{test} (coefficient or model parameter) is not significant, if $LR(\eta_{test})$ is less than chi-squared statistic $\chi^2_{v,1-\alpha}$.
- H_1 (*Alternative Hypothesis*): Effect η_{test} that is represented by γ_{test} (coefficient or model parameter) is significant, if $LR(\eta_{test})$ is larger than chi-squared statistic $\chi^2_{v,1-\alpha}$.

With a certain confidence level $C = 1 - \alpha$, which is assumed to be a two-sided confidence bound on the estimated parameter, the inequality equation based on the LR method is then given as follows:

$$LR(\eta_{test}) \leq \chi^2_{v,\,1-\alpha/2} \tag{8.5.2}$$

where

$\alpha = 1 - C$, while C is the confidence level
ν is the DOFs associated with the χ^2 distribution; $\nu = 1$ for LR

If the null hypothesis is true, $LR(\eta_{test})$ is small. $LR(\eta_{test})$ follows the chi-squared distribution with one DOF ($\nu = 1$). When $LR(\eta_{test})$ is not statistically significant, the impact of effect η_{test} is numerically small or close to zero whether or not the related coefficient γ_{test} is removed from the full model of Equation (8.4.19). It means that the LR because of effect η_{test} will not affect the LR significantly and the related coefficient γ_{test} is not relevant. On the other hand, if $LR(\eta_{test})$ is significantly large, effect η_{test} is statistically significant.

8.5.1. Statistical Inference by χ^2 Distribution

The χ^2 distribution, namely chi-squared distribution, is one of the most widely used probability distributions in inferential statistics, notably in hypothesis testing such as the LR algorithm for MLE or in the construction of confidence intervals. χ^2 is the notation of an individual statistic, not a "squared value," and it is a special case of the gamma distribution.

How to define a new random variable χ^2? One may take a sample from variable Z (independent standard normal random variable) and square the value. The distribution of the squared values is given by a new random variable called χ^2, as $\chi^2 = Z^2$. The distribution of random variable χ^2 is then a chi-squared distribution: χ_1^2, which is a variable having one DOF in chi-squared distribution. If one takes a sample from ν independent standard normal random variables, then the sum of squares is

$$\chi^2_v = \sum_{n=1}^{v} Z_n^2 \tag{8.5.3}$$

that defines the chi-squared distribution with ν DOFs. The DOF, which is the number of Z_n values, becomes the sole parameter of a chi-squared distribution.

Next, consider the sampling of the variable that fits a normal distribution. When a random sample of size N is selected from a normally distributed population that has a population standard deviation equal to σ. Assume that the sample standard deviation is equal to S that is derived from N samples, and then, the chi-squared statistic is written [Blischke and Murthy 2000].

$$\chi^2_{N-1} = (N-1)\, S^2 / \sigma^2 \tag{8.5.4}$$

where

χ^2_{N-1} is the chi-squared statistic with N – 1 DOFs
N is the sample size
S is the sample standard deviation
σ is the population standard deviation

The equation given above follows the definition of a distribution function, generally written as χ^2 distribution, with ν (= N – 1) DOFs. As defined for the distribution of a sum of the squares of ν independent standard normal random variables, the PDF of χ^2 distribution is

$$f(\chi^2) = \frac{(\chi^2)^{\nu/2-1}}{2^{\nu/2}\,\Gamma(\nu/2)} \exp(-\tfrac{1}{2}\chi^2) \qquad \text{if } \chi^2 > 0$$

$$\text{and} \quad f(\chi^2) = 0 \qquad \text{if } \chi^2 \le 0 \tag{8.5.5}$$

of which Γ is the gamma function. The cumulative function of a chi-squared distribution is formed by integrating Equation (8.5.5),

$$F(\chi^2) = \frac{\Gamma_L(\nu/2, \chi^2/2)}{\Gamma(\nu/2)} \qquad \text{if } \chi^2 > 0 \tag{8.5.6a}$$

of which $\Gamma_L(\nu/2, \chi^2/2)$ is the lower incomplete gamma function. The PDF of the χ^2 distribution is plotted in Figure 8.5.1, and the CDF is tabulated in Table 8.5.1.

FIGURE 8.5.1 PDF of the χ^2 distribution.

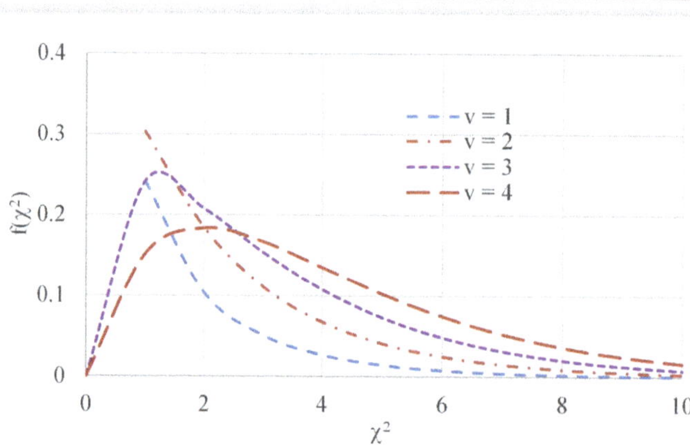

TABLE 8.5.1 Probability points of the accumulative χ^2 distribution with ν DOFs.

ν	95%	90%	80%	50%	20%	10%	5%	1%	0.1%
1	0.004	0.02	0.06	0.46	1.64	2.71	3.84	6.63	10.83
2	0.10	0.21	0.45	1.39	3.22	4.61	5.99	9.21	13.82
3	0.35	0.58	1.01	2.37	4.46	6.25	7.81	11.34	16.27
4	0.71	1.06	1.65	3.36	5.99	7.78	9.49	13.28	18.47
5	1.14	1.61	2.34	4.35	7.29	9.24	11.07	15.09	20.52
6	1.63	2.20	3.07	5.35	8.56	10.64	12.59	16.81	22.46
7	2.17	2.83	3.82	6.35	9.80	12.02	14.07	18.48	24.32
8	2.73	3.46	4.59	7.34	11.03	13.36	15.51	20.09	26.12
9	3.32	4.17	5.38	8.34	12.24	14.68	16.92	21.67	27.88
10	3.94	4.87	6.18	9.34	13.44	15.99	18.31	23.21	29.59

8.5.2. Acceptance Sampling Plans by Chi-Squared Statistic

Acceptance sampling for inspecting a normally distributed data set can be assessed using χ^2 distribution [Dodge 1947], without going through time-consuming 100% inspection. With a predetermined number of failures allowed, the sample size corresponding to a certain confidence level with a given reliability value can be formulated as follows:

$$N = \chi^2_{v,(1-C)} / [2\,(1-R)] \tag{8.5.6b}$$

and

$$v = 2\,(Q+1) \tag{8.5.7}$$

where

N is the sample size
R is the reliability
Q is the number of failures
C is the confidence level, which frequently goes by "confidence" in alias

Chi-squared value, χ^2_v, for a given confidence level for 2(Q+1) DOFs is calculated. This approach can be utilized for a general case study, including no failure [Wang 1991]. LRTs commonly provide the highest statistical power to reject the null hypothesis, as revealed by the Neyman–Pearson lemma [Neyman and Pearson 1933]. However, the normal and chi-squared approximations are only valid asymptotically. This is the reason why the t-distribution (used for DOE-t) is preferred to the normal approximation (used for DOE-F) or the chi-squared approximation for a small sample size.

EXAMPLE 8.5.1

A product under inspection is expected to demonstrate its qualification of R = 90% with C = 95%. What is the minimum sample size required to ensure that the reliability criterion is met when zero, one, two, three, or four failures occur?

Solution:

The calculation procedure of the chi-squared value for assessing the sample size is demonstrated here using the case with two failures as an example:

1. The cumulative probability of failure is expected to be $\alpha = 1 - C = 1 - 0.95 = 0.05$
2. The number of DOFs is calculated as $\nu = 2(Q + 1) = 2(2 + 1) = 6$
3. The chi-squared value is obtained from **Table 8.5.1** as $\chi^2_{6,0.05} = 12.59$
4. Sample size $N = \chi^2_{\nu,(1-C)}/[2(1 - R)] = 12.59/[2(1 - 0.9)] = 62.95 \approx 63$

This procedure can be repeated for Q = 0, 1, 3, and 4. In order to ensure that the validation meets the risk-based acceptance criterion: R = 90% with C = 95%, the minimum sample sizes for 0, 1, 2, 3, and 4 failures are listed as follows:

- When R = 90% with C = 95%
  - Q = 0, then N = 30
  - Q = 1, then N = 48
  - Q = 2, then N = 63
  - Q = 3, then N = 78
  - Q = 4, then N = 92

If the risk-based acceptance criterion is revised for R = 90% and C = 50%, the sample sizes as a minimum required for 0, 1, 2, 3, and 4 failures are then

- When R = 90% with C = 50%
  - Q = 0, then N = 7
  - Q = 1, then N = 17
  - Q = 2, then N = 27
  - Q = 3, then N = 37
  - Q = 4, then N = 48

The above calculations for R = 90% with C = 50% have been meant sometimes for B_{10} (F = 1 − R = 10%) that appears paradoxical in some engineering design textbooks without phrasing the confidence level.

In practice, the accept-on-zero (AOZ) sampling plan has been favored in the automotive engineering practice. Thus, DOF $\nu = 2$ is the choice and Equation (8.5.6b) reduces to

$$N = \chi^2_{2,(1-C)} / [2(1-R)] \tag{8.5.8}$$

With regard to acceptance sampling plans for variables, the reliability and confidence levels are based on success-run theorem and statistical theories such as normal, lognormal, and Weibull distributions.

The acceptance sampling criteria are justified by the reliability and its associated confidence. It is recognized as a (R, C, N) trio.

It is stated in TS16949-8.2.3.1, "The organization shall maintain manufacturing process capability or performance as specified by the customer part approval process requirements." Two general practices in the automotive industry are (R = 90%, C = 90%) and (R = 95%, C = 75%), while testing phases vary from one company to another. As being commonly used in the automotive industry, there are three major product development phases:

1. Design verification (DV)
2. Product validation (PV)
3. Continuous improvement and continuing conformance testing

8.6. Reduction in Product Variability

Most of DOE-t and DOE-F applications allow an analyst to determine the optimal design that is expected to produce a better overall performance. The improvement of performance often is to present that either the average or the variation is better off. How to make an improvement in average has been introduced in the last two chapters, and the improvement in product variability is addressed here. Reduction in product variability, as a part-assembly strategy, is to reduce product variation and make a product more robust or fit to use, e.g., easing the assembly and meeting the product performance requirements regardless of variation. When the new design derived from DOEs is put into practice, it is expected to reduce scrap and warranty costs. Effects frequently act multiplicatively, and variability often increases with the size of the measurement. A logarithmic transformation is explicitly recommended when the standard deviation of the collected data is proportional to the mean value. Generally speaking, there are three sources of variation: manufacturing variation [Steiner and MacKay 2005], environmental/deterioration variation, and usage variation. The approach to reducing the variability can be achieved via process capability study based on DOEs as shown in Figure 8.6.1. After all, the expected improvement in variability from the DOE is expressed in terms of capability improvement indexes such as C_p and C_{pk}.

FIGURE 8.6.1 Variability reduction via process capability study using DOEs.

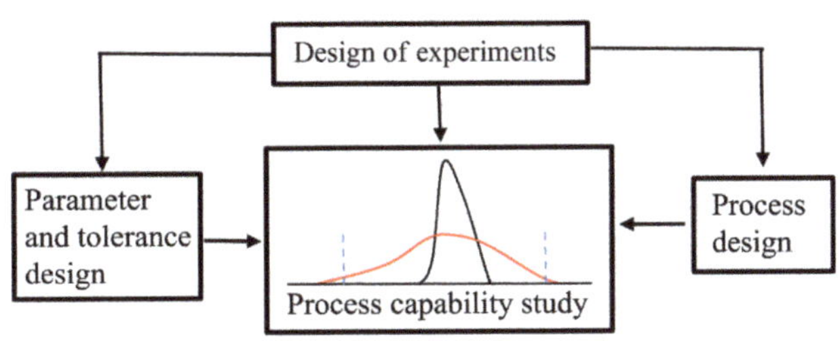

How to determine the optimal settings of decision variables such as the inspection frequency, the sample size, and the termination time for each run, which are influential to the precision of identifying significant factors and the experimental cost was addressed by [Yu and Chiao 2002]. Identification of best nominal values of design parameters to reduce variance and putting output on target (with minimal cost increase), as shown in Figure 8.6.2, means to improve product reliability through variation reduction.

FIGURE 8.6.2 Reduction in product variability with higher C_{pk} (LSL—lower specified limit and USL—upper specified limit).

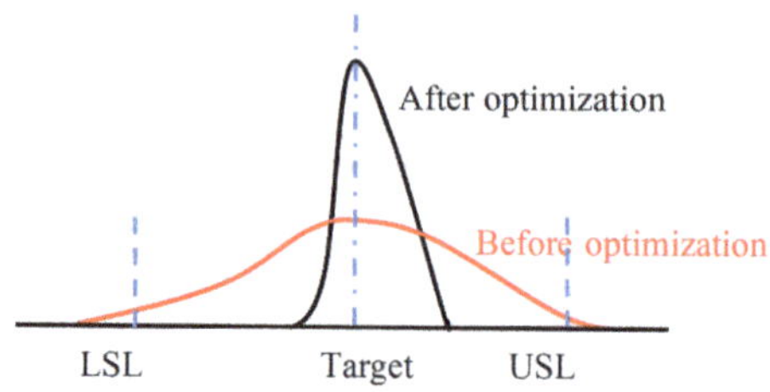

8.6.1. Robust Engineering Design

A robust product is the one that works as intended regardless of variation in a product's manufacturing process, variation resulting from degradation (e.g., deterioration), and variation in use. DOEs can be used to countermeasure various sources of variation: product design, process design, production, degradation, and usage (in service) as described below [Taylor 1990]:

1. Understanding the statistical capability of a process (process capability) can help to either design within the capability of that process or determine when an improved capability is required (e.g., new equipment, outsourcing).
2. By using statistical process control, special cause variations can be identified and cured.
3. Degradation, especially material degradation, is to be predicted based on the service environment in the design stage.
4. In the long run, process variability from common causes can be systematically reduced through process optimization, operator training, preventive maintenance, tool monitoring, standardization of machine settings, climate control, power conditioning, and so on.
5. Variability in usage can be countered by mistake-proofing (e.g., design for assembly and manufacturing techniques), warning labels, easy-to-understand operating manuals and controls, and so on.

The concept of robust design can be realized primarily using DOEs to determine which product and process parameters are sensitive to variation or noise and which factor level settings, i.e., parameter values, minimize the variability in the functional requirement and desired reliability. A framework for robust engineering process using DOEs in a manufacturing environment includes

$$\text{System design} \rightarrow \text{Parameter design} \rightarrow \text{Tolerance design} \rightarrow \text{Process capability}$$

$$\rightarrow \text{Variability reduction} \rightarrow \text{Robust design and process}$$

8.6.2. Logarithmic Transformation of Sample Variances for Variability Study

When studying the product variability, either the ln (i.e., natural log) or log ($\log_{10}$) of the sample variance of each individual treatment (run) can be taken. For the example given in the next section, it is desired for predicting a non-negative variance. Suppose that $z \approx \ln(S^2)$, where z is the predicted value of $\ln(\sigma^2)$. Then, exp(z) becomes the predicted sample variance of σ^2, and it is always non-negative. The applicable working domain for S^2 is $(0, \infty)$. On the other hand, the working domain for $\ln(S^2)$ is $(-\infty, \infty)$, which bears a nearly normal PDF and can be tested using the student t-distribution function with a limited number of tests, or F-distribution function if regression for RSM is applied.

Let y_{qn}, where subscript q is the treatment and subscript n is the number of replications within treatment q of a design matrix, represent an individual piece of data obtained from physical tests. Then,

$$S_q^2 = [\sum_{n=1}^{N_q} (y_{qn} - y_{n,ave})^2] / (N_q - 1) \sim \sigma_q^2 [\chi^2_{vq} / (N_q - 1)] \quad (8.6.3)$$

and $$\nu_q = N_q - 1 \quad (8.6.4)$$

where

y_{qn} is the individual data at replication n within treatment q
$y_{q,ave}$ is the average of replicated results from treatment q
S_q^2 is the sample variance of treatment q
σ_q^2 is the variance of treatment
ν_q is the DOFs
χ^2 is the chi-squared PDF with ν DOFs
N_q is the total number of replications within treatment q

The above equation is given following the definition of chi-squared distribution function (usually written as χ^2 distribution) with ν DOFs, which is the distribution of a sum of the squares of ν independent standard normal random variables. The mean value and variance of a χ^2 distribution function are ν (DOFs) and 2ν, respectively. The variance of $\ln(\chi^2/\nu)$ can be decomposed into two parts as follows:

$$\mathrm{Var}[\ln(\chi^2 / \nu)] \approx \{\frac{d\{\ln[E(\chi^2 / \nu)]}{d[E(\chi^2 / \nu)]}\}^2 \mathrm{Var}[\chi^2 / \nu] \quad (8.6.5)$$

i.e., $$\mathrm{Var}[\ln(\chi^2 / \nu)] \approx \{1\}^2 (2 / \nu) = 2 / \nu \quad (8.6.6)$$

where

$\mathrm{Var}[\ln(\chi^2/\nu)]$ is the variance of $\ln(\chi^2/\nu)$
$E(\chi^2/\nu)$ is the expected value of χ^2/ν
$\mathrm{Var}(\chi^2/\nu)$ is the variance of χ^2/ν

Thus, $$\ln(S_q^2) \approx N[\ln(\sigma_q^2), 2 / (N_q - 1)] \quad (8.6.7)$$

The above equation means that $\ln(S_q^2)$ is normally distributed with mean $\ln(\sigma_q^2)$ and variance $2/(N_q - 1)$. The variance of $\ln(S_q^2)$ is $2/(N_q - 1)$, which gets smaller and smaller as N_q grows and is gradually stabilized as $N_q \geq 10$.

8.6.3. Noises

Factors are classified into two categories in robust engineering design. They are control factors and noise factors. Control factors can be controlled, but noise factors are either difficult or impossible to control during normal operating conditions. There are two different noise factors embedded in a reliability experiment:

(a) Product Noise: Product noises are the factors that vary from product to product. For example, the wear resistance of tires may vary from unit to unit even under the same working condition.

(b) Environmental Noise: Environmental noises are the factors that vary during the usage of the product. For example, temperature, humidity, and road roughness working against the tires can vary during its usage.

To compute the noise, a first assumption is made: having a "zero" mean. The noise should be independently and randomly distributed with respect to the response and factors. If not, then a source of systematic variation is embedded in what was wrongly perceived to be pure noise. If that is the case, one should extract assorted terms from the noise that is related to any factor. This can be figured out by comparing the average of replicated responses with the average noise within the treatment. For each set of replications on a treatment, the following rule can be used for calculating the noise within each treatment:

$$\text{Noise} = \{\, \Sigma\, [\text{Measurement} - \text{Average of Measurements}]\} \,/\, \text{Replication} \tag{8.6.8}$$

Since white noise is normally distributed with mean zero, the standard deviation of the noise using the equation given above should be small enough, as compared to the differences between any two replicated responses within the same treatment.

8.7. MLE for DOE-LN Parameters

The proposed optimality algorithm applied to the MLE is to minimize the asymptotic variances and covariances generated using some percentile life at a specified item or load level. As demonstrated in Table 8.7.1, it is an example given to illustrate the procedure for formulating the equation for product life prediction using MLE. The test results listed in Table 8.7.1 are obtained from full factorial design 2^2 [Reliawiki 2020].

TABLE 8.7.1 Example data for life prediction with two factors.

(a) Time t:				
Run	**A (x_1)**	**B (x_2)**	**AB (x_1x_2)**	**t (time to failure)**
1	−1 (x_{11})	−1 (x_{21})	1	t_1 = 27 days
2	1 (x_{12})	−1 (x_{22})	−1	t_2 = 25 days
3	−1 (x_{13})	1 (x_{23})	−1	t_3 = 50 days
4	1 (x_{14})	1 (x_{24})	1	t_4 = 55 days
Contrast	1.5	26.5	3.5	
Effect	<u>0.75</u>	13.25	<u>1.75</u>	$t_{average}$ = 39.25
(b) y = ln(t):				
Run	**A (x_1)**	**B (x_2)**	**AB (x_1x_2)**	**y = ln(t)**
1	−1 (x_{11})	−1 (x_{21})	1	$\ln(t_1)$ = 3.2958
2	1 (x_{12})	−1 (x_{22})	−1	$\ln(t_2)$ = 3.2189
3	−1 (x_{13})	1 (x_{23})	−1	$\ln(t_3)$ = 3.9120
4	1 (x_{14})	1 (x_{24})	1	$\ln(t_4)$ = 4.0073
Contrast	0.00917	0.7024		0.0861
Effect	<u>0.00459</u>	0.3512	<u>0.04305</u>	$y_{ave} = \ln(t)_{ave}$ = 3.6085

Three different ways of calculating the design contrasts and effects are exhibited for demonstration:

1. DOE-t in the time domain
2. DOE-t after the logarithmic transformation of data
3. MLE

8.7.1. Lognormal Distribution Solution by DOE-t

It is shown in Table 8.7.1 that factor A and two-factor interaction AB are not statistically significant. The predictive equations derived from the test data based on DOE-t in the time domain and after logarithmic transformation are, respectively, as follows:

$$t = 39.25 + 13.25\, x_2 \qquad \text{(Time Domain)} \tag{8.7.1}$$

and $$y = 3.6085 + 0.3512\, x_2 \qquad \text{(After Transformation)} \tag{8.7.2}$$

Note that the statistical significance of each coefficient, i.e., γ_0, γ_1, γ_2, or γ_{12}, is evaluated using the random error made of the pooled small effects (i.e., γ_1 and γ_{12}) based on t-distribution.

8.7.2. Lognormal Distribution Solution by MLE

The same solution to the example problem as presented by Equation (8.7.2) in the logarithmic domain can be obtained using the MLE. As the maximum-likelihood method is applied to estimating the lognormal parameters for DOEs of two factors, Equation (8.4.16) is rewritten as

$$\mu_{yq} = \gamma_0 + \gamma_1 x_{1q} + \gamma_2 x_{2q} + \gamma_{12} x_{1q} x_{2q} \tag{8.7.3}$$

of which the following three associated hypotheses are to be examined:

$$H_o\text{: } \gamma_1 = 0 \text{ and } H_1\text{: } \gamma_1 \neq 0 \tag{8.7.4}$$

$$H_o\text{: } \gamma_2 = 0 \text{ and } H_1\text{: } \gamma_2 \neq 0 \tag{8.7.5}$$

$$H_o\text{: } \gamma_{12} = 0 \text{ and } H_1\text{: } \gamma_{12} \neq 0 \tag{8.7.6}$$

There are four pieces of data for four unknowns, i.e., γ_0, γ_1, γ_2, and γ_{12} in Equation (8.7.3). Since the life data for the present experiment are complete, i.e., no censored or interval data, and the time to failure in this case follows the lognormal distribution, the likelihood function of Equation (8.4.13) can be written in terms of test data t_q (q = 1, 2, 3, and 4) as

$$L_{ln}(\mu_{yq}, \sigma_y, t_q) = \prod_{q=1}^{Q} f(t_q) = \prod_{q=1}^{4} \frac{1}{t_q\, \sigma_y\, (2\pi)^{1/2}} \exp\left\{-\tfrac{1}{2}\left[\frac{\ln(t_q) - \mu_{yq}}{\sigma_y}\right]^2\right\} \tag{8.7.7}$$

The total number of treatments Q = 4. Substituting Equation (8.7.3) into the above equation leads to

$$L_{ln}(\sigma_y, \gamma_0, \gamma_1, \gamma_2) = \prod_{q=1}^{4} \frac{1}{t_q \sigma_y (2\pi)^{1/2}} \exp\{-½ [\frac{\ln(t_q) - (\gamma_0 + \gamma_1 x_{1q} + \gamma_2 x_{2q} + \gamma_{12} x_{1q} x_{2q})}{\sigma_y}]^2\} \quad (8.7.8)$$

Taking a logarithmic transformation of the above equation, one has the following logarithmic likelihood function:

$$\begin{aligned} \ln[L_{ln}(\sigma_y, \gamma_0, \gamma_1, \gamma_2)] \\ &= \sum_{q=1}^{4} \ln\{\frac{1}{t_q \sigma_y (2\pi)^{1/2}} \exp[-½ (\frac{\ln(t_q) - (\gamma_0 + \gamma_1 x_{1q} + \gamma_2 x_{2q} + \gamma_{12} x_{1q} x_{2q})}{\sigma_y})^2]\} \\ &= \ln[\frac{1}{t_1 t_2 t_3 t_4 (\sigma_y)^4 (2\pi)^2}] - ½ \{\sum_{q=1}^{4} [\frac{\ln(t_q) - (\gamma_0 + \gamma_1 x_{1q} + \gamma_2 x_{2q} + \gamma_{12} x_{1q} x_{2q})}{\sigma_y}]^2\} \\ &= -[\ln(t_1 t_2 t_3 t_4) + 4 \ln(\sigma_y) + 2 \ln(2\pi)] \\ &\quad - ½ \{\sum_{q=1}^{4} [\frac{\ln(t_q) - (\gamma_0 + \gamma_1 x_{1q} + \gamma_2 x_{2q} + \gamma_{12} x_{1q} x_{2q})}{\sigma_y}]^2\} \end{aligned} \quad (8.7.9)$$

By taking partial differentiations of the logarithmic likelihood function given above with respect to the four unknowns (i.e., γ_0, γ_1, γ_2, and γ_{12}), one may maximize the logarithmic likelihood as

$$\frac{\partial\{\ln[L_{ln}()]\}}{\partial \gamma_0} = \frac{1}{\sigma_y^2} \sum_{q=1}^{4} [\ln(t_q) - (\gamma_0 + \gamma_1 x_{1q} + \gamma_2 x_{2q} + \gamma_{12} x_{1q} x_{2q})] = 0 \quad (8.7.10)$$

$$\frac{\partial\{\ln[L_{ln}()]\}}{\partial \gamma_1} = \frac{1}{\sigma_y^2} \sum_{q=1}^{4} \{[\ln(t_q) - (\gamma_0 + \gamma_1 x_{1q} + \gamma_2 x_{2q} + \gamma_{12} x_{1q} x_{2q})] x_{1q}\} = 0 \quad (8.7.11)$$

$$\frac{\partial\{\ln[L_{ln}()]\}}{\partial \gamma_2} = \frac{1}{\sigma_y^2} \sum_{q=1}^{4} \{[\ln(t_q) - (\gamma_0 + \gamma_1 x_{1q} + \gamma_2 x_{2q} + \gamma_{12} x_{1q} x_{2q})] x_{2q}\} = 0 \quad (8.7.12)$$

and

$$\frac{\partial\{\ln[L_{ln}()]\}}{\partial \gamma_{12}} = \frac{1}{\sigma_y^2} \sum_{q=1}^{4} \{[\ln(t_q) - (\gamma_0 + \gamma_1 x_{1q} + \gamma_2 x_{2q} + \gamma_{12} x_{1q} x_{2q})] x_{1q} x_{2q}\} = 0 \quad (8.7.13)$$

Only four fitting coefficients, i.e., γ_0, γ_1, γ_2, and γ_{12}, can be considered, since there are merely four pieces of data available. Substituting the data listed in Table 8.7.1 into Equations (8.7.10)–(8.7.13) yields the following four equations:

$$\begin{aligned} \gamma_0 &= [\ln(t_1) + \ln(t_2) + \ln(t_3) + \ln(t_4)] / 4 = [\ln(27) + \ln(25) + \ln(50) + \ln(55)] / 4 \\ &= 3.6085 \end{aligned} \quad (8.7.14)$$

$$\gamma_1 = [-\ln(t_1) + \ln(t_2) - \ln(t_3) + \ln(t_4)] / 4 = [-\ln(27) + \ln(25) - \ln(50) + \ln(55)] / 4$$

$$= 0.00459 \quad (8.7.15)$$

$$\gamma_2 = [-\ln(t_1) - \ln(t_2) + \ln(t_3) + \ln(t_4)] / 4 = [-\ln(27) - \ln(25) + \ln(50) + \ln(55)] / 4$$

$$= 0.3512 \quad (8.7.16)$$

and

$$\gamma_{12} = [\ln(t_1) - \ln(t_2) - \ln(t_3) + \ln(t_4)] / 4 = [\ln(27) - \ln(25) - \ln(50) + \ln(55)] / 4$$

$$= 0.0431 \quad (8.7.17)$$

Once more, the statistical significance of each coefficient, i.e., γ_0, γ_1, γ_2, or γ_{12}, can be evaluated using the random error made of the pooled small effects (i.e., γ_1 and γ_{12}) based on t-distribution, and its resulting predictive equation is

$$y = 3.6085 + 0.3512\, x_2 \quad (8.7.18)$$

The above predictive equation obtained using the maximum-likelihood estimator is exactly the same as Equation (8.7.2), which is obtained by analyzing the logarithmic-transformed data via DOE-t. Finally, the predictive equation in the time domain is

$$t = \exp(y) = \exp(3.6085 + 0.3512\, x_2) \quad (8.7.19)$$

The next step is to calculate variance σ_y^2. By taking partial differentiations of the logarithmic likelihood function in Equation (8.7.9) with respect σ_y, given that $\gamma_1 = 0$ and $\gamma_{12} = 0$, one has

$$\frac{\partial\{\ln[L_{\ln}()]\}}{\partial \sigma_y} = \frac{-4}{\sigma_y} + \frac{1}{\sigma_y^3} \sum_{q=1}^{4} [\ln(t_q) - (\gamma_0 + \gamma_2\, x_{2q})]^2 = 0 \quad (8.7.20)$$

Thus, $\sigma_y^2 = \{ \sum_{q=1}^{4} [\ln(t_q) - (\gamma_0 + \gamma_2\, x_{2q})]^2 \} / 4 = 0.001874$

and $\sigma_y = 0.0433$ (8.7.21)

8.8. Example: Insulation Aging of Electric Motor Windings

This example is given here to illustrate the procedure of resolving an experimental design, of which the data fit in the domain of logarithmic transformation. Either maximum-likelihood method or DOE-t method can be utilized to estimate all the parameters in the proposed model.

$$y = \text{mean} + \eta_a\, a + \eta_b\, b + \eta_c\, c + \eta_{ab}\, a\, b + \eta_{ac}\, a\, c + \eta_{bc}\, b\, c + \eta_{abc}\, a\, b\, c \quad (8.8.1)$$

Polyetherimide (PEI) and PEI/silica nanocomposites are often used as insulation material for low- and medium-voltage electric motors. Thin steel plates (15 cm × 9 cm) coated with PEI enamel with nanosized silica (90 μm coating; thermal rating: 180°C) are employed to physically simulate the insulation behavior [Lahoud et al. 2013]. Operating factors and their design levels for the insulation of windings are listed in Table 8.8.1. These coatings are subjected to accelerated aging based on full factorial design 2^3 under the external loading conditions given in Table 8.8.2. The objective is to explore the electric insulation material life span of PEI/silica [Lahoud et al. 2013] nanocomposites. There are eight replicated data collected for each treatment. The life span of each sample was measured using a timer (one per sample) that stopped counting as soon as the current increased and crossed a threshold.

TABLE 8.8.1 Operating factors and their design levels for the insulation of windings in electric motor [Lahoud et al. 2013].

Variables	Level (−)	Level (+)	Coded variables
A: Voltage (kV)	1	3	a = (A − 2)/1
B: Frequency (kHz)	5	15	b = (B − 10)/5
C: Temperature (°C)	−55°C	180°C	c = (C − 62.5)/117.5

TABLE 8.8.2 Factorial design 2^3 for electric insulation by PEI/nano-silica-plated steel [Lahoud et al. 2013].

Treatment	a	b	c	T_{ave}	$T_1 - T_8$ (min)
q = 1	−1	−1	−1	607.88	378, 418, 568, 587, 634, 642, 786, 850
q = 2	1	−1	−1	27.125	25, 29, 23, 29, 26, 30, 24, 31
q = 3	−1	1	−1	260.375	169, 187, 268, 268, 343, 364, 162, 322
q = 4	1	1	−1	14.2813	14.5, 14.25, 13.4, 13.8, 14.8, 15, 13, 15.5
q = 5	−1	−1	1	26.125	26, 30, 28, 33, 24, 29, 28, 35
q = 6	1	−1	1	6.2125	6, 6.5, 6.2, 6.8, 5.5, 6.5, 6, 6.3
q = 7	−1	1	1	14.920	14, 16.21, 15.07, 15.5, 15.28, 16, 12.5, 14.8
q = 8	1	1	1	1.6373	2.2, 2.233, 1.133, 1.65, 2.033, 1.516, 0.933, 1.4

8.8.1. Insulation Aging of Electric Motor Windings via DOE-t

Since sampling of the product life comes apart (Table 8.8.2), the logarithmic transformation of each piece of data is taken after they are averaged within each individual treatment; i.e., $\ln(\Sigma t_n/8)$, where n = 1, 2, 3, 4, 5, 6, 7, and 8, is presented in Table 8.8.2 as [ln(T)]. Effects obtained from DOE-t operations are listed in Table 8.8.3. A Pareto plot of these effects is presented in Figure 8.8.1.

TABLE 8.8.3 Analysis of electric insulation by PEI/nano-silica-plated steel based on t-statistics.

Treatment	A	B	C	AB	AC	BC	ABC	[ln(T)]	$[\ln(T)]_p$	$T_{p,mean}$
1	−1	−1	−1	1	1	1	−1	6.3770	6.438	588.98
2	1	−1	−1	−1	−1	−1	1	3.2950	3.325	30.19
3	−1	1	−1	−1	1	1	1	5.5175	5.348	243.31
4	1	1	−1	1	−1	−1	−1	2.6574	2.579	12.47
5	−1	−1	1	1	−1	−1	1	3.3651	3.325	31.93
6	1	−1	1	−1	1	1	−1	1.8249	1.755	4.80
7	−1	1	1	−1	−1	−1	−1	2.6997	2.761	13.19
8	1	1	1	1	1	1	1	0.4513	0.482	1.98
Contrast	−2.433	−0.884	−2.376	−0.092	0.508	−0.136	−0.232			
Effect	−1.216	−0.442	−1.188	*−0.061*	0.269	*−0.068*	*−0.116*			
Error, S_y	0.0853	0.0853	0.0853	0.0853	0.0853	0.0853	0.0853			
t-ratio	14.27	5.1853	13.939	0.7131	3.1577	0.7945	1.3639			
$t_{3,5\%}$	2.353	2.353	2.353	2.353	2.353	2.353	2.353			
Significant?	Yes	Yes	Yes	No	Yes	No	No			

FIGURE 8.8.1 Pareto plot of effects.

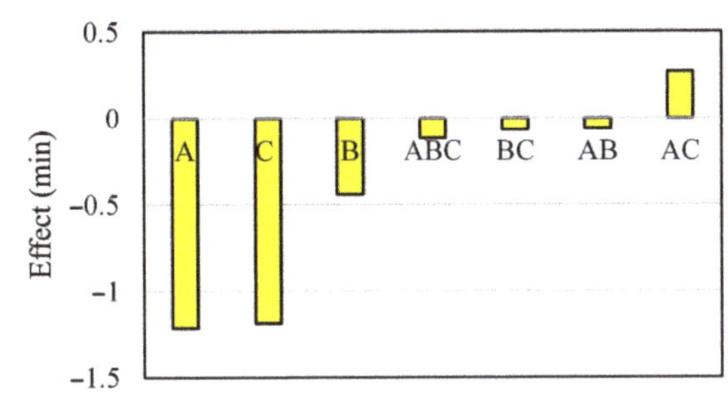

The three small effects, i.e., effects of AB, BC, and ABC given in Table 8.8.3, are taken for estimating random error S_y of each effect.

$$S_y^2 = \{[(-0.061)^2 + (-0.068)^2 + (-0.116)^2] / 3\} = (0.08525)^2$$

$$\text{Thus,} \quad S_y = 0.08525 \tag{8.8.2}$$

which is used to measure the t-statistics associated with the potentially influential effects. By taking a significance level of 5%, one may obtain the following predictive equation of mean value based on t-distribution as

$$\mu_{yq} = 3.2735 - 1.2163\ a - 0.4420\ b - 1.1883\ c + 0.2692\ a\ c \tag{8.8.3}$$

of which a, b, and c are coed variables, ranging from −1 to 1, corresponding to factors A, B, and C, respectively. The calculated main and interactive effects and their statistical significance are given in Table 8.8.3.

Simultaneously, the sample variance for the average value μ_{yq} within each treatment is to be obtained from the three small effects as

$$\sigma_y^2 = (-0.061)^2 + (-0.068)^2 + (-0.116)^2 = 0.0218 = 0.14765^2 \tag{8.8.4}$$

Letting y = ln(T), the mean value of each treatment in the original time domain is to be estimated using Equation (8.2.10) as

$$T_{q,ave} \approx E[T_q] = \exp(\mu_{yq} + \tfrac{1}{2}\sigma_y^2), \qquad q = 1, 2, 3, 4, 5, 6, 7, 8 \tag{8.8.5}$$

Then, one may obtain the predictive equation of the mean value based on t-distribution as follows:

$$\begin{aligned} \mu_{yq} &= \ln(T_q) - \tfrac{1}{2}\sigma_y^2 \\ &= 3.2735 - 1.2163\ a - 0.4420\ b - 1.1883\ c + 0.2692\ a\ c - \tfrac{1}{2}\sigma_y^2 \end{aligned} \tag{8.8.6}$$

of which a, b, and c are coded variables, ranging from −1 to 1, corresponding to factors A, B, and C, respectively.

Predicted mean values using Equations (8.8.6) and (8.8.4) for the eight runs (treatments) are also listed in Table 8.8.3. Model effectiveness and adequacy are further assured by the correlation plot with a correlation of 99.98% and the residual plot with a random pattern, respectively, depicted in Figure 8.8.2(a) and (b). The validity of the predictive equation can be further assured by the normal probability plot shown in Figure 8.8.3, of which the more the deviation stays away from the straight dashed line, the larger the effect.

FIGURE 8.8.2 Diagnostic checking of the predictive equation.

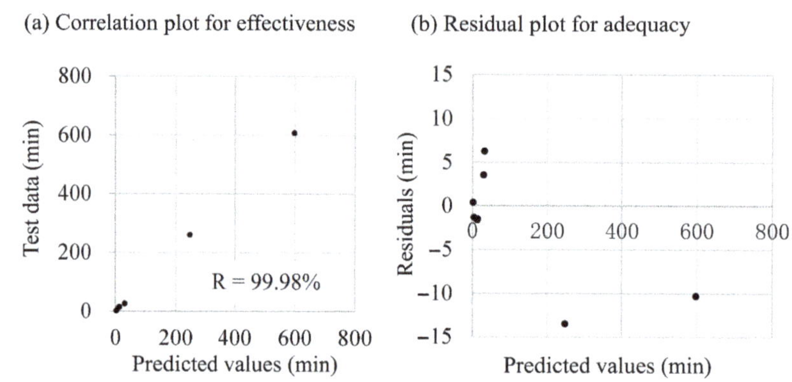

FIGURE 8.8.3 Normal probability plot.

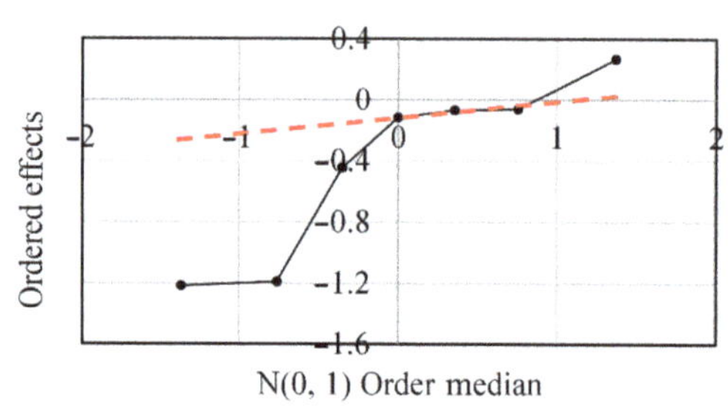

8.8.2. Insulation Aging of Electric Motor Windings via MLE

As the maximum-likelihood method is applied to estimating the lognormal parameters for DOEs of three factors, Equation (8.8.1) is here rewritten as

$$\mu_{yq} = \gamma_0 + \gamma_1 a_q + \gamma_2 b_q + \gamma_3 c_q + \gamma_{12} a_q b_q + \gamma_{13} a_q c_q + \gamma_{23} b_q c_q + \gamma_{123} a_q b_q c_q \tag{8.8.7}$$

Substituting the above equation into Equation (8.7.8) leads to

$$L_{ln}(\sigma_y, \gamma_0, \gamma_1, \gamma_2) = \prod_{q=1}^{8} \frac{1}{t_q \sigma_y (2\pi)^{1/2}}$$

$$\exp\left\{-\tfrac{1}{2}\left[\frac{\ln(t_q) - (\gamma_0 + \gamma_1 a_q + \gamma_2 b_q + \gamma_3 c_q + \ldots + \gamma_{123} a_q b_q c_q)}{\sigma_y}\right]^2\right\}$$

By taking partial differentiations of the logarithmic likelihood function given above with respect to the eight unknowns (i.e., $\gamma_0, \gamma_1, \gamma_2, \ldots,$ and γ_{123}), one may maximize the logarithmic likelihood using the following eight equations:

$$\frac{\partial\{\ln[L_{ln}()]\}}{\partial\gamma_0} = \frac{1}{\sigma_y^2}\sum_{q=1}^{8}[\ln(t_q) - (\gamma_0 + \gamma_1 a_q + \gamma_2 b_q + \ldots + \gamma_{123} a_q b_q c_q)] = 0$$

$$\frac{\partial\{\ln[L_{ln}()]\}}{\partial\gamma_1} = \frac{1}{\sigma_y^2}\sum_{q=1}^{8}\{[\ln(t_q) - (\gamma_0 + \gamma_1 a_q + \gamma_2 b_q + \ldots + \gamma_{123} a_q b_q c_q)]\, a_q\} = 0$$

$$\frac{\partial\{\ln[L_{ln}()]\}}{\partial\gamma_2} = \frac{1}{\sigma_y^2}\sum_{q=1}^{8}\{[\ln(t_q) - (\gamma_0 + \gamma_1 a_q + \gamma_2 b_q + \ldots + \gamma_{123} a_q b_q c_q)]\, b_q\} = 0$$

$$\frac{\partial\{\ln[L_{ln}()]\}}{\partial\gamma_3} = \frac{1}{\sigma_y^2}\sum_{q=1}^{8}\{[\ln(t_q) - (\gamma_0 + \gamma_1 a_q + \gamma_2 b_q + \ldots + \gamma_{123} a_q b_q c_q)]\, c_q\} = 0$$

$$\frac{\partial\{\ln[L_{ln}()]\}}{\partial\gamma_{12}} = \frac{1}{\sigma_y^2}\sum_{q=1}^{8}\{[\ln(t_q) - (\gamma_0 + \gamma_1 a_q + \gamma_2 b_q + \ldots + \gamma_{123} a_q b_q c_q)]\, a_q b_q\} = 0$$

$$\frac{\partial\{\ln[L_{ln}()]\}}{\partial\gamma_{13}} = \frac{1}{\sigma_y^2}\sum_{q=1}^{8}\{[\ln(t_q) - (\gamma_0 + \gamma_1 a_q + \gamma_2 b_q + \ldots + \gamma_{123} a_q b_q c_q)]\, a_q c_q\} = 0$$

$$\frac{\partial\{\ln[L_{ln}()]\}}{\partial\gamma_{23}} = \frac{1}{\sigma_y^2}\sum_{q=1}^{8}\{[\ln(t_q) - (\gamma_0 + \gamma_1 a_q + \gamma_2 b_q + \ldots + \gamma_{123} a_q b_q c_q)]\, b_q c_q\} = 0$$

and

$$\frac{\partial\{\ln[L_{ln}()]\}}{\partial\gamma_{123}} = \frac{1}{\sigma_y^2}\sum_{q=1}^{8}\{[\ln(t_q) - (\gamma_0 + \gamma_1 a_q + \gamma_2 b_q + \ldots + \gamma_{123} a_q b_q c_q)]\, a_q b_q c_q\} = 0$$

Substituting the values of a_q, b_q, and c_q (in Table 8.8.2) into the above eight equations leads to a set of simultaneous equations, which can be solved for the eight unknowns as

$$\gamma_0 = 3.274,\ \gamma_1 = -1.216,\ \gamma_2 = -0.442,\ \gamma_3 = -1.188,\ \gamma_{12} = -0.061,\ \gamma_{13} = 0.2694,\ \gamma_{23} = -0.068,$$

$$\text{and } \gamma_{123} = -0.116$$

The three small effects $\gamma_{12} = -0.061$, $\gamma_{23} = -0.068$, and $\gamma_{123} = -0.116$ are pooled together to assess the sample standard deviation:

$$S_y = [(-0.061)^2 + (-0.068)^2 + (-0.116)^2]^{1/2} = 0.08525$$

The t-ratios for the seven coefficients in Equation (8.8.7) are, respectively,

$$\gamma_1 / S_y = 4.27,\ \gamma_2 / S_y = 5.1853,\ \gamma_3 / S_y = 13.939,\ \gamma_{12} / S_y = 0.7131,\ \gamma_{13} / S_y = 3.1577,$$
$$\gamma_{23} / S_y = 0.7945, \text{ and } \gamma_{123} / S_y = 1.3639$$

Given that $t_{3,95\%} = 2.353$, the valid predictive equation of the mean value is

$$\mu_{yq} = = 3.2735 - 1.2163\ a_q - 0.4420\ b_q - 1.1883\ c_q + 0.2692\ a_q\ c_q - ½\ \sigma_y^2$$

The above is the same as Equation (8.8.6), which is obtained using DOE-t based on logarithmic-transformed data.

8.9. MLE for DOE-LN Parameters with Incomplete Data

For an experimental design involving incomplete data (censored), the maximum-likelihood method combined with the LRT should be applied. An experiment with fluorescent light involving five factors, namely A, B, C, D, and E, is demonstrated for the need by [Wu and Hamada 2009]. Fractional factorial design 2_{III}^{5-2}, with alias equations D = −AC and E = ±BC (E = BC for runs 1–4 and E = −BC for runs 5–8) [Hamada and Tse 1992], is set up to identify the potentially significant factors that can be used for improving the reliability of fluorescent lights [Wu and Hamada 2009]. Each treatment comes with two replicated tests. Inspections are conducted every two days, and the test ends at the 20th day. Test results are exhibited in Table 8.9.1.

TABLE 8.9.1 2_{III}^{5-2} design for evaluating product life with right-censored and interval data [Wu and Hamada 2009].

Treatment	a	b	c	d = −ac	e	T_1 and T_2 (days)	E(T)
1	−1	−1	−1	1	1	(14, 16) and 20⁺	18.2
2	1	−1	−1	−1	−1	(18, 20) and 20⁺	23.2
3	−1	1	−1	1	−1	(8, 10) and (10, 12)	9.0
4	1	1	−1	−1	1	(18, 20) and 20⁺	21.1
5	−1	−1	1	−1	1	20⁺ and 20⁺	39.8
6	1	−1	1	1	−1	(12, 14) and 20⁺	17.0
7	−1	1	1	−1	−1	(16, 18) and 20⁺	19.6
8	1	1	1	1	1	(12, 14) and (14, 16)	15.4

where
T_1 and T_2 (days) is two replicated runs within each treatment
20+ is suspended after 20 days
(18, 20) refers to a failure that occurred at the moment between the 18th and 20th days

As the maximum-likelihood method is applied to estimating the lognormal parameters for fractional factorial design 2_{III}^{5-2} design, Equation (8.8.1) is here rewritten as

$$\mu_{yq} \approx \gamma_0 + \gamma_a\ a + \gamma_b\ b + \gamma_c\ c + \gamma_d\ d + \gamma_e\ e \quad (8.9.1)$$

Subscript q can be s (for suspended data) or i (for interval data). Note that two-factor interactions may be confounded with main effects because fractional factorial design 2_{III}^{5-2} comes with resolution III.

Assume that the shape factor is consistent throughout the experimental tests. In light of Equations (8.4.14) and (8.4.15) for right-suspended and interval data, the ln-likelihood can be written as

$$\ln[L_{\ln}(\beta, \gamma_0, \gamma_a, \gamma_b, \gamma_c, \gamma_d, \gamma_e)] = \ln(L_{suspended}) + \ln(L_{interval}) \tag{8.9.2}$$

where $$L_{suspended} = \prod_{s=1}^{Q_s} \left\{1 - \frac{1}{(2\pi)^{1/2}} \int_{-\infty}^{[\ln(t_s) - \gamma_0 + \gamma_a a + \gamma_b b + \gamma_c c + \gamma_d d + \gamma_e e] / \sigma_y} \exp(-\tfrac{1}{2} \tau^2)\, d\tau\right\} \tag{8.9.3}$$

and $$L_{interval} = \prod_{i=1}^{Q_i} \left\{\frac{1}{(2\pi)^{1/2}} \int_{-\infty}^{[\ln(t_{i,2}) - \gamma_0 + \gamma_a a + \gamma_b b + \gamma_c c + \gamma_d d + \gamma_e e] / \sigma_y} \exp(-\tfrac{1}{2} \tau^2)\, d\tau \right.$$
$$\left. - \frac{1}{(2\pi)^{1/2}} \int_{-\infty}^{[\ln(t_{i,1}) - \gamma_0 + \gamma_a a + \gamma_b b + \gamma_c c + \gamma_d d + \gamma_e e] / \sigma_y} \exp(-\tfrac{1}{2} \tau^2)\, d\tau\right\} \tag{8.9.4}$$

Substituting Equations (8.9.3) and (8.9.4) into Equation (8.9.2) leads to an equation with six unknowns, i.e., γ_0, γ_a, γ_b, γ_c, γ_d, and γ_e, and one may maximize the logarithmic likelihood using the following six equations:

$$\partial\{\ln[L_{\ln}()]\} / \partial\gamma_0 = 0$$
$$\partial\{\ln[L_{\ln}()]\} / \partial\gamma_a = 0$$
$$\partial\{\ln[L_{\ln}()]\} / \partial\gamma_b = 0$$
$$\partial\{\ln[L_{\ln}()]\} / \partial\gamma_c = 0$$
$$\partial\{\ln[L_{\ln}()]\} / \partial\gamma_d = 0$$
and $$\partial\{\ln[L_{\ln}()]\} / \partial\gamma_e = 0$$

The above equations yield 15 (15 = 16 − 1) equations because of treatment 5 that has two suspended data with the same lifetime. Solving these 15 equations for six unknowns leads to [Wu and Hamada 2009]

$$\gamma_0 = 2.939$$
$$\gamma_a = 0.117$$
$$\gamma_b = -0.201$$
$$\gamma_c = -0.051$$
$$\gamma_d = -0.273$$
and $$\gamma_e = -0.153$$

After these parameters are estimated, the LRT is then utilized to justify the goodness of each individual effect coefficient γ_{test} as

$$LR(\gamma_{test}) = -2\, \{\ln[L_w(All - \gamma_{test})] - \ln[L_w(All)]\}$$

of which γ_{test} is one of the following six unknowns: γ_0, γ_a, γ_b, γ_c, γ_d, and γ_e. The results are shown in Table 8.9.2. Removing nonsignificant terms in Equation (8.9.1) according to the p-values listed in the table leads to

$$\mu_{yq} = 2.939 + 0.117\, a - 0.201\, b - 0.273\, d - 0.153\, e \tag{8.9.5}$$

Plugging the above equation into Equation (8.8.5), one can predict the product life within individual treatment q as

$$E[T_q] = \exp(\mu_{yq} + ½\, \sigma_y^2), \qquad q = 1, 2, 3, 4, 5, 6, 7, 8 \tag{8.9.6}$$

Variance σ_y^2 can be obtained by pooling statistically insignificant effects, and it is assumed to be small and negligible in this case. The expected data of life expectancy corresponding to the eight treatments are listed in column E(T) of Table 8.9.1. It can be seen that the predicted values correlate well with test data.

TABLE 8.9.2 Likelihood test for reliability of fluorescent lights based on Weibull MLE [Wu and Hamada 2009].

Effect	DOF	Lognormal MLE	LR	p-Value
γ_0	1	2.939	2089	0.0%
γ_1	1	0.117	3.6	6%
γ_2	1	−0.210	0.7	0.0%
γ_3	1	−0.051	60.76%	41%
γ_4	1	0.273	13.2237	0.0%
γ_5	1	−0.153	3.8504	2%

8.10. Alternate Viable Transformations

Response data are transformed to allow an experimenter to have valid applications using proper statistical inferences, which depend on the assumptions of independence, normality, and even constant variance (or not). One can use a non-normal distribution to calculate process capability, or alternatively, one may try to transform data to follow a normal distribution after a proper transformation [Hernández and Johnson 1980]. For example, the ln(Y) transformation has been proven to be effective in an examination of matrix effects in atomic absorption spectrometry by [Koscielniak and Parczewski 1983]. After transforming the response using ln(Y) or log(Y), one may examine the residual plot and normal probability plot corresponding to the transformed data. Note that if the assumptions appear to be satisfied, one may draw conclusions from the analysis of the transformed data. If not, more alternate transformations are available for the next tryout.

8.10.1. Individual Distribution Identification

Here is the procedure for how to use Minitab (Version 18) to identify the suitable statistical distribution for a given set of data:

Stat → Quality Tools → Individual Distribution Identification

When fitting statistical distributions, a lot of output in both the Minitab Session Window and Graphics will pop up with the following three measures:

1. ***Anderson-Darling (AD) Statistic (Adjusted)***: A lower AD value indicates a better fit [Anderson and Darling 1952], as it measures a squared distance that is weighted heavier in the tails of the distribution. The AD statistic makes use of observations to obtain a random sample of values from a given PDF, to (a) compute CDF, (b) take random samples from the PDF of a uniform distribution

over the range [0, 1], and (c) then look up the random sample values using the selected CDF [Marsaglia 2004]. However, to compare how well different distributions fit the data, one should assess the p-value, as described below.

2. ***P-value***: A high p-value, which is the probability value at the extrema of a statistical distribution, means that a good fit as the null hypothesis prescribes no difference between the data set and the fitted statistical distribution function. It is generally valid to compare p-values between distributions and go with the highest. A low p-value such as 10% or lower indicates that the data do not follow that distribution. This is also a common practice in applying DOEs. Because statistical significance is often declared at a mean difference of 1.96 standard deviations (p-value = 5%) based on normal distribution, it introduces enough noise to conceal an effect in comparisons between the mismatched treatments or to severely skew the apparent magnitude and statistical significance of a larger effect. Nevertheless, that p-value between 5% and 10% is generally applied to automotive engineering DOEs. For some three-parameter distributions, the p-value is impossible to calculate and is represented by asterisks.
3. ***LRT***: This is for examining three-parameter distributions only, and a low value indicates that adding the third parameter is a significant improvement over the two-parameter version when fitting a distribution function using collected sample data. A higher value suggests that the two-parameter version is better. LRT has also been in use for DOEs with product life prediction and others, which have extremely skewed statistical distributions.

Nevertheless, when evaluating the model fit, one should always make sense of the visual results on the probability plots, e.g., quantile–quantile (Q–Q) plots for CDFs. One solution to this problem is to simply increase the sample size of each treatment until variability, namely, discrepancies between treatments, decreases to an acceptable level [Bertsimas et al. 2015].

EXAMPLE 8.10.1

Twenty-three test-to-failure data of a mechanical component subjected to tensile stress (MPa) are listed in ascending order as follows:

291.2, 304.1, 327.6, 332.3, 337.5, 347.0, 353.8, 354.8,
355.1, 356.8, 359.4, 360.4, 366.7, 368.2, 369.2, 372.3,
377.9, 378.7, 382.5, 383.7, 388.0, 394.4, 396.8

What would be the statistical distribution function concerning the above test data?

Solution:

After typing the data in the Minitab, one begins with Stat ➔ Quality Tools ➔ Individual Distribution Identification. The following three important messages are generated:

(A) Goodness-of-Fit Test

Distribution (for the better)	AD (lower)	P (higher)	LRT P (lower)
Normal	0.477	0.215	*
Box-Cox transformation	***0.143***	***0.965***	*
Lognormal	0.636	0.085	*
Three-parameter lognormal	0.477	*	0.192
Exponential	9.138	<0.003	*
Three-parameter exponential	4.195	<0.010	0.000
Weibull	***0.161***	***>0.250***	*

Distribution (for the better)	AD (lower)	P (higher)	LRT P (lower)
Three-parameter Weibull	***0.139***	**>0.500**	***0.641***
SEV	***0.139***	***>0.250***	*
Largest extreme value	1.253	<0.010	*
Gamma	0.578	0.150	*
Three-parameter gamma	9.925	*	1.000
Logistic	0.319	>0.250	*
Log-logistic	0.403	>0.250	*
Three-parameter log-logistic	0.319	*	0.290

Conclusion

1. Box-Cox transformation with $\lambda = 5$ leads to the most fit model.
2. SEV function also fits well.
3. Both two- and three-parameter Weibull distributions fit well. Since the LRT of the three-parameter Weibull distribution is high, there is no need for the third parameter ($t_0 = 0$).

These observations are evidenced by the Q-Q plots.

(B) MLE of Distribution Parameters

Distribution	Location	Shape	Scale	Threshold
Normal*	359.06087	—	26.83130	—
Box-Cox Transformation*	6.26891E+12	—	2.04939E+12	—
Lognormal*	5.88068	—	0.07777	—
Three-parameter Lognormal	11.91367	—	0.00018	-1.48934E+05
Exponential	—	—	359.06087	—
Two-parameter Exponential	—	—	70.94520	288.11543
Weibull	—	17.77070	370.36640	—
Three-parameter Weibull	—	8137.00018	1.65450E+05	-1.65079E+05
SEV	370.99801	—	20.33208	—
Largest extreme value	344.91640	—	30.20827	—
Gamma	—	177.80434	2.01942	—
Three-parameter Gamma	—	7475.34345	0.30458	-1940.27345
Logistic	361.60167	—	14.45728	—
Log-logistic	5.88939	—	0.04104	—
Three-parameter Log-logistic	11.91284	—	0.00010	-1.48808E+05

(C) Q-Q plots for statistical distribution identification are given in Figure 8.10.1.

FIGURE 8.10.1 Example Q-Q plots for statistical distribution identification.

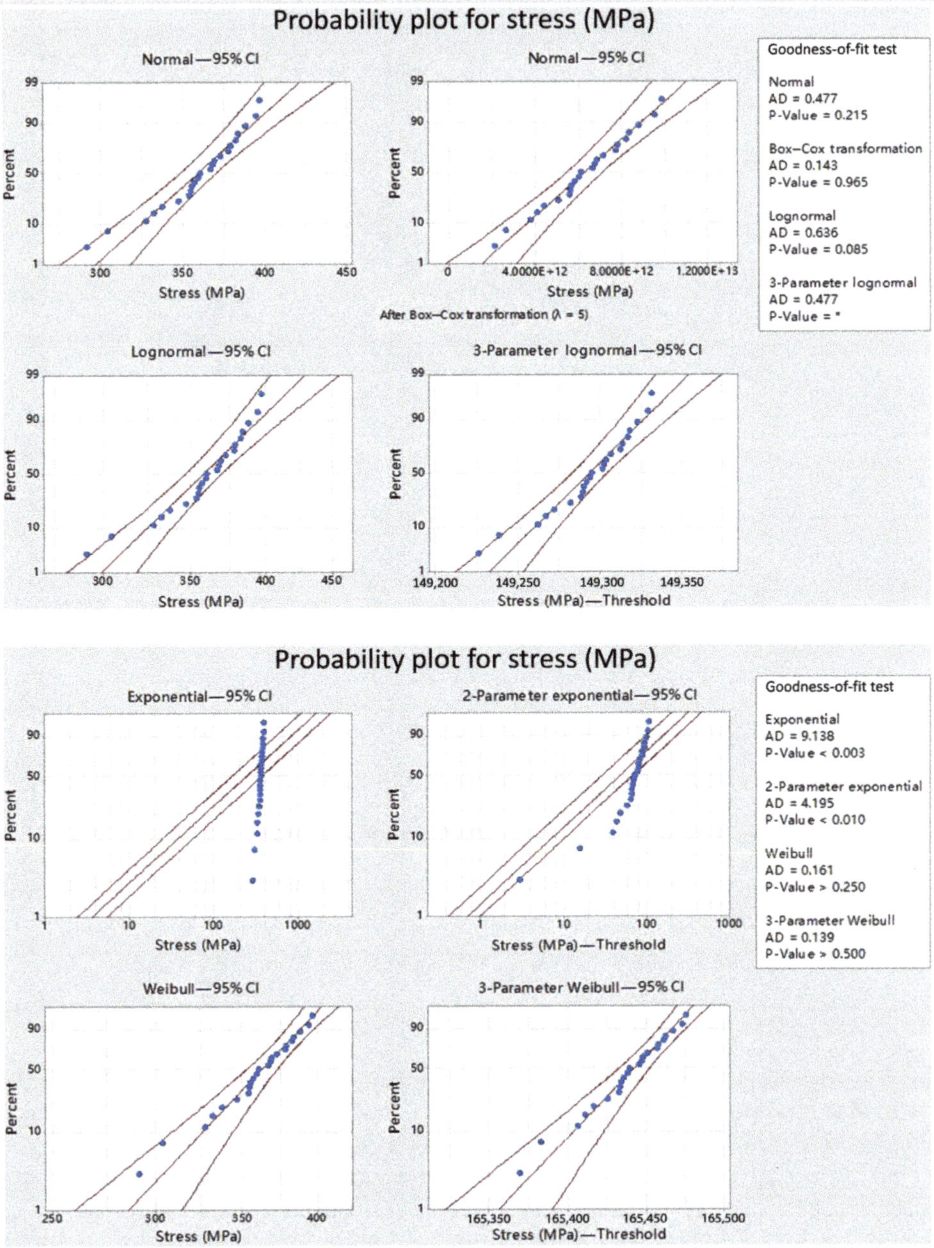

(Continued)

FIGURE 8.10.1 (Continued) Example Q-Q plots for statistical distribution identification.

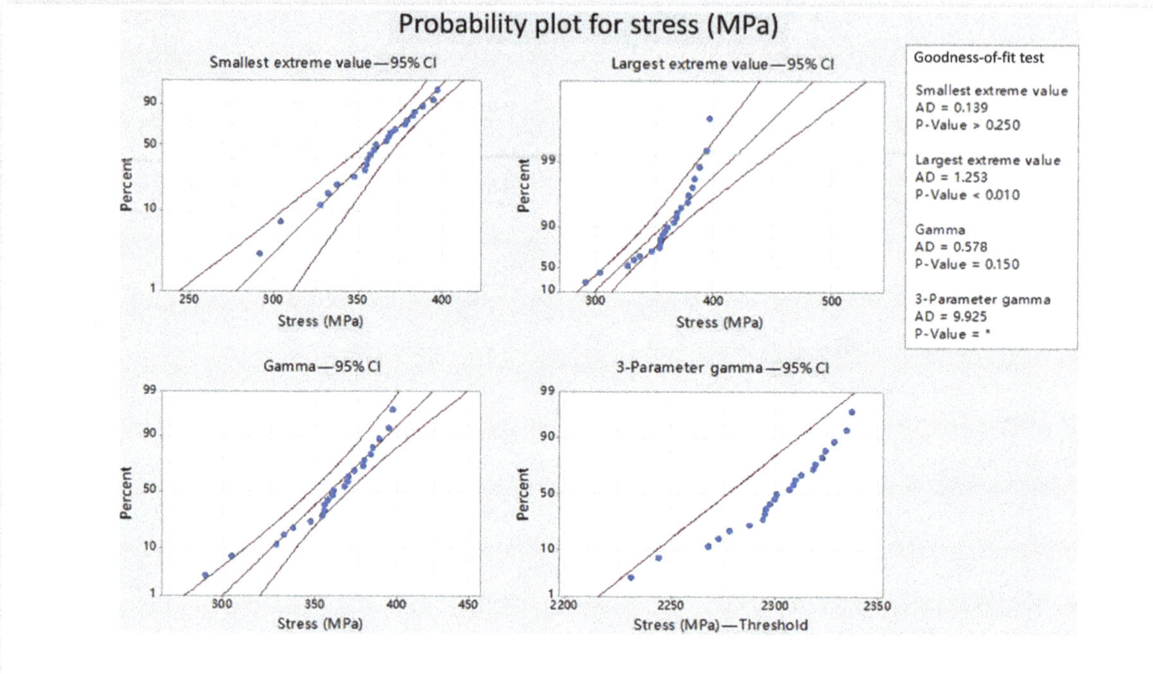

8.10.2. Box-Cox Transformations

At the core of the Box-Cox transformation [Box and Cox 1964] is an exponent, i.e., λ (lambda). The transformation of Y has the form:

$$y = (Y^{\lambda} - 1) / \lambda \tag{8.10.1}$$

of which λ is the power, to which all data should be raised. When λ → 0,

$$y = \ln(Y) \tag{8.10.2}$$

The Box-Cox transformation searches from λ = −5 to λ = +5 until the best value is found, i.e., the one which results in the best approximation of a normally distributed curve. Since ln(Y) = 2.3025 log(Y), the only real difference between these two functions is a scale factor 2.3025.

It is not guaranteed against a risk, because the transformation is justified only by the lowest standard deviation after fitting the data. It is necessary to check the transformed data for normality using a normal probability plot. The Box-Cox power transformation only works if all the data are positive and greater than 0. For Weibull distributed data, the Box-Cox transformation outperforms three other methods including the error function transformation, the dual power transformation, and the exponential transformation [Chortirat et al. 2011].

The case study in Example 8.10.1 shows that $y = (Y^5 - 1)/5$ is the transformation that leads to the best fit among all the models. Although the best value can be a real number, the process works better if this value is rounded to a whole number. This makes it easier to transform the data back and forth. The best whole-number values here are −1 and −2, i.e., Y^{-1} and Y^{-2}, respectively.

Normal-distributed data are required for drawing statistical inferences from a lot of statistical applications, such as statistical process control charts, C_p and C_{pk} analyses, t-tests, and ANOVA [Osborne 2010].

8.10.3. Adjustment to ln or log Transformation

A small constant can be added to all the observations prior to taking the logarithmic transformation. It means that the new transformed variable becomes ln(Y + c) or log(Y + c), given that constant c is a positive value. The adjustment to response Y makes an ln or log transformation feasible. This is also applicable to Box–Cox transformations.

8.10.4. $Y^{1/2}$ Transformation

A $Y^{1/2}$ transformation, i.e., sqrt(Y), may be just right for response data, if ln(Y) or log(Y) is an "over-correction." On the example of tool life subjected to turning operations given in Section 8.8: Table 8.8.2, the experimental assessment of the tool life based on fractional factorial design 3^{3-1} is here further examined using the $Y^{1/2}$ transformation, of which the ANOVA is shown in Table 8.10.1.

TABLE 8.10.1 ANOVA on tool life subjected to turning operations (Table 8.8.2) using $Y^{1/2}$ transformation.

Source of variation	DOF	SS_{Adj}	MS_{Adj}	F-ratio	P-value
Regression	4	965.7	241.4	68.57	0.1%
A	1	331.2	331.2	94.06	0.1%
B	1	83.59	83.59	23.74	0.8%
C	1	19.82	19.82	5.63	7.7%
BC	1	30.52	30.52	8.67	4.2%
Error	4	14.08	3.521	—	—
Subtotal		8	979.8		
Average		1			
Total		9			

Consider the example of modeling tool life Y with the data given in Table 8.8.2. As sample data of the tool life of the nine treatments come apart, the $Y^{1/2}$ transformation of each piece of treatment data (average of the data within each treatment) is taken here instead of ln(Y). Based on F-statistics (Chapters 3 and 4), the calculated main and interactive effects and their validity are demonstrated in Table 8.10.1. By taking a significance level of 10%, one may obtain the following predictive equation for tool life T:

$$Y^{1/2} = 80.92 - 0.1253\ A - 110.9\ B - 10.06\ C + 58.2\ B\ C$$

$$(R = 99.4\%,\ R_{adj} = 98.39\%,\ \text{and}\ R_{pred} = 95.48\%)$$

As effective as Equation (8.3.5) derived using ln(t) data, the above equation is a valid predictive equation for the tool life subjected to turning operations.

8.10.5. $\sin^{-1}(Y^{\delta})$ Transformation

Another useful transformation is arcsine square-root transformation $\sin^{-1}(Y^{\delta})$, of which the popular choice is exponent $\delta = 1/2$. The $\sin^{-1}(Y^{1/2})$ transformation may be taken if the response variable is a proportion (fraction) that happens to fall between 0 and 1, or saying $0\% < Y < 100\%$, which constrains the limit of variability, especially within a range of small percentages (<30%) or a range of large percentages (>70%). The $\sin^{-1}(Y^{1/2})$ transformation is a widely used transformation, and it can be carried out using "asin(sqrt(Y))" in Excel or Minitab.

Percentages usually follow a binomial distribution that violates basic assumptions of a normal distribution for drawing statistical inferences when doing DOEs by F- or t-distribution. The $\sin^{-1}(Y^{1/2})$ transformation provides the variance stabilizing mechanism, as exhibited in Table 8.10.2(i) and (ii). Given that both calculations are carried out using DOE-t procedure:

1. For original data Y, the random error generated is not able to detect the effect of two-factor interaction AC.
2. For transformed data $\sin^{-1}(Y^{1/2})$, the random error generated is able to detect the effect of two-factor interaction AC.

TABLE 8.10.2 Data transformation of factorial design 2^3 by $\sin^{-1}(Y^{1/2})$.

(i) Analysis with original data Y

Treatment	A	B	C	AB	AC	BC	ABC	Y
1	−1	−1	−1	1	1	1	−1	0.1
2	1	−1	−1	−1	−1	1	1	0.0
3	−1	1	−1	−1	1	−1	1	0.9
4	1	1	−1	1	−1	−1	−1	0.4
5	−1	−1	1	1	−1	−1	1	0.7
6	1	−1	1	−1	1	−1	−1	0.0
7	−1	1	1	−1	−1	1	−1	1.0
8	1	1	1	1	1	1	1	0.6
Contrast	−0.425	0.525	0.225	−0.025	−0.125	−0.075	0.175	
Effect	−0.213	0.263	0.113	−0.013	−0.063	−0.038	0.088	0.46 (average)
Error	0.0280	0.0280	0.0280	0.0280	0.0280	0.0280	0.0280	
t-Ratio	7.603	9.391	4.025	0.447	2.236	1.342	3.130	
$t_{2,95\%}$	2.92	2.92	2.92	2.92	2.92	2.92	2.92	
Significant?	Yes	Yes	Yes	No	No	No	Yes	

(ii) Analysis with original data $\sin^{-1}(Y^{1/2})$

Treatment	A	B	C	AB	AC	BC	ABC	$\sin^{-1}(Y^{1/2})$
1	−1	−1	−1	1	1	1	−1	0.322
2	1	−1	−1	−1	−1	1	1	0.000
3	−1	1	−1	−1	1	−1	1	1.249
4	1	1	−1	1	−1	−1	−1	0.685
5	−1	−1	1	1	−1	−1	1	0.991
6	1	−1	1	−1	1	−1	−1	0.000
7	−1	1	1	−1	−1	1	−1	1.571
8	1	1	1	1	1	1	1	0.886
Contrast	−0.640	0.769	0.298	0.016	−0.197	−0.037	0.137	
Effect	−0.320	0.385	0.149	0.008	−0.099	−0.018	0.069	0.71 (average)
Error	0.0141	0.0141	0.0141	0.0141	0.0141	0.0141	0.0141	
t-Ratio	22.70	27.27	10.57	0.566	6.997	1.296	4.864	
$t_{2,95\%}$	2.92	2.92	2.92	2.92	2.92	2.92	2.92	
Significant?	Yes	Yes	Yes	No	Yes	No	Yes	

8.10.6. Sinh Transformation

A transformation from the original data to the standard normal distribution is proposed by [Johnson 1949] as

$$z = \gamma + \delta \sinh^{-1}[(Y - \xi) / \lambda] \quad (8.10.3)$$

where

$z \sim N(0, 1)$
Y is the original data
$\gamma, \delta, \xi, \lambda$ are the constants

Since it has never been easy to identify four constants simultaneously, the simplest derivative model of the above equation is often applied instead.

$$z = \sinh^{-1}(Y) \quad (8.10.4)$$

8.10.7. Sinh-Arcsinh Transformation

An extension of sinh transformation is the sinh–arcsinh transformation proposed by [Jones and Pewsey 2009] as

$$z = \sinh[\gamma + \delta \sinh^{-1}(Y)] \quad (8.10.5)$$

where

$z \sim N(0, 1)$
Y is the original data
γ, δ are the constants

8.10.8. Signal-to-Noise Ratio

Both DOE-t and DOE-F methods, used for estimating factorial effects and thereby determining the desirable factorial levels as an optimal working condition, merely examine the response averages of the test results. It may not be able to capture the variability of results within a trial condition [Taguchi 1987], because the average calculation is relatively simple. A method is proposed to compare the population mean with the mean squared deviation, which combines the effects of both average E[Y] and variance Var(Y) of the test results within each treatment. The measure is guided by the following rules [Tsui and Li 1994]:

1. The larger the better, e.g., mechanical strength; the mission is twofold—to find factor settings that maximize E(Y) and to find other factor settings that minimize Var(Y).
2. The smaller the better, e.g., material rust; the mission is also twofold—to find factor settings that minimize E(Y) and to find other factor settings that minimize Var(Y).
3. On target the better, e.g., the nominal value of a given geometric dimension that is the target of interest in the study.

In pursuit of the convenience of linearity and accommodation of wide-ranging data, a logarithmic transformation, called the signal-to-noise ratio, is recommended. When the signal-to-noise ratio is used for analysis, the optimal condition identified from such an analysis is more likely to produce consistent performance, as argued by [Taguchi 1987]. The algorithm to minimize the variation from the target leads to exploiting measures of the process output that incorporate both the location and deviation of the output data. One of such measures is called Taguchi's signal-to-noise ratio [Ross 2005, Roy 2001], of which "signal" means the targeted experimental response value resulting from tests. Noise factors can be either internal or external to the system according to the following [Phadke 1989]:

(a) Inner Noise: It refers to any noise developed in the system or by the system itself.
(b) Outer Noise: It is the variation subjected to service circumstances.

Noise factors refer to the factors that cannot be controlled in real life. The signal-to-noise ratio is to check whether the performance depends on service circumstances.

The signal-to-noise ratio is thus defined to provide a measure of the impact of noise factors on the objective function (performance). Once the signal-to-noise ratio is statistically significant, the system is accordingly called being robust. Nevertheless, both [Hunter 1987, Montgomery 2019] discourage the use of signal-to-noise ratios and state that a more effective approach is to model the mean and variance separately. The larger the signal-to-noise ratio, the more robust the product is against noise. The definition of the signal-to-noise ratio depends on the experimental objective that could be the larger the better, the smaller the better, and on target the better:

(a) The Larger the Better: Positive logarithmic likelihood values are recommended for pursuing "the larger the better" objective. The signal-to-noise ratio for the larger the better is defined as

$$(S/N)_q \text{ ratio} \equiv -10 \log_{10}[(\sum_{n=1}^{N_q} y_{qn}^{-2}) / N_q] \tag{8.10.6}$$

where
y_{qn} is the data obtained from tests according to the DOE design matrix
q is the subscript q is the treatment
n is the subscript n is the number of the replicated run within the qth treatment
N_q is the total number of data available within the qth treatment

(b) The Smaller the Better: Negative logarithmic likelihood values are recommended for pursuing "the smaller the better" objective. Then, the signal-to-noise ratio for the smaller the better is

$$(S/N)_q \text{ ratio} \equiv -10 \log_{10}[(\sum_{n=1}^{N_q} y_{qn}^{2}) / N_q] \tag{8.10.7}$$

(c) Nominal the Better: The signal-to-noise ratio for the nominal the better is

$$(S/N)_q \text{ ratio} \equiv -10 \log_{10}(\mu_q^2 / S_q^2) \tag{8.10.8}$$

where
μ_q is the target (nominal) value
S_q is the sample standard deviation of all the data obtained within treatment q

The above transformations and some DOE design matrices given by Dr. S. Taguchi constitute a DOE methodology, which is called Taguchi's methods. The central theme of Taguchi's methods is to redeem a robust design [Byrne and Taguchi 1987], but the embedded technical robustness is debatable [Hunter 1987].

8.11. Finite Lognormal Mixture Distribution

Assume that each treatment in the DOEs has its own distinct inferential mechanism with its standardized variate as

$$z_q = [\ln(t) - \mu_{yq}] / \sigma_{yq} \tag{8.11.1}$$

where

t is the data, e.g., time, stress
μ_{yq} is the sample mean within the qth treatment
σ_{yq} is the sample standard deviation within the qth treatment

If μ_{yq} and σ_{yq} given in the above equation are not independent of treatment conditions, the DOE modeling becomes a problem-solving procedure involving a finite mixture distribution. Again, as long as an observation (t_{yq}) is censored, the corresponding residual is also to be censored in the residual plot when involved with a finite mixture distribution.

A finite lognormal mixture distribution appears when one deals with a heterogeneous population that consists of a finite number of "more homogeneous subpopulation" such as each treatment q in a DOE design matrix. The finite lognormal mixture distribution that consists of Q-weighted PDFs can be written as

$$f(t) = \sum_{q=1}^{Q} w_q f_q(t) \tag{8.11.2}$$

and $$f_q(t) = \frac{1}{t\,\sigma_{yq}\,(2\pi)^{1/2}} \exp\{-\tfrac{1}{2}\,[\frac{\ln(t)-\mu_{yq}}{\sigma_{yq}}]^2\} \quad \text{if } t > 0 \tag{8.11.3}$$

where

w_q is the mixing weight attached to treatment q, which ranges from 1 to Q
$f_q(t)$ is the lognormal function associated with treatment q

Then, the combined reliability function of the finite lognormal mixture distribution becomes

$$R(t) = \sum_{q=1}^{Q} w_q R_q(t) \tag{8.11.4}$$

In light of Equation (8.2.8b), individual reliability function $R_q(t)$ can be expressed as

where $$R_q(t) = 1 - F_q(t) = 1 - \frac{1}{(2\pi)^{1/2}} \int_{-\infty}^{[\ln(t)-\mu_{yq}]/\sigma_{yq}} \exp(-\tfrac{1}{2}\,\tau^2)\, d\tau\} \tag{8.11.5}$$

or $$R_q(t) = 1 - \Phi[\frac{\ln(t)-\mu_{yq}}{\sigma_{yq}}] \tag{8.11.6}$$

Note that Φ[ ] is the standard normal distribution function. The hazard rate of the finite lognormal mixture distribution is defined as the ratio of f(t) to R(t). Thus,

$$h(t) = \frac{f(t)}{R(t)} = [\sum_{q=1}^{Q} w_q f_q(t)] / R(t) = \sum_{q=1}^{Q} w_q h_q(t) \quad (8.11.7)$$

$$\text{where } h_q(t) \equiv \frac{f_q(t)}{R(t)} = \frac{\dfrac{1}{t\,\sigma_{yq}\,(2\pi)^{1/2}} \exp\{-\tfrac{1}{2}[\dfrac{\ln(t)-\mu_{yq}}{\sigma_{yq}}]^2\}}{1-\Phi[\dfrac{\ln(t)-\mu_{yq}}{\sigma_{yq}}]} \quad (8.11.8)$$

According to Equation (8.2.10), the related mean time to failure (MTTF) of the finite lognormal mixture distribution can be derived as

$$\text{MTTF} = \text{Mean}\,[t] = E[t] = \sum_{q=1}^{Q} w_q \exp[\mu_{yq} + \sigma_{yq}^2 / 2] \quad (8.11.9)$$

Given that means and random effects are incorporated into the model through MTTF as shown in the above equation, it is feasible to improve the product reliability with maximizing the MTTF.

8.12. Statistical Inference from Fisher Information

Fisher information is able to quantify the relationship between the response to the input working variables and model parameters by sensitivity analysis. The FIM of a finite mixture distribution function plays a significant role in the follow-up statistical inference by unraveling the related variances and covariances. The steps for conceptual realization of Fisher information are detailed in [Wolpert 2021].

Consider the Fisher information with parameter θ, denoted as I(θ), of a continuous random variable y sampled from a PDF f(y | θ). With any given PDF, it can be shown that

$$\int f(y \mid \theta)\, dy = 100\% \quad (8.12.1)$$

Assume that the orders of differentiations and integrations are exchangeable between variable y and parameter θ. Taking a partial differentiation of the above equation with respect to a model parameter θ yields

$$\frac{\partial[\int f(y \mid \theta)\, dy]}{\partial\theta} = 0 \quad (8.12.2)$$

$$\text{Thus, } \int \frac{\partial f(y \mid \theta) / \partial\theta}{f(y \mid \theta)} f(y \mid \theta)\, dy = 0 \quad (8.12.3)$$

This is formulated for a continuous random variable y. For a discrete distribution, just replace the integrals below with sums.

A score function is defined as the differential change in the PDF with respect to the variation of a model parameter divided by the PDF itself, i.e.,

$$\lambda'(y \mid \theta) \equiv \partial\{[\lambda(y \mid \theta)]\} / \partial\theta \equiv \frac{\partial\{[f(y \mid \theta)]\} / \partial\theta}{f(y \mid \theta)} \tag{8.12.4}$$

of which $\lambda'(t \mid \theta)$ is called the score function. Plugging the above equation into Equation (8.12.3) leads to

$$\int \lambda'(y \mid \theta)\, f(y \mid \theta)\, dy = 0 \tag{8.12.5}$$

which can be rewritten as

$$E_\theta[\lambda'(y \mid \theta)] = 0 \tag{8.12.6}$$

Differentiating (8.12.5) with respect to model parameter θ, using the product rule, lends

$$\partial\,[\int \lambda'(y \mid \theta)\, f(y \mid \theta)\, dy] / \partial\,\theta = \int \lambda''(y \mid \theta)\, f(y \mid \theta)\, dt + \int \lambda'(y \mid \theta)\, f'(y \mid \theta)\, dy = 0$$

$$\text{i.e.,} \quad \int \lambda''(y \mid \theta)\, f(y \mid \theta)\, dy + \int \lambda'(y \mid \theta) \frac{f'(y \mid \theta)}{f(y \mid \theta)} f(y \mid \theta)\, dy = 0 \tag{8.12.7}$$

$$E_\theta\{[\lambda''(y \mid \theta)]\} + E_\theta\, \{[\lambda'(y \mid \theta)]^2\} = 0$$

Fisher information, namely I(θ), is defined as the variance of the score function, i.e., the second term in the above equation.

$$I(\theta) \equiv E_\theta\{[\lambda'(y \mid \theta)]^2\} \tag{8.12.8}$$

$$\text{or} \quad I(\theta) = \int \{\frac{\partial[f(y \mid \theta)]}{\partial\theta}\}^2 f(y \mid \theta)\, dy \tag{8.12.9}$$

of which Fisher information I(θ) is an expectation of the score function taken over variable y. Substituting Equation (8.12.8) into Equation (8.12.7), one has [Frieden and Gatenby 2010]

$$I(\theta) = E_\theta\{[\lambda'(y \mid \theta)]^2\} = -E_\theta\{[\lambda''(y \mid \theta)]\} \tag{8.12.10}$$

Since y =ln(t) is also a continuous variable as long as variable t is a continuous variable, Equation (8.12.9) can be rewritten as

$$I(\theta) = \int (\frac{\partial\{\ln[f(t \mid \theta)]\}}{\partial\theta})^2 f(t \mid \theta)\, dt \tag{8.12.11}$$

The above equation fits the derivative of the log-likelihood with respect to parameter θ when applying the log-likelihood method.

When there are two parameters of interest, the information between these two parameters is formed as

$$I(\theta_1, \theta_2) = -E\left[\frac{\partial^2\{\ln[f(t \mid \theta_1 \text{ and } \theta_2)]\}}{\partial\theta_1\, \partial\theta_2}\right] \quad (8.12.12)$$

How to form a FIM is illustrated using Example 6.1.1, based on the normal distribution.

8.12.1. FIM for DOEs in Lognormal Distribution

When the fisher information is used for measuring a set of parameters with a certain amount of test data available in DOEs, it becomes a FIM that leads to how well can the parameters of interest be estimated. These parameters may include the statistical parameters for a specified statistical distribution and the intercept (constant) and coefficients of the polynomial equation that would fit the DOE data. With a series of response measurements over time instants with no replication, one may put these data as follows:

$$\{y_1, y_2, \ldots, y_Q\} = \{\ln(t_1), \ln(t_2), \ldots, \ln(t_Q)\} \quad (8.12.13)$$

where

$t_1, t_2, \ldots, t_Q$ are the response
$y_1, y_2, \ldots, y_Q$ are the $y_q = \ln(t_q)$, of which $q = 1, 2, 3, \ldots,$ and Q
Q is the total number of treatments, i.e., the total number of observations without replication

Consider finite mixture distribution f_Q, which is defined as the joint probability function of $y_1, y_2, \ldots, y_Q$,

$$f_Q(Y \mid \theta) = \prod_{q=1}^{Q} f(y_q \mid \theta) \quad (8.12.14)$$

The corresponding joint likelihood function is then

$$L_Q(Y \mid \theta) = \ln\left[\prod_{q=1}^{Q} f(y_q \mid \theta)\right] = \sum_{q=1}^{Q} \ln[f(y_q \mid \theta)] = \sum_{q=1}^{Q} L(y_q \mid \theta) \quad (8.12.15)$$

The derivatives of the joint likelihood function and individual likelihood are, respectively,

$$L_Q'(Y \mid \theta) \equiv \partial[L_Q(Y \mid \theta)] / \partial\theta$$
$$= \{\partial[f_Q(Y \mid \theta)] / \partial\theta\} / f_Q(Y \mid \theta) = f_Q'(Y \mid \theta) / f_Q(Y \mid \theta) \quad (8.12.16)$$

and $$L'(Y \mid \theta) \equiv \partial[L(Y \mid \theta)] / \partial\theta \quad (8.12.17)$$

According to Equation (8.12.6), the likelihood of the joint PDF is

i.e., $$E_\theta[L_Q'(Y \mid \theta)] = 0 \quad (8.12.18)$$

The Fisher information given in Equation (8.12.10) can then be expanded to accommodate generalized variance (including covariances and variances), involved in the joint PDF, namely, FIM, as

$$[I] = Var_\theta\{[L_Q'(Y \mid \theta)]^2\} = -E_\theta[L_Q''(Y \mid \theta)] \quad (8.12.19)$$

$$\text{Thus,} \quad |[I]| \equiv \det[I] = -E_\theta[\sum_{q=1}^{Q} L''(Y \mid \theta)] = -\sum_{q=1}^{Q} E_\theta[L''(Y \mid \theta)] = Q\, I(\theta) \quad (8.12.20)$$

According to Equation (8.12.12) that is derived for a single normal distribution function, and letting $|[I]| = \sigma_\theta^{-2}$ for the finite mixture distribution, the above equation leads to

$$\text{i.e.,} \quad \sigma_\theta^2 \approx [Q\, I(\theta)]^{-1} \quad (8.12.21)$$

where

σ^2 is the variance for one of its constituents, i.e., a single normal distribution

σ_θ^2 is the variance for the finite mixture distribution

When Equation (8.12.19) is expanded to cover multidimensional analysis, the FIM used for the MLE with a set of given test data, e.g., $\{y_1 = \ln(t_1), y_2 = \ln(t_3), y_2 = \ln(t_3), \ldots, y_Q = \ln(t_Q)\}$, can be practically related to the log-likelihood function using the following equation [Lehmann and Casella 1998, Lindsay and Yao 2012, Lin and Stefanopoulou 2015]:

$$[I] = -E[\frac{\partial^2 \ln[f(y \mid \{\theta\})]}{\partial\{\theta\}\, \partial\{\theta\}^T}] \quad (8.12.22)$$

$$\text{i.e.,} \quad [I] = -E[\begin{Bmatrix} \partial/\partial\theta_1 \\ \partial/\partial\theta_2 \\ \cdots \\ \partial/\partial\theta_{G-1} \\ \partial/\partial\theta_G \end{Bmatrix}_{G\times 1} \{\partial y/\partial\theta_1\ \partial y/\partial\theta_2 \ldots \partial y/\partial\theta_{G-1}\ \partial y/\partial\theta_G\}_{1\times G}] \quad (8.12.23)$$

where

y: $y \in \{y\} = \{y_1\ y_2 \ldots y_Q\} = \{\ln(t_1)\ \ln(t_2) \ldots \ln(t_Q)\}$, where $t_1, t_2, \ldots$ and t_Q are test data

$\{\theta\} = \{\theta_1\ \theta_2 \ldots \theta_G\}$ are the model parameters, e.g., mean μ of a normal distribution

G is the total number of model parameters of interest, i.e., unknowns to be estimated

The above equation yields the explicit FIM that can be written as

$$[I] = E[\begin{matrix} -\partial^2 y/\partial\theta_1^2 & -\partial^2 y/\partial\theta_1\partial\theta_2 & \cdots & -\partial^2 y/\partial\theta_1\partial\theta_G \\ -\partial^2 y/\partial\theta_2\partial\theta_1 & -\partial^2 y/\partial\theta_2^2 & \cdots & -\partial^2 y/\partial\theta_2\partial\theta_G \\ \cdots & \cdots & \cdots & \cdots \\ -\partial^2 y/\partial\theta_{G-1}\partial\theta_1 & -\partial^2 y/\partial\theta_{G-1}\partial\theta_2 & \cdots & -\partial^2 y/\partial\theta_{G-1}\partial\theta_G \\ -\partial^2 y/\partial\theta_G\partial\theta_1 & -\partial^2 y/\partial\theta_G\partial\theta_2 & \cdots & -\partial^2 y/\partial\theta_G^2 \end{matrix}] \quad (8.12.24)$$

For G model parameters of interest, [I] is a $G \times G$ symmetric matrix [Cordeiro and Andrade 2011]. Although FIM [I] is a function of {y} and {θ}, it is merely decided on the basis of test data $\{y_1, y_2, y_3, \ldots, y_Q\}$. Furthermore, FIM is not a function of a particular observation, as the random variable $y = \ln(t)$ has been averaged out, as given by Equation (8.12.11).

DOEs are balanced and orthogonal if the FIM [I] is diagonal and vice versa. A balanced orthogonal design usually implies that the coefficients have a minimum generalized variance as derived from multiple parameters. When a linear model is fit with a balanced orthogonal design, e.g., each one of 2^{K-P} and 3^{K-P}, the parameter estimates are uncorrelated and they are independent of one another in the model. The variance induced by one single random variable is σ^2 according to Equation (8.12.12). When the sample size Q is significantly large with every treatment having the same statistical distribution, the random error for estimating a certain parameter, namely σ_θ, is assumedly normal-distributed [Ly et al. 2017]. Thus, in light of Equation (8.12.21),

$$(\theta - \Theta) \sim N(0, \sigma_\theta^2) \approx N(0, [Q\ I(\theta)]^{-1}) \tag{8.12.25}$$

where

θ is the estimated parameter
Θ is the true value
σ_θ^2 is the sample variance

The variance of the estimation by an unbiased estimator in DOEs is exactly the inverse of FIM.

With an extension to a general case hereon for a multiparametric problem with random errors in the experiment, the generalized variance of unknown parameter θ_g being sought is bounded by the inverse of FIM [I] evaluated at the actual value of θ_g [Cramer 1999] as

$$-|[I(\theta_g)]|^{-1} \geq \mathrm{Var}([\theta_g]) \geq |[I(\theta_g)]|^{-1} \tag{8.12.26}$$

The above equation is called Cramer–Rao bounds. Diagonal elements of $[I]^{-1}$ prescribe the lower bound of the variance of θ_g, where g = 1, 2, …, and G. The Cramer–Rao bound is able to predict the best-case generalized variance based on the knowledge of the variance of the output measurement, denoted by σ_y^2, and the correlation between the output and the state/parameter estimates specified by the related partial derivatives.

8.12.2. Bounds on Estimated Lognormal Parameters by FIM

Each of the lognormal parameter (i.e., σ_y) and the parameters in the approximating equation, i.e., Equation (8.4.16), has been in the lognormal "domain" when they are estimated using the maximum-likelihood method. It is therefore asymptotically normal-distributed. It unveils that the distribution of each estimated parameter derived from the same population would be close to the normal distribution given that the sample size is statistically significant [Meeker et al. 1998]. If the sample size is small in practice, one has to apply the LR method instead of the FIM.

Assume that θ_g is the estimated parameter of population parameter Θ_g in a lognormal distribution, then $\ln(\theta_g)$ is the estimated parameter of $\ln(\Theta_g)$ in the domain of a normal distribution, as evidenced by Equation (8.4.4). With a confidence level C = 1 − α, which is assumed to provide two-sided confidence bounds on the estimated parameter, the inequality equation is given as follows:

$$\theta_g - z_{\alpha/2}\, \sigma_{\theta g} < \Theta_g < \theta_g + z_{1-\alpha/2}\, \sigma_{\theta g} \tag{8.12.27}$$

where

Θ_g is the population parameter, which can be anyone of σ_y, γ_0, γ_i, γ_{ij}, and γ_{ijm}
θ_g is the estimated parameter that refers to anyone of σ_y, γ_0, γ_i, γ_{ij}, and γ_{ijm}
g is the index of parameters, where g = 1, 2, 3, …, and G
G is the total number of parameters
$z_{\alpha/2}$, $z_{1-\alpha/2}$ is the lower and upper bounds of standard normal-distributed variable z
$\sigma_{\theta g}$ is the sample standard deviation of θ_g

The point of the FIM formalism is to predict how well the DOE will be able to constrain the model parameters. When applying FIM bounds for a function, one must determine the mean and variance of the statistical function (e.g., lognormal, Weibull, or exponential) of concern in advance. For a reliability function with multiple parameters, such as $\theta_1, \theta_2, \theta_3, \ldots$, and θ_G, the generalized variance, also called covariance matrix, is defined as a collection of all the variances and covariances of these parameters.

$$[Var] = \begin{bmatrix} \sigma^2_{\theta 1} & Cov(\theta_1, \theta_2) & \ldots & \ldots & Cov(\theta_1, \theta_G) \\ Cov(\theta_2, \theta_1) & \sigma^2_{\theta 2} & \ldots & \ldots & Cov(\theta_2, \theta_G) \\ \ldots & \ldots & \ldots & \ldots & \\ Cov(\theta_{G-1}, \theta_1) & Cov(\theta_{G-1}, \theta_2) & \ldots & \sigma^2_{0\,G-1} & Cov(\theta_{G-1}, \theta_G) \\ Cov(\theta_G, \theta_1) & Cov(\theta_G, \theta_2) & \ldots & \ldots & \sigma^2_{0G} \end{bmatrix} \quad (8.12.28)$$

where

[Var] is the sample variance–covariance matrix

$\sigma^2_{\theta g}$ is the sample variance of θ_g

$Cov(\theta_i, \theta_j)$ is the covariance of variables θ_i and θ_j

Sample standard deviation $\sigma_{\theta g}$, or sample variance $\sigma^2_{\theta g}$, in the above equation for estimating the bounds on each estimated lognormal parameter can be obtained by inverting the FIM. For example, let us consider likelihood function $L_{ln}(\sigma_y, \gamma_0, \gamma_i, \gamma_{ij}, \gamma_{ijm})$ with multiparameters $\sigma_y, \gamma_0, \gamma_i, \gamma_{ij}$, and γ_{ijm}, as described by Equation (8.4.19). According to Equation (8.12.24), the FIM that accommodates the logarithmic likelihood of transformed data can be calculated as [Cramer 1999] follows:

$$[I] = \begin{bmatrix} -\partial^2\{\ln L_{ln}()\}/\partial\theta_1^2 & -\partial^2\{\ln L_{ln}()\}/\partial\theta_1\partial\theta_2 & \ldots & -\partial^2\{\ln L_{ln}()\}/\partial\theta_1\partial\theta_G \\ -\partial^2\{\ln L_{ln}()\}/\partial\theta_2\partial\theta_1 & -\partial^2\{\ln L_{ln}()\}/\partial\theta_2^2 & \ldots & -\partial^2\{\ln L_{ln}()\}/\partial\theta_2\partial\theta_G \\ \ldots & \ldots & \ldots & \ldots \\ -\partial^2\{\ln L_{ln}()\}/\partial\theta_{G-1}\partial\theta_1 & -\partial^2\{\ln L_{ln}()\}/\partial\theta_{G-1}\partial\theta_2 & \ldots & -\partial^2\{\ln L_{ln}()\}/\partial\theta_{G-1}^2 \\ -\partial^2\{\ln L_{ln}()\}/\partial\theta_G\partial\theta_1 & -\partial^2\{\ln L_{ln}()\}/\partial\theta_G\partial\theta_2 & \ldots & -\partial^2\{\ln L_{ln}()\}/\partial\theta_G^2 \end{bmatrix} \quad (8.12.29)$$

where $\ln L_{ln}()$ is referring to $\ln[L_{ln}(\sigma_y, \gamma_0, \gamma_i, \gamma_{ij}, \gamma_{ijm})]$.

Then, the determination of variances and covariances of these parameters in the regression analysis can be accomplished via the FIM.

8.12.3. Bounds on Estimated Lognormal Parameter σ_y by FIM

Random error σ_y, a lognormal parameter, is to be estimated using Equations (8.4.20a, 8.4.20b)–(8.4.24). The two-sided confidence bound on estimated parameter σ_y is then given as follows:

$$\sigma_y - z_{\alpha/2}\,[var(\sigma_y)]^{1/2} < \Sigma_y < \sigma_y + z_{1-\alpha/2}\,[var(\sigma_y)]^{1/2} \quad (8.12.30)$$

of which Σ_y is the corresponding population parameter for the estimated σ_y. Here, assume that σ_y is positive. Since $y = \ln(t)$, sample random error σ_t is related to sample random error σ_y by

$$var(\sigma_y) = [\partial \ln(\sigma_t) / \partial\sigma_t]^2\, var(\sigma_t) = (1/\sigma_t)^2\, var(\sigma_t) \quad (8.12.31)$$

The confidence bounds on σ_y can then be determined using the above equation, where lognormal parameter σ_y is obtained from the MLEs using Equations (8.4.20a, 8.4.20b)–(8.4.24) and its standard deviation from the covariance matrix in Equation (8.12.28). Substituting Equation (8.12.31) into Equation (8.12.30) leads to

$$\ln(\sigma_t) - (z_{\alpha/2} / \sigma_t)\,[\mathrm{var}(\sigma_t)]^{1/2} < \ln(\Sigma_t) < \ln(\sigma_t) + (z_{\alpha/2} / \sigma_t)\,[\mathrm{var}(\sigma_t)]^{1/2} \quad (8.12.32)$$

$$\text{i.e.,}\quad \sigma_t \exp\{(z_{\alpha/2} / \sigma_t)\,[\mathrm{var}(\sigma_t)]^{1/2}\} < \Sigma_t < \sigma_t \exp\{(z_{\alpha/2} / \sigma_t)\,[\mathrm{var}(\sigma_t)]^{1/2}\} \quad (8.12.33)$$

of which subscript α is the specified p-value for both sides.

8.12.4. Confidence Bounds on Function $\ln L_{ln}()$

The variance of function $\ln L_{ln}()$ can be estimated using the equation for the variance, after the values for the variance and covariance of the parameters are estimated from the FIM equation. Once they have been obtained, the approximate confidence bounds on the function are given as

$$E[\ln L_{ln}()] - z_{\alpha/2}\,[\mathrm{var}(\ln L_{ln}())]^{1/2} < [\ln L_{ln}()]_{population} < E[\ln L_{ln}()] + z_{1-\alpha/2}\,[\mathrm{var}(\ln L_{ln}())]^{1/2} \quad (8.12.34)$$

Variate $z_{\alpha/2}$ is negative and variate $z_{1-\alpha/2}$ is positive in general, and both fall on the abscissa of the standard normal distribution.

8.12.5. Bounds on Estimated Functional Coefficients and Parameters

Consider the example data for the life prediction with two factors (Table 8.7.1). According to Equation (8.7.18), there are only two statistically significant parameters, i.e., intercept γ_0 and coefficient γ_2.

$$y = \gamma_0 + \gamma_0\, x_2 + \sigma_y = 3.6085 + 0.3512\, x_2 + \text{random error} \quad (8.12.35)$$

Bounds on these two effects and the related random error σ_y can be estimated using the FIM, of which matrix elements can be calculated as

$$[I] = -[\begin{matrix} \partial^2[\ln L_{ln}()]/\partial\gamma_0^2 & \partial^2[\ln L_{ln}()]/\partial\gamma_0\partial\gamma_2 & \partial^2[\ln L_{ln}()]/\partial\gamma_0\partial\sigma_y \\ \partial^2[\ln L_{ln}()]/\partial\gamma_2\partial\gamma_0 & \partial^2[\ln L_{ln}()]/\partial\gamma_2^2 & \partial^2[\ln L_{ln}()]/\partial\gamma_2\partial\sigma_y \\ \partial^2[\ln L_{ln}()]/\partial\sigma_y\partial\gamma_0 & \partial^2[\ln L_{ln}()]/\partial\sigma_y\partial\gamma_2 & \partial^2[\ln L_{ln}()]/\partial\sigma_y^2 \end{matrix}] \quad (8.12.36)$$

which is an expansion of Equation (8.12.29) to include random error σ_y.

By the same token of derivations of Equations (8.7.10) and (8.7.12), one has

$$-\partial^2[\ln L_{ln}()] / \partial\gamma_0^2 = 4 / (\sigma_y)^2 = 2134$$

$$-\partial^2[\ln L_{ln}()] / \partial\gamma_0\partial\gamma_2 = -\partial^2[\ln L_{ln}()] / \partial\gamma_2\partial\gamma_0 = (\Sigma\, x_{q2}) / (\sigma_y)^2 = 0$$

$$-\partial^2[\ln L_{ln}()] / \partial\gamma_0\partial\sigma_y = -\partial^2[\ln L_{ln}()] / \partial\sigma_y\partial\gamma_0 = 2\,\{\Sigma\,[\ln(t_q) - \mu_{yq})]\} / (\sigma_y)^3 \approx 0$$

$$-\partial^2[\ln L_{ln}()] / \partial\gamma_2^2 = (\Sigma\, x_{q2}^2) / (\sigma_y)^2 = 2136$$

$$-\partial^2[\ln L_{ln}()] / \partial\gamma_2\partial\sigma_y = -\partial^2[\ln L_{ln}()] / \partial\sigma_y\partial\gamma_2 = (2\,\Sigma\, x_{q2}\,[\ln(t_q) - \mu_{yq}]) / (\sigma_y)^3 \approx 0$$

$$\text{and}\quad -\partial^2[\ln L_{ln}()] / \partial\sigma_y^2 = 4 / (\sigma_y)^2 - 3\,\{\Sigma\,[\ln(t_q) - \mu_{yq}]^2\} / (\sigma_y)^4 = 4272$$

Substituting the above equations into Equation (8.12.36) yields the corresponding FIM [I]:

$$[I] = -\begin{bmatrix} 2134 & 0 & 0 \\ 0 & 2136 & 0 \\ 0 & 0 & 4272 \end{bmatrix}$$

Inverting the above FIM leads to the generalized variance matrix:

$$[Var] = \begin{bmatrix} \sigma^2_{\gamma 0} & Cov(\gamma_0, \gamma_2) & Cov(\gamma_0, \sigma_y) \\ Cov(\gamma_2, \gamma_0) & \sigma^2_{\gamma 2} & Cov(\gamma_2, \sigma_y) \\ Cov(\sigma_y, \gamma_0) & Cov(\sigma_y, \gamma_2) & \sigma^2_{\sigma y} \end{bmatrix} = [I]^{-1}$$

$$= [I]^{-1} = \begin{bmatrix} 0.000469 & 0 & 0 \\ 0 & 0.000468 & 0 \\ 0 & 0 & 0.000234 \end{bmatrix} \tag{8.12.37}$$

Thus, $\sigma^2_{\gamma 0} = 4.69 \times 10^{-4}$

$\sigma^2_{\gamma 2} = 4.68 \times 10^{-4}$

and $\sigma^2_{\sigma y} = 2.34 \times 10^{-4}$ (8.12.38)

i.e., $\sigma_{\gamma 0} = 0.02165$

$\sigma_{\gamma 2} = 0.02164$

and $\sigma_{\sigma y} = 0.0153$

The confidence intervals for γ_0 and γ_2 at $\alpha = 10\%$ (i.e., C = 90%) are, respectively,

$$3.6085 + z_{5\%}\,\sigma_{\gamma 0} < \gamma_0 < 3.6085 + z_{95\%}\,\sigma_{\gamma 0} \tag{8.12.39}$$

and $$0.3512 + z_{5\%}\,\sigma_{\gamma 2} < \gamma_2 < 0.3512 + z_{95\%}\,\sigma_{\gamma 2} \tag{8.12.40}$$

which follow some normal distributions. Parameter t (lifetime) is always positive and the operations for fitting lognormal function y = ln(t), which ranges from $-\infty$ to ∞, can be approximately treated as being normally distributed. Given that standard normal-distributed variable $z_{5\%} = -1.645$ and $z_{95\%} = 1.645$ (as read from the standard normal distribution chart), the inequality relationships for γ_0 and γ_2 in Equations (8.12.35) and (8.12.36) become

$$3.6085 + 1.645\,(0.02165) < \gamma_0 < 3.6085 + 1.645\,(0.02165)$$

and $$0.3512 + 1.645\,(0.02164) < \gamma_2 < 0.3512 + 1.645\,(0.02164) \tag{8.12.41}$$

Thus, $3.573 < \gamma_o < 3.644$

and $$0.3156 < \gamma_2 < 0.387 \tag{8.12.42}$$

8.12.6. Bounds on Estimated Model Parameter σ_y for Ball Grid Array Packaging

If random error σ_y of variable y can take only a positive value, it is assumed that the $\ln(\sigma_y)$ follows the normal distribution [Meeker et al. 1998]. The bounds on population standard deviation Σ_y for sample standard error σ_y can be obtained from

$$\ln(\sigma_y) + z_{5\%} \ln(\sigma_y) < \ln(\Sigma_y) < \ln(\sigma_y) + z_{95\%} \ln(\sigma_y) \tag{8.12.43}$$

$$\text{Then,}\quad \exp[\ln(\sigma_y) + (z_{5\%} / \sigma_y)\, \sigma_{\sigma y}] < \Sigma_y < \exp[\ln(\sigma_y) + z_{95\%} / \sigma_y)\, \sigma_{\sigma y}] \tag{8.12.44}$$

$$\text{i.e.,}\quad \sigma_y + \exp[(z_{5\%} / \sigma_y)\, \sigma_{\sigma y}] < \Sigma_y < \sigma_y + \exp[(z_{95\%} / \sigma_y)\, \sigma_{\sigma y}] \tag{8.12.45}$$

Plugging in the respective numerical data, i.e., $z_{5\%} = -1.645$, $z_{95\%} = 1.645$, $\sigma_y = 0.0433$, and $\sigma_{\sigma y} = 0.0153$ (from Equation 8.12.38), into Equation (8.12.44) yields

$$0.0433 \exp\left[\left(\frac{-1.645}{0.0433}\right) 0.0153\right] < \Sigma_y < 0.0433 \exp\left[\left(\frac{1.645}{0.0433}\right) 0.0153\right]$$

$$\text{i.e.,}\quad 0.0242 < \Sigma_y < 0.0774$$

This is the confidence interval for the estimated random error σ_y.

References

Anderson, T. W. and Darling, D. A. (1952), "Asymptotic Theory of Certain 'Goodness-of-Fit' Criteria Based on Stochastic Processes," *Annals of Mathematical Statistics*, 23, pp. 193-212.

Bertsimas, D. et al. (2015), "The Power of Optimization Over Randomization in Designing Experiments Involving Small Samples," *Operations Research*, 63(4), pp. 868-876.

Blischke, W. and Murthy, D. (2000), *Reliability. Modeling, Prediction and Optimization*, John Wiley & Sons Inc., New York.

Box, G. E. P. and Cox D. R. (1964), "An Analysis of Transformation," *Journal of the Royal Statistical Society, Series B (Methodological)*, 26(2), pp. 211-252.

Brezina, J. (1963), "Kapteyn's Transformation of Grain Size Distribution," *Journal of Sedimentary Petrology*, 33(4), pp. 931-937.

Byrne, D. M. and Taguchi, S. (1987), "The Taguchi Approach to Parameter Design," *Quality Progress*, pp. 19-26.

Chortirat, T. et al. (2011), "Comparison of Four Data Transformation Methods for Weibull Distributed Data," *Kasetsart Journal (Natural Science)*, 18(45), pp. 366-383.

Cordeiro, G. M. and Andrade, M. G. (2011), "Transformed Symmetric Models," *International Journal of Statistical Modelling*, 11(4), pp. 1-13.

Cramer, H. (1999), *Mathematical Methods of Statistics*, Princeton University Press, 575 pages.

Das, R. N. and Park, J. S. (2012), "Discrepancy in Regression Estimates between Log-Normal and Gamma: Some Case Studies," *Journal of Applied Statistics*, 39(1), pp. 97-111.

Das, R. N. et al. (2017), "Correlated Log-normal Composite Error Models for Different Scientific Domains," *Model Assisted Statistics and Applications*, 12, pp. 39-53.

Dodge, H. F. (1947), "Sampling Plans for Continuous Production," *Industrial Quality Control*, 14, pp. 5-9.

Frieden, B. R. and Gatenby, R. A. (2010), *Exploratory Data Analysis Using FI*, Springer, New York; ISBN: 978-1846285066.

Guo, H. and Mettas, A. (2010), "Design of Experiments and Data Analysis," *IEEE Reliability and Maintainability Symposium*, San Jose, January 2010.

Hamada, M. (1995), "Using Statistically Designed Experiments to Improve Reliability and to Achieve Robust Reliability," *IEEE Transactions on Reliability*, 44, pp. 206-215.

Hamada, M. and Tse, S. K. (1992), "On Estimability Problems in Industrial Applications with Censored Data," *Statistica Sinica*, 2, pp. 381-391.

Hernández, F. and Johnson, R. A. (1980), "The Large-Sample Behavior of Transformations to Normality," *Journal of the American Statistical Association*, 75(372), pp. 855-861.

Holgate, P. (1989), "The Lognormal Characteristic Function," *Communications in Statistics - Theory and Methods* 18(12), pp. 4539-4548.

Hunter, J. S. (1987), "Signal-to-Noise Ratio Debated," *Quality Progress*, 20(5), pp. 7-9.

Johnson, N. L. (1949), "Systems of Frequency Curves Generated by Methods of Translation," *Biometrika*, 36(1-2), pp. 149-176.

Jones, M. C. and Pewsey, A. (2009), "Sinh-arcsinh Distributions," *Biometrika*, 96(4), pp. 761-780.

Keene, O. N. (1995), "The Log Transformation Is Special," *Statistics in Medicine*, 14, pp. 811-819.

Koscielniak, P. and Parczewski, A. (1983), "Nonlinear Transformation of Factors in the Design of Experiments," *Analytica Chimica Acta*, 163, pp. 103-109.

Lahoud, N. et al. (2013), "Electrical Aging of the Insulation of Low Voltage Machines: Model Definition and Test with the Design of Experiments," *IEEE Transactions on Industrial Electronics*, 60, pp. 4147-4155.

Lehmann, E. L. and Casella, G. (1998), *Theory of Point Estimation*, 2nd Edition, Springer, New York.

Lin, X. and Stefanopoulou, A. G. (2015), "Analytic Bound on Accuracy of Battery State and Parameter Estimation," *Journal of the Electrochemical Society*, 162(9), pp. A1879-A1891.

Lindsay, B. G. and Yao, W. (2012), "Fisher Information Matrix: A Tool for Dimension Reduction, Projection Pursuit, Independent Component Analysis, and More," *The Canadian Journal of Statistics*, 40(4), pp. 712-730.

Liu, B. et al. (2013), "Lifetime Prediction and Failure Analysis of IGBT Module Based on Accelerated Lifetime Test," *Jiangsu Daxue Xuebao (Ziran Kexue Ban)/Journal of Jiangsu University*, 34(5).

Ly, A. et al. (2017), "A Tutorial on Fisher Information," www.alexander-ly.com/; https://jasp-stats.org/.

Makuch, R.W., Freeman, D. H., and Johnson, M. F. (1979), "Justification for the Lognormal Distribution as a Model for Blood Pressure," *Journal of Chronic Diseases*, 32(3), pp. 245-250.

Marsaglia, G. (2004), "Evaluating the Anderson-Darling Distribution," *Journal of Statistical Software*, 9(2), pp. 730-737.

Meeker, W. Q. (1984), "A Comparison of Accelerated Life Test Plans for the Weibull and Lognormal Distributions," *Technometrics*, 26, pp. 157-172.

Meeker, W. Q., Escobar, L. A., and Pascual, F. G. (1998), *Statistical Methods for Reliability Data*, 2nd Edition, John Wiley & Sons, New York, 704 pages; ISBN: 978-1-118-11545-9.

Montgomery, D. C. (2019), *Design and Analysis of Experiments*, 10th Edition, John Wiley & Sons, Inc., Hoboken, 688 pages; ISBN: 978-1-119-49244-3.

Nelson, W. and Kielpinski, T. J. (1976), "Theory for Optimized Censored Accelerated Tests for Normal and Lognormal Life Distributions," *Technometrics*, 18, pp. 105-114.

Neyman, J. and Pearson, E. S. (1933), "IX. On the Problem of the Most Efficient Tests of Statistical Hypotheses," *Philosophical Transactions of the Royal Society of London, Series A*, 231(694-706), pp. 289-337.

Osborne, J. W. (2010), "Improving your Data Transformation: Applying Box-Cox Transformation," *Practical Assessment, Research & Evaluation*, 15(12), pp. 1-7.

Phadke, M. S. (1989), *Quality Engineering Using Robust Design*, Prentice Hall, Englewood Cliffs.

Reliawiki (2020), "Reliability DOE for Life Tests," Retrieved August 2, 2020, https://reliawiki.org/index.php/Reliability_DOE_for_Life_Tests.

Ross, P. J. (2005), *Taguchi Techniques for Quality Engineering*, McGraw-Hill, New York.

Roy, R. K. (2001), *Design of Experiments Using the Taguchi Approach: 16 Steps to Product and Process Improvement*, John Wiley & Sons, New York.

Schmee, J. and Hahn, G. J. (1979), "A Simple Method for Regression Analysis with Censored Data," *Technometrics*, 21, pp. 417-434.

Scholz, F. W. (1981), "On the Uniqueness of Roots of the Likelihood Equations," Technical Report No. 14, Department of Statistics, University of Washington, Seattle, WA.

Silvestri, S. et al. (2008), "Statistical Analysis towards the Identification of Accurate Probability Distribution Models for the Compressive Strength of Concrete," *The 14th World Conference on Earthquake Engineering*, Beijing, China, October 12–17, 2008.

Steiner, S. H. and MacKay, R. J. (2005), "*Statistical Engineering: An Algorithm for Reducing Variation in Manufacturing Processes*," ASQ Quality Press, Milwaukee, WI.

Stovig, J. G. (2014), "Censored Weibull Distributed Data in Experimental Design," MS thesis, Norwegian University of Science and Technology, Trondheim.

Taguchi, G. (1987), *System of Experimental Design*, UNIPUB, Kraus International Publications, New York.

Taylor, W. A. (1990), *Optimization & Variation Reduction in Quality*, McGraw-Hill, New York.

Tsui, K. L. and Li, A. (1994), "Analysis of Smaller-and Larger-the Better Robust Design Experiments," *Int'l Journal of Industrial Engineering*, 1, pp. 193-202.

Vander Wiel, S. A. and Meeker, W. Q. (1990), "Accuracy of Approximate Confidence Bounds Using Censored Weibull Regression Data from Accelerated Life Tests," *IEEE Transactions on Reliability*, 39(3), pp. 346-351.

Vining, G. G., Freeman, L. J., and Kensler, J. L. K. (2015), "An Overview of Designing Experiments for Reliability Data," in *Frontiers in Statistical Quality Control 11*, Springer, Cham, pp. 321-326; ISBN: 978-3319355450.

Wang, C. J. (1991), "Sample Size Determination of Bogey Tests without Failures," *Quality and Reliability Engineering International*, 7(1), pp. 35-38.

Wolpert, R. L. (Last edited: May 11, 2021), "Fisher Information & Efficiency," Department of Statistical Science, Duke University, Durham, NC.

Wu, C. F. J. and Hamada, M. S. (2009), *Experiments: Planning, Analysis and Parameter Design Optimization*, 2nd Edition, Wiley, New York; ISBN: 978-0471699460.

Yang, T. and Pan, R. (2013), "A Novel Approach to Optimal Accelerated Life Test Planning with Interval Censoring," *IEEE Transactions on Reliability*, 62(2), pp. 527-536.

Yang, W. H. and Tarng, Y. S. (1998), "Design Optimization of Cutting Parameters for Turning Operations Based on the Taguchi Method," *Journal of Materials Processing Technology*, 8, pp. 122-129.

Yu, H. F. and Chiao, C. H. (2002), "An Optimal Designed Degradation Experiment for Reliability Improvement," *IEEE Transactions on Reliability*, 51(4), pp. 427-433.

Zhang, Z. H. (2000), "Moment Estimation of Weibull Distribution under Type I Random Censorship," *Chinese Journal of Engineering Mathematics*, 17(3), pp. 43-49.

Problems

P8.1: Resumed from P4.6, the formula for the maximum deflection of a centrally loaded simply supported beam with a rectangular cross section is given as

$$\delta = \frac{1}{4\,E}\left(\frac{D\,C^3}{B\,A^3}\right)$$

where E is the Young's modulus of the material. Fit a model of this form to the data from Problem P4.6. Compare the performance of the two models (P4.6 and P8.1). Why do some of the exponents of variables deviate from those indicated by the formula? Given the implications of the theoretical formula for deflection, how would one design a new experiment to study this process?

9

Weibull Reliability and DOEs

Reliability DOEs is the DOEs involved with reliability goals, such as rubber cure and fatigue life cycles. The Weibull distribution is able to model hazard functions that are decreasing, constant, or increasing, allowing it to describe any phase of a component's fatigue life. Acceptance sampling is an inspection procedure that is then provided in the test plan and used to determine whether to accept or reject a specific quantity of products. DOE-W is an important branch of statistical life prediction under reliability DOE discipline. The exponential distribution is a special case of the Weibull distribution.

9.1. Product Reliability as Measured by Noncompliance

Reliability DOEs are built upon creating proper stochastic models based on reliability statistics in combination with physical laws, which provide a system-level estimate of the output engineering parameters. When the product life data, such as those obtained from mechanical failure tests or material degradation observations, are suspended or interval data are identified because of the nature of experiments, the transformation schemes followed by ANOVA are no longer statistically proper to be applied directly to the "adjusted" data [Rigdon et al. 2022]. Modeling based on standard DOE-t and DOE-F solution procedures assumes the following about the test data:

(a) Data are normally distributed.

(b) Data have constant variances across all treatments.

If data transformations (Chapter 8, Section 8.9) cannot provide the answer for resolving the above two requirements simultaneously, DOE-W or DOE-LN using the maximum-likelihood method is often applied to characterizing such product life data, because of its capability in modeling skewed data and versatility in fitting various mechanical failure modes and material degradation.

9.1.1. Hazard Rate

Reliability statistics are the foundation for clarifying the functional requirements of products and assessing long-term product performance, as applied in the customer's service environment. Product reliability is generally measured by noncompliance. Consider the fatigue life testing of a structural component. Let the cumulative probability of damage done to the structural component be represented as function F(t), of which variable t is an independent random variable and $t \geq 0$. For example, variable t is the time for a fatigue test. The conversion rate is defined as the failure rate divided by the balance of the survivor portion.

$$h(t) = \frac{dF(t) / dt}{R(t)} = \frac{f(t)}{R(t)} \tag{9.1.1}$$

where

- h(t) is the hazard rate or conversion rate
- F(t) is the CDF of failure at time t, also called unreliability
- R(t) is the CDF of success at time t, also called reliability
- f(t) is the PDF, also called failure rate

Conversion rate h(t) is also called the hazard rate when applied to a failure mode. Statistically, the hazard rate is also called the instantaneous failure rate, which is different from the failure rate f(t).

9.1.2. Reliability and Unreliability

The denominator on the right side of Equation (9.1.1) is a cumulative probability, by which the structural component may survive the testing. When applied to the study on a failure mode, it is called reliability.

$$R(t) = 1 - F(t) \tag{9.1.2}$$

Thus, reliability for a non-repairable item can be defined as the probability that an item will fulfill a prescribed function without failure under given conditions for a stated period of time. The accumulated failure, F(t), is also called unreliability. The numerical value of either reliability or unreliability is expressed as a probability ranging from 0 to 1 and has no unit. Reliability is defined over a very long time period and using a very large sample size, or infinite in theory. However, the practice in the real world is somehow biased. The unreliability of a system at time t, F(t), is simply the number of systems that failed up to time t divided by the total number of samples tested. For example, 22 drive shafts have gone through a 1000-h fatigue test and one broke. Then, F(t) = 1/22 = 4.545%.

Substituting Equation (9.1.2) into Equation (9.1.1) yields

$$h(t) = \frac{dF(t) / dt}{1 - F(t)}$$

which leads to the CDF of failure

$$F(t) = 1 - \exp[-\int_0^t h(\tau)\, d\tau] = 1 - \exp[-H(t)] \tag{9.1.3}$$

where H(t) is the cumulative hazard function and CDF of the hazard rate (or conversion rate).

$$H(t) = \int_0^t h(\tau)\, d\tau \tag{9.1.4}$$

Nevertheless, the physical meaning of the conversion rate h(t) depends on the application. For example, when Equations (9.1.1)–(9.1.3) are applied to tire curing in a mold, h(t) is the instantaneous success rate (instantaneous curing rate) for a chemical reaction. Consequently, F(t) becomes the cumulative probability of successful curing and R(t) denotes the cumulative probability of unsuccessful curing.

Each statistical distribution function related to reliability can be derived using its corresponding h(t) function. Weibull distribution function, lognormal distribution function, and normal distribution function have most fits to mechanical engineering and structural applications. The ultimate tensile strength of structural material usually meets the Weibull distribution, while automotive tire wear fits into a lognormal distribution. Note that the exponential distribution function is a special case of the Weibull distribution function.

9.2. Product Reliability Growth Based on the Weibull Distributions

In terms of conversion rate, i.e., h(t) in Equation (9.1.1), the three-parameter Weibull distribution function is defined as [Weibull 1939, 1951, 1967]

$$h(t) = \left(\frac{\beta}{\eta}\right)\left(\frac{t - t_0}{\eta}\right)^{(\beta - 1)} \tag{9.2.1a}$$

where

h(t) is the PDF of conversion, namely, conversion rate or hazard rate

t_0 is the location factor, also called guaranteed value, setting a limit that $(t - t_0) \geq 0$

η is the scale factor or characteristic life, at which 63.212% of the population will fail

β is the shape parameter that has the same unit as t

The related CDF of conversion, also called the cumulative hazard function, is then

$$H(t) = \int_0^t h(\tau)\, d\tau = \frac{(t - t_0)^{\beta}}{\eta} \tag{9.2.1b}$$

It can be understood that the cumulative hazard function H(t) refers to the change in status (conversion) with respect to variable t, raised with power β while measured by the scale η.

Substituting Equation (9.2.1b) into Equation (9.1.3), one obtains the three-parameter Weibull CDF as

$$F(t) = 1 - \exp\left[-\left(\frac{t - t_0}{\eta}\right)^{\beta}\right] \tag{9.2.2}$$

of which F(t) is the CDF for the failed fraction. It is named after Swedish mathematician Waloddi Weibull, who addressed the statistics in detail in 1951, but it might have been first identified by French mathematician Maurice Fréchet in 1927.

Taking a natural logarithmic transformation of $1 - F(t) = \exp\{-[(t - t_0)/\eta)]^{\beta}\}$, one has

$$\ln\left\{\ln\left[\frac{1}{1 - F(t)}\right]\right\} = \beta \ln(t - t_0) + \ln\left(\frac{1}{\eta}\right)$$

$$\text{or} \quad \ln\{-\ln[1 - F(t)]\} = \beta \ln(t - t_0) + \ln(1/\eta) \tag{9.2.3}$$

$$\text{i.e.,} \quad \ln\{-\ln[R(t)]\} = \beta \ln(t - t_0) + \ln(1/\eta)$$

which results in the form of a linear relationship on the logarithmic scale. These three parameters can be estimated using the maximum-likelihood method, percentile estimator, method of moments, or linear regression model. A schematic Weibull CDF plot of a set of sample data is exhibited in Figure 9.2.1. Note that $\eta > 0$, $\beta > 0$, and $t \geq t_0$. Shape parameter β is the slope in the ln-ln scale, while scale factor η occurs at the moment when the cumulative probability of the population is at 63.212% of the horizontal axis, i.e., axis ln(t).

FIGURE 9.2.1 Cumulative failure probability (%) of a three-parameter Weibull distribution.

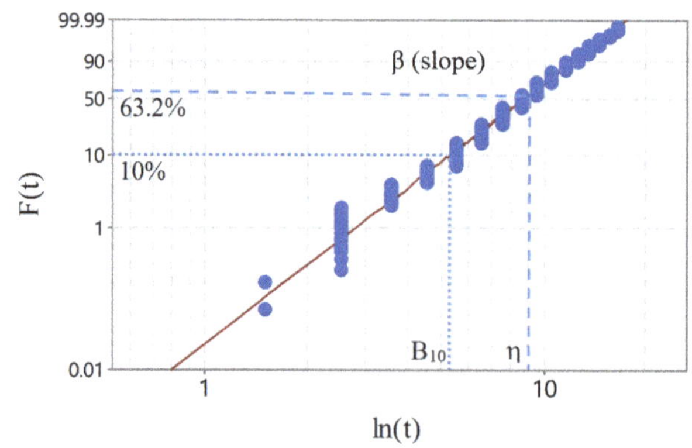

The reliability function, denoted by R(t), of the three-parameter Weibull distribution is thus

$$R(t) = 1 - F(t) = \exp\left[-\left(\frac{t - t_0}{\eta}\right)^{\beta}\right] \tag{9.2.4}$$

Since R(t) is reliability, F(t) is called unreliability. The Weibull PDF can be determined by taking the time derivative of F(t) as

$$f(t) = \frac{dF(t)}{dt} = \frac{\beta}{\eta}\left(\frac{t-t_0}{\eta}\right)^{(\beta-1)} \exp\left[-\left(\frac{t-t_0}{\eta}\right)^{\beta}\right] \tag{9.2.5}$$

Ever since, the Weibull distribution is the most popular distribution for modeling time-to-event (e.g., time-to-failure) data.

Reliability predictions are important for failure prevention. The Weibull distribution is able to model both mechanical and electric devices with a decreasing failure rate, a constant failure rate, or an increasing failure rate. This versatility of the Weibull distribution lends itself to a wide range of applications, and it is one of the most widely used lifetime distributions in reliability engineering.

9.2.1. Shape Factor β

Parameter β is called the shape factor (or shape parameter) as it depicts the shape of the Weibull cumulative failure probability curve as demonstrated in Figure 9.2.2. It is the slope of the curve on a Weibull cumulative failure probability paper as exhibited in Figure 9.2.1 and can be estimated in such a manner. The cumulative failure probability plot is particularly useful, because it not only allows for a simple check of whether a linear fit matches the data, but can also indicate other potential solutions when the fit is not linear. If the distribution does not provide an accepted linear fit, the Weibull plot may reveal a more appropriate distribution or suggest a better interpretation of the data.

The influence of the shape factor on unreliability (cumulative failure probability) is demonstrated in Figure 9.2.3. In consideration of the useful life of a mechanical or electric component, some statistical inferences from the product durability-based Weibull distribution function can be drawn as follows:

(A) $\beta < 1$: It shows infancy mortality. The product is extremely dangerous if it is a durable product such as a ground vehicle or an airplane. Problems are because of special causes such as strength deficiency (e.g., over-stressed and inferior material) or production defects (e.g., mal-assembly).

(B) $\beta = 1$: It shows random failures. The Weibull distribution function becomes an exponential distribution function statistically. The system has a constant instantaneous conversion rate as time increases. The failure modes may be maintenance errors, such as human errors or failure because of nature (e.g., lightning).

(C) $\beta = 2$: The Weibull PDF becomes the Rayleigh distribution. The failure rate grows linearly with respect to time.

(D) $1 < \beta < 4$: It turns out to be a normal or near-normal wear-out. For example, a component fails because of fatigue, corrosion, and erosion.

(E) $\beta \approx 3.4$: Data that fit into a normal distribution also generate good Weibull plots, as presented in Figure 9.2.2(a).

(F) $4 > \beta$: It turns out to be an aged wear-out. Potentially, wear, corrosion, erosion, and creep are causes.

(G) $\beta = \infty$: It is a perfect product as designed, manufactured, and used.

FIGURE 9.2.2 Weibull PDFs with varying shape factors.

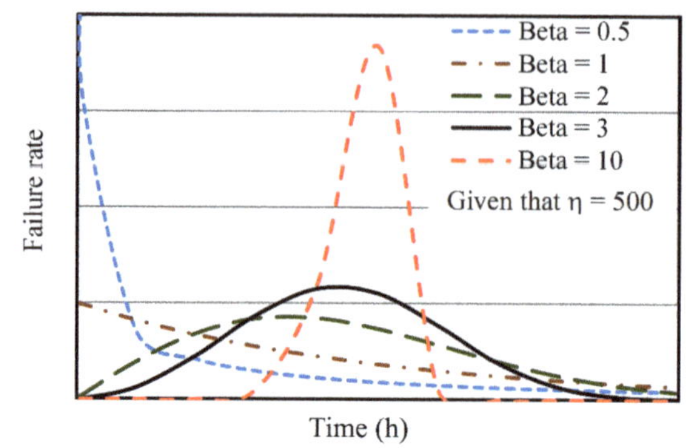

FIGURE 9.2.3 Influence of shape factor on a Weibull failure distribution (unreliability).

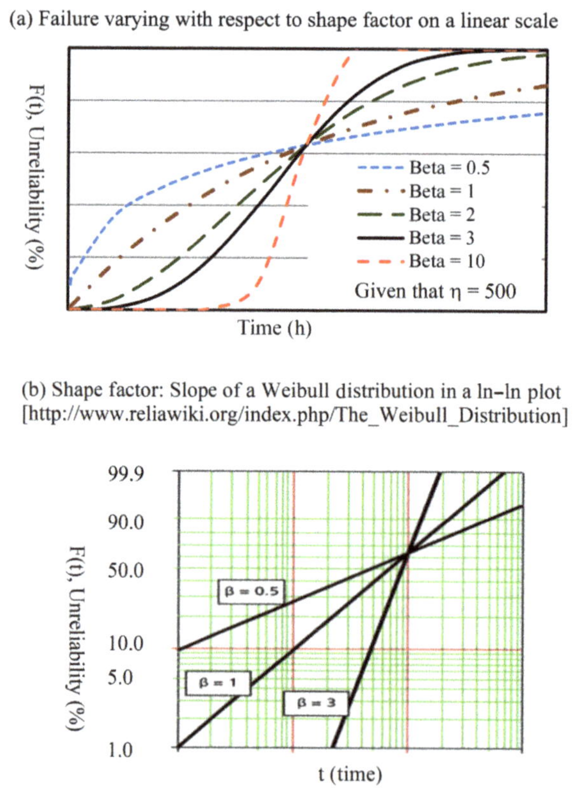

9.2.2. Scale Factor η

Parameter η is called the scale factor in the Weibull reliability relationship because it scales the value of variable t. As indicated in Figure 9.2.4, when the area under a PDF curve is a constant value of one, the "peak" of the PDF curve decreases with increasing η with β being held constant. It means that the increase in η value has the effect of stretching or contracting the failure distribution along the variable t-axis. When t = β, Equation (9.2.2) reduces to

$$F(t) = 1 - \exp[-(t/\eta)^{\beta}] = 1 - \exp(1) = 63.212\%$$

Thus, scale factor η is also interpreted as the quantile, at which 63.212% of the units fail, as shown in Figure 9.2.1. It determines how the hazard rate changes over variable t.

FIGURE 9.2.4 Spread-up of the PDF with increasing scale factor.

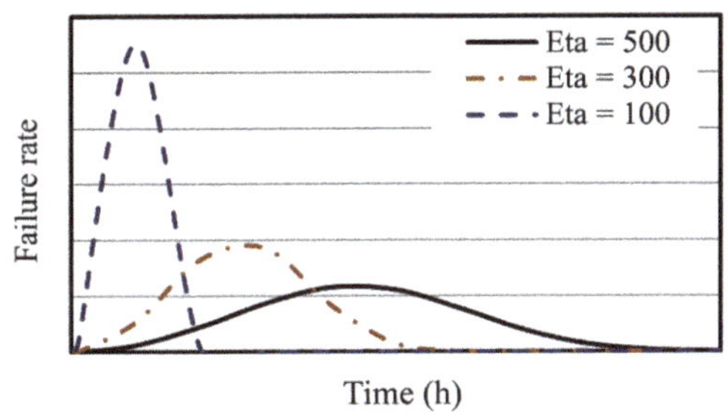

In the event of no failures, one technique is to add a failure at the end. Then, the estimates reflect a type of lower bound on the η value, rather than a real estimate.

9.2.3. Guaranteed Value t_0

Parameter t_0 is the guaranteed value, also called the locator. In general, $-\infty < t_0 < \infty$. By definition, there is a zero probability of failure for $t < t_0$. It means that all data must be greater than the threshold. Statistically speaking, the guaranteed value can be negative and it implies that units failed prior to testing, although this is unusual.

The guaranteed value cannot be zero ($t_0 \neq 0$) for durable mechanical components such as bearings, which do not fail instantaneously. It is also called threshold parameters, which refers to the distribution's shift away from 0. A negative threshold means that the distribution shifts to the left of 0, and a positive threshold means that the distribution shifts to the right of 0.

EXAMPLE 9.2.1

Following the steps of the bootstrap technique, one is able to finalize 500 data for evaluating the ultimate tensile strength of 2000 bolts in the same batch using the micro-indentation equipment. When rounded to a whole number with an interval of 5 MPa apart and expressed as (strength, number of bolts), they are classified into 16 groups (from **Table 5.6.1**) as follows:

(590, 1), (595, 6), (600, 9), (605, 21), (610, 41), (615, 54), (620, 69), (625, 76), (630, 69), (635, 65), (640, 41), (645, 25), (650, 14), (655, 6), (660, 2), (665, 1)

Should it fit a three-parameter Weibull distribution, please identify its shape factor, scale factor, and guaranteed value.

Solution:

The first step is to identify the statistical distribution of the given data set. Leaving the calculations to Minitab (Stat ➔ Reliability/Survival ➔ Distribution ID Plot (Right Censoring) ➔ Distribution ID Plot), one is able to get the following information.

Distribution	AD statistic (adjusted)
Weibull	7.633
Lognormal	3.123
Exponential	219.737
Log-logistic	3.673
Three-parameter Weibull	3.049
Three-parameter lognormal	3.115
Two-parameter exponential	93.395
Three-parameter log-logistic	3.671
SEV	8.287
Normal	3.117
Logistic	3.674

It can be seen that the three-parameter Weibull distribution is the leading candidate, followed by the normal distribution. The AD statistic (adjusted) measures the area between the fitted line based on the fitting distribution (e.g., Weibull distribution) and the empirical distribution function that is based on the data points. Thus, it is the smaller the better.

One is able to calculate two additional columns, i.e., ln(Y) and ln(ln(1/R)) displayed in **Table 9.2.1**, in light of Equation (9.2.3) that is here rewritten in terms of stress as

$$\ln\{\ln[1 / R(Y)]\} = \beta \ln(Y - Y_0) + \ln(1 / \eta)$$

Substituting the data of Y and R available in the table into the above equation yields

$$\ln\{\ln[1/0.999]\} = \beta \ln(590 - Y_0) + \ln(1/\eta)$$

$$\ln\{\ln[1/0.992]\} = \beta \ln(595 - Y_0) + \ln(1/\eta)$$

$$\ldots$$

and $$\ln\{\ln[1/0.001]\} = \beta \ln(665 - Y_0) + \ln(1/\eta)$$

The 16 equations given above can be solved for three unknowns, i.e., β, η, and Y_0, using two frequently used approaches, i.e., (a) maximum-likelihood method and (b) least squares method.

(a) Maximum-likelihood Method

Using Minitab (Stat → Reliability/Survival → Distribution ID Plot (Right Censoring) → Parametric Distribution Analysis) to estimate the three parameters by MLEs, one is able to obtain the three parameters as

$$\beta = 3.55476,\ \eta = 45.8383,\ \text{and } Y_0 = 584.487$$

$$\text{Thus,}\quad F(t) = 1 - \exp[-(\frac{Y - Y_0}{\eta})^{\beta}] = 1 - \exp[-(\frac{Y - 584.487}{45.8383})^{3.55476}]$$

A plot of the above equation with 95% CI is shown in Figure 9.2.1(a).

(b) Least Squares Method

Using Minitab (Stat → Reliability/Survival → Distribution ID Plot (Right Censoring) → Parametric Distribution Analysis) to estimate the three parameters and changing the method from MLEs to least squares estimates (LSXY), one is able to obtain the following three parameters:

$$\beta = 3.93118,\ \eta = 49.0235,\ \text{and } Y_0 = 625.799$$

$$\text{Thus,}\quad F(t) = 1 - \exp[-(\frac{Y - Y_0}{\eta})^{\beta}] = 1 - \exp[-(\frac{Y - 625.799}{49.0235})^{3.93118}]$$

A plot of the above equation with 95% CI is shown in Figure 9.2.1(b).

The maximum-likelihood method and least squares method for Weibull statistics are to be elaborated in detail in Sections 9.4 and 9.7, respectively. Based on the AD statistics (adjusted) presented in Figure 9.2.5, it can be seen that the maximum-likelihood method is more robust.

TABLE 9.2.1 Bootstrap technique for identification of three-parameter Weibull distribution.

q	Y	n	f = n/(Nw)	f w	F	R	ln(Y)	ln(ln(1/R))
1	590	1	0.0004	0.002	0.001	0.999	6.380	-6.907
2	595	6	0.0024	0.012	0.008	0.992	6.389	-4.824
3	600	9	0.0036	0.018	0.023	0.977	6.397	-3.761
4	605	21	0.0084	0.042	0.053	0.947	6.405	-2.910
5	610	41	0.0164	0.082	0.115	0.885	6.413	-2.102
6	615	54	0.0216	0.108	0.210	0.790	6.422	-1.445
7	620	69	0.0276	0.138	0.333	0.667	6.430	-0.904
8	625	76	0.0304	0.152	0.478	0.522	6.438	-0.431
9	630	69	0.0276	0.138	0.623	0.377	6.446	-0.025
10	635	65	0.026	0.13	0.757	0.243	6.454	0.347
11	640	41	0.0164	0.082	0.863	0.137	6.461	0.687
12	645	25	0.01	0.05	0.929	0.071	6.469	0.973
13	650	14	0.0056	0.028	0.968	0.032	6.477	1.236
14	655	6	0.0024	0.012	0.988	0.012	6.485	1.487
15	660	2	0.0008	0.004	0.996	0.004	6.492	1.709
16	665	1	0.0004	0.002	0.999	0.001	6.500	1.933

where
q is the class
Y (MPa) is the ultimate tensile strength
n is the number of specimens in the class
N is the total number of specimens; N = 500
w (MPa) is 5 MPa
F is the statistical distribution function of failure
R is the reliability

FIGURE 9.2.5 Weibull probability with 95% CI for ultimate tensile strength of steel bolts.

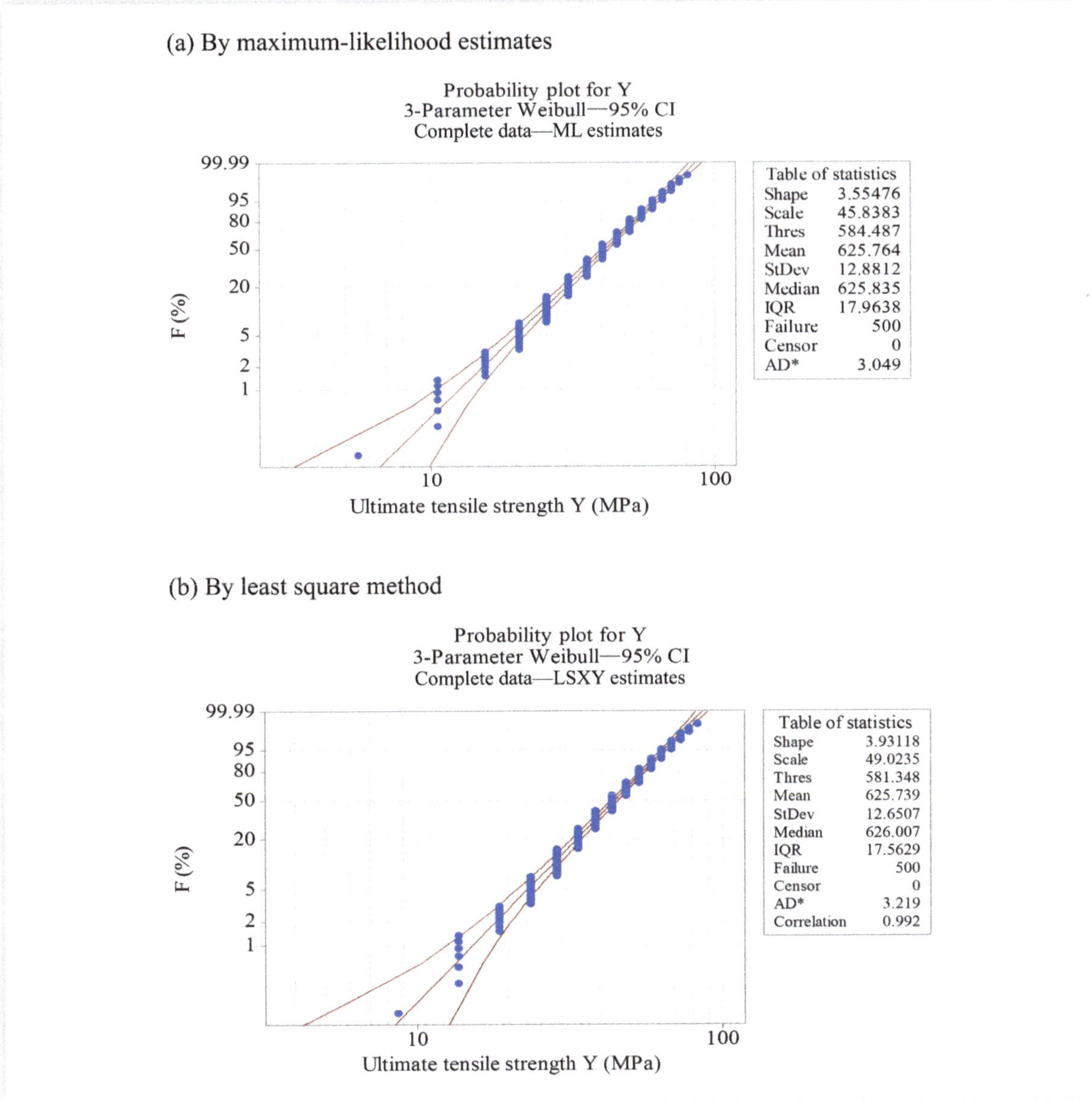

9.2.4. Mean and Standard Deviation

The mean of a three-parameter Weibull distribution function is the expected value that can be expressed as

$$E(t) = \mu_t = t_0 + (\eta - t_0)\,\Gamma(1 + 1/\beta) \tag{9.2.6}$$

in which Γ( ) is the gamma function, i.e.,

$$\Gamma(1 + 1/\beta) = \int_0^{\infty} e^{-y}\, y^{1/\beta}\, dy \tag{9.2.7}$$

Note that mean μ_t is the mean time between failures (MTBFs) for the Weibull distribution function if variable t is time. When $\beta = 1$, it turns into an exponential function and MTBF = $\mu_t = \eta$ since $\Gamma(2) = 1! = 1$.

The standard deviation of the Weibull distribution function is

$$\sigma_t = t_0 + (\eta - t_0)\,[\Gamma(1 + 2/\beta) - \Gamma^2(1 + 1/\beta)]^{1/2} \tag{9.2.8}$$

9.2.5. Median and Mode

The median for the Weibull distribution can be obtained by setting 50% for the CDF, i.e., MTTF, as

$$MTTF = median = B_{50} = t_{50\%} = t_0 + (\eta - t_0)\,[\ln(2)]^{1/\beta} \tag{9.2.9}$$

MTTF describes the expected time to failure for a non-repairable product, as an important and well-accepted measure for product reliability. If $\beta = 1$ and $t_0 = 0$, the Weibull distribution becomes an exponential distribution with only one parameter, i.e., η, and $t_{50\%}$ = MTTF = $[\ln(2)]\,\eta = 0.693\,\eta$.

Note that the MTTF of a set of sample data that follows the Weibull distribution can be also obtained from the following integration:

$$MTTF = \int_0^{\infty} R(t)\, dt \tag{9.2.10}$$

The mode for the Weibull distribution can be obtained by setting the first differentiation of its PDF to zero. In terms of time t, it is

$$t_{mode} = t_0 + (\eta - t_0)\left(\frac{\beta - 1}{\beta}\right)^{(1/\beta)} \tag{9.2.11}$$

Two special cases of t_{mode}:

(a) If $\beta = 1$, $t_{mode} = t_0$.

(b) As $\beta \rightarrow +\infty$, the Weibull distribution degenerates to $t_{mode} = \eta$. It means that when a Weibull distribution has a large β its corresponding Weibull PDF consequently exhibits a sharp peak at the mode.

9.2.6. Two-Parameter Weibull Distribution

Assume that $t_0 = 0$, or say that the Weibull distribution function with two parameters is adequate to describe the converting behavior. For the two-parameter Weibull distribution, the conversion (hazard) rate given by Equation (9.2.1a) reduces to

$$h(t) = \left(\frac{\beta}{\eta^{\beta}}\right) t^{(\beta - 1)} \quad (9.2.12)$$

The two-parameter Weibull CDF, PDF, and reliability function become, respectively,

$$F(t) = 1 - \exp\left[-\left(\frac{t}{\eta}\right)^{\beta}\right] \quad (9.2.13)$$

$$f(t) = \frac{\beta}{\eta}\left(\frac{t}{\eta}\right)^{(\beta-1)} \exp\left[-\left(\frac{t}{\eta}\right)^{\beta}\right] \quad (9.2.14)$$

and

$$R(t) = 1 - F(t) = \exp\left[-\left(\frac{t}{\eta}\right)^{\beta}\right] \quad (9.2.15)$$

The two-parameter Weibull distribution is used extensively in reliability applications to model failure times.

9.2.7. Standard Weibull Distribution

When $t_0 = 0$ and $\eta = 1$, the Weibull distribution becomes a standard Weibull distribution. The corresponding standard Weibull CDF, PDF, and reliability function become, respectively,

$$F(t) = 1 - \exp(-t^{\beta}) \quad (9.2.16)$$

$$f(t) = \beta\, t^{(\beta - 1)} \exp(-t^{\beta}) \quad (9.2.17)$$

and

$$R(t) = 1 - F(t) = \exp(-t^{\beta}) \quad (9.2.18)$$

Note that the general form of probability functions can be expressed in terms of the standard distribution.

9.2.8. SEV Distribution

The relationship of the Weibull distribution to the extreme-value distribution is just like that of the normal distribution to the lognormal distribution. After the likelihood function is formulated on the Weibull distribution, its logarithmic transformation becomes a new distribution function, namely, extreme-value distribution. Extreme-value theory or extreme-value analysis is a branch of statistics dealing with the extreme deviations from the median of probability distributions. For example, let $x = \ln(t)$, of which variable t follows a Weibull distribution. Then, variable x follows a SEV distribution. This SEV distribution is called type I limiting distribution, as long as the limiting distribution is referred to the minimum of many unbounded identically distributed random variables [Murthy et al. 2004].

The SEV distribution is skewed to the left, providing the limiting distribution for the minimum (e.g., number of defects) of a very large collection of random observations from the same arbitrary distribution such as the Weibull distribution. For example, the distribution of the mechanical strength of a gear train is skewed to the left, because the chain breaks when the weakest gear breaks.

Assume that variable t follows the two-parameter Weibull distribution. Then, variable x has the PDF and CDF for the minimum, respectively [Coles 2001]:

$$f(x) = \frac{1}{\sigma_x} \exp\left(\frac{x-\mu_x}{\sigma_x}\right) \exp\left[-\left(\frac{x-\mu_x}{\sigma_x}\right)\right] \tag{9.2.19}$$

and
$$F(x) = 1 - \exp\left[-\exp\left(\frac{x-\mu_x}{\sigma_x}\right)\right] \equiv \Phi_{sev}\left(\frac{x-\mu_x}{\sigma_x}\right) \tag{9.2.20}$$

where

x is the variable (e.g., life cycles) following the Weibull distribution
μ_x is the mean of variable x, i.e., the location parameter of the extreme-value distribution
σ_x is the standard deviation of variable x, i.e., the scale parameter of the extreme-value distribution
Φ_{sev} is the standard SEV distribution

Note that $-\infty < \mu_x < \infty$ and $\sigma_x > 0$. Equations (9.2.19) and (9.2.20) exactly fit the definition of the SEV distribution that is also called the "Gumbel distribution-minimum."

The hazard rate is

$$h(x) = \exp(x) \tag{9.2.21}$$

The mean value and variance of the Gumbel distribution-minimum are, respectively,

$$E(x) = \mu_x - 0.\ 577215665\ \sigma_x \tag{9.2.22}$$

and
$$Var(x) = (\pi^2 / 6)\ \sigma_x^2 \tag{9.2.23}$$

The formula for the MTTF of the SEV distribution is thus obtained from Equation (9.2.22) as

and
$$MTTF = \exp(\mu_x - 0.\ 577215665\ \sigma_x) \tag{9.2.24}$$

which is expressed in the time domain, as obtained from Equation (9.2.2).

Equations (9.2.19) and (9.2.20) exactly fit the definition of the SEV distribution that is also named Gumbel distribution-minimum, of which the variate is calculated as

$$f(x) = \frac{1}{\sigma_x} \exp\left(\frac{x-\mu_x}{\sigma_x}\right) \exp\left[-\left(\frac{x-\mu_x}{\sigma_x}\right)\right] \tag{9.2.25}$$

and
$$F(x) = 1 - \exp\left[-\exp\left(\frac{x-\mu_x}{\sigma_x}\right)\right] \equiv \Phi_{sev}\left(\frac{x-\mu_x}{\sigma_x}\right) \tag{9.2.26}$$

9.2.9. Standard SEV Distribution

Given that $\mu_x = 0$ and $\sigma_x = 1$, the SEV distribution is called standard SEV distribution or standard Gumbel distribution-minimum. The PDF and CDF of the standard SEV distribution can be obtained from Equations (9.2.19) and (9.2.20) as, respectively [Genschel and Meeker 2010],

$$f(x) = \exp(x)\exp[-\exp(x)] \tag{9.2.27}$$

and

$$F(x) = 1 - \exp[-\exp(x)] \equiv \Phi_{sev}(x) \tag{9.2.28}$$

Because of this relationship, computational tactics designed for the SEV distribution can be used to analyze logarithmic-transformed Weibull distribution data. The situation resembles using normal distribution programs to analyze lognormal data, after first taking natural logarithms of the original data.

For a standardized SEV distribution, there exist the following two delimits:

$$\mu_x = \ln(\eta) \tag{9.2.29}$$

and

$$\sigma_x = \beta^{-1} \tag{9.2.30}$$

The standard variable of the two-parameter Weibull distribution is thus

Let
$$z = \frac{x - \mu_x}{\sigma_x} = \frac{\ln(t) - \ln(\eta)}{\beta^{-1}} = \beta\,[\ln(t) - \ln(\eta)] \tag{9.2.31}$$

For example, let $\eta = 160{,}000$ cycles and $\beta = 3.2$ for a two-parameter Weibull distribution. Then, $\mu_x = \ln(\eta) = \ln(160{,}000) = 11.983$ and $\sigma_x = 1/\beta = 1/3.2 = 0.3125$, which are the counterparts for the corresponding SEV distribution. Once a significant number of random numbers (e.g., 100 points) are generated from this extreme-value distribution, a probability chart resembling the one shown in Figure 9.2.6 can be plotted. It is similar to the normal probability plot for a lognormal distribution after the logarithmic transformation. Data derived from the SEV distribution will line up approximately along a straight line. The slope of the line is an estimate of σ_x and the vertical axis value on the line corresponding to "zero point" on the horizontal axis is an estimate of μ_x.

FIGURE 9.2.6 SEV plot of a series of Weibull random data [NIST 2022].

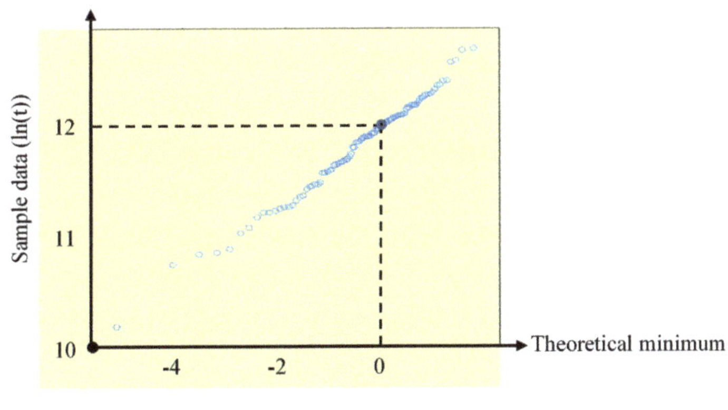

9.2.10. Largest Extreme-Value Distribution

A largest extreme-value distribution is skewed to the right, providing the limiting distribution for the maximum (e.g., number of defects) of a very large collection of random observations from the same arbitrary distribution such as the Weibull distribution. The largest extreme-value distribution and the SEV distribution are closely related. For example, if x has the SEV distribution, –x has a largest extreme-value distribution and vice versa.

9.3. Reliability Demonstration Testing (RDT)

An automotive manufacturer, as well as its sub-tier suppliers, will have to demonstrate that each part, assembly, system, and completed product have met a goal value of reliability with a specific confidence level at a given time under a prescribed procedure. This activity is called RDT. It is employed toward the end of the product reliability growth period for verifying if the specific reliability level has been achieved. It may be based on one individual comprehensive experimental treatment after the completion of DOEs.

RDT is conducted to prove that an acceptable amount of the population will survive a known period of field use under service conditions. Statistical distributions that have been hypothesized to help engineers do reliability demonstration tests include cumulative binomial, nonparametric binomial, exponential chi-squared, nonparametric Bayesian, and the Weibull distribution. A work task on RDT necessitates the following quantified data:

1. Number of units to be tested.
2. Stop moment, i.e., the moment when either the amount of tested parts passes or the number of failures that must occur.
3. Measure of failure (or success), by which the number of failures allowed in a passing test, e.g., every unit runs for the specified amount of time and there are no failures.

Reliability means the percentage of population surviving desired life; e.g., 90% reliability means nine out of ten exhibiting the specified life. The symbol B_α is used to designate the load or time to failure, exceeded by $(1 - \alpha)$% of a sample population of a part or system in the automotive industry, as "B" stands for bearing life. For example, B_{10} is used to designate that 90% of a sample population of test units survive in a test-to-failure experiment. In other words, B_{10} is also interpreted to demonstrate reliability at R = 90%. If the confidence level is not requested, B_{10} is generally calculated on the basis of the median rank, i.e., having confidence C = 50%. Testing can be classified into two categories:

1. Reliability Testing: To demonstrate whether a reliability requirement has been met.
2. Substantiation Testing: To demonstrate whether a refined system (part) has eliminated a known failure mode or mitigated its influence significantly.

9.3.1. Statistics of RDT

Two widely applied criteria for designing reliability demonstration tests for the purpose of quality control are based on binomial equations for α-risk and β-risk allowing for Q failures:

(a) α-risk: Given an incoming lot of sample size N, a confidence level calculated for each sample refers to that the proportion of the confidence level will include the true population parameter is 1 – α.

$$1-\alpha = \sum_{q=0}^{Q} \frac{N!}{(N-q)!\,q!}[F_{incoming}(t)]^q\,[1-F_{incoming}(t)]^{N-q} \tag{9.3.1}$$

where

N is the sample size

q is the failure per lot of sample size N

Q is the number of failures allowed during the test

α is the alpha risk (type I error), also called producer's risk, ideally $5\% \leq \alpha \leq 10\%$

$F_{incoming}(t)$ is the cumulative failure probability of incoming lot

The acceptable quality limit (AQL) is defined as a percent defective that is the requirement for the quality of the producer's product in a lot acceptance sampling plan (LASP). It is the type I error rate (i.e., α-risk) that is associated with producer's risk, accepting the lot while it has a defect level less than or equal to the AQL. It also gives how many units out of how many test units will exhibit the reliability performance specified when multiple units of samples are tested.

(b) β-risk: Given an outgoing lot of sample size N, the β-risk calculated for the lot refers to that the proportion of the probability will include the true population parameter is

$$\beta = \sum_{q=0}^{Q} \frac{N!}{(N-q)!\,q!} [F_{outgoing}(t)]^q [1 - F_{outgoing}(t)]^{N-q} \tag{9.3.2}$$

where

β is the beta risk (type II error), also called consumer's risk, ideally $10\% \leq \beta \leq 20\%$

$F_{outgoing}(t)$ is the cumulative failure probability of outgoing lot

The rejectable quality limit (RQL) is a maximum percent defective that would be unacceptable to the consumer. It is the type II error rate (i.e., β-risk), also called the consumer's risk, accepting a lot while it has a defect level as high as the RQL in a LASP.

It is assumed that binomial sampling is valid above. Nevertheless, the Weibull distribution is also widely applied for assessing these risks [Kim and Yum 2011]. Operating characteristic (OC) curves are the primary tool for determining LASPs.

9.3.2. Substantiation Test

OC curves with zero failure are often used for identifying specified levels of alpha risk and beta risk in RDT in the automotive industry. Zero-failure design, as practiced in the automotive industry, means to eliminate critical failures by design. The confidence level, also called CI that is defined as $1 - \alpha$, is expressed in percentage. The confidence level expresses how often the reliability status will be expected to be true. The relationship between reliability and confidence level (type I error) without any failure after completing N test specimens can be obtained from Equation (9.3.1) by setting Q = 0 as

$$1 - \alpha = [R(t)]^N = [1 - F(t)]^N \tag{9.3.3}$$

$$\text{i.e.,} \quad R = (1 - \alpha)^{1/N} \tag{9.3.4}$$

$$\text{Thus,} \quad N = \frac{\ln(1 - \alpha)}{\ln(R)} \tag{9.3.5}$$

where

F(t) is the cumulative failure probability

R(t) or R is the reliability

$1 - \alpha$ is the confidence level corresponding to α-risk, also called confidence for short

Zero-failure reliability testing is also often called substantiation testing, which provides statistical evidence that a designed or redesigned unit has suppressed or significantly reduced a known cause of failure. It is a test conducted to substantiate that a given design exceeds a requirement or is better than a previous design. A steeper OC curve exhibits a greater discriminatory power, since its delimited alpha and beta errors are smaller.

The life specifications for testing, i.e., L for life, R for reliability, and C for confidence, form an assurance index for a finite number of sample tests (i.e., N). Life usually refers to the test duration measured in terms of kilometers, cycles, hours, etc. The three frequently asked reliability dilemmas associated with no (zero) failure in automotive engineering tests are as follows:

(a) L = ?: On a certain level of reliability (R) with a specified confidence (C), are the test results from the sample size (N) evident of passing the life requirement?

(b) N = ?: How many samples are required to test, for a given reliability (R), confidence (C), and life specification (L) with zero failure?

(c) What is the alternative life level (L_1), greater than the given engineering life specification (L_0), should be, for a given reliability (R), confidence (C), and sample size (N) with zero failure?

A comprehensive accelerated life test (ALT) plan with built-in statistics in a substantiation test can be invaluable, because it reveals not only the loading conditions to apply but also the contingencies at times of failure, required test equipment, definitions of failures, and alternative fixtures to use to continue testing after experiencing a failure in the test plan [Allmen and Lu 1994].

9.3.3. Cumulative Fault Frequency (CFF)

CFF, which is obtained by dividing the number of failure (i.e., faults) by the total number of samples, has also been used in demonstration tests. For example, if a dealer sold 3000 cars and only 45 cars of the reported for repair fall between 0 and 20,000 km, CFF = 45/3000 = 0.015, or 15 out of 1000 units during 0–20,000 km for this vehicle. The range of life is an important factor and must be used as a qualifier when doing the warranty analysis.

9.3.4. Weibull RDT

The following two equations that demonstrate the relationship between confidence level and lifetime are obtained by substituting Equations (9.2.2) and (9.2.13) into Equation (9.3.4), respectively,

$$(1-C)^{1/N} = \exp\{-[(t-t_0)/(\eta - t_0)]^{\beta}\} \qquad \text{(3-parameter)} \qquad (9.3.6)$$

and

$$(1-C)^{1/N} = \exp\{-(t/\eta)^{\beta}\} \qquad \text{(2-parameter)} \qquad (9.3.7)$$

A demand of 90% reliability at 160,000 km (100,000 miles) with 90% confidence, i.e., R = 90% and C = 90%, is a general practice in the automotive industry. This reliability goal has to be stretched forth for new vehicle design from the early conceptual design stage.

EXAMPLE 9.3.1

Replicated with the same treatment, the fatigue life cycles of ten cables under a flex test are given as follows: Seven of them failed at 36,657; 58,111; 78,014; 109,032; 124,859; 169,316; and 204,050 cycles, respectively, and the last three were suspended at 218,457 cycles. How much are the life expectancies at (A) R = 90% and C = 95% and (B) R = 95% and C = 95%. Also, the sample mean, standard deviation, median, and mode are estimated.

Solution:

1. Arranging these data in three columns (Minitab).

Cycles	Censored	Frequency
218,457	1	3
78,014	0	1
124,859	0	1
36,657	0	1
109,032	0	1
58,111	0	1
169,316	0	1
204,050	0	1

2. Weibull parameters can be calculated using the maximum-likelihood method (to be discussed in Section 9.4). Hereon, following directory Path in Minitab: Stat ➔ Reliability/Survival ➔ Parametric Distribution Analysis ..., one will get scale factor η = 191,314 cycles and shape factor β = 1.62135.
3. Putting "Cycles" in the "Variables" and "Frequency" in the "Frequency Columns"; clicking "Censor" icon; and putting "Suspended" in the "Use Censoring Column" with censoring value "1," one can obtain **Figure 9.3.1**. When the mouse cursor points over the curve of interest, the corresponding life cycles of interest can be read from the pop-up data. Specifically, for the case study,

$$8984 < \text{Life} < 104428 \qquad (R = 95\%, C = 95\%)$$

$$18161 < \text{Life} < 125541 \qquad (R = 90\%, C = 95\%)$$

and

$$\text{Life} = 47749 \qquad (R = 90\%, C = 50\%)$$

4. The mean of the Weibull distribution function can be calculated using Equation (9.2.6) with $t_0 = 0$.

$$E(t) = \mu_t = \eta\, \Gamma(1 + 1/\beta) = 191314\, \Gamma(1 + 1/1.62135) = 171324.3$$

5. The sample standard deviation of the Weibull distribution function can be calculated using Equation (9.2.8) with $t_0 = 0$.

$$\sigma_t = \eta\, [\Gamma(1 + 2/\beta) - \Gamma^2(1 + 1/\beta)]^{1/2}$$

$$= 191314\, [\Gamma(1 + 2/1.62135) - \Gamma^2(1 + 1/1.6235)]^{1/2} = 108341.7$$

6. The median (MTTF) of the Weibull distribution function can be calculated using Equation (9.2.9) with $t_0 = 0$.

$$\text{MTTF} = \text{median} = B_{50} = t_{50\%} = \eta\,[\ln(2)]^{1/\beta} = 191314\,[\ln(2)]^{1/1.62135} = 152606.4$$

7. The mode of the Weibull distribution function can be calculated using Equation (9.2.11) with $t_0 = 0$.

$$t_{mode} = \eta\left(\frac{\beta-1}{\beta}\right)^{(1/\beta)} = 191314\left[\left(\frac{1.62135-1}{1.62135}\right)^{1/1.62135}\right] = 105885.6$$

8. Notes:

(a) Interquartile Range (IQR) (in Figure 9.3.1) is a measure of the spread of the distribution. Quartiles represent the value for which 25% of the data are below quartile 1 (Q1) and the value for which 25% of the data are above quartile 3 (Q3). The IQR is the difference between these two quartiles (IOQ = Q3 − Q1). A major advantage of using the IQR to estimate variability is that it is less sensitive to outliers than the sample standard deviation-based (or the sample variance-based) summary statistics.

(b) AD Value: The AD goodness-of-fit value (AD value) measures the area between the fitted line based on a Weibull distribution and the empirical distribution function that is based on the data points. It is also called AD statistic, which is a squared distance that is weighted more heavily in the tails of the distribution. A smaller AD value indicates a better fit [Marsaglia 2004].

FIGURE 9.3.1 Quantile plot of the Weibull distribution for fatigue cycles under cable flex test.

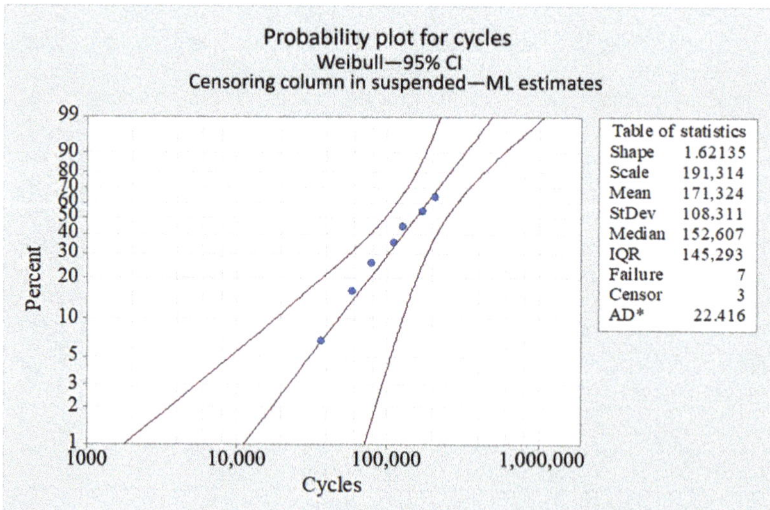

Assume that the AOZ sample plan is applied; i.e., no failure is allowed (Q = 0). Equation (9.3.4) can be used for estimating the sample size as [Durivage and Mehta 2016]

$$N = \frac{\ln(1-C)}{\ln(R)} \tag{9.3.8}$$

The sample size required for inspection without any failure based on the two-parameter Weibull distribution is given in Table 9.3.1, following the equation given above. In general, R = 90% and C = 90% are accepted by most automotive manufacturers.

TABLE 9.3.1 Confidence, reliability, and sample size in the Weibull distribution with no failure.

Confidence (C)	Reliability (R)	Required sample size (N)
95%	95%	58
95%	90%	28
95%	85%	18
95%	80%	13
95%	75%	10
95%	70%	8
90%	95%	45
90%	90%	22
90%	85%	14
90%	80%	10
90%	75%	8
90%	70%	6
0.8	95%	31
0.8	90%	15
0.8	85%	10
0.8	80%	7
0.8	75%	6
0.8	70%	5
0.5	95%	14
0.5	90%	7
0.5	85%	4
0.5	80%	3
0.5	75%	2
0.5	70%	2

9.3.5. Reliability, Confidence, and Sample Size Based on the Weibull Distribution

One situation encountered frequently in testing involves a trade-off between sample size and testing time. If testing is inexpensive, the sample size can be increased to reduce the time required for validation. An experimenter usually cannot afford a huge number of test specimens, because

(a) Cost constraints: financial, effort, time

(b) Complex items are hard to come by

(c) Durable product may take a long time to acquire data

The Weibull distribution is one of the most used distributions in life data analysis. The values of its shape factor (β) and scale factor (α) have great influence on the failure rate. Product life tests can be utilized to accurately estimate the value of β. Substitution of Equation (9.2.15) into Equation (9.3.7) yields

$$\ln(1 - C) = N \ln(R) = N\,[-(t/\eta)^{\beta}] \tag{9.3.9}$$

i.e.,

$$(1 - C)^{1/N} = \exp[-(t/\eta)^{\beta}] \tag{9.3.10}$$

or

$$N = -(t/\eta)^{-\beta} \ln(1 - C) \tag{9.3.11a}$$

A random failure without special causes means that β = 1. Life tests for durable devices such as electric motors or relays can be run much closer to their total expected life, resulting in better predictions for η. If high-cycle counts can be achieved in the life test, the ratio of test cycles to the expected life can be closer to one, reducing the need for huge sample sizes, as shown in Table 9.3.2. It is shown that high confidence levels can be obtained for total run cycles within 1 to 100% of the characteristic life, or saying 1% < t/η < 100%.

TABLE 9.3.2 N (number of tests) required for a zero-failure test at C = 90% with respect to different dimensionless test time (t/η) and shape factors.

	β = 0.1	β = 0.5	β = 1	β = 2	β = 3	β = 4	β = 10	β = 20
t/η = 0.001	5	73	2303	—	—	—	—	—
t/η = 0.002	5	52	1152	—	—	—	—	—
t/η = 0.004	4	37	576	—	—	—	—	—
t/η = 0.006	4	30	384	—	—	—	—	—
t/η = 0.008	4	26	288	—	—	—	—	—
t/η = 0.01	4	24	231	—	—	—	—	—
t/η = 0.02	4	47	116	5757	—	—	—	—
t/η = 0.04	4	12	58	1440	—	—	—	—
t/η = 0.06	4	10	39	640	—	—	—	—
t/η = 0.08	3	9	29	360	4498	—	—	—
t/η = 0.1	3	8	23	231	2303	23,026	—	—
t/η = 0.2	3	6	12	58	288	1440	—	—
t/η = 0.3	3	5	8	26	86	285	—	—
t/η = 0.4	3	4	6	15	36	90	—	—
t/η = 0.5	3	4	5	10	19	37	2358	—
t/η = 0.6	3	3	4	7	11	18	381	—
t/η = 0.7	3	3	4	5	7	10	82	2886
t/η = 0.8	3	3	3	4	5	6	22	200
t/η = 0.9	3	3	3	3	4	4	7	19
t/η = 1.0	3	3	3	3	3	3	3	3

Note: A random failure without special causes means that β = 1.

However, the relationship between two sample sizes and their related test time with a fixed confidence and the same scale factor, e.g., cycles for fatigue testing and strength for material rigidity, can be derived from Equation (9.3.10) as

$$\frac{N_1}{N_2} = \left(\frac{T_1}{T_2}\right)^{\beta} \tag{9.3.11b}$$

where

N_1 and N_2 are the two sample sizes of interest

T_1 and T_2 are the lifetime expected for the two sample sizes to justify the demanded confidence

Note that the exponent β has to be estimated from actual historical laboratory or field test results.

"A system is supposed to pass the life expectancy of 80,000 km (averaged mileage for five-year driving in US), demonstrating reliability at R = 90% with a confidence C = 90%. How many specimens (or tests) should be tested up to how many kilometers?" This question consists of the following key elements:

1. Goal: lifetime = 80,000 km
2. Reliability: R = 90%
3. Confidence: C = 90%
4. Number of tests or specimens: N = ?
5. Type of tests?
   (i) No failure
   (ii) One or zero failure
   (iii) Sudden death
6. Suspended (censored) or not?

   If yes, each specimen should be tested up to how many kilometers.
   This is to be further interpreted using Example 9.3.2.

EXAMPLE 9.3.2

There are ten passenger vehicles. How many kilometers must each accumulate to demonstrate that B_{10} life is 80,000 km (90% confidence level) based on the zero-failure criterion? Assume that the shape factor β = 3.0 for the Weibull distribution function.

Solution:

A. Find η (Characteristic Life): Given that the time-to-failure distribution is the Weibull distribution function with β = 3 and demonstrates reliability at R = 90% with a life expectancy of 80,000 km (i.e., t = B_{10}), the expected characteristic life (η) is obtained from Equation (9.2.15),

$$R(t) = \exp\left[-(t/\eta)^{\beta}\right]$$

i.e., 90% = $\exp[-(80{,}000/\eta)^3]$

Thus, η = 169,381 km

B. Designing Test Plans: Given that ten passenger vehicles are available for testing and the expected confidence level is C = 90%, one can substitute the time-characteristic life ratio into Equation (9.3.10) as

$$(1-90\%)^{1/10} = \exp\left[-(t/\eta)^{3}\right]$$

Thus, t = 0.613 and η = 103,817 km

Thus, the zero-failure test plan for demonstrating 90% reliability at 80,000 km with 90% confidence requires testing the ten product units for 103,817 km with no failure. Note that shape factor β can be obtained from test physical tests, and β = 3 is assumedly hypothetical.

EXAMPLE 9.3.3

Given that each spring can accumulate 1,200,000 cycles. How many product units are required for testing to demonstrate that B_{10} life = 1,000,000 cycles (90% confidence level), using the zero-failure criterion? Assume that the shape factor β = 3.0 based on the two-parameter Weibull distribution function.

Solution:

A. Find η (Characteristic Life): Given that the time-to-failure mode fits the two-parameter Weibull distribution function with β = 3 and demonstrates reliability at R = 90% with a life cycle of 1,000,000 cycles, the expected characteristics life (η) can be calculated using Equation (9.2.15):

$$R(t) = \exp\left[-(t/\eta)^{\beta}\right]$$

$$\text{or} \quad 90\% = \exp\left[-(1{,}000{,}000/\eta)^{3}\right]$$

Thus, η = 2,117,260 cycles.

B. Designing Test Plans: Given that each product can accumulate 1,200,000 cycles and will be suspended exactly at 1,200,000 cycles, then the test time-to-characteristic life ratio is

$$t/\eta = 1{,}200{,}000/2{,}117{,}260 = 0.567$$

Hence, one can plug the time-to-characteristic life ratio into Equation (9.3.10) to calculate the required total number of test units at a confidence level of 90% (i.e., C = 90%) as

$$(1-C)^{1/N} = \exp\left[-(t/\eta)^{\beta}\right]$$

$$\text{or} \quad (1-90\%)^{1/N} = \exp\left[-(0.567)^{3}\right]$$

Thus, N = 12.65.

It is concluded that the zero-failure test plans to demonstrate 90% reliability at 1,000,000 cycles with 90% confidence requires testing 13 product units for at least 1,200,000 cycles.

One more fundamental question remains to be answered. How to identify if the test data fit the Weibull distribution? Many methods of estimating Weibull parameters have been suggested [ASTM D7846-21 2021]. Identification of a homogenous Weibull distribution using a set of test data can be accomplished using one of the following methods:

(a) MLE
(b) MME
(c) Simple percentile estimator [e.g., median rank regression (MRR)]
(d) Regression models (e.g., least squares method)

The MRR is not a convenient tool for censored data, the method of moments has a low efficiency with a biased β value, and the method of maximum likelihood is nearly universally superior to the least squares method [Evans et al. 2019]. Therefore, the method of MLE has become a universal tool for the evaluation of the Weibull parameters.

9.3.6. Decision-Making Rule with Statistical Thinking

One management guideline on the confidence level for reliability is the so-called golden rule in general practice for deciding the product service life. The decision-making rule is described as follows:

(a) Green Zone: It writes the design inputs and test acceptance criteria such that the lower 90% confidence bound is the limit for a clear pass.

(b) Red Zone: The nominal estimate is the limit for a clear fail.

(c) Yellow Zone: It is a caution zone that triggers a management review and pass/fail decision.

9.4. Calculating Weibull Parameters by the Maximum-Likelihood Method

The prior justification of MLE is its modeling efficiency with a large sample. Under the condition of mild regularity, the MLE of a single parameter from singly censored samples has been shown to converge to the true parameter value with a given large sample size [Halperin 1952]. Many of these properties can be extended to the case of several parameters and more general censoring [Evans et al. 2019]. This estimator for the parameters of a statistical function is important in the stochastic world, because it is asymptotic unbiasedness resulting from minimal variance.

9.4.1. MLE of Weibull Parameters with Test-to-Failure Data

A likelihood function is first defined by treating Weibull parameters as unknown values and finding the joint probability density of all observations, which are assumed to be IID. If the data points are all highly likely under specific parameter values, their product will be the "most likely" outcome, called the maximum likelihood. Once the likelihood function is formed, the maximum of that function can be found by partial differentiations with respect to these related parameters.

Assume that the lifetime data of all the test units, i.e., t_q, q = 1, 2, …, and Q, come with no censoring and follow a three-parameter Weibull distribution, the likelihood function and logarithmic likelihood function based on a given set of data can be written, respectively [Freeman 2010],

$$L_w(\beta, \eta, t_0) = \prod_{q=1}^{Q} f(t_q) \tag{9.4.1}$$

and

$$\ln[L_w(\beta, \eta, t_0)] = \sum_{q=1}^{Q} \ln[f(t_q)] \tag{9.4.2}$$

where

$L_w(\beta, \eta, t_0)$ is the likelihood function with three-Weibull parameters
$\ln[L_w(\beta, \eta, t_0)]$ is the logarithmic transformation of $L_w(\beta, \eta, t_0)$
$f(t_q)$ is the Weibull PDF evaluated at t_q
q is the number of data
Q is the total number of test-to-failure data

Substituting Equation (9.2.5) into the above equation leads to

$$\ln[L_w(\beta, \eta, t_0)] = Q \ln(\beta) - Q\,\beta \ln(\eta) + (\beta - 1) \sum_{q=1}^{Q} \ln(t_q - t_0) - \sum_{q=1}^{Q} \left(\frac{t_q - t_0}{\eta}\right)^{\beta} \tag{9.4.3}$$

As enlightened by the rule of maximizing the likelihood, Weibull parameters β, η, and t_0 are to be estimated by taking partial differentiations with respect to these three parameters and setting them to zeros, respectively,

$$\frac{\partial\{\ln[L_w(\beta, \eta, t_0)]\}}{\partial \eta} = \frac{-Q\,\beta}{\eta} + \frac{\beta}{\eta^{\beta+1}} \sum_{q=1}^{Q} (t_q - t_0)^{\beta} = 0 \tag{9.4.4}$$

$$\frac{\partial\{\ln[L_w(\beta, \eta, t_0)]\}}{\partial \beta} = \frac{Q}{\beta} - Q \ln(\eta) + \sum_{q=1}^{Q} \ln(t_q - t_0) + \sum_{q=1}^{Q} \left(\frac{t_q - t_0}{\eta}\right)^{\beta} \ln\left(\frac{t_q - t_0}{\eta}\right) \tag{9.4.5}$$

and

$$\frac{\partial\{\ln[L_w(\beta, \eta, t_0)]\}}{\partial t_0} = (\beta - 1) \sum_{q=1}^{Q} (t_q - t_0)^{-1} + \beta\,\eta^{-\beta} \sum_{q=1}^{Q} (t_q - t_0)^{\beta-1} = 0 \tag{9.4.6}$$

Equation (9.4.4) gives an equation, revealing that η is a function of β and t_0.

$$\eta = \left[\frac{1}{Q} \sum_{q=1}^{Q} (t_q - t_0)^{\beta}\right]^{1/\beta} \tag{9.4.7}$$

Plugging the above equation into Equations (9.4.5) and (9.4.6) yields two simultaneous equations as

$$\frac{Q}{\beta} + \sum_{q=1}^{Q} (t_q - t_0) - Q \left\{\sum_{q=1}^{Q} (t_q - t_0)^{\beta} \ln(t_q - t_0)\right] / \left[\sum_{q=1}^{Q} (t_q - t_0)^{\beta}\right]\} = 0 \tag{9.4.8}$$

and

$$\frac{\beta}{1-\beta} + \frac{1}{Q} \sum_{q=1}^{Q} \frac{1}{(t_q - t_0)} \left[\sum_{q=1}^{Q} (t_q - t_0)^{\beta}\right] / \left[\sum_{q=1}^{Q} (t_q - t_0)^{\beta-1}\right] = 0 \tag{9.4.9}$$

The above two equations can be solved for the two unknowns (β and t_0), although they are generally nonlinear. It is an iterative process of solving these two equations for the fit curve that most likely regenerates accurate data values. The Newton–Raphson method is one of the proper ways to do this. After β and t_0 are obtained, they can be substituted into Equation (9.4.7) for η.

Next, consider the Weibull distribution with two parameters, having $t_0 = 0$ and neglecting Equation (9.4.6). Equations (9.4.7) and (9.4.8) reduce to

$$\eta = \left[\frac{1}{Q} \sum_{q=1}^{Q} (t_q)^{\beta}\right]^{1/\beta} \tag{9.4.10}$$

and

$$\frac{Q}{\beta} + \sum_{q=1}^{Q} \ln(t_q) - Q \left[\sum_{q=1}^{Q} t_q^{\beta} \ln(t_q)\right] / \left[\sum_{q=1}^{Q} t_q^{\beta}\right] = 0 \tag{9.4.11}$$

The above equation has only one unknown parameter that is β, but it is still quite nonlinear. A numerical procedure, such as the Newton–Raphson method and expectation–maximization algorithm [McCool 2012], may be required for identifying the root. After the value of parameter β is obtained, it can be substituted back into Equation (9.4.10) to obtain parameter η.

9.4.2. MLE of Weibull Parameters with Incomplete Data

Incomplete data refer to censored and interval data, while test-to-failure data are complete data. During a fatigue life test, the test may be censored (suspended) after the specimen runs significantly higher than the required life cycles. This belongs to type I right censoring. In other words, the time to failure of each censored unit is unknown, but it is known that the time to failure has some positive value. As long as there are censored data, the maximum-likelihood method provides a valid way of estimating Weibull parameters, instead of the least squares method or method of moments. This pertains to the traditional type I right-censoring scheme; i.e., N specimens are placed on test first and the number of uncensored data (N – S) by test to failure is predetermined. Instead of continuing experiments until all N items have failed, the censored experiments are terminated when the $(N-S)^{th}$ unit fails [Lawless 2003]. The remaining S test units are regarded as censored data. Most product validation tests of automotive parts fall into this category.

Assume that the lifetime of the test units follows a two-parameter Weibull PDF. The likelihood function and logarithmic likelihood function with N_f test-to-failure data and N_s right-suspended data can be written as, respectively [Kim and Yum 2011],

$$L_w(\beta, \eta) = \{\prod_{f=1}^{N_f} f(t_f)\}\ \{\prod_{s=1}^{N_s} [S(t_s)]\} = \{\prod_{f=1}^{N_f} f(t_f)\}\ \{\prod_{s=1}^{N_s} [R(t_s)]\} \tag{9.4.12}$$

and $$\ln[L_w(\beta, \eta)] = \{\sum_{f=1}^{N_f} \ln[f(t_f)]\}\ \{\sum_{s=1}^{N_s} \ln[R(t_s)]\} \tag{9.4.13}$$

where

- N_s is the total number of censored (suspended) test units
- N_f is the total number of failed test units
- t_f is the life of unit f, up to failure
- t_s is the life of unit s, up to suspension (censoring)
- $f(t_f)$ is the PDF of failure evaluated at time t_f
- $R(t_s)$ is the reliability function evaluated at time t_s

Note that reliability function R(t) is equal to survival function S(t) for the Weibull distribution when dealing with discrete suspended data. Substituting Equations (9.2.4)–(9.4.5) into the above equation yields

$$\ln[L_w(\beta, \eta, t_0)] = N_f \ln(\beta) - N_f\,\beta \ln(\eta) + (\beta-1) \sum_{f=1}^{N_f} \ln(t_f - t_0) - \sum_{f=1}^{N_f} \left(\frac{t_f - t_0}{\eta}\right)^{\beta} - \sum_{s=1}^{N_s} \left(\frac{t_s - t_0}{\eta}\right)^{\beta} \tag{9.4.14}$$

When guaranteed value $t_0 = 0$, the above equation reduces to

$$\ln[L_w(\beta, \eta)] = N_f \ln(\beta) - N_f\,\beta \ln(\eta) + (\beta-1) \sum_{f=1}^{N_f} \ln(t_f) - \sum_{f=1}^{N_f} \left(\frac{t_f}{\eta}\right)^{\beta} - \sum_{s=1}^{N_s} \left(\frac{t_s}{\eta}\right)^{\beta} \tag{9.4.15}$$

In order to maximize the likelihood, it is necessary to take partial derivatives of $\ln[L_w(\beta, \eta)]$ in the above equation with respect to η and β, respectively, and equal each of them to zero as shown below:

$$\partial\{\ln[L_w(\beta, \eta)]\} / \partial\eta = 0 \tag{9.4.16}$$

and

$$\partial\{\ln[L_w(\beta, \eta)]\} / \partial\beta = 0 \tag{9.4.17}$$

Substituting Equation (9.4.5) into the above two equations yields

$$\frac{\partial\{\ln[L_w(\beta, \eta, t_0)]\}}{\partial\eta} = \frac{-N_f\,\beta}{\eta} + \frac{\beta}{\eta^{\beta+1}} \sum_{f=1}^{N_f} (t_f)^\beta - \frac{\beta}{\eta^{\beta+1}} \sum_{s=1}^{N_s} (t_s)^\beta = 0 \tag{9.4.18}$$

and

$$\frac{\partial\{\ln[L_w(\beta, \eta, t_0)]\}}{\partial\beta} = \frac{N_f}{\beta} - N_f \ln(\eta) + \sum_{f=1}^{N_f} \ln(t_f) + \sum_{f=1}^{N_f} \left(\frac{t_f}{\eta}\right)^\beta \ln\left(\frac{t_f}{\eta}\right)$$

$$+ \sum_{s=1}^{N_s} \left(\frac{t_s}{\eta}\right)^\beta \ln\left(\frac{t_s}{\eta}\right) = 0 \tag{9.4.19}$$

General speaking, an iterative process can be applied for solving these two simultaneous nonlinear equations. Again, other proper ways to obtain the roots include the Newton–Raphson method and expectation–maximization algorithm [McCool 2012].

9.4.3. MLE of Weibull Parameters with Interval Data

Assume that the lifetime of the test specimens follows a two-parameter Weibull PDF, and the likelihood function and logarithmic likelihood function with both right-censored and interval-censored data can be written as

$$L_w(\beta, \eta) = \left\{\prod_{f=1}^{N_f} f(t_f)\right\} \left\{\prod_{s=1}^{N_s} [R(t_s)]\right\} \left\{\prod_{i=1}^{I} [F(t_{i,2}) - F(t_{i,1})]\right\} \tag{9.4.20}$$

Substituting Equations (9.2.4), (9.4.5), and (9.2.13) into the above equation yields and taking a natural logarithmic transformation yield

$$\ln[L_w(\beta, \eta)] = N_f \ln(\beta) - N_f\,\beta \ln(\eta) + (\beta - 1) \sum_{f=1}^{N_f} \ln(t_f) - \sum_{f=1}^{N_f} \left(\frac{t_f}{\eta}\right)^\beta - \sum_{s=1}^{N_s} \left(\frac{t_s}{\eta}\right)^\beta$$

$$+ \sum_{i=1}^{I} \ln\left\{\exp\left[-\left(\frac{t_{i,2}}{\eta}\right)^\beta\right] - \exp\left[-\left(\frac{t_{i,1}}{\eta}\right)^\beta\right]\right\} \tag{9.4.21}$$

In order to maximize the likelihood, it is necessary to take partial derivatives of $\ln[L_w(\beta, \eta)]$ in the above equation with respect to β and η, respectively, and equal each of them to zero as shown below:

$$\frac{\partial\{\ln[L_w(\beta, \eta, t_0)]\}}{\partial\eta} = \frac{-N_f\,\beta}{\eta} + \frac{\beta}{\eta^{\beta+1}}\sum_{f=1}^{N_f}(t_f)^\beta - \frac{\beta}{\eta^{\beta+1}}\sum_{s=1}^{N_s}(t_s)^\beta + \left(\frac{1}{\eta}\right)^{\beta+1}\sum_{i=1}^{I}\frac{(t_{i,1})^\beta \exp[-(t_{i,1}/\eta)^\beta] - (t_{i,2})^\beta \exp[-(t_{i,2}/\eta)^\beta]}{\exp[-(t_{i,2}/\eta)^\beta] - \exp[-(t_{i,1}/\eta)^\beta]} \quad (9.4.22)$$

and

$$\frac{\partial\{\ln[L_w(\beta, \eta, t_0)]\}}{\partial\beta} = \frac{N_f}{\beta} - N_f\ln(\eta) + \sum_{f=1}^{N_f}\ln(t_f) + \sum_{f=1}^{N_f}\left(\frac{t_f}{\eta}\right)^\beta \ln\left(\frac{t_f}{\eta}\right) + \left(\frac{1}{\eta}\right)^\beta\sum_{i=1}^{I}\frac{(t_{i,1})^\beta \exp[-(t_{i,1}/\eta)^\beta]\ln(t_{i,1}) - (t_{i,2})^\beta \exp[-(t_{i,2}/\eta)^\beta]\ln(t_{i,2})}{\exp[-(t_{i,2}/\eta)^\beta] - \exp[-(t_{i,1}/\eta)^\beta]} \quad (9.4.23)$$

9.4.4. Unbiased Estimate of Shape Factor

It has been unveiled that the shape parameter (β) obtained from the MLE can be biased, especially when the sample size is small. Nevertheless, they can be corrected using the following modifications:

(a) The value of β estimated by maximum-likelihood method is biased if the sample size is small. When there are only complete data (i.e., no censored or interval data), the following equation is suggested to calculate the unbiased shape factor, from the shape factor estimated using Equation (9.4.11), as [Hirose 1999, Zhang et al. 2007]:

$$\beta_{unbiased} = \beta / (1.01155 + 1.278 / N_f + 2.001 / N_f^2 + 20.35 / N_f^3 - 46.98 / N_f^4) \quad (9.4.24)$$

where

$\beta_{unbiased}$ is the unbiased estimate of β

β is the β estimated from estimate, such as Equation (9.4.11)

N_f is the total number of test-to-failure data

(b) When there exist right-censored data, it is suggested to calculate the unbiased shape factor, from the shape factor estimated using Equations (9.4.18) and (9.4.19), using [Ross 2005, Zhang et al. 2007].

$$\beta_{unbiased} = \beta / \{1 + 1.37\,[(N_f + N_s) / N_f]^{1/2} / (N_f - 1.92)\} \quad (9.4.25)$$

where

$\beta_{unbiased}$ is the unbiased estimate of β

β is the β estimated from the MLE, such as Equations (9.4.18) and (9.4.19)

N_s is the total number of suspended data

$N_f + N_s$ are the total number of data

(c) One may use the table of "unbiasing multipliers" for data with a sample size, ranging from five to 120, given by [Thoman et al. 1969]. A shape factor calculated using MLE can be multiplied by the corresponding multiplier to redeem the unbiased estimate.

(d) Another way to do it with a small sample size is to calculate shape factor β with the method of MRR or method of moments. Then, scale factor η and guaranteed value t_0 are calculated using the maximum-likelihood method.

(e) Use expectation–maximization algorithm [McCool 2012].

9.5. Confidence Level for Product Life Prediction

Reliability study of a product has been used for predicting the failure rates of its related components and the overall system as designed. These predictions are used to evaluate design feasibility, compare design alternatives, identify potential failure areas, assess trade-off system design factors, and track reliability growth. Nevertheless, reliability is the cumulative probability that a system performs its intended function satisfactorily for a specific interval of time when operating in the manner and environment intended, as defined by Equation (9.2.4). A prediction interval is defined as a range that is likely to contain the response value of an individual new observation under specified settings of the applicable predictors with a certain confidence level [Nordman and Meeker 2002], e.g., 95%.

The reliability trio, which consists of reliability (R), confidence level (C), and sample size (N), has to be identified to uniquely justify the robustness of each prediction, specifically fitness of the statistics. Confidence level is the cumulative probability that the interval estimate contains the true value being estimated with a given reliability, especially when the number of test data is limited. It is an ideal case for characterizing the failure mode, if all the test units can be tested to failure.

It would be desired to assess the reliability at the certain desired confidence level after estimating the mean from a sample (often a series of test data). The sample is usually obtained from a limited sample size, while the population standard deviation is not available. For example, a typical question in the automotive industry is "Do we satisfy the requirement of R = 95% and C = 90%?" In a fatigue-cycling test, it means that the experimenter has 90% confidence level that only or less than 5% of product units will fail at the customer's wishful number of durability cycles in the ten years' service life.

The following standards provide failure rate and MTBF data for electronic and mechanical parts and their related equipment:

1. MIL-HDBK-217: electronics reliability prediction
2. Bellcore/Telcordia: electronics reliability prediction
3. Naval Surface Warfare Center (NSWC): mechanical reliability prediction

9.5.1. Tolerance Interval

A tolerance interval is defined as range that is likely to include a prescribed proportion of a sample population with an accountable probability, i.e., for containing this prescribed proportion at a certain predetermined confidence level. The tolerance interval, so defined using (p, 1−α), based on a set of limited number data collected from a given sample population is constructed so that it would include at least a proportion p of the sampled population with confidence 1−α [Young 2010]. It means that a tolerance interval provides a tolerance limit or tolerance limits (i.e., ending points of tolerance interval), within which at least a certain proportion (p) of the population falls with a given level of confidence (1 − α). Tolerance limits, assigned to a tolerance interval, can be classified into the following three types:

1. Lower and Upper Bounds: Between what tolerance limits will the tolerance interval contain *p* percent of the population measurements?
2. Lower Bound: What will guarantee that *p* percent of population measurements will not fall below a lower tolerance limit?
3. Upper Bound: What will guarantee that *p* percent of population measurements will not exceed an upper tolerance limit?

As described above, the tolerance interval bounds the range of data values that includes a specific proportion of the population [Vardeman 1992], while a CI bounds a single-valued population parameter such as the mean or the variance with some confidence. For example, if a population is normally distributed with known population mean and standard deviation σ, the interval $\mu \pm 3\sigma$ includes 99.73% of the population (3 is the z-score for 99.73% coverage of a normally distributed population). However, if there is merely a set of finite sample data available, one can know only the sample mean A and sample standard deviation S, which are estimates of the true parameters. In that case, $A \pm 3S$ will not necessarily include 99.73% of the population, because of sampling variance in these estimates. A tolerance interval bounds this sampling variance by introducing a confidence level $(1 - \alpha)$, which is the confidence with which this interval actually includes the specified proportion of the population.

9.5.2. Lower and Upper Bounds of Lifetime by MLE

Consider a scenario that fit-in prevails. For example, the control of one critical dimension of a part shall fall within 10.00 ± 0.05. After measuring 50 parts in the trial run, the calculated reliability is 90%, or saying R = 90%. It could indicate that the engineer is $(100 - \alpha)\%$ confident that at least 90% of the dimensions meet the requirement. However, the statistics are still not well-defined without figuring out the CI regulated by the desired $\alpha\%$. This is representative of a lower bound and an upper bound that are required to demonstrate reliability R (e.g., 90%) with confidence level $C = (100 - \alpha)\%$.

For a regression model, the typical method of computing CIs relies on the maximum-likelihood theory that applies. Consider Equation (9.2.3) and let

$$u(t) = \ln(t) \tag{9.5.1}$$

$$\text{Then,}\quad u(t) = \ln(\eta) + \ln\{-\ln[R(t)]\} / \beta \tag{9.5.2}$$

Then, the lower and upper bounds of u(t) with $(100 - \alpha)\%$ confidence, i.e., $C = (100 - \alpha)\%$, are, respectively,

$$u_{lb} = \hat{u} - w_{\alpha/2\%}\, \sigma_{\hat{u}} \tag{9.5.3}$$

$$\text{and}\quad u_{ub} = \hat{u} + w_{(100-\alpha/2)\%}\, \sigma_{\hat{u}} \tag{9.5.4}$$

where

$\hat{u}$ is the sample mean or called estimator $\hat{u}$, which is the estimated value of u

$\sigma_{\hat{u}}$ is the sample standard deviation of estimator $\hat{u}$, also called sample standard error

$w_{\alpha/2\%}$ and $w_{(100-\alpha/2)\%}$ are the percentage points of the Weibull distribution

α is the confidence level, or confidence in short

As the estimated parameters are asymptotically normal, $w_{\alpha/2\%}$ and $w_{(100-\alpha/2)\%}$ of the Weibull distribution are counterparts of $z_{\alpha/2\%}$ and $z_{(100-\alpha/2)\%}$ of the normal distribution function. The lower and upper bounds in the domain of lifetime are then obtained from Equations (9.5.3) and (9.5.4), respectively,

$$t_{lb} = \exp(u_{lb}) \tag{9.5.5}$$

$$\text{and}\quad t_{ub} = \exp(u_{ub}) \tag{9.5.6}$$

The sample standard deviation in Equations (9.5.3) and (9.5.4) can be estimated by two methods: (a) the inverse of the Fisher information matrix and (b) the LR method.

The inverse of the Fisher information matrix yields the generalized variance (variance–covariance) matrix, which provides the variances of the parameters. These CIs are approximate, while the methodological exactness gets more precise with an increasing sample size. Consider the two-parameter Weibull distribution with test-to-failure and right-censored data. The related sample standard error given in Equations (9.5.3) and (9.5.4), i.e., point estimator $\sigma_{\hat{u}}$, is calculated as

$$\sigma_{\hat{u}}^2 = (\partial u / \partial\beta)^2 \sigma_\beta^2 + (\partial u / \partial\eta)^2 \sigma_\eta^2 + 2 (\partial u / \partial\beta)(\partial u / \partial\eta) \mathrm{Cov}(\beta, \eta) \tag{9.5.7}$$

where

σ_β^2 is the variance of shape factor β
σ_η^2 is the variance of scale factor η
$\mathrm{Cov}(\beta, \eta)$ is the covariance of shape factor β and scale factor η

Substituting Equation (9.5.3) into the above equation yields

$$\sigma_{\hat{u}}^2 = \beta^{-4} \{\ln[-\ln(R)]\}^2 \sigma_\beta^2 + \eta^{-2} \sigma_\eta^2 - 2 \beta^{-2} \eta^{-1} \{\ln[-\ln(R)]\} \mathrm{Cov}(\beta, \eta) \tag{9.5.8}$$

The variances and covariance given above, i.e., σ_β^2, σ_η^2, and $\mathrm{Cov}(\beta, \eta)$, can be calculated according to the FIM as [Escobar and Meeker 1999]

$$\begin{bmatrix} \sigma_\beta^2 & \mathrm{Cov}(\beta, \eta) \\ \mathrm{Cov}(\beta, \eta) & \sigma_\eta^2 \end{bmatrix} = \mathrm{Inv}\begin{bmatrix} -\partial^2[\ln(L_w)] / \partial\beta^2 & -\partial^2[\ln(L_w)] / (\partial\beta\, \partial\eta) \\ -\partial^2[\ln(L_w)] / (\partial\beta\, \partial\eta) & -\partial^2[\ln(L_w)] / \partial\eta^2 \end{bmatrix} \tag{9.5.9}$$

The variances and covariances given above are assumedly calculated from their corresponding sample variances and covariance of the estimated parameters as

$$\sigma_\beta^2 \approx S_\beta^2 = -\partial^2[\ln(L_w)] / \partial\beta^2$$

$$= \frac{-\partial^2[\ln(L_w)]}{\partial\beta^2} = \frac{N_s}{\beta} + \sum_{f}^{N_f} (t_f / \eta)^\beta \ln^2(t_f / \eta) \tag{9.5.10}$$

$$\sigma_\eta^2 \approx S_\eta^2 = -\partial^2[\ln(L_w)] / \partial\eta^2$$

$$= \frac{-\partial^2[\ln(L_w)]}{\partial\eta^2} = \frac{-N_s\, \beta}{\eta^2} + \frac{\beta(\beta+1)}{\eta^2} \sum_{f=1}^{N_f} (t_f / \eta)^\beta \tag{9.5.11}$$

and $\sigma_{\beta\eta}^2 \approx S_{\beta\eta}^2 = -\partial^2[\ln(L_w)] / (\partial\beta\, \partial\eta)$

$$= \frac{-\partial^2[\ln(L_w)]}{\partial\beta\, \partial\eta} = \frac{N_s}{\eta} - \frac{1}{\eta} \sum_{f=1}^{N_f} \{(t_f / \eta)^\beta [1 + \beta \ln(T_f / \eta)]\} \tag{9.5.12}$$

It means that the calculated sample variances and covariances of β and η based on the above three equations can be used for estimating the corresponding true variances and covariances of β and η. Note that $\ln(L_w) = \ln[L_w(\beta, \eta)]$, as described by Equation (9.4.15).

9.5.3. Lower Bound of Lifetime

Consider a scenario that the more the better. For example, the ultimate tensile strength of a material has to be at least 200 MPa. Assume that the calculated reliability is 93%, or say R = 93%. It could indicate that at least 93% of the material's ultimate tensile strength is above 200 MPa with (100 – α)% confidence. This is representative of a lower one-sided bound with 100 – α% confidence level. The lower bound of u(t) with C = (100 – α)% confidence is

$$u_{lb} = \hat{u} - w_{\alpha\%}\,\sigma_{\hat{u}} \tag{9.5.13}$$

Then, the lower bound lifetime is

$$t_{lb} = \exp(u_{lb}) \tag{9.5.14}$$

EXAMPLE 9.5.1

The following 12 test data are obtained from durability tests of side-door hinge systems for a sedan: 230,000; 135,000; 240,000 (suspended); 240,000 (suspended); 210,000; 240,000 (suspended), 160,000; 150,000; 240,000 (suspended); 240,000 (suspended); 200,000 (suspended); and 190,000. The eleventh specimen was censored (suspended) because its fixture failed, and the final five hinge systems were censored after they reach three-lifetime as cost mandates. Please use the maximum-likelihood method to estimate the lifetimes, given that (a) reliability R = 90% and confidence C = 50% and (b) reliability R = 90% and confidence C = 90%.

Solution:

(a) Lower bound with R = 90% and C = 50%

Plugging the data into the Minitab (➔ Stat ➔ Reliability/Survival ➔ Distribution Analysis (Right Censoring) ➔ Parametric Distribution Analysis) and adjusting the confidence to "C = 50%" and "LB (Lower Bound)," one is able to obtain Figure 9.5.1. Since the test-to-failure data are the more the better, the lower bound is taken. There are six test-to-failure data, i.e., blue circular dots. It can be read from the figure that t = 146,777 cycles when R = 90%, as located by the red triangle. The side-door hinge system will survive 146,777 cycles with R = 90% and C = 50%. This is a reference number and is not to be used for the purpose of design.

FIGURE 9.5.1 Weibull probability plot with C = 50% using MLE.

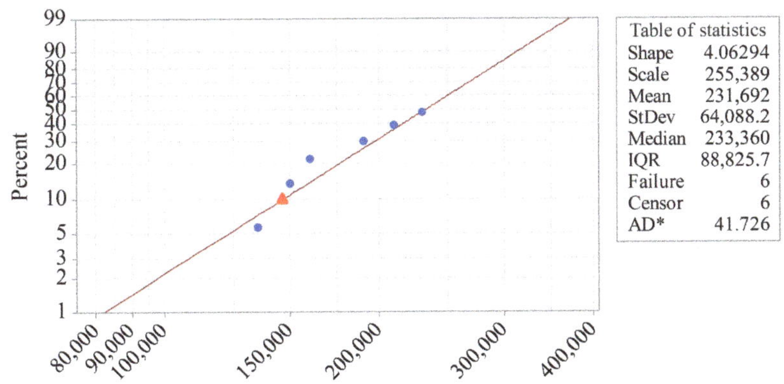

(b). Lower bound with R = 90% and C = 90%

Plugging the data into the Minitab (➔ Stat ➔ Reliability/Survival ➔ Distribution Analysis (Right Censoring) ➔ Parametric Distribution Analysis) and adjusting the confidence to "C = 90%" and "Lower Bound (LB)," one is able to obtain Figure 9.5.2. Since the test-to-failure data are the more the better, the lower bound is taken. There are six test-to-failure data, i.e., blue circular dots.

It can be read from the figure that t = 115,632 cycles when R = 90% and C = 90%, as located by the red triangle. The side-door hinge system will survive 115,632 cycles with R = 90% and C = 90%. The requirement of R = 90% and C = 90% is generally demanded for durability tests in the automotive industry. If the driver uses the front door hinges ten cycles per day, 115,632 cycles mean to survive 31.68 (=115,632/10/365) years that well exceed the general expectation of an automobile life (15 years).

FIGURE 9.5.2 Weibull probability plot with 90% lower bound using MLE.

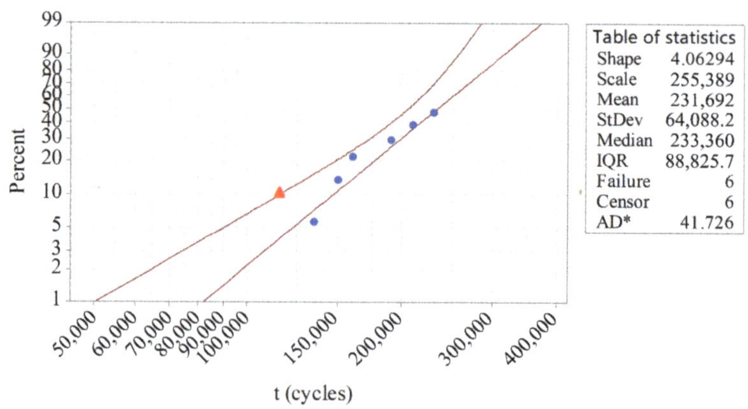

9.5.4. Upper Bound of Lifetime

Consider another scenario that the less the better. For example, the pore volume (content of porosities) of cast iron has to be less than 1% for a specific application. Assume that the calculated reliability is 93%, or say R = 93%. It could indicate that one is (100 – α)% confident that at least 93% of the material's content of voids is less than 1%. This is representative of an upper one-sided bound with α% confidence level. Then, the upper bounds of u(t) with (100 – α)% confidence, i.e., C = (100 – α)%, respectively, are

$$u_{ub} = \hat{u} + w_{(1-\alpha)\%}\, \sigma_{\hat{u}} \tag{9.5.15}$$

and

$$t_{ub} = \exp(u_{ub}) \tag{9.5.16}$$

9.5.5. Upper and Lower Bounds of Estimated Parameters

Assume that the independent variable is t that can be time, load, or other factors. The bounds of an estimated statistic can be estimated by solving the reliability equation as discussed in [Nelson 2000]. The relative uncertainty of the estimated β-value (shape factor) in the Weibull distribution based on the maximum-likelihood method happens to be the coefficient of variation given as follows:

$$\text{Uncertainty} = (S_\beta^2 / \beta_{ml})^{1/2} \tag{9.5.17}$$

where

S_β^2 is the sample variance of β_{ml} derived from MLE

β_{ml} is the β-value estimated using the maximum-likelihood method

It can be verified that the uncertainty is constant for all the beta values for a given sample size of homogeneous data and that it is independent of the true value of shape factor β. Therefore, Equation (9.5.17) can be used to simulate a life test without having to find out the value of β in the Weibull statistics.

Let β_{ml} and η_{ml} be the two parameters obtained from the maximum-likelihood method. The lower and upper bounds of the CI for shape factor β can be calculated using the following formula [Abernethy 2006]:

$$\beta_{ml}\left[\exp\left(\frac{\mu_\alpha S_\beta^2}{\beta_{ml}}\right)\right]^{-1} < \beta_{ml} < \beta_{ml}\exp\left(\frac{\mu_\alpha S_\beta^2}{\beta_{ml}}\right) \tag{9.5.18}$$

where S_β^2 can be obtained from Equation (9.5.10). By the same token, the lower and upper bounds of the CI for scale factor β are

$$\text{and} \quad \eta_{ml}\left\{\exp\left[\frac{\mu_\alpha S_\eta^2}{\eta_{ml}}\right]\right\}^{-1} < \eta_{ml} < \eta_{ml}\exp\left(\frac{\mu_\alpha S_\eta^2}{\eta_{ml}}\right) \tag{9.5.19}$$

where S_η^2 can be obtained from Equation (9.5.11).

9.5.6. CI for the Predicted Lifetime by the Bootstrap Method

The bootstrap method is a nonparametric technique for making statistical inferences. The idea of the bootstrap method is to estimate the sampling variability that would occur by taking repeated samples from the population. This is done by taking a random sample from the estimation sample, i.e., computing the statistics of interest and repeating this for some number of replications. The bootstrap method is due to the assumption that bootstrapped statistics approximately fit the normal distribution subject to the central limit theorem [Aerts et al. 1994], as detailed in Section 5.5 (Volume 1, Chapter 5). When both shape and scale factors of the SEV distribution, e.g., ln(t) where t fits a Weibull distribution, vary over experimental factors, a bootstrap-based methodology for identifying significant factors was proposed by [Wang et al. 2017].

9.6. Estimating Weibull Parameters by Method of Moments

The MME uses the statistical moments of the component distributions and the first and second partial derivatives of the desired function to find the first four moments of the statistical distribution, e.g., Equations (7.1.6)–(7.1.9). These four moments are used to find the parameters of a general distribution such as the Pearson system [Pearson 1894], the Johnson system, or the Lambda distribution. After the parameters of a statistical distribution are determined, the fraction outside of the possible limits is assessed using numerical integrations, tables, or algebraic equations in some cases. The MME can be utilized to estimate the two parameters as the first approximation, although the MME is superseded by Fisher's MLE concerning asymptotic unbiasedness and minimal variance in many cases.

Assume that the data set of interest fits for a two-parameter Weibull distribution. The MME can be derived by hand calculation. The MME begins with the following first two moments:

$$\mu_1 = \frac{1}{Q} \sum_{q=1}^{Q} t_q = \mu_t \tag{9.6.1}$$

and $$\mu_2 = \frac{1}{Q} \sum_{q=1}^{Q} (t_q)^2 \approx \mu_t^2 + \sigma_t^2 \tag{9.6.2}$$

where

μ_k is the moments, where k = 1 or 2, i.e., first and second moments estimated from test data
Q is the total number of data available
μ_t is the average of test data t, where $t \in (t_1, t_2, \ldots, t_Q)$
σ_t^2 is the sample variance of test data t, where $t \in (t_1, t_2, \ldots, t_Q)$

Simultaneously, the moment generating function for the two-parameter Weibull distribution is [Casella and Berger 2002]

$$m_K(t) = E[t^k] = \eta^k \, \Gamma(1 + k / \beta) \tag{9.6.3}$$

and $$\Gamma(1 + k / \beta) = \int_0^{\infty} t^{(k/\beta)} \exp(-t)\, dt \tag{9.6.4}$$

where

$m_K(t)$ is the kth theoretical moment, k = 1, 2, 3, or 4
$E[t^k]$ is the expectation of t^k
Γ() is the gamma function

Consider the first two moments only. One may set these moments derived numerically equal to the theoretical moments from the moment generating function, i.e., $\mu_1 = m_1$ and $\mu_2 = m_2$, which leads to

$$\mu_1 = \eta \, \Gamma(1 + 1 / \beta) \tag{9.6.5}$$

and $$\mu_2 = \eta^2 \, \Gamma(1 + 2 / \beta) \tag{9.6.6}$$

It is not encouraged to use $\mu_t^2 + \sigma_t^2$ instead of μ_2 when the sample size is limited. Substituting $\eta = \mu_1/\Gamma(1 + 1/\beta)$, obtained from Equation (9.6.5) into Equation (9.6.6), yields

$$\mu_2 [\Gamma(1 + 1 / \beta)]^2 - \mu_1^2 \Gamma(1 + 2 / \beta) = 0 \tag{9.6.7}$$

The above equation can be solved for parameter β since it is the only unknown. The least squares method is a good option.

After parameter η is obtained, the following equation can be used for calculating shape factor β based on Equation (9.6.5) as

$$\eta = \mu_1 / \Gamma(1 + 1 / \beta) \tag{9.6.8}$$

When estimating parameters of a known family of probability distributions, the MME is superseded by Dr. Fisher's method of maximum-likelihood, because the MLE has a higher probability of being close to the quantities to be estimated and is often less biased [Bowman and Shenton 1998].

EXAMPLE 9.6.1

Twelve assembly robots went through a durability test, and they failed at 315, 386, 487, 509, 512, 613, 660, 725, 753, 811, 849, and 872 h, respectively. Assume that the two-parameter Weibull distribution is of interest. Please estimate Weibull parameters using the MME.

Solution:

Following Equations (9.6.1) and (9.6.2), one has

$$\mu_1 = \frac{1}{Q} \sum_{q=1}^{Q} t_q = 624.25 \tag{a}$$

and

$$\mu_2 = \frac{1}{Q} \sum_{q=1}^{Q} (t_q)^2 = 420652 \tag{b}$$

Substituting the above two equations into Equation (9.6.6) leads to

$$420652 [\Gamma(1 + 1 / \beta)]^2 - (624.25)^2 \Gamma(1 + 2 / \beta) = 0$$

The above equation can be solved iteratively for β,

$$\beta = 2.264209 \tag{c}$$

Plugging Equations (a) and (c) into Equation (9.6.8) yields

i.e., $\beta = \mu_1 / \Gamma(1 + 1 / \eta) = 624.25 / \Gamma(1 + 1 / 2.264209) = 704.75$

9.7. Assessing Weibull Parameters by MRR

Given the t_q (n = 1, 2, 3, …, Q) values, how to identify the probability p_q to approximate $F(t_q)$ for making a p_q-t_q plot, or saying $F(t_q)$-t_q plot? As one of the percentile estimators, MRR is a genuine procedure for estimating Weibull parameters based on the regression method through the points on a probability plot [Lloyd 1952].

A simple but effective way to estimate the Weibull parameters is to utilize Equation (9.2.3) that exhibits a linear relationship on the logarithmic scale.

$$\ln(t - t_0) = \ln(\eta) + \ln\{-\ln[1 - F(t)]\} / \beta \tag{9.7.1}$$

or

$$\ln(t - t_0) = \ln(\eta) + \ln(\ln\{1 / [1 - F(t)]\}) / \beta \tag{9.7.2}$$

Let $p \equiv F(t)$, Equation (9.7.2) can be rewritten as

$$\ln(t - t_0) = \ln(\eta) + \ln\{\ln[1 / (1 - p)]\} / \beta \quad \text{(3-parameter Weibull)} \tag{9.7.3}$$

or

$$\ln(t) = \ln(\eta) + \ln\{\ln[1 / (1 - p)]\} / \beta \quad \text{(2-parameter Weibull)} \tag{9.7.4}$$

For a given two-parameter Weibull CDF F(t), Equation (9.7.4) yields a straight line, as illustrated by Figure 9.2.5, of which the slope is $1/\beta$ and the constant is $\ln(\eta)$.

Given the t_q (n = 1, 2, 3, …, Q) values and their corresponding $F(t_q)$, Equation (9.7.3) or Equation (9.7.4) can be solved for β and η by regression, the least squares method is usually the choice because of its simplicity [Zanakis 1979]. The least squares method-based median rank algorithm is also known as the MRR method because the estimation of the Weibull parameters is dependent upon regressing some form of logarithmic transformation and rank positioning of a given dataset.

As extended from Equation (9.7.4), the inverse of cumulative probability function of standard SEV distribution, i.e., $\Phi_{sev}^{-1}(p)$, and percent point function can be established. In light of Equation (9.7.4), the inverse function of SEV distribution is

$$\Phi_{sev}^{-1}(p) = \ln\{-\ln[1 - p]\} \tag{9.7.5}$$

Following Equation (9.2.8), one has

$$\ln(t) = \ln(\eta) + \Phi_{sev}^{-1}(p)\,\sigma \tag{9.7.6}$$

The analytical motivation for MRR is that the log of the Weibull p quantile is a linear function of $\Phi_{sev}^{-1}(p)$, which is the inverse of cumulative probability function of standard SEV distribution.

9.7.1. Estimating Weibull Parameters with MRR

The median ($t_{50\%}$) rank regression is a least squares fit or linear regression to get the least error between time-to-failure data and fit curve [Lehmann 1975]. The mean order number is used to denote the most likely position that a failure occupies in an increasing order. One implemented operation to obtain the median rank of test data is given as follows [Abernethy 2006, Genschel and Meeker 2010]:

1. List all data (including test-to-failure and suspended units) in the ascending order (from the smallest to the largest). The mean order number is here used to denote the most likely position that test-to-failure data occupy in an increasing order in terms of variable t, such that $t_1 \le t_2 \le t_3 \le \cdots \le t_Q$.
2. If there are no suspended data, the increasing order is the mean order number for each test-to-failure data, with increment $I_q = 1$ by default.
3. When a suspension is encountered increment, I_q is calculated using

$$I_q = \frac{Q + 1 - \text{Previous Mean Order Number}}{1 + Q_{More}} \tag{9.7.5a}$$

where
Q is the sample size
Q_{More} is the number of test-to-failure and suspended data beyond the present suspended

The mean order number of the test-to-failure data right after the suspended set is given by adding the updated increment (I_q) to the mean order number of the previous test-to-failure data. If the very first piece of data is suspended, then the mean order number of "the previous mean order number" is zero.

4. The adjusted mean order number q_{adj} can be calculated by adding the adjusted increment to the previous mean order number.
5. Median ranks are inherent with at a specific confidence level (50%), the ordinate for the plot can be assessed using the following equation in terms of binomial distribution function:

$$\sum_{q=1}^{Q} \binom{Q}{q} p_q^{\,q} (1 - p_q)^{Q-q} = 50\%$$

$$\text{or} \quad F(t_q) = p_q = \frac{1}{1 + [(Q - q + 1) / q]\, F_{50\%,\, 2(Q-q+1),\, 2q}} \tag{9.7.6a}$$

which can be approximated by

$$F(t_q) = p_q = (q_{adj} - 0.3) / (Q + 0.4) \tag{9.7.7}$$

6. The abscissa for the plot will be t_i, where i = 1, 2, …, N are the data for the test result.

Equation (9.7.7) is an approximation based on the MMR, and it redeems the curve fitting with the confidence level of 50% (C = 50%).

9.7.2. MRR (Confidence = 50%)

MRR, which utilizes a best-fit straight line, is a reasonable method to estimate Weibull parameters, even when the sample size is small. Consider the Weibull distribution with three parameters, i.e., Equation (9.7.3). By putting $1 - F(t_f - t_0)$ on one side of the "=" sign and the exponential term on the other side, and then taking natural logarithm twice, one has

$$\ln\{\ln[\frac{1}{1 - F(t_f - t_0)}]\} = \beta \ln(t_f - t_0) - \beta \ln(\eta - t_0) \quad (9.7.8)$$

or

$$\ln\{\ln[\frac{1}{1 - p_f}]\} = \beta \ln(t_f - t_0) - \beta \ln(\eta - t_0) \quad (9.7.9)$$

Note that $\ln\{\ln[1 - F(t_f - t_0)]^{-1}\} = \ln\{-\ln[1 - F(t_f - t_0)]\}$. Substituting the N_f data and their corresponding F(t) into the above equation yields N_f equations, which in general can be solved numerically for β, η, and t_0. Two possible outcomes may appear.

A plot of $\ln\{-\ln[1 - F(t_f - t_0)]\}$ versus $\ln(t_f - t_0)$ is a straight line with β as the slope, as long as variable t fits the Weibull distribution.

If it is not a straight line, then choose a statistical distribution other than the Weibull distribution.

A simple least squares regression [Snee 2001] can yield a reasonable approximate solution to Equation (9.7.8). As Equation (9.7.8) is rearranged in a linear form,

$$Y = \beta X - \beta A \quad (9.7.10)$$

i.e.,

$$X = \beta^{-1} Y + A \quad (9.7.11)$$

where

$$X = \ln(t - t_0) \quad (9.7.12)$$

$$Y = \ln\{-\ln[1 - F(t - t_0)\} \quad (9.7.13)$$

and

$$A = \ln(\eta - t_0) \quad (9.7.14)$$

The fit of data to a Weibull distribution can be visually assessed using a Weibull plot that is representative of the empirical CDF of data on special axes in a type Q–Q plot following Equation (9.7.9). Equation (9.7.11) can be solved for 1/β and A using the least squares method as follows [Cook 1974]:

$$\beta = \frac{(\Sigma X_i Y_i) - (\Sigma X_i)(\Sigma Y_i) / N}{\Sigma X_i^2 - (\Sigma X_i)^2 / N} \quad (9.7.15)$$

and

$$\eta = \exp(X_{ave} - Y_{ave} / \beta) \quad (9.7.16)$$

of which summation sign Σ denotes the sum of all data, ranging from 1 to N. Parameters X_{ave} and Y_{ave} are averages of all X_i and Y_i, respectively. The correlation is calculated using

$$\rho = (\frac{\sigma_{XY}^2}{\sigma_X^2 \sigma_Y^2})^{1/2} = \frac{\Sigma (X_i - X_{ave})(Y_i - Y_{ave})}{\{[\Sigma (X_i - X_{ave})^2][\Sigma (Y_i - Y_{ave})^2]\}^{1/2}} \quad (9.7.17)$$

where

σ_{XY}^2 is the covariance of X and Y
σ_X^2 is the variance of X
σ_Y^2 is the variance of Y

Of course, the correlation follows the rule that the higher the better. MRR redeems the curve fitting with the confidence level of 50%. This method can be extended to a general rank regression that redeems the curve fitting with an arbitrary confidence level, e.g., 5, 10, 90, or 95%.

9.7.3. Weibull Plot

The Weibull plot is a plot of the empirical CDF of data on special axes in a type of Q–Q plot. In light of Equation (9.2.3), the Weibull probability plot is formed by

(a) Vertical Axis: Weibull cumulative probability expressed as a percentage

(b) Horizontal Axis: ordered failure times (in a logarithmic scale)

The vertical scale is $\ln\{-\ln[1 - F(t_q)]\}$, where $F(t_q) = p_q = (q_{adj} - 0.3)/(Q + 0.4)$ and q_{adj} is the adjusted rank of an individual test-to-failure observation. This scale is chosen in order to linearize the resulting plot for Weibull data.

In order to let the starting value of the ordinate be zero at F(t) = 0.1%, a Weibull probability plot is usually constructed using the following equation:

$$Y - 6.90826 = \beta\, X - (\beta\, A + 6.90826) \tag{9.7.18}$$

The above equation is derived from the transformation of Equation (9.7.10) by subtracting constant 6.90826 as illustrated using the data in Table 9.7.1. Note that $\ln(\ln(1.001) = -6.90826$. The mean order number is used to denote the most likely position that a failure occupies in an increasing order. If a plot of $Y = \ln(\ln\{1/[1 - F(t)] - 1\}) + 6.90826$ versus $X = \ln(t - t_0)$ is a straight line, the data set fits the Weibull distribution.

TABLE 9.7.1 MRR for the fitted coordinates in the Weibull probability plot.

Ordinate			Abscissa	
F(t)	Y	Y + 6.90826	t	X = ln(t)
0.1%	−6.90826	0 Y-units	1	0 X-units
1.0%	−4.60	2.31	2	0.60
5.0%	−2.97	3.94	3	1.10
10%	−2.25	4.66	4	1.39
20%	−1.50	5.41	5	1.61
30%	−1.03	5.88	10	2.30
40%	−0.67	6.24	20	3.00
50%	−0.37	6.54	100	4.605
60%	−0.09	6.82	200	5.30
70%	0.19	7.10	500	6.215
80%	0.48	7.39	1000	6.91
90%	0.83	7.74	10,000	9.21
95%	1.10	8.01	100,000	11.51
99%	1.53	8.44	...	...
99.9%	1.93	8.84	...	...

EXAMPLE 9.7.1

Nine automotive latch systems went through a life-cycle test and they failed at 70,560; 71,252; 71,267; 71,275; 71,297; 72,587; 91,011; 91,786; and 97,187 cycles, respectively. Assume that the two-parameter Weibull distribution is of interest. Estimate Weibull parameters using MRR.

Solution:

(a) Two-Parameter Weibull Probabilities: (q_{adj} = q for complete data)

Life (t_f)	q	$F(t_f) \approx (q - 0.3)/(N + 0.4)$	$R(t_f) = 1 - F(t_f)$
70,560	1	7.45%	92.55%
71,252	2	18.09%	81.91%
71,267	3	28.72%	71.28%
71,275	4	39.36%	60.64%
71,297	5	50.00%	50.00%
72,587	6	60.64%	39.36%
91,011	7	71.28%	28.72%
91,786	8	81.91%	18.09%
97,187	9	92.55%	7.45%

(b) Accessing Weibull Parameters: Let

$$X = \ln(t_s)$$

$$\text{and} \quad Y = \ln\{\ln[\frac{1}{1 - F(t_s)}]\}$$

Following Equations (9.7.15) and (9.7.16), one has

$$\text{Then,} \quad \beta = \frac{(\Sigma X_i Y_i) - (\Sigma X_i)(\Sigma Y_i)/N}{\Sigma X_i^2 - (\Sigma X_i)^2/N} = 10.452$$

$$\eta = \exp(X_{ave} - Y_{ave}/\beta) = 82007$$

EXAMPLE 9.7.2

Twelve automotive latch systems went through a life-cycle test. Nine of them fail at 70,560; 71,252; 71,267; 71,275; 71,297; 72,587; 91,011; 91,786; and 97,187 cycles, respectively, one of them was suspended at 90,000 cycles, and the other two were suspended at 100,000 cycles. Assume that the two-parameter Weibull distribution is of interest. Estimate Weibull parameters using MRR.

Solution:

(a) Two-Parameter Weibull Probabilities: Since the first six data are obtained from test-to-failure units, $I_1 = I_2 = I_3 = I_4 = I_5 = I_6 = 1$. The increment for q = 7 has to be adjusted using Equation (9.7.5), i.e.,

$$I_7 = (12 + 1 - 6)/(1 + 5) = 1.167$$

Update I_8, I_9, and I_{10} values by the same token. Derive the adjusted mean order (q_{adj}) according to the increment for individual data q. Then, calculate $F(t_q)$ and $R(t_q)$.

Life (t_f)	Q	I_q	q_{adj}	$F(t_q) \approx (q_{adj} - 0.3)/(N + 0.4)$	$R(t_q) = 1 - F(t_q)$
70,560	1	1	1	5.645%	94.35%
71,252	2	1	2	13.71%	86.29%
71,267	3	1	3	21.77%	78.23%
71,275	4	1	4	29.84%	70.16%
71,297	5	1	5	37.90%	62.10%
72,587	6	1	6	45.97%	54.03%
90,000⁺	7	1.167	—	—	—
91,011	8	1.167	7.167	55.38%	44.62%
91,786	9	1.167	8.333	64.79%	35.22%
97,187	10	1.167	9.5	74.19%	25.81%
100,000⁺	11	—	—	—	—
100,000⁺	12	—	—	—	—

where 90,000⁺ and 100,000⁺ are suspended at 90,000 and 100,000 cycles, respectively.

$R_q = Q - q_{adj} + 1$ is applicable to the test-to-failure data only.

(b) Accessing Weibull Parameters: Consider the test-to-failure data only. Let

$$X = \ln(t_q)$$

and $$Y = \ln\{\ln[\frac{1}{1 - F(t_q)}]\}$$

Following Equations (9.7.15) and (9.7.16), one has

$$\beta = \frac{(\Sigma X_i Y_i) - (\Sigma X_i)(\Sigma Y_i)/N}{\Sigma X_i^2 - (\Sigma X_i)^2/N} = 9.807$$

and $$\eta = \exp(X_{ave} - Y_{ave}/\beta) = 85725$$

9.7.4. Upper and Lower Limits of General Ranking

Order statistics are sample values placed in ascending order with a given number of sample values; i.e., number Q in Table 9.7.2 can be used for generating upper and lower limits of general ranking. On average, the order statistics in a sample tend to divide the area under the curve y = F(x) into Q = n + 1 regions, each with area 1/(n + 1). The requirement that meets reliability R = 90% and confidence C = 90% is a general practice in the automotive industry. Similar to Equation (9.7.6), for a general rank regression the upper and lower limits can be obtained from the two following equations:

$$\sum_{q-1}^{Q} \binom{Q}{q} p_q^{\,q} (1 - p_q)^{Q-q} = C \tag{9.7.19}$$

$$\text{and} \quad \sum_{q=1}^{Q} \binom{Q}{q} p_q^{\,q} (1 - p_q)^{Q-q} = 1 - C \qquad (9.7.20)$$

For the given upper and lower bounds, the generalized ranking probability p_q can be solved using the above equation for any combination of sample size Q and confidence C. The two extremities demonstrated by these two equations are the accumulative probabilities of no failures and total failure, respectively. Tabulated % ranks listed in Table 9.7.2 are given for an accumulative probability ranging from 1 to 99.9%, i.e., percentile points of order statistics in random samples of a uniform distribution [Romig 1947].

Percentile points of ordered statistics in samples from beta, normal, and chi (1 DOF) populations are given in [Govindarajulu and Hubacker 1962].

TABLE 9.7.2 Percentile points of order statistics in random samples of a uniform distribution [Romig 1947].

Order	1%	5%	10%	25%	50%	75%	90%	95%	99%	99.9%
Q = 1										
1	1.0	5	10	25	50	75	90	95	99	99.9
Q = 2										
1	0.501	2.532	5.132	13.40	29.29	50.0	68.38	77.64	90.0	96.84
2	10.0	22.36	31.62	50	70.71	86.60	94.87	98.74	99.50	99.95
Q = 3										
1	0.334	1.695	3.451	9.144	20.63	37.00	53.58	63.16	90.00	96.84
2	5.850	13.54	19.58	32.635	50.00	67.365	80.42	86.47	94.11	98.16
3	21.54	36.84	46.42	63.00	79.37	90.86	96.55	98.31	99.67	99.97
Q = 4										
1	0.251	1.274	2.60	6.94	15.91	29.29	43.77	52.71	68.38	82.22
2	4.2	9.76	14.26	24.3	38.57	54.37	67.95	75.14	85.91	93.6
3	14.09	24.86	32.05	45.63	61.43	75.7	85.74	90.24	95.8	98.7
4	31.62	47.29	46.23	70.71	84.09	93.06	97.4	98.73	99.75	99.98
Q = 5										
1	0.201	1.021	2.085	9.591	12.95	24.21	36.90	45.07	60.19	74.88
2	2.288	7.644	11.22	19.38	31.38	45.42	58.39	65.74	77.79	87.80
3	10.56	18.93	24.66	35.94	50.00	64.06	75.34	81.07	89.44	95.25
4	22.21	34.3	41.61	54.58	68.62	80.62	88.78	92.36	96.73	98.99
5	39.81	54.93	63.1	75.79	87.06	94.41	97.92	98.98	99.80	99.98
Q = 6										
1	0.167	0.851	1.741	4.682	10.91	20.63	31.17	39.30	53.58	68.38
2	2.676	6.285	9.26	16.12	26.45	38.95	51.03	58.18	70.57	81.86
3	8.473	15.32	20.09	29.69	42.14	55.32	66.68	72.87	82.69	90.61
4	17.31	27.13	33.32	44.68	57.86	70.31	79.91	84.68	91.53	96.21
5	29.43	41.112	48.97	61.05	73.56	83.88	90.74	93.72	97.32	99.17
6	46.42	60.70	68.13	79.37	89.09	95.32	98.26	99.15	99.92	99.83

(Continued)

TABLE 9.7.2 (Continued) Percentile points of order statistics in random samples of a uniform distribution [Romig 1947].

Order	1%	5%	10%	25%	50%	75%	90%	95%	99%	99.9%
Q = 7										
1	0.143	0.730	1.494	4.026	9.428	17.97	28.03	34.82	48.21	62.72
2	2.267	5.338	7.882	13.80	22.85	34.07	45.26	52.07	64.34	76.25
3	7.080	12.876	16.96	25.31	36.41	48.61	59.62	65.87	76.37	85.62
4	14.23	22.53	27.86	37.89	50.00	62.12	72.14	77.47	85.77	92.34
5	23.63	34.13	40.38	51.39	63.59	74.69	83.04	87.12	92.92	96.84
6	35.66	47.93	54.74	65.93	77.15	86.20	92.12	94.66	97.73	99.30
7	51.76	65.18	71.97	82.03	90.57	95.97	98.51	99.27	99.86	99.99
Q = 8										
1	0.126	0.639	1.308	3.532	8.3	15.91	25.01	31.23	43.77	57.83
2	1.966	3.185	6.863	12.06	20.11	30.27	40.63	47.07	58.99	71.13
3	4.746	6.084	8.523	22.06	32.05	43.32	53.82	59.97	70.68	80.73
4	12.1	19.29	14.69	32.91	44.02	55.55	65.54	71.08	80.18	88.04
5	19.82	28.92	34.46	44.45	55.98	67.09	76.03	80.71	87.91	93.52
6	29.32	40.03	46.18	56.68	67.95	77.94	85.32	88.89	93.92	97.3
7	41.01	52.93	59.38	69.73	79.89	87.94	93.14	95.36	98.03	99.4
8	56.23	68.77	74.99	84.09	91.7	96.47	98.69	99.36	99.87	99.99
Q = 9										
1	0.112	0.568	1.164	3.146	7.413	14.28	22.57	28.31	40.05	53.58
2	1.736	4.102	6.077	10.72	17.96	27.23	36.84	42.91	54.40	66.51
3	5.335	9.775	12.95	19.55	28.62	39.05	49.01	54.96	65.63	76.12
4	10.53	16.88	21.04	29.10	39.31	50.20	59.94	65.51	75.00	83.71
5	17.10	25.14	30.10	39.20	50.00	60.84	69.90	74.86	82.90	89.75
6	25.00	34.49	40.06	49.80	60.69	70.90	78.96	83.13	89.47	94.38
7	34.37	45.04	50.99	60.95	71.38	80.45	87.05	90.23	94.67	97.63
8	45.60	57.09	63.16	72.77	82.04	89.28	93.92	95.90	98.26	99.47
9	59.95	71.69	77.43	85.72	92.59	96.85	98.84	99.43	99.89	99.99
Q = 10										
1	0.1	0.512	1.048	2.836	6.697	12.945	20.567	25.89	36.90	49.88
2	1.554	3.677	5.453	9.64	16.23	24.74	33.69	39.42	50.44	62.37
3	4.751	8.726	11.58	17.56	25.86	35.54	44.96	50.69	61.17	71.85
4	9.321	15.0	18.76	26.09	35.51	45.77	55.17	60.66	70.29	79.54
5	15.04	22.24	26.73	35.07	45.17	55.55	64.58	69.65	78.17	85.87
6	21.113	30.35	35.42	44.45	54.83	64.93	73.27	77.76	84.96	91.02
7	29.71	39.34	44.83	54.23	64.49	73.92	81.24	85.0	90.68	95.04
8	38.83	49.31	55.04	64.46	74.14	82.44	88.42	91.27	95.25	97.9
9	49.57	60.58	66.32	75.27	83.77	90.36	94.55	96.32	98.45	99.52
10	63.10	74.11	79.43	87.06	93.30	97.16	98.95	99.49	99.90	99.99

(Continued)

TABLE 9.7.2 (Continued) Percentile points of order statistics in random samples of a uniform distribution [Romig 1947].

Order	1%	5%	10%	25%	50%	75%	90%	95%	99%	99.9%
Q = 11										
1	0.091	0.465	0.953	2.581	6.107	11.84	18.89	23.84	34.21	46.63
2	0.982	3.332	4.945	8.761	14.80	22.66	31.02	36.44	46.98	58.66
3	4.282	7.882	10.48	15.93	23.58	32.61	41.52	47.01	57.32	67.93
4	8.366	13.51	23.64	31.74	41.19	42.05	51.08	56.44	66.04	75.60
5	13.44	19.96	24.05	31.74	41.19	51.11	59.95	65.02	73.78	82.06
6	19.36	27.13	31.77	40.16	50.00	59.84	68.23	72.88	80.60	87.51
7	26.22	34.98	40.05	48.89	58.81	68.27	75.95	80.04	86.56	92.01
8	33.96	43.56	48.93	57.95	67.62	76.36	83.08	86.49	91.63	95.56
9	42.77	52.99	58.48	67.39	76.42	84.07	89.52	92.12	95.72	98.11
10	53.02	63.56	68.98	77.34	85.20	91.24	95.06	96.67	98.59	99.57
11	65.75	76.16	81.11	88.16	93.89	97.42	99.05	99.54	99.91	99.991
Q = 12										
1	0.084	0.427	0.874	2.369	5.613	10.91	17.46	22.09	31.117	43.77
2	1.285	3.046	4.524	8.029	13.60	20.91	28.75	33.87	43.95	55.34
3	3.858	7.187	9.565	14.59	21.67	30.12	38.55	43.81	53.74	64.36
4	7.589	12.29	15.42	21.62	29.76	38.88	47.53	52.73	62.22	71.92
5	12.15	18.10	21.117	28.99	37.85	47.31	55.90	60.91	69.76	78.41
6	17.46	24.53	28.82	36.63	45.95	55.47	63.77	68.48	76.51	84.01
7	23.49	31.52	36.23	44.53	54.05	63.37	71.18	75.47	82.54	88.80
8	30.24	39.09	44.10	52.69	62.15	71.02	78.13	81.90	87.85	92.79
9	37.78	47.27	52.47	61.12	70.24	78.38	84.58	87.72	92.41	95.98
10	46.27	56.19	61.45	69.88	78.33	85.42	90.44	92.81	96.10	98.28
11	56.05	66.13	71.25	79.09	86.40	91.97	95.48	96.95	98.72	99.61
12	68.13	77.91	82.54	89.09	94.39	97.63	99.13	99.57	99.92	99.992
Q = 13										
1	0.077	0.394	0.807	2.189	5.192	10.12	16.23	20.58	29.83	41.22
2	1.182	2.805	4.169	7.410	12.58	19.41	26.78	31.63	41.28	52.34
3	3.578	6.605	8.800	13.45	20.04	27.98	35.98	41.01	50.62	61.10
4	6.946	11.27	14.16	19.91	27.53	36.15	44.43	49.47	58.78	68.51
5	11.08	16.57	20.05	26.68	35.02	44.03	52.34	57.26	66.09	74.97
6	15.88	22.40	26.37	33.68	42.51	51.57	59.82	64.52	72.71	80.62
7	21.29	28.71	33.09	40.90	50.00	59.10	66.91	71.30	78.71	85.57
8	27.29	35.48	40.18	48.33	57.49	66.32	73.63	77.60	84.12	89.43
9	33.91	42.74	47.66	55.97	64.98	73.32	79.95	83.43	88.92	93.44
10	41.22	50.54	55.57	63.85	72.47	80.09	85.84	88.73	93.05	96.32
11	49.38	58.99	64.02	72.02	79.96	86.55	91.20	93.40	96.42	98.42
12	58.72	68.37	73.22	80.60	87.42	92.59	85.83	97.20	98.82	99.64
13	70.17	79.42	83.77	89.89	94.81	97.81	99.19	99.61	99.92	99.992

(Continued)

TABLE 9.7.2 (Continued) Percentile points of order statistics in random samples of a uniform distribution [Romig 1947].

Order	1%	5%	10%	25%	50%	75%	90%	95%	99%	99.9%
Q = 14										
1	0.072	0.366	0.75	2.034	4.83	9.428	15.17	19.26	28.03	38.95
2	1.095	2.600	3.866	6.879	11.70	18.10	25.07	29.67	38.91	49.64
3	3.306	6.110	8.148	12.48	18.65	26.12	33.72	38.54	47.83	58.12
4	6.403	10.41	13.09	18.46	25.61	33.77	41.70	46.57	55.67	65.37
5	10.19	15.27	18.51	24.71	32.58	41.17	49.20	54.00	62.74	71.73
6	14.57	20.61	24.32	31.17	39.54	48.35	56.31	60.96	69.20	77.38
7	19.47	26.36	30.45	37.83	46.52	55.35	63.09	67.50	75.12	82.41
8	24.88	32.05	36.91	44.65	53.49	62.18	69.55	73.64	80.53	86.85
9	30.80	39.04	43.69	51.65	60.46	68.83	75.68	79.39	85.43	90.71
10	37.26	46.00	50.80	58.83	67.43	95.29	81.49	84.73	89.81	93.98
11	44.33	53.43	58.30	6.23	74.39	81.54	86.91	89.60	93.60	94.62
12	52.17	61.46	66.28	73.88	81.35	87.53	91.85	93.89	96.69	98.54
13	61.09	70.33	74.94	81.90	88.30	93.12	96.13	97.40	98.91	99.66
14	71.97	80.74	84.83	90.57	95.17	97.97	99.25	99.63	99.93	99.993
Q = 22										
1	0.046	0.233	0.478	1.299	3.102	6.107	9.937	15.44	18.89	26.95
2	0.689	1.640	2.444	4.374	7.512	11.78	16.56	19.81	26.58	34.95
3	2.057	3.822	5.117	7.904	11.97	17.05	22.42	25.95	33.05	41.51
4	3.943	6.460	8.175	11.66	16.44	22.12	27.89	31.59	38.87	47.30
5	6.214	9.411	11.49	15.56	20.91	27.05	33.10	36.91	44.26	52.83
6	8.789	12.60	15.00	19.57	25.38	31.87	38.12	41.98	49.33	57.47
7	11.62	15.99	18.67	23.66	29.86	36.60	42.97	46.85	54.12	62.02
8	14.68	19.57	22.48	27.83	34.33	41.26	47.68	51.55	58.68	66.30
9	17.93	23.27	26.42	32.07	38.81	45.85	52.28	56.09	63.04	70.33
10	21.38	27.13	30.46	36.37	43.29	50.39	56.75	60.48	67.21	74.12
11	25.01	31.13	34.62	40.73	47.76	54.86	61.12	64.75	71.19	77.69
12	28.81	35.25	38.88	45.14	52.24	59.27	65.38	68.87	74.99	81.04
13	32.80	39.52	43.25	49.62	56.71	63.63	69.54	72.87	78.62	84.18
14	36.96	43.91	47.73	54.15	61.19	67.93	73.58	76.73	82.07	87.11
15	41.32	48.45	52.32	58.74	65.67	72.17	77.52	80.44	85.32	89.80
16	45.88	53.15	57.03	63.40	70.14	76.34	81.33	84.01	88.38	92.26
17	50.67	58.02	61.88	68.13	74.62	80.44	85.00	87.40	91.21	94.46
18	55.74	63.09	66.90	72.95	79.09	84.44	88.51	90.59	93.79	96.36
19	61.13	68.41	72.11	77.88	83.56	88.34	91.83	93.54	96.06	97.93
20	66.95	74.05	77.58	82.95	88.03	92.10	94.88	96.18	97.943	99.10
21	73.42	80.19	83.44	88.22	92.49	95.63	97.56	98.36	99.31	99.79
22	81.11	87.27	90.06	93.89	96.90	98.70	99.52	99.77	99.95	99.995

where Q is the total number of data.

EXAMPLE 9.7.3

The following 12 test data are obtained from durability tests of side-door hinge systems for a sedan: 230,000; 135,000; 240,000 (suspended); 240,000 (suspended); 210,000; 240,000 (suspended); 160,000; 150,000; 240,000 (suspended); 240,000 (suspended); 200,000 (suspended); and 190,000. The 11th specimen was censored (suspended) because its fixture failed, and the final five hinge systems were censored after they reach three-lifetime as cost mandates. What is the lifetime level having (a) reliability R = 90% and confidence C = 50% and (b) reliability R = 90% and confidence C = 90%?

Solution:

(a) Two-Parameter Weibull Probabilities: Since the first four data are obtained from test-to-failure units, $I_1 = I_2 = I_3 = I_4 = 1$. The increment for q = 5 has to be adjusted using Equation (9.7.5), i.e.,

$$I_5 = (12+1-4)/(1+7) = 1.125$$

Update I_6 and I_7 values accordingly. Derive the adjusted mean order (q_{adj}) according to the increment for the follow-up individual data q. Then, calculate $F(t_q)$ and $R(t_q)$.

Life (t_f)	q	I_q	q_{adj}	$F(t_q) \approx (q_{adj} - 0.3)/(N + 0.4)$	$R(t_q) = 1 - F(t_q)$
135,000	1	0	1	5.65%	94.35%
150,000	2	1	2	13.71%	86.29%
160,000	3	1	3	21.77%	78.23%
190,000	4	1	4	29.84%	70.16%
200,000⁺	5	1.125	—	—	—
210,000	6		5.125	38.91%	61.09%
230,000	7	1.125	6.25	47.98%	52.02%
240,000⁺	8	—	—	—	—
240,000⁺	9	—	—	—	—
240,000⁺	10	—	—	—	—
240,000⁺	11	—	—	—	—
240,000⁺	12	—	—	—	—

where 200,000⁺ and 240,000⁺ are suspended at 200,000 and 240,000 cycles, respectively.

$R_q = Q - q_{adj} + 1$ is applicable to the test-to-failure data only.

(b) Accessing Weibull Parameters: Consider the test-to-failure data only. Let

$$X = \ln(t_q)$$

and

$$Y = \ln\{\ln[\frac{1}{1-F(t_q)}]\}$$

Following Equations (9.7.15) and (9.7.16), one has

$$\beta = \frac{(\Sigma X_i Y_i) - (\Sigma X_i)(\Sigma Y_i)/N}{\Sigma X_i^2 - (\Sigma X_i)^2/N} = 4.464$$

and $$\eta = \exp(X_{ave} - Y_{ave}/\beta) = 240257$$

(c) Lower Bound with R = 90% and C = 50%

Plugging the data into the Minitab (→ Stat → Reliability/Survival → Distribution Analysis (Right Censoring) → Parametric Distribution Analysis) and adjusting the confidence to "C = 50%" and "Lower Bound (LB)," one is able to obtain Figure 9.7.1. Since the test-to-failure data are the more the better, the lower bound is taken. There are six test-to-failure data, i.e., blue circular dots. It can be read from the figure that t = 145,125 cycles when R = 90%, as located by the red triangle. The side-door hinge system will survive 145,125 cycles with R = 90% and C = 50%.

FIGURE 9.7.1 Weibull probability plot with 50% lower bound using LSXY estimates.

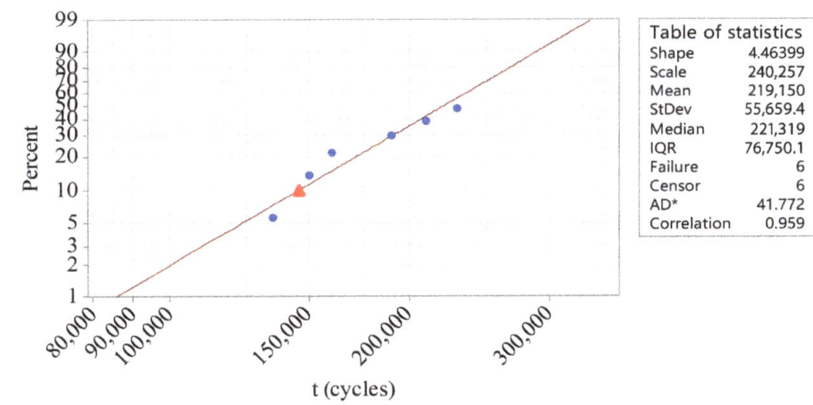

(d) Lower Bound with R = 90% and C = 90%

Plugging the data into the Minitab (→ Stat → Reliability/Survival → Distribution Analysis (Right Censoring) → Parametric Distribution Analysis) and adjusting the confidence to "C = 90%" and "Lower Bound (LB)," one is able to obtain Figure 9.7.2. Since the test-to-failure data are the more the better, the lower bound is taken. There are six test-to-failure data, i.e., blue circular dots. It can be read from the figure that t = 115,030 cycles when R = 90%, as located by the red triangle. The side-door hinge system will survive 115,030 cycles with R = 90% and C = 90%. The requirement of R = 90% and C = 90% is generally demanded for durability tests in the automotive industry. If the driver uses the front door hinges ten cycles per day, 115,030 cycles mean to survive 31.515 (=115,030/10/365) years that well exceed the general expectation of an automobile life (15 years).

FIGURE 9.7.2 Weibull probability plot with 90% lower bound using LSXY estimates.

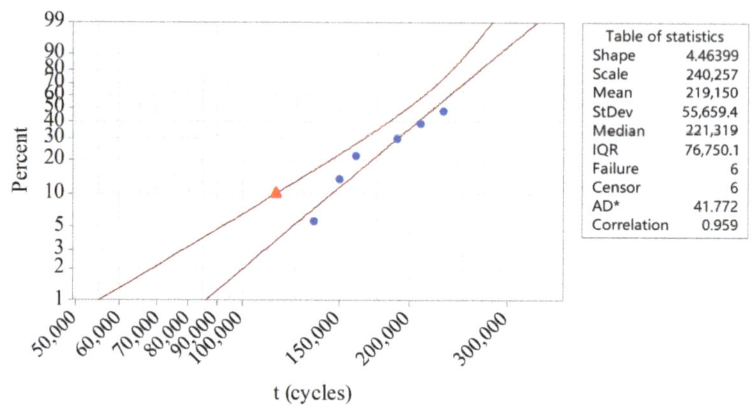

9.7.5. Comparison of the Estimators

The three frequently used methods for estimating the Weibull parameters are maximum-likelihood method, method of moments, and MRR [Nwobil and Ugomma 2014]. Assume that the product life data (time-to-failure) collected from 15 test specimens are 0.438, 2.413, 3.073, 3.079, 3.137, 3.198, 3.918, 4.287, 4.508, 4.981, 5.115, 5.592, 5.848, 5.958, and 6.013. The Weibull parameters estimated using these three methods are listed as follows [Littell and Rao 1978]:

Method	β (shape)	η (scale)	Mean square error
Maximum-likelihood method	2.923	4.553	3.57×10^{-3}
Method of moments	2.941	4.587	3.17×10^{-3}
MRR	1.852	4.756	8.40×10^{-3}

It can be seen that both the maximum-likelihood method and method of moments are correlated to each other quite well for the estimated shape factor. The shape factor estimated using the MRR has a larger mean square error. Given a set of test data of a durable system, its value of shape factor β is frequently hard to come by.

9.8. Reliability DOE-W

Consider the two-parameter Weibull statistics, whose CDF and PDF are given by Equations (9.2.13) and (9.2.14), respectively. Let t_{qn} be the observed lifetime for the nth observation under treatment q test condition. Assume that the failure time within treatment q follows a two-parameter Weibull distribution. Equations (9.2.13) and (9.2.14) become

$$F(t_{qn}) = 1 - \exp\left[-\left(\frac{t_{qn}}{\eta_q}\right)^{\beta_q}\right] \quad \text{(Weibull CDF)} \tag{9.8.1}$$

$$\text{and} \quad f(t_{qn}) = \frac{\beta_q}{\eta_q}\left(\frac{t_{qn}}{\eta_q}\right)^{(\beta_q-1)} \exp\left[-\left(\frac{t_{qn}}{\eta_q}\right)^{\beta_q}\right] \quad \text{(Weibull PDF)} \tag{9.8.2}$$

where

t_{qn} is the life data; subscript q means the qth treatment and n is the nth replication within it
β_q is the shape factor within treatment q
η_q is the scale factor within treatment q

The mean of the Weibull distribution is carried over from the expectation as described by Equation (9.2.6) with the assumption that guaranteed value $t_0 = 0$.

$$E(t_{qn}) = \eta_q \Gamma(1 + 1 / \beta_q) \tag{9.8.3}$$

9.8.1. Coupling between DOE Model and Weibull Distribution

If there are N data that result from the qth treatment in the DOE being addressed, the average is expected to approach μ_{yq} when sample size N is large. The difference among all the data within treatment is then due to the stochastic property within treatment q, and it is assumedly governed by the Weibull distribution. In the meanwhile, the difference among the Q treatments is governed by an acceleration model, e.g., the deterministic approximation of a polynomial in the DOE like Equation (2.3.1). In reliability DOE based on the Weibull distribution (DOE-W), the scale parameter (η_q) activated by each run condition q (as a combination of acceleration factors) is assumed to be

$$\mu_{tq} \approx E(t_q) = \eta_q \Gamma(1 + 1 / \beta_q)$$

$$\text{Or} \quad \ln(\mu_{tq}) \approx \ln[E(t_q)] = \ln(\eta_q) + \ln[\Gamma(1 + 1 / \beta_q)] \tag{9.8.4}$$

It means that the influences of η_q and β_q on the expected lifetime can be accounted for separately.

9.8.2. Finite Mixture Distribution in Two-Parameter Weibull Statistics

When each treatment in the DOEs has its own distinct Weibull distribution. In light of Equation (9.2.31), the standardized variable is

$$z_q = \beta_q [\ln(t_q) - \mu_{yq}] = \beta_q [\ln(t_q) - \ln(\eta_q)] \tag{9.8.5}$$

where

t_q is the data, the qth
β_q is the shape factor of treatment q
η_q is the scale factor of treatment q

If β_q and/or η_q given in the above equation depend on the treatment condition (q), the DOE modeling may become a problem-solving procedure involving a finite Weibull mixture distribution. A finite mixture distribution in Weibull statistics appears when one deals with a heterogeneous population that consists of finite number of "more homogeneous Weibull subpopulation" within each treatment q. The finite Weibull mixture distribution that consists of Q-weighted heterogeneous Weibull PDFs can be written as [Tarum 1999a, 1999b]

$$f(t) = \sum_{q=1}^{Q} w_q f_q(t) \tag{9.8.6}$$

and

$$f_q(t) = \left(\frac{\beta_q}{\eta_q}\right) \left(\frac{t}{\eta_q}\right)^{(\beta_q - 1)} \exp\left[-\left(\frac{t}{\eta_q}\right)^{\beta_q}\right] \tag{9.8.7}$$

where

w_q is the mixing weight attached to treatment q, which ranges from 1 to Q
$f_q(t)$ is the Weibull function associated with treatment q

Then, the related reliability function becomes

$$R(t) = \sum_{q=1}^{Q} w_q R_q(t) \tag{9.8.8}$$

where

$$R_q(t) = \exp\left[-\left(\frac{t}{\eta_q}\right)^{\beta_q}\right] \tag{9.8.9}$$

Since the hazard rate is defined as the ratio of $f_q(t)$ to $R_q(t)$, one has

$$h(t) = \frac{f(t)}{R(t)} = \sum_{q=1}^{Q} w_q h_q(t) \tag{9.8.10}$$

and

$$f_q(t) = \left(\frac{\beta_q}{\eta_q}\right) \left(\frac{t}{\eta_q}\right)^{(\beta_q - 1)} \tag{9.8.11}$$

Finally, the summarized MTTF can be expressed in terms of the MTTFs of heterogeneous Weibull distributions given by Equation (9.8.3) as

$$MTTF = \sum_{q=1}^{Q} w_q \eta_q \Gamma\left(1 + \frac{1}{\beta_q}\right) \tag{9.8.12}$$

It can be seen that scale factor η_q is in effect as an individual contributor to the overall MTTF of a finite mixture distribution in two-parameter Weibull statistics.

9.8.3. MLE for a Homogeneous Two-Parameter Weibull Distribution

If the null hypothesis is true, the test data $\ln(\eta_q)$ resulting from treatment q can be approximated by Equation (9.8.4).

$$\ln(\eta_{yq}) \approx \gamma_0 + \sum_{i=1}^{k} \gamma_i x_{iq} + \sum_{i=1}^{k} \sum_{j=1}^{k} \gamma_{ij} x_{iq} x_{jq} + \sum_{i=1}^{k} \sum_{j=1}^{k} \sum_{m=1}^{k} \gamma_{ijm} x_{iq} x_{jq} x_{mq} \tag{9.8.13}$$

where

η_{yq} is the lognormal mean value of a Weibull distribution under treatment q

γ_0, γ_i, γ_{ij}, and γ_{ijm} are the effects (coefficients) to be estimated using the maximum-likelihood method

x_{iq}, x_{jq}, and x_{mq} are the test variates; for example, each takes (–1, 0, +1) for factorial design 3^K

Similar to Equation (8.4.12) for the lognormal distribution, the likelihood for the two-parameter Weibull distribution that accommodates all three types of test data is

$$L_w(\beta, \gamma_0, \gamma_i, \gamma_{ij}, \gamma_{ijm}) = L_{failure}\, L_{suspended}\, L_{interval}$$

$$= \{\prod_{f=1}^{Q_f} f(t_f)\}\ \{\prod_{s=1}^{Q_s} [R(t_s)]\}\ \{\prod_{i=1}^{Q_i} [F(t_{i,2}) - F(t_{i,1})]\} \tag{9.8.14}$$

where

Q_f is the total number of test-to-failure specimens; subscript f for number order

Q_s is the total number of suspended (censored) test specimens; subscript s for number order

Q_i is the total number of interval data; subscript n for number order

t_f is the life of specimen, up to failure

t_s is the life of specimen, up to the moment of suspension (censoring)

$t_{i,2}$ and $t_{i,1}$ are the time interval between time "1" and time "2," and $t_{i,1} < t_{failure} < t_{i,2}$

Shape factor $\beta_q = \beta$ is assumedly a constant throughout the entire model, while η_{yq} varies with respect to treatment q (run condition). As all the treatments have a common shape factor, it is called homogeneous Weibull distribution, and so, Equation (9.8.3) reduces to

$$E(t_{qn}) = \eta_q\, \Gamma(1 + 1 / \beta) \tag{9.8.15}$$

i.e., the expected value of test data given that the shape factor is assumedly consistent throughout the entire model.

Plugging Equations (9.8.1) and (9.8.2) into the three terms on the right side of Equation (9.8.14), one has the following three terms [Meeker et al. 1998, Wu 2002]:

1. Test-to-Failure Data

$$L_{failure} = \prod_{f=1}^{Q_f} f(t_f, \eta_q) = \prod_{f=1}^{Q_f} \{\frac{\beta}{\eta_f} (\frac{t_f}{\eta_f})^{(\beta-1)} \exp[-(\frac{t_f}{\eta_f})^{\beta}]\} \tag{9.8.16}$$

2. Right-Suspended Data (Type I Censoring)

$$L_{suspended} = \prod_{s=1}^{Q_s} [R(t)] = \prod_{s=1}^{Q_s} \exp\left[-\left(\frac{t_s}{\eta_s}\right)^{\beta}\right] \tag{9.8.17}$$

3. Interval Data ($t_{i,1} < t_{failure} < t_{i,2}$)

$$L_{interval} = \prod_{i=1}^{Q_i} \{F(t_{i,2}) - F(t_{i,1})\} = \prod_{i=1}^{Q_i} \left(\left\{1-\exp\left[-\left(\frac{t_{i,2}}{\eta_i}\right)^{\beta}\right]\right\} - \left\{1-\exp\left[-\left(\frac{t_{i,1}}{\eta_i}\right)^{\beta}\right]\right\}\right)$$

$$= \prod_{i=1}^{Q_i} \left\{\exp\left[-\left(\frac{t_{i,1}}{\eta_i}\right)^{\beta}\right] - \exp\left[-\left(\frac{t_{i,2}}{\eta_i}\right)^{\beta}\right]\right\} \tag{9.8.18}$$

When only the first three terms in Equation (9.8.18) is considered, for simplicity but without losing its generality, $L_w(\beta, \mu_q, t_q)$ is approximated by $L_w(\sigma_y, \gamma_0, \gamma_i, \gamma_{ij}, \gamma_{ijm})$ as

$$L_w(\beta, \mu_q, t_q) \approx L_w(\beta, \gamma_0, \gamma_i, \gamma_{ij}, \gamma_{ijm}) \tag{9.8.19}$$

where

β is the shape factor to be estimated
μ_q is the mean of y, where $y = \ln(t_q)$; thus, $\mu_q = \ln(\eta_q)$
γ_0, γ_i, γ_{ij}, and γ_{ijm} are the parameters to be estimated

Weibull parameters of the model are to be obtained by maximizing $[L_w(\beta, \eta_q, t_q)]$, i.e., Equation (9.8.14), as

$$\ln[L_w(\beta, \gamma_0, \gamma_i, \gamma_{ij}, \gamma_{ijm})] = \ln(L_{failure}) + \ln(L_{suspended}) + \ln(L_{interval}) \tag{9.8.20}$$

where
$$\ln(L_{failure}) = \sum_{f=1}^{Q_f} \ln\left\{\frac{\beta}{\eta_f}\left(\frac{t_f}{\eta_f}\right)^{(\beta-1)} \exp\left[-\left(\frac{t_f}{\eta_f}\right)^{\beta}\right]\right\} \tag{9.8.21}$$

$$\ln(L_{suspended}) = \sum_{s=1}^{Q_s} \left[-\left(\frac{t_s}{\eta_s}\right)^{\beta}\right] \tag{9.8.22}$$

and
$$\ln(L_{interval}) = \sum_{i=1}^{Q_i} \ln\left\{\exp\left[-\left(\frac{t_{i,1}}{\eta_i}\right)^{\beta}\right] - \exp\left[-\left(\frac{t_{i,2}}{\eta_i}\right)^{\beta}\right]\right\} \tag{9.8.23}$$

9.8.4. Hypotheses in DOEs

Consider a general case up to the third-order terms with no replication, as given by Equation (9.8.13). The lognormal mean value of treatment q can be reasonably assumed to be approximated by the following series of expansion. Note that three-factor and higher-order interactions are usually not included in the analysis for most engineering case studies in practice. To test out the significance of each effect (coefficient), the following hypotheses come with the above equation:

$$H_o: \gamma_0 = 0 \qquad H_1: \gamma_0 \neq 0 \tag{9.8.24}$$

$$H_o: \gamma_i = 0 \qquad H_1: \gamma_i \neq 0 \quad (i = 1, 2, 3, \ldots) \tag{9.8.25}$$

$$H_o: \gamma_{ij} = 0 \qquad H_1: \gamma_{ij} \neq 0 \quad (i = 1, 2, 3, \ldots \;\&\; j = 1, 2, 3, \ldots) \tag{9.8.26}$$

and

$$H_o: \gamma_{ijm} = 0 \qquad H_1: \gamma_{ijm} \neq 0 \quad (i = 1, 2, 3, \ldots, j = 1, 2, 3, \ldots, \;\&\; m = 1, 2, 3, \ldots) \tag{9.8.27}$$

The likelihood function is a function of unknowns β, γ_0, γ_i, γ_{ij}, and γ_{ijm}, which are to be evaluated against observations t_q, where $q = 1, 2, \ldots, Q$.

As prescribed by the maximum-likelihood method, unknown parameters β, γ_i, γ_{ij}, and γ_{ijm} are to be calculated using the following first differentials:

$$\frac{\partial\{\ln[L_w(\beta, \gamma_0, \gamma_i, \gamma_{ij}, \gamma_{ijm})]\}}{\partial\beta} = 0 \tag{9.8.28}$$

$$\frac{\partial\{\ln[L_w(\beta, \gamma_0, \gamma_i, \gamma_{ij}, \gamma_{ijm})]\}}{\partial\gamma_0} = 0 \tag{9.8.29}$$

$$\frac{\partial\{\ln[L_w(\beta, \gamma_0, \gamma_i, \gamma_{ij}, \gamma_{ijm})]\}}{\partial\gamma_i} = 0 \tag{9.8.30}$$

$$\frac{\partial\{\ln[L_w(\beta, \gamma_0, \gamma_i, \gamma_{ij}, \gamma_{ijm})]\}}{\partial\gamma_{ij}} = 0 \tag{9.8.31}$$

and

$$\frac{\partial\{\ln[L_w(\beta, \gamma_0, \gamma_i, \gamma_{ij}, \gamma_{ijm})]\}}{\partial\gamma_{ijm}} = 0 \tag{9.8.32}$$

After parameter β, constant γ_0, and parameters γ_i, γ_{ij}, and γ_{ijm} are obtained from the simultaneous equations given above, one may proceed with examining the statistical significance of each unknown using the LR test for formulating the predictive equation. The procedure is to be demonstrated using an example in the next section (Section 9.9).

9.8.5. LR Test

For the Weibull distribution, the LR test is put into practice as the goodness of fit that determines the model parameters. The following LR test can be used [Guo and Mettas 2010] to justify the validity of each individual effect coefficient γ_{test}, which can be anyone of γ_o, γ_i, γ_{ij}, and γ_{ijm} in Equation (9.8.13):

$$LR(\gamma_{test}) = -\ln\left[\frac{L_w(All-\gamma_{test})}{L_w(All)}\right]^2$$

$$\text{i.e.,} \quad LR(\gamma_{test}) = -2\,\{\ln[L_w(All-\gamma_{test})] - \ln[L_w(All)]\} \tag{9.8.33}$$

where

$LR(\gamma_{test})$ is the LR for effect x that corresponds to coefficient γ_{test}
$L_w(All)$ is the likelihood value with all γ_o, γ_i, γ_{ij}, and γ_{ijm}
$L_w(All - \gamma_{test})$ is the likelihood value with all γ_o, γ_i, γ_{ij}, and γ_{ijm}, except γ_{test}
γ_{test} is the one of γ_o, γ_i, γ_{ij}, and γ_{ijm}, as its statistical significance is of interest for the test

Because $\ln[L_w(All - \gamma_{test})]$ is usually smaller than $\ln[L_w(All)]$, a negative sign is given in the above equation to assure that there is a positive LR. Two possibilities are as follows:

(a) If effect γ_{test} is not relevant, $LR(\gamma_{test})$ is numerically very small or close to zero whether or not it is removed from the full model of Equation (9.8.13). It means that the likelihood value of effect γ_{test} will not affect the LR significantly. Statistically, $LR(\gamma_{test})$ follows a chi-squared distribution with one DOF, if effect γ_{test} is not significant.
(b) On the other hand, if the $LR(\gamma_{test})$ is large, effect γ_{test} is statistically significant. The null hypothesis is rejected at a significance level, α, when $LR(\gamma_{test})$ is greater than the critical value $\chi^2_{1,\alpha}$.

9.8.6. Standardized Residual for Weibull Distribution

Residual plots are widely used in linear regression analyses, in which residuals are random and supposedly normal-distributed. The probability plot is used to check whether the residuals follow the SEV distribution with a mean of 0. By examining the pattern of residual plots, one can identify whether there are additional variables that should be included or excluded in the regression model. Again, as long as an observation (t_q) is censored, the corresponding residual is also to be censored in the residual plot.

The standard deviation is defined by Equation (9.8.5), and the SEV distribution is defined by Equation (9.2.23). When a set of data follows the Weibull distribution, the standardized residuals can be obtained from the following equation:

$$\begin{aligned}\varepsilon_q &= \beta\,[\ln(t_q) - \ln(\eta_q)] \\ &= \beta\,[\ln(t_q) - (\gamma_0 + \gamma_1 x_{1q} + \gamma_2 x_{2q} + \gamma_3 x_{3q} + \ldots)]\end{aligned} \tag{9.8.34}$$

In linear regression, the standardized residual resulting from the logarithmic-transformed Weibull distribution is expected to follow the standard SEV distribution. The suspended data do not constitute one DOF. When an observation, i.e., t_q, is censored, the corresponding residual is also to be censored in the residual plot.

9.9. Product Life Prediction Examples Using DOE-W

Three categories of product life data are frequently encountered in a product life prediction, i.e., time-to-failure data, right-suspended (censored) data, and interval data, when DOE-W is applied. The data may be measured in hours, miles, or any other unit that constitutes the product life in practice.

9.9.1. Product Life Prediction Using 2_V^{5-1} Design with Right-Suspended Data

DOEs are set up to study the reliability of an electric motor. The experiment mimics the operating conditions that an electric motor is likely to encounter in the real world, including shock and vibration, thermal changes, humidity, and power cycling, while varying the levels of five design factors as shown in Table 9.9.1.

TABLE 9.9.1 Design factors and relevant design levels for life prediction of an electric motor [Reliasoft 2008].

Factor	Level (−)	Level (+)	Coded variable
A: Wire material	Al	Cu	a = −1 for Al and a = 1 for Cu
B: Wire diameter (mm)	36	50	b = (B − 43)/7
C: Wire tension (N)	60	90	c = (C − 75)/15
D: Winding method	D−	D+	d = −1 for D− and d = 1 for D+
E: Coating material	Old	New	e = −1 for Old and d = 1 for New

The design matrix of fractional factorial design 2_V^{5-1} with alias equation E = ABCD is set up to identify the significant factors that may have influenced the product life of the electric motor, as shown in Table 9.9.2. There are 16 motors. Each test motor is constructed according to the run condition prescribed by the assigned individual treatment, respectively. Test results are given in column Y (hours) of the table. The motor tested within the 10th treatment is not broken but suspended after 120 h (5 days).

TABLE 9.9.2 Fractional factorial design 2_V^{5-1} with E = ABCD for testing electric motors [Reliasoft 2008].

Run	a	b	d	d	e	Y (h)	ln(η)	η	β	E[Y]
1	−1	−1	−1	−1	1	9.53	2.636	13.96	2.566	12.40
2	1	−1	−1	−1	−1	4.65	2.491	12.08	2.566	10.72
3	−1	1	−1	−1	−1	3.5	1.568	4.80	2.566	4.260
4	1	1	−1	−1	1	35.54	3.299	27.07	2.566	24.04
5	−1	−1	1	−1	−1	28.53	3.096	22.11	2.566	19.63
6	1	−1	1	−1	1	32.65	4.826	124.76	2.566	110.8
7	−1	1	1	−1	1	6.19	1.804	6.08	2.566	5.39
8	1	1	1	−1	−1	9.09	2.355	10.54	2.566	9.36
9	−1	−1	−1	1	−1	10.29	3.096	22.11	2.566	19.63
10	1	−1	−1	1	1	120+	4.826	124.76	2.566	110.8
11	−1	1	−1	1	1	4.13	1.804	6.08	2.566	5.39
12	1	1	−1	1	−1	13.52	2.355	10.54	2.566	9.36
13	−1	−1	1	1	1	4.59	1.941	6.96	2.566	6.18
14	1	−1	1	1	−1	13.51	2.491	12.08	2.566	10.72
15	−1	1	1	1	−1	5.61	1.568	4.80	2.566	4.26
16	1	1	1	1	1	8.65	3.299	27.07	2.566	24.04

where
Y is life in hours
120+ is the data right-suspended (censored) at the end of 120 h

Equation (9.8.4) can be expanded to accommodate fractional factorial design 2_V^{5-1} as

$$\ln(\eta_q) \approx \gamma_0 + \gamma_a a + \gamma_b b + \gamma_c c + \gamma_d d + \gamma_e e + \gamma_{ab} a b + \gamma_{ac} a c + \gamma_{ad} a d + \gamma_{ae} a e + \gamma_{bc} b c + \gamma_{bd} b d + \gamma_{be} b e + \gamma_{cd} c d + \gamma_{ce} c e + \gamma_{de} d e \qquad (9.9.1)$$

Subscript q can be f (for test-to-failure data) or s (for suspended data). Note that the two-factor interactions may be confounded with three-factor interactions since 2_V^{5-1} is an experimental design with resolution V. For a general mixture of various distributions [Jiang and Murphy 1995], each treatment has its own distinct Weibull parameters, β and η. With test-to-failure and right-suspended data, the ln-likelihood can be written in light of Equations (9.8.20)–(9.8.22) as

$$\ln[L_w(\beta, \gamma_0, \gamma_i, \gamma_{ij}, \gamma_{ijm})] = \ln(L_{failure}) + \ln(L_{suspended})$$
$$= \sum_{f=1}^{Q_f} \ln\{\frac{\beta_f}{\eta_f} (\frac{t_f}{\eta_f})^{(\beta_f - 1)} \exp[-(\frac{t_f}{\eta_f})^{\beta_f}]\} + \sum_{s=1}^{Q_s} [-(\frac{t_s}{\eta_s})^{\beta_f}] \qquad (9.9.2)$$

The above equation is rewritten as

$$\ln[L_w(\beta, \gamma_0, \gamma_i, \gamma_{ij}, \gamma_{ijm})] = \sum_{f=1}^{Q_f} \ln(\beta_f) - \sum_{f=1}^{Q_f} \beta_f \ln(\eta_f) + \sum_{f=1}^{Q_f} (\beta_f - 1) \ln(t_f) - \sum_{f=1}^{Q_f} (\frac{t_f}{\eta_f})^{\beta_f} - \sum_{s=1}^{Q_s} (\frac{t_s}{\eta_f})^{\beta_s} \qquad (9.9.3)$$

Assume that shape factor β is consistent throughout all the experimental treatments, then the above equation reduces further to

$$\ln[L_w(\beta, \gamma_0, \gamma_i, \gamma_{ij}, \gamma_{ijm})] = Q_f \ln(\beta) - \sum_{f=1}^{Q_f} \beta \ln(\eta_q) + (\beta - 1) \sum_{f=1}^{Q_f} \ln(t_f) - \sum_{f=1}^{Q_f} (\frac{t_f}{\eta_q})^{\beta} - \sum_{s=1}^{Q_s} (\frac{t_s}{\eta_q})^{\beta} \qquad (9.9.4)$$

Substituting Equation (9.9.1) into the above equation renders an equation with 16 unknowns, i.e., β, γ_0, γ_a, γ_b, γ_c, γ_d, γ_e, γ_{ab}, γ_{ac}, γ_{ad}, γ_{ae}, γ_{bc}, γ_{bd}, γ_{be}, γ_{cd}, γ_{ce}, and γ_{de}, as

$$\ln[L_w(\beta, \gamma_0, \gamma_i, \gamma_{ij}, \gamma_{ijm})] = Q_f \ln(\beta) - \sum_{f=1}^{Q_f} \beta \ln(\gamma_0 + \gamma_a a + \gamma_b b + \gamma_c c + \gamma_d d + \gamma_e e + \ldots)$$
$$+ (\beta - 1) \sum_{f=1}^{Q_f} \ln(t_f) - \sum_{f=1}^{Q_f} [t_f / \exp(\gamma_0 + \gamma_a a + \gamma_b b + \gamma_c c + \gamma_d d + \gamma_e e + \ldots)]^{\beta}$$
$$- \sum_{s=1}^{Q_s} [t_s / \exp(\gamma_0 + \gamma_a a + \gamma_b b + \gamma_c c + \gamma_d d + \gamma_e e + \ldots)]^{\beta} \qquad (9.9.5)$$

Since there is only one right-suspended run condition (i.e., run 10), $Q_f = 15$ and $Q_s = 1$. Note that the values of coded variables, i.e., a, b, c, d, and e, take the respective run condition (treatment) given in Table 9.9.2. For example, a = −1, b = −1, c = −1, d = −1, and e = 1 for the first run condition (treatment) and a = 1, b = −1, c = −1, d = 1, and e = 1 for run 10.

In order to maximize the ln-likelihood given above, one may take a partial differentiation with respect to each individual unknown (i.e., $\beta, \gamma_0, \gamma_a, \gamma_b, \gamma_c, \gamma_d, \gamma_e, \ldots$) and set it equal to zero. This yields 16 equations to be solved for the 16 unknowns: $\beta, \gamma_0, \gamma_a, \gamma_b, \gamma_c, \gamma_d, \gamma_e, \gamma_{ab}, \gamma_{ac}, \gamma_{ad}, \gamma_{ae}, \gamma_{bc}, \gamma_{bd}, \gamma_{be}, \gamma_{cd}, \gamma_{ce}$, and γ_{de}.

Based on Equation (9.8.33), the LR test is then utilized to justify the goodness of each individual effect coefficient γ_{test}

$$LR(\gamma_{test}) = -2\ \{\ln[L_w(All - \gamma_{test})] - \ln[L_w(All)]\} \tag{9.9.6}$$

of which γ_{test} is one of the 15 unknowns: $\gamma_0, \gamma_a, \gamma_b, \gamma_c, \gamma_d, \gamma_e, \gamma_{ab}, \gamma_{ac}, \gamma_{ad}, \gamma_{ae}, \gamma_{bc}, \gamma_{bd}, \gamma_{be}, \gamma_{cd}, \gamma_{ce}$, and γ_{de}.

Equations (9.9.5) and (9.9.6), with the test data given in Table 9.9.2 (Column Y), are analyzed using the maximum-likelihood method available in ReliaSoft Weibull++ software. Results for the two-parameter Weibull distribution are given in Table 9.9.3.

The p-values derived from the LR test based on the two-parameter Weibull distribution are then assessed. At the significance level of 10% (the p-value of LR), the statistically significant effects are A, B, E, AE, and CD. Thus, Equation (9.9.1) reduces to

$$\ln(\eta_q) \approx \gamma_0 + \gamma_a\, a + \gamma_b\, b + \gamma_e\, e + \gamma_{ae}\, a\, e + \gamma_{cd}\, c\, d \tag{9.9.7}$$

TABLE 9.9.3 Significances of main and interactive effects on the reliability of electric motors [Reliasoft 2008] by Weibull MLE.

Effect	DOF	ln(likelihood value)	LR	p-Value
A	1	-47.156	13.446	***0.02%***
B	1	-43.969	7.072	***0.78%***
C	1	-40.555	0.244	62.14%
D	1	-40.433	0.0006	98.01%
E	1	-41.940	3.014	***8.25%***
AB	1	-40.543	0.220	63.92%
AC	1	-41.760	2.655	10.32%
AD	1	-41.613	2.360	12.45%
AE	1	-46.284	11.701	***0.06%***
BC	1	-40.479	0.093	76.1%
BD	1	-40.757	0.648	42.08%
BE	1	-40.595	0.327	56.95%
CD	1	-44.834	8.802	***0.3%***
CE	1 (not accounted for)			
CF	1 (not accounted for)			
Y_{ave}	1	—		
Total	16	—		

Note that E = A B C D, and Run 10 is suspended (Table 9.9.2).

Because design 2_V^{5-1} is of resolution V and both factors C and D are not statistically significant, it is here assumed that interaction CD is not statistically significant. It means that interaction CD may be replaced by interaction ABE in light of the alias equation that I = ABCDE. Thus, Equation (9.9.7) turns into

$$\ln(\eta_q) \approx \gamma_0 + \gamma_a a + \gamma_b b + \gamma_e e + \gamma_{ae} a e + \gamma_{cd} a b e \tag{9.9.8}$$

The calculations, involving Equations (9.9.2)–(9.9.5), are repeated in search of seven unknowns, i.e., β, γ_0, γ_a, γ_b, γ_e, γ_{ae}, and γ_{cd}. The final predictive equations for β and $\ln(\eta_q)$ are, respectively,

$$\beta = 2.5657 \tag{9.9.9}$$

and $$\ln(\eta_q) = 2.6726 + 0.5702 a - 0.416 b + 0.2949 e + 0.5248 a e - 0.3479 a b e \tag{9.9.10}$$

Thus, $$\eta_q = \exp(2.6726 + 0.5702 a - 0.416 b + 0.2949 e + 0.5248 a e - 0.3479 a b e) \tag{9.9.11}$$

Plugging the above equation into Equation (9.8.15) with n = 1 (i.e., no replication within each run condition), one can obtain the predictive life within individual treatment q as

$$E(Y) = \eta_q \Gamma(1 + 1/\beta) = \eta_q \Gamma(1 + 1/2.5657) \tag{9.9.12}$$

The expected life corresponding to each run condition calculated using the above equation is listed in column E[Y] in Table 9.9.2.

Next, consider the optimization of product life against the wire material and diameter. The individual main effect of switching coating material from the old to the new (factor E) generates a positive gradient 0.2949 as indicated by Equation (9.9.10). Given that e = +1, Equation (9.9.10) reduces to

$$\ln(\eta_q) = 2.9675 + 1.095 a - 0.416 b - 0.3479 a b \tag{9.9.13}$$

or $$\eta_q = \exp(2.9675 + 1.095 a - 0.416 b - 0.3479 a b) \tag{9.9.14}$$

Plugging the above equation into Equation (9.9.12) yields the life expectancy:

$$E[Y] = \exp(2.9675 + 1.095 a - 0.416 b - 0.3479 a b) \Gamma(1 + 1/2.5657) \tag{9.9.15}$$

The response surface of life expectancy E[Y] as a function of wire material and diameter is plotted in Figure 9.9.1. It can be seen that the wire made of copper with a small diameter will improve life expectancy.

FIGURE 9.9.1 Response surface of the scale factor for life prediction versus wire material and diameter.

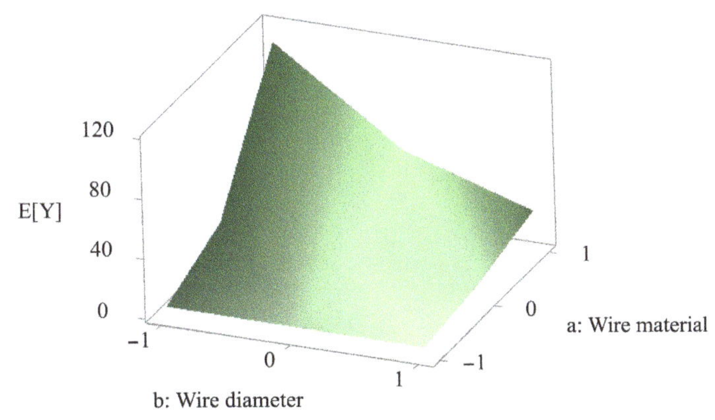

9.9.2. Product Life Prediction for Design with Right-Censored Data and Interval Data

This is a continuation of study on the reliability of fluorescent lights [Wu and Hamada 2009] with the data presented in Table 8.9.1. The test data are rephrased in Table 9.9.4. The coming examination is to inspect the impact of these five proposed factors on the product life prediction using the maximum-likelihood method based on the Weibull distribution.

TABLE 9.9.4 2_{III}^{5-2} design for evaluating product life with right-censored and interval data [Wu and Hamada 2009].

Treatment	a	b	c	d = −ac	e	Y_1 and Y_2 (days)	E(Y)
1	−1	−1	−1	1	1	(14, 16) and 20+	22.5
2	1	−1	−1	−1	−1	(18, 20) and 20+	29.3
3	−1	1	−1	1	−1	(8, 10) and (10, 12)	11.4
4	1	1	−1	−1	1	(18, 20) and 20+	23.5
5	−1	−1	1	−1	1	20+ and 20+	45.6
6	1	−1	1	1	−1	(12, 14) and 20+	22.0
7	−1	1	1	−1	−1	(16, 18) and 20+	23.0
8	1	1	1	1	1	(12, 14) and (14, 16)	17.7

where
Y_1 and Y_2 (days) is two replicated runs within each treatment
20+ is suspended after 20 days
(18, 20) is referring to a failure that occurred at the moment between the 18th and 20th days

Because the resolution of 2_{III}^{5-2} design is 3, the scale factor on the logarithmic scale can be written as

$$\ln(\eta_q) \approx \gamma_0 + \gamma_a a + \gamma_b b + \gamma_c c + \gamma_d d + \gamma_e e \tag{9.9.16}$$

Subscript q can be s (for suspended data) or i (for interval data). Note that two-factor interactions may be confounded with main effects because fractional factorial design 2_{III}^{5-2} comes with resolution III.

Assume that the shape factor is consistent throughout the experimental tests. With right-suspended and interval data, the ln-likelihood can be written in light of Equations (9.8.22) and (9.8.23) as

$$\ln[L_w(\beta, \gamma_0, \gamma_a, \gamma_b, \gamma_c, \gamma_d, \gamma_e)] = \ln(L_{suspended}) + \ln(L_{interval})$$

$$= \sum_{s=1}^{Q_s} [-(\frac{t_s}{\eta_s})^{\beta}] + \sum_{i=1}^{Q_i} \ln\{\exp[-(\frac{t_{i,1}}{\eta_i})^{\beta}] - \exp[-(\frac{t_{i,2}}{\eta_i})^{\beta}]\} \quad (9.9.17)$$

Substituting Equation (9.9.14) into the above equation renders an equation with seven unknowns, i.e., β, γ_0, γ_a, γ_b, γ_c, γ_d, and γ_e, as

$$\ln[L_w(\beta, \gamma_0, \gamma_a, \gamma_b, \gamma_c, \gamma_d, \gamma_e)] = -\sum_{s=1}^{Q_s} [\, t_s / \exp(\gamma_0 + \gamma_a\, a + \gamma_b\, b + \gamma_c\, c + \gamma_d\, d + \gamma_e\, e)]^{\beta}$$

$$+ \sum_{i=1}^{Q_i} \ln(\exp\{-[t_{i,1} / \exp(\gamma_0 + \gamma_a\, a + \gamma_b\, b + \gamma_c\, c + \gamma_d\, d + \gamma_e\, e)]^{\beta}\} \quad (9.9.18)$$

$$- \exp\{-[t_{i,2} / \exp(\gamma_0 + \gamma_a\, a + \gamma_b\, b + \gamma_c\, c + \gamma_d\, d + \gamma_e\, e)]^{\beta}\})$$

The values of coded variables, i.e., a, b, c, d, and e, take the respective run condition (treatment) given in Table 9.9.1. For example, "a = −1, b = −1, c = −1, d = 1, and e = 1" will be applied twice for the first run condition (treatment), one ends up with suspension at "20 days" and the other with interval data (14, 16). Therefore, there will be 16 equations to be solved for the six unknowns.

In order to maximize the ln-likelihood given above, one may take a partial differentiation with respect to each individual unknown (i.e., β, γ_0, γ_a, γ_b, γ_c, γ_d, and γ_e) and set it equal to zero. This yields 15 (15 = 16 − 1) equations because of treatment five that has two suspended data with the same lifetime. Solving these 15 equations for seven unknowns leads to

$$\beta = 7.27$$
$$\gamma_0 = 2.9959$$
$$\gamma_a = 0.1052$$
$$\gamma_b = -0.2256$$
$$\gamma_c = -0.0294$$
$$\gamma_d = 0.2477$$
$$\text{and} \quad \gamma_e = -0.1166$$

Note that the data given above are slightly different from the coefficients of Equation (22) presented in [Guo and Mettas 2010]. Based on Equation (9.8.33), the LR test is then utilized to justify the goodness of each individual effect coefficient γ_{test}.

$$LR(\gamma_{test}) = -2\ \{\ln[L_w(All - \gamma_{test})] - \ln[L_w(All)]\}$$

of which γ_{test} is one of the following five unknowns: γ_0, γ_a, γ_b, γ_c, γ_d, and γ_e. The results are shown in Table 9.9.5. Removing nonsignificant terms in Equation (9.9.16) according to Table 9.9.5 leads to

$$\ln(\eta_q) = 2.9959 + 0.1052\, a - 0.2256\, b + 0.2477\, d - 0.1166\, e \quad (9.9.19)$$

Plugging the above equation into Equation (9.8.15) yields the predictive life within individual treatment q as

$$E(Y) = \eta_q \Gamma(1 + 1/\beta) = \eta_q \Gamma(1 + 1/7.27) \tag{9.9.20}$$

The expected data of life expectancy corresponding to the eight treatments are listed in column E(Y) of Table 9.9.4.

For comparison, the same data set has been analyzed using both the lognormal distribution (Chapter 8, Section 8.9.1) and the Weibull distribution (Chapter 9, Section 9.9.2). In comparison of data of column E(T) in Table 8.9.1 with those in column E(Y) of Table 9.9.4, it shows that the lognormal approach is more conservative than Weibull statistics for this case study while both are acceptable.

TABLE 9.9.5 Likelihood test for reliability of fluorescent lights based on the Weibull MLE [Guo and Mettas 2010].

Effect	DOF	ln(Likelihood value)	LR	p-Value
γ_a	1	-20.7181	3.1411	7.63%
γ_b	1	-24.6436	10.9922	0.09%
γ_c	1	-19.2794	0.2638	60.76%
γ_d	1	-25.7594	13.2237	0.03%
γ_e	1	-21.0727	3.8504	4.97%

9.10. Multi-Objective Optimization with Pareto Optimality Based on DOE-W

This is an example of how to use reliability DOEs to seek out the optimal solution in the criterion domain of product life and product performance. After a screen design 2^{K-P} study, an experiment was set up to finalize the effects of the following three factors on the product life prediction of a part: thickness (factor A), turning time (factor B), and width (factor C). The design levels are listed in Table 9.10.1. As extended from factorial design 2^3, a central composite design with three factors is adopted for factorial analysis because of its high efficiency for problem optimization by means of the RSM. This case study was reported in [ReliaSoft 2014], but its optimal solution does not fall within the experimental ranges of all the design factors given in Table 9.10.1. Extrapolation of any design factor outside its experimental range is dubious, especially unreliable when both responses to the assigned design factors are quite nonlinear. Here, the multi-objective particle swarm optimization with Pareto optimality is applied and it leads to a different optimal solution.

TABLE 9.10.1 Design factors and relevant design levels for exploring product life of an automotive part [ReliaSoft 2014].

	Level					
Factor	(−1.682)	(−1)	(0)	(1)	(1.682)	Coded variable
A: Thickness (in.)	2.318	3	4	5	5.682	a = (A − 4)/1
B: Tuning time (min)	13.182	20	30	40	46.818	b = (B − 30)/10
C: Width (in.)	8.2955	10	12.5	15	16.7045	c = (C − 12.5)/2.5

The spherical composite design, extended from full factorial design 2^3 with $d = 2^{K/4} = 2^{3/4} \approx 1.682$ (Volume 1, Chapter 5), is taken with six replicated runs at the center point. The 34 experimental data of product lifetime resulting from these 15 test treatments (run conditions), of which each either ends up with failure or is suspended, are exhibited in column Y of Table 9.10.2. Specimens, which do not fail but are suspended at 2000 h, are specified as 2000^+ in the table. Of course, the life (Y) is the longer the better. Simultaneously, the part is supposed to support a flow rate of fluid between 75 and 85 gal/h of fluid, and thus, it forms a constraint that $75 \le V \le 85$. This is a problem that can be resolved with the Pareto frontier.

TABLE 9.10.2 Central composite design for evaluating product life with censored data [ReliaSoft 2014].

Run	a	b	c	A	B	C	Y (cycles)	V	E(Y)	E(V)
1	−1	−1	−1	3	20	10	561, 428	56, 52	549	54.2
2	−1	−1	1	3	20	15	814, 1158	75, 70	976	72.2
3	−1	1	−1	3	40	10	1406, 1608	53, 51	1525	50.3
4	−1	1	1	3	40	15	2000^+, 2000^+	66, 70	2712	68.3
5	1	−1	−1	5	20	10	125, 167	31, 33	165	32.4
6	1	−1	1	5	20	15	267, 383	93, 93	293	94.7
7	1	1	−1	5	40	10	278, 319	31, 26	457	29.5
8	1	1	1	5	40	15	588, 1155	92, 89	812	90.8
9	−1.682	0	0	2.318	30	12.5	2000^+, 2000^+	47, 46	2956	47.4
10	1.682	0	0	5.682	30	12.5	239, 505	45, 54	389	48.0
11	0	−1.682	0	4	13.182	12.5	183, 283	56, 56	208	55.3
12	0	1.682	0	4	46.818	12.5	1036, 1479	49, 46	1159	48.7
13	0	0	−1.682	4	30	8.2955	298, 587	50, 47	378	49.2
14	0	0	1.682	4	30	16.705	650, 961	115, 121	996	116.8
15	0	0	0	4	30	12.5	726, 842, 964, 1185, 1327, 1555	53, 53	1073	52.0

Notes:
2000^+ means that the piece of data is suspended at 2000 h
Y (cycle) is the lifetime data; the longer the better
V (gal/h) is the flow rate of fluid, while it is expected that 75 < V < 85
E(Y) is the life expectancy, calculated from the productive equation, i.e., Equation (9.9.20)
E(V) is the expected volume flow
Run 15 is the six replicates for product life prediction to provide the random error information

9.10.1. Central Composite Design for Evaluating Product Life with Right-Censored Data

Assume that the lifetime data given in column Y of Table 9.10.2 follow the two-parameter Weibull distribution. The characteristic life used in DOE-W analysis is the scale parameter, i.e., η, which is assumed to be a function of main effects and two-factor interactions, as well as quadratic terms. After the logarithmic transformation of η_q, the acceleration model is equipped with a polynomial up to the second order as

$$\ln(\eta_q) = \gamma_0 + \gamma_1\, a + \gamma_2\, b + \gamma_3\, c + \gamma_{12}\, a\, b + \gamma_{13}\, a\, c + \gamma_{23}\, b\, c + \gamma_{11}\, a^2 + \gamma_{22}\, b^2 + \gamma_{33}\, c^2 \tag{9.10.1}$$

As the generalized method to approach a problem with lifetime data, it is necessary for the time to failure to follow a certain statistical distribution and the life characteristic of the distribution may be affected by the levels of design factors, i.e., factors A, B, and C or coded variables a, b, and c in this case. Since suspended data are involved in the lifetime data, they do not fit the "test to failure" category and a regular regression model based on F-distribution is not applicable. After some tryouts of different statistical models, it is encouraged to use the more suitable two-parameter Weibull distribution. In light of Equation (9.9.5), one has

$$\ln[L_w(\beta, \gamma_0, \gamma_1, \gamma_2, \gamma_3, \gamma_{12}, \gamma_{13}, \gamma_{23}, \gamma_{11}, \gamma_{22}, \gamma_{33})] = Q_f \ln(\beta) -$$

$$- \sum_{f=1}^{Q_f} \beta \ln(\gamma_0 + \gamma_1 a + \gamma_2 b + \gamma_3 c + \ldots + \gamma_{33} c^2)$$

$$+ (\beta - 1) \sum_{f=1}^{Q_f} \ln(t_f) - \sum_{f=1}^{Q_f} [t_f / \exp(\gamma_0 + \gamma_1 a + \gamma_2 b + \gamma_3 c + \ldots + \gamma_{33} c^2)]^\beta$$

$$- \sum_{s=1}^{Q_s} [t_s / \exp(\gamma_0 + \gamma_1 a + \gamma_2 b + \gamma_3 c + \ldots + \gamma_{33} c^2)]^\beta \quad (9.10.2)$$

There are 34 equations, and they are to be solved numerically for 11 unknowns, i.e., β, γ_0, γ_1, γ_2, γ_3, γ_{12}, γ_{13}, γ_{23}, γ_{11}, γ_{22}, and γ_{33}. The calculations are done using the maximum-likelihood method available in ReliaSoft Weibull++ software.

Conduct the work procedure as applied for Example 9.9.1. After removing the most insignificant term one by one and repeating the calculation again and again, the final model resulting from the LR test as summarized in Table 9.10.3 is

$$\beta = 4.1168 \quad (9.10.3)$$

$$\text{and} \quad \ln(\eta_q) = 7.0745 - 0.6027\, a + 0.5108\, b + 0.2877\, c - 0.2763\, b^2 - 0.1973\, c^2 \quad (9.10.4)$$

Plugging the above equation into Equation (9.8.15), one can obtain the predictive life within individual treatment q as

$$E(Y) = \eta_q \Gamma(1 + 1/\beta) = \eta_q \Gamma(1 + 1/4.1168) \quad (9.10.5)$$

The expected lifetimes corresponding to the 15 treatments are listed in column E(Y) in Table 9.10.2. If suspended data (i.e., with sign "+") are omitted, the correlation between expected life E(Y) and the "$Y_{average}$" for the other 13 treatments is 96.6%.

TABLE 9.10.3 LR test for central composite design with test-to-failure and suspended data [ReliaSoft 2014].

Term	Contrast	Effect	Error	z-Value	p-Value
β	—	4.1168	0.5726	—	—
Intercept	—	7.0745	0.0824	85.89	0
a	-1.2055	-0.6027	0.0635	-9.496	0
b	1.0217	0.5108	0.0486	10.5107	0
c	0.5753	0.2877	0.0554	5.1911	2.09×10^{-7}
b^2	-0.5526	-0.2763	0.0481	-5.7492	8.97×10^{-9}
c^2	-0.3946	-0.1973	0.0512	-3.8564	1×10^{-4}

9.10.2. ANOVA for Examining Flow Rate

Because there are no censored data for the flow rate, ANOVA will be applied to detect the significant effects. The data analysis is done with Minitab (Stat → Regression → Regression → Fit Regression Model). The "most insignificant" factor or interaction that comes with the largest p-value is weeded out one by one until all p-values are less than 10% that is here defined as the parting significance level based on F-distribution. In light of the ANOVA presented in Table 9.10.4, the final predictive equation for the life expectancy based on the factorial design is summarized as follows:

$$V = 427.6 - 43.05\,A - 0.1962\,B - 53.47\,C - 1.510\,A^2 + 1.7521\,C^2 + 4.425\,A\,C$$

$$(R = 99.6\%,\ R_{adj} = 99.4\%,\ \text{and}\ R_{pred} = 99.2\%;\ \text{p-value of constant}\ \gamma_0 = 0.0\%) \quad (9.10.6)$$

High correlations and low p-value of constant γ_0 given above, as well as high p-value of lack of fit (Table 9.10.4), confirm the effectiveness and adequacy of the above predictive equation.

TABLE 9.10.4 ANOVA table for flow rate.

Source	SS_{adj}	DOF	MS_{adj}	$F_{u,v}$	p-Value
Regression	16,210.7	6	2701.78	424.8	0.0%
A	382.8	1	382.79	60.19	0.0%
B	105.1	1	105.15	16.53	0.0%
C	2927.3	1	2927.27	460.3	0.0%
A^2	45.0	1	45.00	7.07	1.4%
C^2	2366.7	1	2366.70	372.1	0.0%
AC	1958.1	1	1958.06	307.9	0.0%
Error	146.3	23	6.36		
Lack of fit	28.8	8	3.60	0.46	86.6%
Pure error	117.5	15	7.83		
Subtotal	16,357.0	29			
Grand ave.	—		1		
Total		—	30		

9.10.3. Pareto Frontier for Design Optimization

The Pareto frontier with two objective functions, i.e., lifetime versus flow rate as shown in Figure 9.10.1, can be figured out and observed easily using the regression models based on DOEs. Equations (9.10.5) and (9.10.6) are used to generate feasible criterion points (swarm particles) using Excel, as denoted by the 125 (5 × 5 × 5) swarm particles (dark circular spots) in the figure that are generated using five design levels for each factor. The two vertical straight dashed lines underlie feasible criterion window for the flow rate range between 75 and 85 gal/h. Criterion point (78.91, 3470) is obviously the sole nondominated point (on the Pareto frontier), while its corresponding design point (A = 2.318 in., B = 40 min, and C = 16.7045 in.) can be read out from the Excel file.

FIGURE 9.10.1 The Pareto frontier of lifetime versus flow rate.

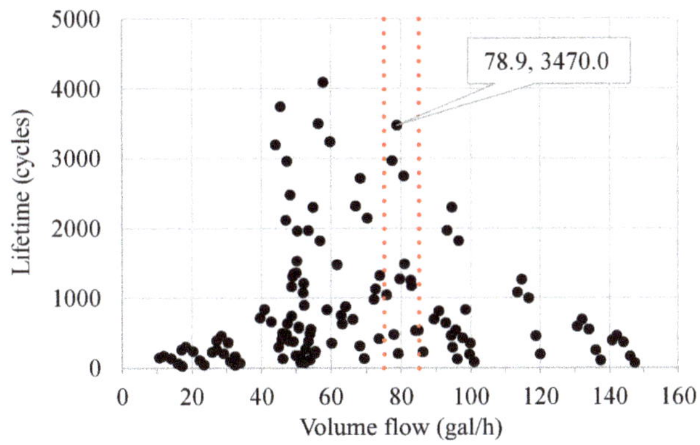

9.11. Exponential Distribution

By statistics, life cycles of some automotive electronics tend to fail according to the exponential distribution. Exponential distribution-based DOEs (DOE-E), as a special case of DOE-W, is another important statistical discipline in reliability DOEs. As a natural extension of an exponential distribution, the two-parameter Weibull CDF with $\beta = 1$ reduces to an exponential distribution function,

$$F(t) = 1 - \exp(-\lambda t) \qquad (\text{if } \lambda > 0) \tag{9.11.1a}$$

and $$F(t) = 0 \qquad (\text{if } \lambda < 0) \tag{9.11.1b}$$

where $$\lambda = 1/\eta \geq 0 \tag{9.11.2}$$

The PDF of an exponential distribution describes the time between events in a Poisson point process, i.e., a process in which events occur continuously and independently at an average rate λ. Thus, λ is also called rate parameter, while η is also called survival rate in the sense that if a random variable t is the duration of time that a given mechanical, electronic, or biological system manages to survive. The corresponding reliability function is

$$R(t) = 1 - F(t) = \exp(-\lambda\, t) \tag{9.11.3}$$

Choice of second-order response surface designs for logistic and Poisson regression models is discussed in [Johnson and Montgomery 2009], as extended from [Box and Wilson 1951]. Equation (9.11.1a) is the exponential distribution function and its corresponding PDF can be obtained by taking a differentiation with respect to time.

$$f(t) = \lambda \exp(-\lambda\, t) \tag{9.11.4}$$

The PDF of an exponential distribution is shown in Figure 9.11.1. The associated hazard rate for an exponential distribution is

$$h(t) \equiv \frac{f(t)}{R(t)} = \lambda \tag{9.11.5}$$

Conceptually, this is the same as $h(t) \approx N_f/N_{risk}$ delineated by Equation (8.1.3). According to Equations (8.1.1)–(8.1.3), the cumulative hazard rate and survival rate can be calculated as, respectively,

$$H(t) = \int h(t)\,dt = \lambda\, t \tag{9.11.6}$$

and $$S(t) = \exp\{\int [-h(t)]\,dt\} = \exp[-H(t)] = \exp(-\lambda\, t) \tag{9.11.7}$$

In comparison of the above equation with Equation (9.11.3), $S(t) = R(t) = \exp(-\lambda t)$ for an exponential distribution. The mean and variance of the exponential distribution are

$$\mu = E(t) = 1/\lambda = \eta \tag{9.11.8}$$

and $$\sigma^2 = 1/\lambda^2 = \lambda^{-2} \tag{9.11.9}$$

In other words, the expected duration of survival of the system is η units of time and its standard deviation is equal to the mean. Note that variance λ^{-2} also leads to Fisher information.

FIGURE 9.11.1 PDF of exponential distribution.

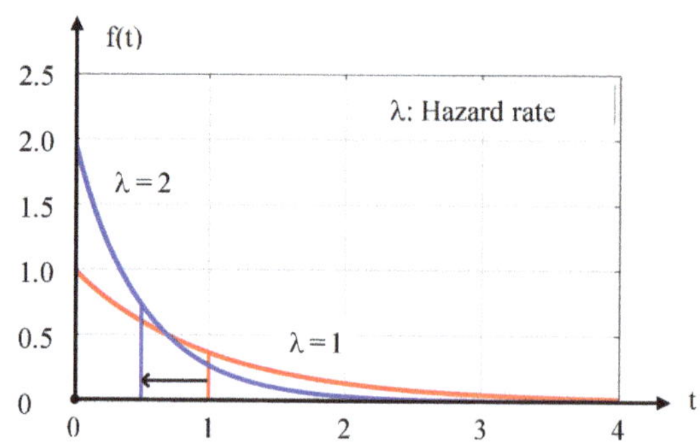

For a system that fits the exponential function, its mean is the MTBFs for a repairable system or MTTF for an un-repairable system. Therefore,

$$MTBF = 1/\lambda \quad \text{(Repairable system)} \tag{9.11.10}$$

and $$MTTF = 1/\lambda \quad \text{(Un-repairable system)} \tag{9.11.11}$$

A typical electronic component's failure follows an exponential distribution, having a constant hazard rate, i.e., constant MTTF, as shown by the second stage of hazard rate in Figure 9.11.2. Note that a failure in the break-in period with a higher hazard rate, i.e., the first stage, is usually because of uncommon causes. If all the system components' failure times are exponential and they are in a series configuration, the system's failure time is still exponential. This makes the exponential distribution more convenient for fault tree analysis (FTA) than others.

FIGURE 9.11.2 Hazard model of typical mechanical and electronic components.

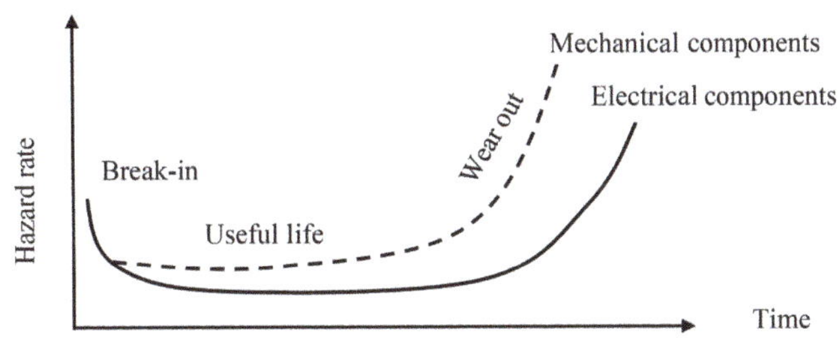

EXAMPLE 9.11.1

One of two suppliers is to be selected for jobs on the result of competitive cost analysis. It takes P dollars per unit to acquire the product from supplier 1, but nP (n > 1) dollars per unit from supplier 2. Assume that the products produced by both suppliers follow the exponential distribution with the following hazard rates: $\lambda_1 = 10^{-4}$/h and $\lambda_2 = 10^{-5}$/h. On a warranty clause of three years (≈1000 h in service for this product), each failed product would cost W dollars per unit (including installation, downtime, and others). Assume that W = 2P, which supplier is more cost-effective?

Solution:

The cost per unit from suppliers 1 and 2 are, respectively,

$$\text{Cost}_1 = P \qquad \text{If } T \geq 1000 \text{ hours}$$

$$= P + W \qquad \text{If } T < 1000 \text{ hours}$$

$$\text{Cost}_2 = n\,P \qquad \text{If } T \geq 1000 \text{ hours}$$

$$= n\,P + W \qquad \text{If } T < 1000 \text{ hours}$$

The expected costs are

$$E[Cost_1] = (P + W) \int_0^T \lambda_1 \exp(-\lambda_1 t)\, dt + P \int_T^\infty \lambda_1 \exp(-\lambda_1 t)\, dt$$

$$= (P + W) \int_0^{1000} 10^{-4} \exp(-10^{-4} t)\, dt + P \int_{1000}^\infty 10^{-4} \exp(-10^{-4} t)\, dt$$

$$= (P + W)\, [-\exp(-0.1) + 1] + P\, [-0 + \exp(-0.1)]$$

$$= P + W\, [1 - \exp(-0.1)]$$

$$E[Cost_2] = (n\, P + W) \int_0^T \lambda_1 \exp(-\lambda_2 t)\, dt + n\, P \int_T^\infty \lambda_1 \exp(-\lambda_2 t)\, dt$$

$$= (n\, P + W) \int_0^{1000} 10^{-5} \exp(-10^{-5} t)\, dt + n\, P \int_{1000}^\infty 10^{-5} \exp(-10^{-5} t)\, dt$$

$$= (n\, P + W)\, [-\exp(-0.01) + 1] + n\, P\, [-0 + \exp(-0.01)]$$

$$= n\, P + W\, [1 - \exp(-0.01)]$$

Thus, $E[Cost_2] - E[Cost_1] = (n - 1)\, P + W\, [-\exp(-0.01) + \exp(-0.1)]$

$$= (n - 1)\, P - 0.08521\, W$$

If $n > 1 + 0.08521\, W/P = 1.17042$, supplier 1 is the choice. In other words, if supplier 2 charges 17.042% more than supplier 1, supplier 1 is the choice.

9.12. MLE of Hazard Rate of Exponential Distribution

MLE can be used for assessing hazard rate λ of an exponential PDF with a given set of data, which fit into an exponential distribution. The algorithm is to maximize the likelihood function.

9.12.1. MLE of λ with No Censored Data

Assume that the lifetime data of the test units, i.e., t_q, $q = 1, 2, \ldots, Q$, follow an exponential distribution with no censor. The likelihood function and logarithmic likelihood function for λ based on the Q IID sample $t = (t_1, \ldots, t_Q)$ drawn from the random variable can be written as, respectively,

$$L_c(\lambda) = \prod_{q=1}^{Q} f(t_q) \qquad (9.12.1)$$

$$\text{and} \quad \ln[L_e(\lambda)] = \sum_{q=1}^{Q} \ln[f(t_q)] \tag{9.12.2}$$

where

$L_e(\lambda)$ is the likelihood function with parameter λ
$\ln[L_e(\lambda)]$ is the logarithmic transformation of $L_e(\lambda)$
$f(t_q)$ is the exponential density function evaluated at t_n
q is the number of data
Q is the total number of data

Substituting Equation (9.11.4) into the above equation leads to

$$L_e(\lambda) = \prod_{q=1}^{Q} f(t_q) = \prod_{q=1}^{Q} \lambda \exp(-\lambda\, t) = \lambda^Q [\exp(-\lambda \sum_{q=1}^{Q} t_q)] \tag{9.12.3}$$

$$\text{Thus,} \quad \ln[L_e(\lambda)] = Q \ln(\lambda) - (\lambda \sum_{q=1}^{Q} t_q) \tag{9.12.4}$$

As enlightened by the maximum-likelihood rule, parameter λ is to be estimated by taking partial differentiation of $\ln[L_w(\lambda)]$ with respect to this parameter and setting it to zero.

$$\frac{\partial\{\ln[L_e(\lambda)]\}}{\partial\lambda} = \frac{Q}{\lambda} - \sum_{q=1}^{Q} t_q = 0 \tag{9.12.5}$$

The above equation can be solved for parameter λ as

$$\lambda = Q / (\sum_{q=1}^{Q} t_q) \tag{9.12.6}$$

Hazard rate λ can be calculated using the test data directly when there are no censored data.

9.12.2. MLE of Hazard Rate λ with Type I Right-Censored Data

Assume that the lifetime of the test specimens follows an exponential distribution statistically, the likelihood function can be written as

$$L_e(\lambda) = \{\prod_{f=1}^{N_f} f(t_f)\} \{ \prod_{s=1}^{N_s} [S(t_s)]\} = \{\prod_{f=1}^{N_f} f(t_f)\} \{ \prod_{s=1}^{N_s} [R(t_s)]\} \tag{9.12.7}$$

where

N_s is the total number of censored (suspended) test specimens
N_f is the total number of failed test specimens
t_f is the life of specimen, up to failure
t_s is the life of specimen, up to suspension (censor)

Note that reliability function R(t) is equal to survival function S(t) for the Weibull distribution, including exponential distribution. Substituting Equations (9.11.3) and (9.11.4) into the above equation yields

$$L_e(\lambda) = \{\lambda^{N_f} \exp(-\lambda \sum_{f=1}^{N_f} t_f)\} [\exp(-\lambda \sum_{s=1}^{N_s} t_s)] \quad (9.12.8)$$

$$\text{Thus,} \quad \ln[L_e(\lambda)] = N_f \ln(\lambda) - \lambda \sum_{f=1}^{N_f} t_f - \lambda \sum_{s=1}^{N_s} t_s \quad (9.12.9)$$

As enlightened by the maximum-likelihood rule, parameter λ is to be estimated by taking partial differentiation of $\ln[L_e(\lambda)]$ with respect to this parameter and setting it to zero.

$$\frac{\partial\{\ln[L_e(\lambda)]\}}{\partial\lambda} = \frac{N_f}{\lambda} - (\sum_{f=1}^{N_f} t_f + \sum_{s=1}^{N_s} t_s) = 0$$

The above equation can be solved for parameter λ as

$$\lambda = N_f / (\sum_{f=1}^{N_f} t_f + \sum_{s=1}^{N_s} t_s) \quad (9.12.10)$$

It can be concluded from Equations (9.12.10) and (9.12.6) that the random process that fits an exponential distribution is memoryless. In other words, the chance of product failure in the future remains the same regardless of its past experience. This fact lends the statistical probabilities based on exponential distributions to the traditional FTA.

9.12.3. Standardized Residual for Exponential Distribution

Residual plots are widely used in linear regression analyses, in which residuals are assumed to be normally distributed. By examining the pattern of residual plots, one can identify whether there are additional variables that should be included in the regression model. In linear regression, the standardized residual for an exponential distribution, which is a special case of Weibull statistics, follows the standard SEV distribution, and it is defined as

$$\Phi_{sev}(z_q) = \Phi_{sev}\{[\ln(t_q) - \ln(1/\lambda_q)]\} \quad (9.12.11)$$

$$\text{and} \quad z_q = [\ln(t_q) - \ln(1/\lambda_q)] \quad (9.12.12)$$

where

t_q is the life data, where subscript i means the qth treatment

Φ_{sev} is the standard SEV distribution function

λ_q is the scale factor corresponding to treatment q

When the data follow the Weibull distribution, the standardized residuals can be obtained from the following equation:

$$\varepsilon_q = \ln(t_q) - \ln(1/\lambda_q)$$
$$= \ln(t_q) - (\gamma_0 + \gamma_1 x_{1q} + \gamma_2 x_{2q} + \gamma_3 x_{3q} + \ldots + \gamma_{12} x_{1q} x_{2q} + \ldots.) \quad (9.12.13)$$

The probability plot is used to check whether the residuals follow the SEV distribution with a mean of 0. If an observation is censored, its corresponding residual is also to be censored in the residual plot.

9.12.4. CIs for the Exponential Hazard Rate

The exponential hazard rate can be obtained from an experiment with the practice of type II censoring. It means that the experiment continues until a predetermined number of test-to-failure units, i.e., N_f, is reached. As the first approximation, exponential hazard rate λ based on such an experiment can be calculated using Equation (9.12.10) based on the maximum-likelihood method as

$$h = N_f / \left(\sum_{f=1}^{N_f} t_f + \sum_{s=1}^{N_s} t_s \right) \quad (9.12.14)$$

where h is the estimated λ.

Type II censoring means that the experiment is run until a preset number of failures, i.e., N_f, occur. For a two-sided bound, an exact 100(1 – α)% CI for λ is given as

$$P\left(\frac{h}{2\,N_f} \chi^2_{\alpha/2} \le \lambda \le \frac{-h}{2\,N_f} \chi^2_{1-\alpha/2}\right) = 1 - \alpha \quad (9.12.15)$$

of which $\chi^2_{\alpha/2}$ and $\chi^2_{1-\alpha/2}$ are the values of the corresponding chi-square distribution with $2N_f$ DOFs. Probability α/2 is applied to the left side and 1 – α/2 to the right side, respectively. One may replace α/2 by α for one-sided lower bound, while replacing 1 – α/2 by 1 – α for one-sided upper bound.

9.12.5. Product Life Prediction Based on Exponential Distribution

An exponential distribution can be effectively incorporated into reliability analysis only if the assumption of a constant failure rate can be justified. Although a constant hazard rate is generally inadequate to describe a failure mode, piecewise exponential functions under different working conditions can be piece-wisely put together to model the entire deterioration process like applying the finite element method to a space continuum. Advanced D-optimal designs for lifetime experiments with exponential distribution and censoring are demonstrated in [Müller 2013].

References

Abernethy, R. B. (2006), *The New Weibull Handbook*, 5th Edition, North Palm Beach, FL.

Aerts, M., Janssen, P., and Veraverbeke, N. (1994), "Bootstrapping Regression Quantiles," *Journal of Nonparametric Statistics*, 4, pp. 1-20.

Allmen, C. R. and Lu, M. W. (1994), "A Reduced Sampling Approach for Reliability Verification," *Quality and Reliability Engineering International*, 10(1), pp. 71-77.

ASTM D7846-21 (2021), Standard Practice for Reporting Uniaxial Strength Data and Estimating Weibull Distribution Parameters for Advanced Graphites, Last Updated: February 2, 2022.

Bowman, K. O. and Shenton, L. R. (1998), "Estimator: Method of Moments," in *Encyclopedia of Statistical Sciences*, Wiley & Sons, Inc., New York, pp. 2092-2098.

Box, G. E. P. and Wilson, K. B. (1951), "On the Experimental Attainment of Optimum Conditions," *Journal of the Royal Statistical Society, Series B*, 13, pp. 1-45.

Bullington, R. G., Lovin, S., Miller, D. M., and Woodall, W. H. (1993), "Improvement of an Industrial Thermostat Using Designed Experiments," *Journal of Quality Technology*, 25, pp. 262-270.

Casella, G. and Berger, R. L. (2002), *Statistical Inference*, 2nd Edition, Duxbury, Pacific Grove, CA.

Coles, S. (2001), *An Introduction to Statistical Modeling of Extreme Values*, Springer-Verlag, London.

Cook, R. D. (1974), "Influential Observations in Linear Regression," *Journal of the American Statistician Association*, 74, pp. 169-174.

Durivage, M. A. and Mehta, B. (2016), *Practical Process Validation*, ASQ Quality Press, Milwaukee, WI.

Escobar, L. A. and Meeker, W. Q. (1999), "Statistical Prediction Based on Censored Life Data," *Technometrics*, 41(2), pp. 113-124.

Evans, J. W. et al. (February 2019), "Procedures for Estimation of Weibull Parameters," General Technical Report FPL-GTR-264, Forest Product Laboratory, USDA, Madison, WI.

Freeman, L. J. (2010), "Statistical Methods for Reliability Data from Designed Experiments," PhD dissertation in Statistics, Virginia Polytechnic Institute and State University.

Genschel, U. and Meeker, W. K. (2010), "A Comparison of Maximum Likelihood and Median-Rank Regression for Weibull Estimation," *Quality Engineering*, 22, pp. 236-255.

Govindarajulu, Z. and Hubacker, N. W. (1962), "Percentile Points of Ordered Statistics in Samples from Beta, Normal, and Chi (1 D.F.) Populations," Unclassified AD 410083, Defense Documentation Center for Scientific and Technical Information, Cameron Station, Alexandria, VA.

Guo, H. and Mettas, A. (2010), "Design of Experiments and Data Analysis," *IEEE Reliability and Maintainability Symposium*, San Jose, 2010.

Halperin, M. (1952), "Maximum Likelihood Estimation in Truncated Samples," *Annals of Mathematical Statistics*, 23, pp. 226-238.

Hirose, H. (1999), "Bias Correction for Maximum Likelihood Estimates in the Two Parameter Weibull Distribution," *IEEE Transactions on Dielectrics and Electrical Insulation*, 6, pp. 66-68.

Jiang, R. and Murphy, D. N. P. (1995), "Reliability Modeling Involving Two Weibull Distributions," *Reliability Engineering & System Safety*, 47(3), pp. 187-198.

Johnson, R. T. and Montgomery, D. C. (2009), "Choice of Second-Order Response Surface Designs for Logistic and Poisson Regression Models," *Int'l Journal of Experimental Design and Process Optimisation*, 1(1), pp. 1-23.

Kim, M. and Yum, B. (2011), "Life Test Sampling Plans for Weibull Distributed Lifetimes under Accelerated Hybrid Censoring," *Statistical Papers*, 52(2), pp. 327-342.

Lawless, J. F. (2003), *Statistical Models and Methods for Lifetime Data*, 3rd Edition, John Wiley & Sons, New York.

Lehmann, E. L. (1975), *Nonparametrics: Statistical Methods Based on Ranks*, Holden-Day, San Francisco.

Littell, R. C. and Rao, P. V. (1978), "Confidence Regions for Location and Scale Parameters Based on the Kolmogorov Smirnov Goodness of Fit Statistic," *Technometrics*, 20(1), pp. 23-27.

Lloyd, E. H. (1952), "Least-Squares Estimation of Location and Scale Parameters Using Order Statistics," *Biometrika*, 39, pp. 88-95.

Marsaglia, G. (2004), "Evaluating the Anderson-Darling Distribution," *Journal of Statistical Software*, 9(2), pp. 730-737.

McCool, J. I. (2012), *Using the Weibull Distribution, Reliability, Modeling and Inference*, John Wiley & Sons, New York.

Meeker, W. Q., Escobar, L. A., and Pascual, F. G. (1998), *Statistical Methods for Reliability Data*, 2nd Edition, John Wiley & Sons, New York, 704 pages; ISBN: 978-1-118-11545-9.

Müller, C. H. (2013), *Advances in Model-Oriented Design and Analysis, Contributions to Statistics*, Edited by Ucinski, D. et al., Springer, New York.

Murthy, D. P., Xie, M., and Jiang, R. (2004), *Weibull Models*, John Wiley & Sons, Inc., Hoboken.

Nelson, W. B. (2000), "Weibull Prediction of a Future Number of Failures," *Quality and Reliability Engineering International*, 16, pp. 23-26.

NIST 8.1.6.3 (2022), "8.1.6.3. Extreme Value Distributions," Engineering Statistics Handbook, Department of Commerce, National Institute of Standards and Technology, Gaithersburg, Retrieved November 1, 2022.

Nordman, D. J. and Meeker, W. Q. (2002), "Weibull Prediction Intervals for a Future Number of Failures," *Technometrics*, 44, pp. 15-23.

Nwobil, F. N. and Ugomma, C. A. (2014), "A Comparison of Methods for the Estimation of Weibull Distribution Parameters," *Metodološki zvezki*, 11(1), pp. 65-78.

Pearson, K. (1894), "Contributions to the Mathematical Theory of Evolution. II. On the dissection of Symmetrical Frequency Curves," *Philosophical Transactions of the Royal Society of London, Series A*, 185, pp. 71–110.

ReliaSoft 2008 (2008), "Hot Topics: An Example of Using Reliability DOE for Life Testing," Reliability Hot Wire, Issue 88, ReliaSoft Publishing, Tucson, AZ; https://www.weibull.com/hotwire/issue88/hottopics88.htm.

ReliaSoft 2014 (2014), "Improving Product Reliability Using Design of Experiments," Reliability Hotwire, Issue 155, ReliaSoft Publishing, Tucson, AZ.

Rigdon, S. E. Pan, R., Montgomery, D. C., and Borror, C. M. (2022), *Experiments for Reliability Achievement*, Wiley, Hoboken, 416 pages; ISBN: 978-1119237693.

Romig, H. G. (1947), *50-100 Binominal Tables*, John Wiley & Sons, New York.

Ross, P. J. (2005), *Taguchi Techniques for Quality Engineering*, McGraw-Hill, New York.

Snee, R. D. (2001), "My Process Is Too Variable—Now What Do I Do? How to Produce and Use a Successful Multi-Vari Study," *Quality Progress*, 34, pp. 65-68.

Sundaresan, S. G. and Kalivoda, F. E. (1982), "Reliability Life Testing and Evaluation of 3-Phase Motors," *Int'l Compressor Engineering Conference*, West Lafayette, IN, 1982, Paper 377.

Tarum, C. D. (1999a), "Determination of the Critical Correlation Coefficient to Establish a Good Fit for Weibull and Log-Normal Failure Distributions," *SAE Technical Paper 1999-01-0057*, https://doi.org/10.4271/1999-01-0057.

Tarum, C. D. (1999b), "Classification and Analysis of Weibull Mixtures," *SAE Technical Paper 1999-01-0055*, https://doi.org/10.4271/1999-01-0055.

Thoman, D. L., Bain, L. L., and Antle, C. E. (1969), "Inferences on the Parameters of the Weibull Distribution," *Technometrics*, 11, pp. 445-460.

Vardeman, S. B. (1992), "What about the Other Intervals," *The American Statistician*, 46(3), pp. 193-197.

Wang, G. et al. (2017), "Bootstrap Analysis of Designed Experiments for Reliability Improvement with a Non-constant Scale Parameter," *Reliability Engineering and System Safety*, 160, pp. 114-121.

Weibull, W. (1939), "A Statistical Theory of the Strength of Materials," Ingeniorsvetenskapsakademiens Handlingar, Number 151, Stockholm.

Weibull, W. (1951), "A Statistical Distribution Function of Wide Applicability," *Journal of Applied Mechanics*, 18, pp. 293-305.

Weibull, W. (1967), "Estimation of Distribution Parameters by a Combination of the Best Linear Order Statistic Method and Maximum Likelihood," USAF AFML-TR-67-105.

Wu, S.-J. (2002), "Estimations of the Parameters of the Weibull Distribution with Progressively Censored Data," *Journal of the Japan Statistical Society*, 32(99), pp. 155-163.

Wu, C. F. J. and Hamada, M. S. (2009), *Experiments: Planning, Analysis and Parameter Design Optimization*, 2nd Edition, Wiley, New York; ISBN: 978-0471699460.

Young, D. S. (2010), "Tolerance: An R Package for Estimating Tolerance Intervals," *Journal of Statistical Software*, 36(5), pp. 1-39.

Zanakis, S. H. (1979), "A Simulation Study of Some Simple Estimators for the Three-Parameter Weibull Distribution," *Journal of Statistical Computation and Simulation*, 9, pp. 101-116.

Zhang, L., Xie, M., and Tang, L. (2007), "A Study of Two Estimation Approaches for Parameters of Weibull Distribution Based on WPP," *Reliability Engineering and System Safety*, 92, pp. 360-368.

Problems

P9.1: There are 1000 identical test specimens. Derived from the experiment, the ultimate tensile strengths (σ_{uts}) versus the corresponding numbers of failure data (No) are given in order as follows:

- σ_{uts}: 56.5, 57.5, 58.5, 59.5, 60.5, 61.5, 62.5, 63.5, 64.5, 65.5, 66.5, 67.5, 68.5, 69.5, 70.5, 71.5
- No: 2, 18, 23, 31, 83, 109, 138, 151, 139, 130, 82, 49, 28, 11, 4, 2

Please make a plot of three-parameter Weibull distribution using computer software packages.

P9.2: The "cycles to failure" data from the plug reversal life testing lend itself to a simple but valid method of quantitative analysis of electric motors. Eight sample test data of a 35 hp three-phase BLDC motor are given as follows: 96,000; 127,000; 138,000$^+$; 170,000; 182,000$^+$; 230,000; 280,000; and 390,000 [Sundaresan and Kalivoda 1982]. Note that superscript + means the respective test is suspended, while others are tested to failure. Please identify the corresponding Weibull distribution. Make a plot of the cycles to failure at a reliability of 90%, with a given confidence of 90%.

P9.3: Five data were obtained from an up-to-failure life-cycle test for an automotive hood latch system: 6100 (latch failure), 7000 (latch failure), 9000 (latch failure), 9500 (latch failure), and 22,000 (latch failure). Does this latch system meet the requirement of 5000 life cycles (less than one cycle per day for 15 years), demonstrating reliability at R = 90% with confidence C = 90%?

P9.4: Five data were obtained from a life-cycle test for an automotive hood latch system: 6100 (latch failure), 7000 (latch failure), 9000 (latch failure), 9500 (suspended because of a broken fixture), and 22,000 (suspended on purpose to reduce the cost and time). Does this latch system meet the requirement of 5000 cycles (less than one cycle per day for 15 years), demonstrating reliability at R = 90% and confidence C = 90%?

P9.5: An experimental design was conducted by employees of Eaton Controls Division at the Athens plant (Alabama) to improve the reliability of an industrial thermostat subjected to 11 process factors [Bullington et al. 1993]. Please identify the factors that have influences on the product life in terms of operating cycles, i.e., Y (k-cycles). Data are listed as follows.

Run	a	b	c	d	e	f	G	h	j	k	m	Y
1	–	–	–	–	–	–	–	–	–	–	–	Y_1
2	–	–	–	–	–	+	+	+	+	+	+	Y_2
3	–	–	+	+	+	–	–	–	+	+	+	Y_3
4	–	+	–	+	+	–	+	+	–	–	+	Y_4
5	–	+	+	–	+	+	–	+	–	+	–	Y_5
6	–	+	+	+	–	+	+	–	+	–	–	Y_6
7	+	–	+	+	–	–	+	+	–	+	–	Y_7
8	+	–	+	–	+	+	+	–	–	–	+	Y_8
9	+	–	–	+	+	+	–	+	+	–	–	Y_9
10	+	+	+	–	–	–	–	+	+	–	+	Y_{10}
11	+	+	–	+	–	+	–	–	–	+	+	Y_{11}
12	+	+	–	–	+	–	+	–	+	+	–	Y_{12}

where	Y_1 = (957	2846	7342	*	*	*	*	*	*	*)
	Y_2 = (206	284	296	305	313	343	364	420	422	543)
	Y_3 = (63	113	129	138	149	153	217	272	311	402)
	Y_4 = (76	104	113	234	270	364	398	481	517	611)
	Y_5 = (92	126	245	250	390	390	479	487	533	573)
	Y_6 = (490	971	1615	6768	7342	*	*	*	*	*)
	Y_7 = (232	326	326	351	372	446	459	590	597	732)
	Y_8 = (56	71	92	104	126	156	161	167	216	263)
	Y_9 = (142	142	238	247	310	318	420	482	663	672)
	Y_{10} = (259	266	306	337	347	368	372	426	451	510)
	Y_{11} = (381	420	7342	*	*	*	*	*	*	*)
	Y_{12} = (56	62	92	104	113	121	164	232	258	731)

Arrhenius model was adopted by [Bullington et al. 1993] and a bootstrap analysis based on the Weibull distribution was undertaken to solve the problem. Please use t-distribution (DOE-t) using the life expectancy based on the Weibull distribution for each run condition to solve the problem.

10

Product Reliability Growth

DFR embraces orchestrated team-based activities including carrying out test plans for growing product reliability. Reliability growth (RG) is some quantitative thinking, which is used for analyzing the potential improvement made on a product under development or introduced into the market, as well as verifying whether a certain reliability goal has been reached. A test plan consists of a series of experimental activities and analytical techniques that aim to achieve objectives that mainly target market needs. It includes organizing test and analysis teams and fulfilling all the detailed work tasks, such as a specific beta test for validating a software or hardware product.

10.1. DFR Tests

To see is to believe. Product engineers face major challenges over how to design an efficient test plan for reliability demonstration. Should an error occur, the error-tolerant design can gracefully detect it and resolve it. Up to an error-tolerant design in the automotive industry, for example, an objective to demonstrate 90% reliability and 90% confidence (R90/C90), for 240,000 customer equivalent kilometers or more in the field is required for sustainable automotive components. Depending on the product and the responsibility of the organization to which the test plan applies, a test plan may plot a strategy for one or more of the following tests:

1. Compliance DV tests
2. Production (manufacturing) tests
3. Acceptance tests and commissioning measurements
4. Service and repair tests
5. Regression tests

Product reliability testing used to "take too long and cost too much." Traditional MTBF calculations tend to be inaccurate, and FMEA/FTA rarely identifies DFR issues because of the limited focus on the failure mechanism. Significant improvements have been made in recent years through a combination of DOEs, accelerated life test methods, physics-of-failure models, and measurement techniques. As the most priming problem-solving method, DOEs provides a variety of tools for obtaining a fair comparison of design alternatives and meeting various product development needs.

10.1.1. Compliance Tests

Compliance tests are a method of testing a product to assure that it meets all of its design specifications. It is thus also called a DV test. DV determines whether the product is designed right. It is performed during the development or approval stage, typically on a small sample of units. This is usually conducted in the R&D laboratory. Strategies on zero-failure design (i.e., no failure allowed), fault tolerance (e.g., redundancy), derating (i.e., a component is used at a load that is much below its capability rating), durability, and safety margin (based on life cycles) are in order of precedence the designer uses to prevent failures. How long might the product last is assessed under service conditions, and it is called design life.

Aligned with a product development team's responsibility, there is a work plan to verify the designed product through a sequence of testing, including verifying functionality, achieving the reliability goal, and meeting the customer's needs. A compliance test plan includes who to do the test, list of results, and status for design review activities. Design Verification Plan and Report (DVP&R) used to be a Ford and Chrysler approach, but is adopted by most automobile makers nowadays. DVP&R refers to a live configurable worksheet for DV plans and reports. Each plan can be created from scratch or import data. Constraints by costs and timing for completing the planned work tasks drive the need for an efficient way to achieve DV and even product validation, therefore, additional thinking on how and how much testing is needed. Activities involved in compliance tests can be divided into two stages:

(A) ***Product Design***: Activities in the product design stage are intended to assure that the product will work well under a wide variety of field conditions. Usually, the product parameters of an automotive component can be obtained by decomposing it into various mechanical and electronic elements. For example, a bolted fastener can be factorized as a bolt, a nut, a washer, fastened members, applied torque, and fastening force curves. The life-centered design framework that aims to materialize a product by looking at design challenges through the entire ecosystem's lens must be also taken into consideration.

(B) ***Process Design***: Activities in the process design stage are intended to assure that the product will make the production process robust and flexible under a wide variety of manufacturing conditions. Usually, the process parameters of a production process can be obtained by decomposing it into a step-by-step production procedure. The purpose of validating the process design is to mitigate noises in such a manner that engineers have a high degree of confidence that the production process will make reliable high-performance products meeting or exceeding the predetermined customer's requirements.

Automotive quality standards, such as ISO/TS 16949 and QS-9000, demand automotive suppliers to specify a scripted plan for engineering testing that is used for product development, including DV and PV, and, in some cases, continuous conformance. Each automotive company requires a specific format. A DVP&R is usually itemized and tabulated in Excel in two major sections, i.e., test plan and test report. The essential elements are as follows:

(A) Test plan	(B) Test report
Test ID (identification no.)	Actual completion (timing)
Specification and test method	Number (sample)
Test description	Level (sample)
Acceptance criteria	Pass/not pass
Test stage	Actual results
Target requirements	Additional notes
Test responsibility	
Start (timing)	
Completed (timing)	
Quantity (sample)	
Type (sample)	

Each of the above items listed above will be recorded as an individual column in an Excel file or the alike. Inputs are crucial for proper planning of implementation strategy and evaluation techniques, including sample sizes and timing for completing the tests. DVP&R receives inputs from various sources, such as

(a) Requirements and specifications
(b) New contents and QFD [Hauser and Clausing 1988]
(c) DFMEA
(d) Robustness analysis
(e) Part history and legacy of data

One may transfer data from the DFMEA that has been defined for the same item including actions. For example, it is a general practice to use the controls and perhaps some types of recommended activities from the DFMEA as the starting point when developing a new DVP&R. Some robust design guidelines that may help develop DVP&R are given as follows:

1. Conduct conceptual studies based on parametric design and tolerance design.
2. Perform risk assessment and DFMEAs.
3. Identify critical characteristics.
4. Advance system design in light of virtual design tools (e.g., FEA).
5. Get an early involvement of suppliers and the customer.
6. Make prototypes matching to mass production.
7. Consider "tendency to fail" such as degradation analysis, rather than time to fail.
8. Do it right the first time.

For example, programming for each of the 100^{+} central processing units (CPUs) in a passenger vehicle at the design stage may face such a technical challenge. Sneak circuit analysis (SCA) can be conducted on hardware and software to identify latent service conditions that inhibit demanded functions or cause undesired functions because of logical flaws, but not by part failures.

10.1.2. Production Tests

Production (manufacturing) tests are performed during the preparation or assembly of the product in an ongoing manner for purposes of assembly, performance verification, and quality control. This is conducted on the factory floor. Effective production tests redeem product quality, production line yield, cost-effectiveness, and customer acceptance. The finished products comply with the expectations set in the R&D stage. Burn-in tests (i.e., environmental stress screen tests or screening tests) are typical.

Focus on the stability of a production system in light of input–output functions in the early stage, i.e., reducing functional variation to improve reliability, control system design, calibration methods, prognostics, and measurement systems. Burn-in tests (i.e., environmental stress screen tests or screening tests) are typically done at the production stage. PFMEA and statistical process control (SPC) are viable tools for keeping track of potential problems.

10.1.3. Acceptance Test Commissioning Measurements

Acceptance tests are performed at the product delivery or installation. For example, site acceptance tests are performed at the buyer's location to validate that the product's specifications meet the buyer's requirements and standards. Commissioning measurements are thus taken to assure that the right product is built. Therefore, this constitutes an algorithm of product validation and renders a common commitment between the supplier and buyer. Production reliability acceptance tests (i.e., failure rate tests or MTBF tests) typically apply.

Production acceptance tests, i.e., failure rate or MTBF tests, are used to detect any product degradation in order to remove the inherent and latent unreliability of a product over the course of production [Raheja and Allocco 2006]. Acceptance tests are also used to assure that products made from a certain production lot meet the customer's reliability requirements. The products of an individual production lot are then accepted on the basis of a sampling plan.

10.1.4. Service and Repair Tests

These tests are performed as required over the service life of a product, and they are done at a service shop or the customers' site. A systematic approach to failure reporting, analysis, and corrective actions is set up, such as failure reporting, analysis, and corrective action system (FRACAS). How long the product lasts in service is called service life, as evaluated in the field under operational conditions in the working environment.

The objective of service and repair tests is twofold. First, it is essential to have a systematic approach to failure reporting, analysis, and corrective actions, such as FRACAS, in order to provide the data needed to identify deficiencies for correction to ensure that inherent or latent unreliability is not disregarded. The next is the analysis of warranty where unreliability affects the manufacturer's revenue and profit. The data obtained from FRACAS are to detect unreliability treads and to respond in a timely manner as a preventive tool to reduce the warranty cost.

10.1.5. Regression Tests

Regression refers to returning to a former state. Regression tests are performed on an existing operational product to assure that there is no loss in the existing functionality, while other aspects of the environment are changed. This is a testing practice that ensures an application still functions as expected after any change, update, or improvement. For example, tests immediately after upgrading the operating system (software) verify that the existing application software still works.

Regression tests are employed to determine the good distribution to use for making an accurate prediction of the entire product life or components of concern. Many failure mechanisms observed in an automobile, such as wear of brake discs, fractured leaf springs, and broken light bulbs, can be measured and analyzed as their performance degrades, eventually leading to weakness or sudden failures. One can generate pseudo-failure times to predict the corresponding product lifetime using the data collected from regression tests. Regression tests can be further used for growing product reliability.

10.2. Kaplan-Meier Product Limit Estimator

A product limit estimator was proposed to assess the survival function without making any assumption regardless of the distribution of survival time [Kaplan and Meier 1958]. Consider the discrete-time domain of each potential failure time recorded as t_1, t_2, etc., since the test data are generally discrete as given. The cumulative probability function of survival up to time t_q is defined as

$$S(t_q) \equiv P(t > t_q) \tag{10.2.1}$$

The Kaplan–Meier survival curve refers to the probability of surviving in a given length of time that is divided into many small intervals. The probability for the survival up to time t_{q+1} can be estimated using the following conditional probability of instantaneous survival rate:

$$s(t) = P(t > t_{q+1} \mid t > t_q) = \frac{P(t > t_{q+1},\, t > t_q)}{P(t > t_q)} = \frac{P(t > t_{q+1})}{P(t > t_q)} = \frac{S(t_{q+1})}{S(t_q)} \tag{10.2.2}$$

where
- t is the lifetime of the product
- $S(t_q)$ is the instantaneous survival probability, starting with $S(0) = 100\%$

It is a means of product limit estimate, involving computing of probabilities of occurrence of event at a certain time point of interest. This survival time is generally used for type I right censoring.

The Kaplan–Meier curve starts out with $S(t_0) = 1$ for the initial event time at $t_0 = 0$. In light of Equation (8.11.7), the hazard rate is calculated as the "number of subjects failed" divided by the number of "subjects at risk" for each specified interval of time as

$$h(t) = f(t) / R(t) \approx N_f / N_{risk} \tag{10.2.3}$$

The survival rate at time t_i, namely, $s(t_i)$, and the related cumulative survival estimate, namely, $S(t_i)$, are, respectively,

$$s(t_i) = 1 - h(t_i) \tag{10.2.4}$$

and
$$\hat{S}(t_i) = s(t_i)\, \hat{S}(t_{i-1}) \tag{10.2.5}$$

The sample variance, namely, $\sigma_S^2(t)$, can be calculated using the following equation [Peto 1984]:

$$\sigma_S^2(t_i) = [\hat{S}(t_i)]^2\, [1 - \hat{S}(t_i)] / N_{risk} \tag{10.2.6}$$

This is called Peto's method, given that $\sigma_S^2(t_0) = 0$ and $\hat{S}(t_i)$ is the estimated cumulative survival function. The 95% CI for the Kaplan–Meier estimate at the time step t_i is

$$\hat{S}(t_i) - 1.96\,\sigma_S(t_i) < \hat{S}(t_i) < \hat{S}(t_i) + 1.96\,\sigma_S(t_i),\ i = 1, 2, 3, \ldots \tag{10.2.7}$$

The Kaplan–Meier survival estimate is the simplest way of computing the survival over time in spite of potential difficulties associated with the test subjects or processes.

Once the cumulative survival time is calculated, the cumulative hazard can be estimated as [Peterson 1977]

$$\hat{H}(t) = -\ln[\hat{S}(t)] \tag{10.2.8}$$

Both $\hat{S}$ and $\hat{H}$ derived above do not assume specific distributions for survival or hazard curves. If survival plots indicate specific functions, e.g., lognormal and Weibull distributions, more accurate estimates of $\hat{S}$ and $\hat{H}$ might be achieved by the prescribed distribution function.

EXAMPLE 10.2.1

Assume that 10,000 vehicles (N_0 = 10,000) were introduced into the market 15 years ago. It is here intended to assess the survival estimate, with the given failed and suspended numbers of vehicles (N_f and N_s) as follows:

t (years)	N_s	N_{risk}	N_f	h(t)	s(t)	$\hat{S}$ (t)	σ_S^2(t)
$t_0 = 0$	0	1000	0	0	1.0	$\hat{S}(t_0) = 10$	0
$t_1 = 1$	0	1000	1000	0.1	0.9	$\hat{S}(t_1) = 0.9\ S(0) = 90\%$	8.1×10^{-6}
$t_2 = 3$	1000	8000	1000	0.125	0.875	$\hat{S}(t_2) = 0.875\ S(1) = 78.75\%$	16.47×10^{-6}
$t_3 = 8$	0	7000	1000	0.143	0.857	$\hat{S}(t_3) = 0.857\ S(2) = 67.49\%$	21.15×10^{-6}
$t_4 = 15$	0	6000	1000	0.167	0.833	$\hat{S}(t_4) = 0.833\ S(3) = 56.22\%$	23.06×10^{-6}
$t \geq 15$	0	5000	—	—	—	—	—

Solution:

(a) Given the experimental data, i.e., N_S, N_{risk}, and N_f, one is able to estimate the following:

$$h(t_i) = N_f(t_i) / N_{risk}(t_i)$$

$$s(t_i) = 1 - h(t_i)$$

$$\text{and} \quad \hat{S}(t_i) = \hat{S}(t_{i-1})\, s(t_{i-1})$$

where i = 1, 2, 3, and 4. The results are listed in the table. A plot of cumulative survival function in terms of number of years after an example model car has been introduced into the market is plotted in Figure 10.2.1.

(b) $\hat{S}(t \geq 15 \text{ years}) = \hat{S}(t \geq t_4) = 56.22\%$

(c) The sample variance can be calculated using Equation (8.2.6). Taking time step t_i into consideration, one has

$$\sigma_S^2(t_1) = [\hat{S}(t_1)]^2 [1 - \hat{S}(t_1)] / N_{risk} = [0.9]^2 [1 - 0.9] / 10000 = 0.0000081$$

$$\sigma_S^2(t_2) = [\hat{S}(t_2)]^2 [1 - \hat{S}(t_2)] / N_{risk} = [0.7875]^2 [1 - 0.7875] / 8000 = 0.000001647$$

$$\sigma_S^2(t_3) = [\hat{S}(t_3)]^2 [1 - \hat{S}(t_3)] / N_{risk} = [0.6749]^2 [1 - 0.6749] / 7000 = 0.000002115$$

and $$\sigma_S^2(t_4) = [\hat{S}(t_4)]^2 [1 - \hat{S}(t_4)] / N_{risk} = [0.5622]^2 [1 - 0.5622] / 6000 = 0.000002306$$

Thus, $\sigma_S(t_1) = 0.00284605$, $\sigma_S(t_2) = 0.004059$, $\sigma_S(t_3) = 0.004599$, and $\sigma_S(t_4) = 0.004802$

(d) The 95% CI of the Kaplan-Meier estimate for each time step t_i can be calculated using Equation (10.2.7).

$$\hat{S}(t_i) - 1.96\,\sigma_S(t_i) < \hat{S}(t_i) < \hat{S}(t_i) + 1.96\,\sigma_S(t_i)$$

Thus, $89.44\% < \hat{S}(t_1) = \hat{S}(1\text{ year}) < 90.56\%$

$77.95\% < \hat{S}(t_2) = \hat{S}(3\text{ year}) < 79.55\%$

$66.59\% < \hat{S}(t_3) = \hat{S}(8\text{ year}) < 68.39\%$

and $55.28\% < \hat{S}(t_4) = \hat{S}(15\text{ year}) < 57.16\%$

The lower ends of these inequalities can be considered in an automotive warranty policy.

This approach can be applied to each individual failure mode of a vehicle.

FIGURE 10.2.1 Cumulative survival after being introduced into the market (Example 10.2.1).

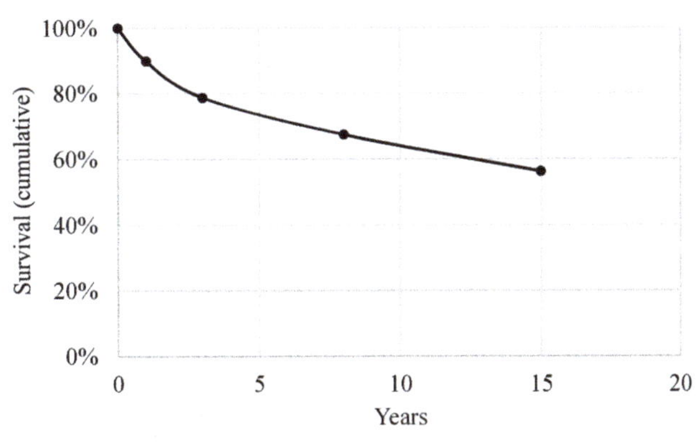

10.3. RG Analysis

A RG program is a well-structured process of improving product reliability by testing, incorporating corrective actions, and monitoring effectiveness of the fixes throughout analysis and test phases. The reliability of the product is improved, or experiences growth, as the design is modified to eliminate failure modes. It is essentially a closed-loop corrective action in the test–analyze–and-fix (TAAF) manner.

By default, the first lot of prototypes assembled in the development stage of a new complex product generally contain design, manufacturing, and/or engineering deficiencies. Because of these deficiencies, the initial reliability of these prototypes usually falls below the product reliability goal. In order to identify and correct these deficiencies, the prototypes are often subjected to a rigorous analysis and testing program. Problems are identified, and appropriate corrective actions are then taken. Product reliability goal can then be accomplished based on proper analytic techniques and RG tests.

Product RG is the improvement in the reliability of a product over a period of time because of changes in the product's design and manufacturing process. It may take place during prototype tests, DVP&R system tests, production tests, and inspecting tests after the customer's feedback from the field. RG is usually performed under normal operating conditions. Accelerated test plans may be employed to speed up improving product reliability prior to conducting RG testing, but they cannot replace it.

10.3.1. Product Failure Rate

RG analysis can keep track of the effectiveness of each design change, as an assessment tool used to decide if a reliability goal has been met and whether, and how much, additional testing is required. A schematic diagram for the failure rate of typical machine elements, e.g., bearings and gears, under normal operation is depicted in Figure 10.3.1, which is called a reliability bathtub curve that describes graphically the failure rate of components versus time, over the life of the product. It is also called life characteristic curve or lambda characteristic curve. In general, a durable product cycle life can be divided into three separate periods as follows:

(a) Break-In Period: It is the early life stage where the failure rate is high initially but decreases to somewhat minimum. This period is characterized by a decreasing failure rate and consists of failures caused by both birth defects and blunders made. Break-in happens to most new products and is called the infant mortality period. Zero-kilometer defects of passenger vehicles belong to this category.

(b) Useful Life Period: In this period, the failure rate tends to become somehow constant if wear-out does not persist in the product. It is represented by the flat bottom of a bathtub. If it is purely flat, failures occur randomly in this period for products like electronic components. Sometimes a slight slope turning upward exhibits in accordance with a minor increase in the failure rate as wear-out kicks as shown in mechanical parts with moving parts. Unreliability in this period consists of random failures typically caused by being "used beyond designed."

(c) Wear-Out Period: It is the stage where the slope starts to rise till the end. Thus, it is also called burn-in period since it typically happens to products when they get old and the failure rate increases. Wear-out is usually caused by breakdown because of various reasons such as physical wear, stress-induced fatigue, corrosion, and/or degradation of materials.

Various statistical distribution functions can be used to characterize different failure modes of products. The exponential distribution function fits electronic components smoothly, while the two-parameter Weibull distribution function is of great practical interest for mechanical components. The three-parameter Weibull distribution function is applied if the two-parameter model is not adequate. The bathtub curve is not intended for characterizing the failure rate of a single item. Instead, the curve describes the relative failure rate of an entire population of products over time. It should be taken into consideration in the cost model as the economic impact is quite dependent on the period of warranty, especially with automobiles that are subject to wear, tear, fatigue, corrosion, and other time-related failure modes.

FIGURE 10.3.1 Product failure rate.

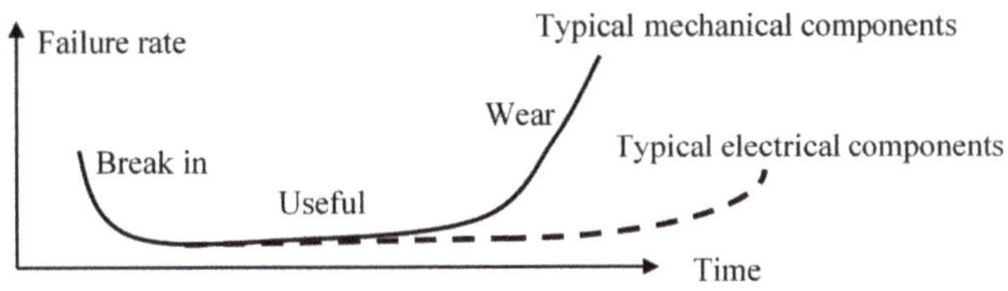

10.3.2. RG: Homogeneous Poisson Process (HPP)

As the product development progresses from the conceptual design stage to development, testing, and manufacturing stages, it is expected that the implementation of every design change will improve the product reliability goal. Typically, a formal TAAF program is implemented to discover design flaws and mitigate them. The gradual product improvement through the elimination of design deficiencies may result in the increase in inter-failure arrival time that is known as RG. Statistically, the fundamental theme is to reduce MTBF. It is addressed in ANSI/GEIA-STD-0009 that the following four objectives shall be met regarding RG:

- Understand customer/user requirements and constraints
- DFR [Rigdon et al. 2022] and re-DFR
- Producing reliable systems
- Monitoring and assessing user's experienced reliability

OC curve analysis should be used to mitigate risks in demonstrating reliability requirement. It can be used to differentiate risks: type I or type II errors, i.e., either the supplier's risk or the buyer's risk in a product acquisition. Comparison across multiple OC curves helps gauge the sample size as a function of allowable failures and risk to define reliability (R) and confidence level (C). For example, R = 90% and C = 90% are the normalcy for Ford Motor Company's request to its suppliers. Two opportunities for assessing the RG of a product are recommended:

1. The first opportunity is to make an RG plan through analysis and design improvements in the product design phase, in which initial reliability, reliability goal, and planned design improvements, along with their expected magnitudes, should be addressed. Reliability allocation has to be applied during the design stage. It applies methods to assigning reliability targets to subsystems and parts based on a product reliability goal. The required equipment for testing, analytical techniques, and even redundancy configurations to meet reliability targets should be also recommended in a reliability allocation project.
2. The second opportunity for assessing the RG comes by the product validation phase. It takes a data set of accumulated test times, at which relevant failures occurred or were observed, and the time of termination of the test if different from the time of the final failure. Basically, the trend of the failures encountered during testing is the growth curve that can be used for assessing the RG rate.

A test effort is generally guided by prior information. The collected data used as the input for the analytic model begins after the completion of preliminary tests, e.g., environmental stress screening (ESS) testing, are intended to stabilize the product's initial failure intensity. Though several RG models are available, the power-law model is used frequently. Assume that F(t) is the cumulative number of failures up to time t (Columns 1 and 2 in Example 10.3.1) as obtained from the test. The corresponding growth rate of failure, namely, cumulative failure rate m(t), can be calculated as [Duane 1964]

$$f(t) = d[F(t)] / dt \approx \Delta F(t) / \Delta t \qquad (10.3.1)$$

As derived from the test data shown in Figure 10.3.2, the discernable linear relationship between ln[f(t)] and ln(t), i.e., f(t) in the natural log scale versus time t in natural log scale, verifies the validity of the power-law model for denoting failure rate. Therefore, failure rate f(t) can be written as [Duane 1964]

$$f(t) = \lambda\, t^{\phi} \tag{10.3.2}$$

where

f(t) is the cumulative failure rate or cumulative failure intensity

λ is the failure rate coefficient

ϕ is the growth rate, and generally, $0 < \phi < 1$ along an actual growth path

t is the time

which is a special form reduced from Equation (9.2.12), i.e., two-parameter Weibull model. The hazard conversion rate, also called instantaneous failure intensity, can be obtained by subtracting the change in cumulative failure rate with respect to time from the cumulative failure rate as [Duane 1964]

$$h(t) = f(t) - d[f(t)]\,/\,dt = \lambda\, t^{\phi} - \phi\, \lambda\, t^{\phi} = \lambda\,(1-\phi)\, t^{\phi} \tag{10.3.3}$$

This is the rate of occurrence of failures (ROCOF), designated as the hazard conversion rate or instantaneous failure intensity. Note that instantaneous failure rate h(t) is less than cumulative failure rate f(t). The difference between the hazard conversion rate (instantaneous failure intensity) and cumulative failure rate is the change in f(t) with respect to time, i.e.,

$$d[f(t)]\,/\,dt = -\lambda\, \phi\, t^{\phi} \tag{10.3.4}$$

Taking a natural log on both sides of the above equation leads to a linear relationship between ln[df(t)/dt] versus t as

$$\ln[df(t)\,/\,dt] = \ln(\lambda\, \phi) + \phi\, \ln(t) = \text{Constant} + \phi\, \ln(t) \tag{10.3.5}$$

Note that parameter ϕ is the growth rate, which is exactly parameter α given in one figure published by [Duane 1964] and recited here as Figure 10.3.2. This is a special case of failure model, and it is called HPP.

FIGURE 10.3.2 Plot of failure rate versus accumulated time [Duane 1964].

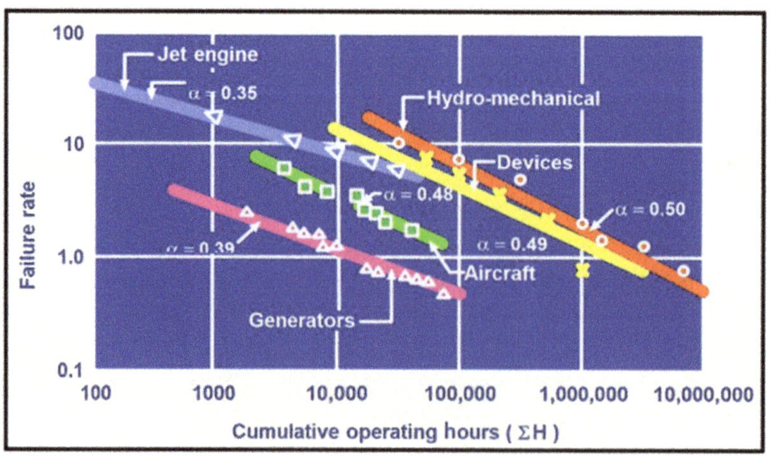

Once the HPP constant repair rate model takes over, the MTBF at time T (the moment when the test is suspended) for the system from then on is the reciprocal of the final repair rate. According to Crow's model [AMSAA 2010], the MLEs for λ and ϕ are estimated based on actual failure times or grouped data. The estimate of the hazard conversion rate, h(t), at time T (the end of test) referring to t = 0 (starting time) can be obtained from Equation (10.3.3).

$$h(T) = \lambda\,(1-\phi)\,T^{-\phi} = \frac{N}{1-\phi}\,(1-\phi)\,T^{-\phi} = \frac{-N\,(1-\phi)}{T} \tag{10.3.6}$$

Since MTBF is the inverse of the hazard conversion rate (instantaneous failure rate),

$$\text{MTBF} = 1\,/\,h(T) = \frac{T}{N\,(1-\phi)} \tag{10.3.7}$$

where

T is the time, at the end of test

N is the total number of test failures observed

The formulae for a 1 – α confidence interval (M_L, M_U) for the MTBF at the end of the test are approximated by

$$F_U = \text{MTBF}\,\frac{N\,(N-1)}{\{N + \tfrac{1}{4}\,(z_{1-\alpha/2})^2 + [\tfrac{1}{2}\,N\,(z_{1-\alpha/2})^2 + (z_{1-\alpha/2})^4\,/\,16]^{1/2}\}^2} \tag{10.3.8}$$

$$F_U = \text{MTBF}\,\frac{N\,(N-1)}{[N - z_{1-\alpha/2}\,(\tfrac{1}{2}\,N)^{1/2}]^2} \tag{10.3.9}$$

where $z_{1-\alpha/2}$ is the 100 (1 – ½α)% percentile point of the standard normal distribution. Note that F_L and F_U are the 100 (1 – ½α)% one-sided lower and upper confidence bounds, respectively.

The power-law model [Duane 1964] has been successful in modeling industrial reliability improvement data that follow the HPP. An example is given below to illustrate the work procedure for calculating the growth rate and MTBF with confidence levels.

EXAMPLE 10.3.1

A test-to-failure experiment is conducted. Ten failures have been observed before the preset test cycle (1500 h) is reached, and the data are listed in the first two columns of the following table:

i	t_i	t_i/i	$\ln(t_i)$	$\ln(t_i/i)$
1	5	5	1.61	1.61
2	40	20	3.69	3.0
3	43	14.33	3.76	2.66
4	175	43.75	5.16	3.78
5	389	77.8	5.96	4.35
6	712	118.7	6.57	4.78
7	747	106.7	6.62	4.67
8	795	99.38	6.68	5.6
9	1299	144.3	7.17	4.97
10	1478	147.8	7.3	5.0
End	1500			

What would be the estimated MTBF after 1500 h, at the moment the test is suspended?

Solution:

1. Calculate t_i/i, $\ln(t_i)$, and $\ln(t_i/i)$. The results are shown in columns 3, 4, and 5.
2. Regression of $\ln(t_i/i)$ to $\ln(t_i)$ leads to the following model:

$$\ln(t_i / i) = 0.5964 + 0.6136 \ln(t_i)$$

   It means that $\phi = 0.6136$ and $\ln(\lambda\, \phi) = 0.5964$. Thus, $\phi = 0.6136$ and $\lambda = 0.338$.

3. The MTBF, at the moment right after 1500 h when the test is suspended, is

$$MTBF = \frac{T}{N(1-\phi)} = \frac{1500}{10(1-0.6136)} = 388 \text{ hours}$$

10.3.3. RG: Non-HPP (NHPP)

A plot on log–log paper of successive MTBF estimates versus system time of fail for reliability improvement test data is called a Duane plot. It was observed that the plotted MTBF estimates versus the times of failure on log–log paper tended to line up following a straight line [Duane 1964], of which the slope line through the points is called the RG slope or the Duane plot slope.

Assume that the RG failure time follows a NHPP. By a change in variable,

$$\phi = 1 - \beta \tag{10.3.10}$$

the hazard function described by Equation (10.3.7) becomes [Crow 2011]

$$h(t) = \lambda\, \beta\, t^{(\beta-1)} \tag{10.3.11}$$

where

λ is the coefficient of hazard conversion rate (instantaneous failure intensity)

β is the shape factor of a Weibull density function; $\beta = 1 - \phi$

It is called crow model [Barringer 2006] or crow-AMSAA model [AMSAA 2010]. It is popular to use crow-AMSAA for system reliability modeling. Nevertheless, it must be applied to only historically identical events for the model to represent the truth and provide believable predictions. If $0 < \beta < 1$, the corrective action is working; if $\beta > 1$, the action makes it worse as shown in Figure 10.3.3. The above equation is the same as Equation (10.3.2), except that the RG is $1 - \beta$ instead of ϕ.

FIGURE 10.3.3 Hazard function ($\lambda = 1$) as a function of β, as denoted by Equation (10.3.11).

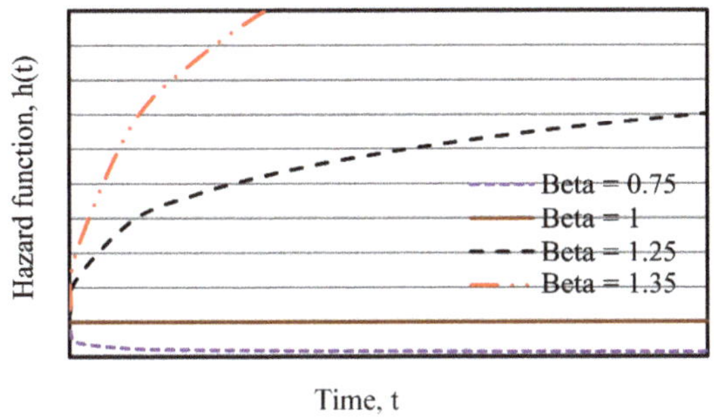

The RG failure times follow a NHPP with an instantaneous failure intensity following Equation (10.3.10), which will allow for statistical procedures based on this process for RG analyses [Crow 2011]. The cumulative number of failures is then

$$N(T) = \int_0^T h(t)\, dt = \int_0^T \lambda\, \beta\, t^{(\beta-1)}\, dt = \lambda\, T^{\beta} \tag{10.3.12}$$

$$\text{Thus,} \quad \lambda = N(T) / T^{\beta} \tag{10.3.13}$$

By taking a natural log on both sides of the above equation, one has a linear relationship between the number of failure times and the failure time as

$$\ln[N(t)] = \ln(\lambda) + \beta \ln(T) \tag{10.3.14}$$

According to [Crow 2011], λ and β can be estimated using the maximum-likelihood method, respectively, as

$$\beta \approx N / \{ \sum_{i=1}^{N} \ln(T / t_i)\} \tag{10.3.15}$$

where

T is the duration of the test

n is the number of failures observed by time T

t_i is the time, at which each corresponding failure is observed

The MTBF is the inverse of the instantaneous failure intensity, i.e.,

$$\text{MTBF} = 1 / h(T) \tag{10.3.16}$$

EXAMPLE 10.3.2

A test-to-failure experiment is conducted for five prototypes (electric motors) at five test stands (S-1, S-2, S-3, S-4, and S-5), respectively. What is the MTBF based on the failure times? Test data (hours) are tabulated as follows:

Failure no.	S-1	S-2	S-3	S-4	S-5	t_i
i = 1	14.3^F	0	0	0	0	14.3
2	55.0^F	19.2	0	0	0	74.2
3	88.4	21.5^F	18.4	15.8	0	144.1
4	140.4^F	46.4	21.3	21.7	31.9	223.9
5	149.5	69.2	32^F	38.7	39.6	329
6	182.3	98.7^F	75.4	51.2	49.8	457.4
7	214.2	149.7	121.5	64.6^F	79	629
8	263	198.6	170.3	113.4	80.4^F	825.7
9	288.1^F	225.4	197.1	140.2	107.2	958
10	314.3	251.6	209.9^F	166.4	133.5	1075.7
11	381.8	262.6^F	277.3	233.9	200.9	1356.5
12	400	334.1	348.4	250.3^F	272.5	1605.7
13	400	382.2	396.9	298.3	293^F	1770.4
14	400	400	400	385.4^F	400	1985.4
End	400	400	400	400	400	2000

where superscript F means that failure occurs and the machine is repaired. The test continues until the accumulated time of each machine reaches 400 h.

Solution:

The two parameters for the instantaneous failure intensity can be calculated using Equations (10.3.14) and (10.3.15) as

$$\beta \approx N / \{ \sum_{i=1}^{N} \ln(T / t_i)\} = 14 / \{ \sum_{i=1}^{14} \ln(2000 / t_i)\} = 0.6839$$

$$\text{and} \quad \lambda \approx N / T^{\beta} = 14 / (2000)^{0.6839} = 0.07734$$

The hazard conversion rate (instantaneous failure intensity) is

$$h(T) = \lambda\, \beta\, T^{(\beta - 1)} = 0.07334\,(0.6839)\,(2000)^{(0.6839 - 1)} = 0.004788$$

$$\text{Thus,} \quad \text{MTBF} = 1 / h(T) = 1 / 0.004788 \approx 209 \text{ hours}$$

10.3.4. RG-Weibull Density Function

Let the coefficient of instantaneous failure intensity (hazard conversion rate), i.e., λ in Equation (10.3.2), be rewritten as

$$\lambda = \eta^{-\beta} \tag{10.3.17}$$

Then, the instantaneous failure intensity in RG model becomes the hazard conversion rate of a Weibull density function as [Lawless 2003]

$$h(t) = \eta^{-\beta}\, \beta\, t^{(\beta - 1)} \tag{10.3.18}$$

Note that the above equation is exactly Equation (10.3.11). Parameter η is the scale factor of a Weibull density function, while parameter β is the shape factor.

10.3.5. Effectiveness of a Corrective Action

Consider here a single repairable equipment, which operates from age zero and fails at age T_1 since it is restored to a satisfactory condition by a corrective action that involves design improvement. It operates again and fails later at age T_2 measured from age zero and then gets restored again to a satisfactory condition by a design improvement and operates again, and the process gets repeated. If more than one unit is involved in the RG process, then T_i is the unit hours of operation accumulated by all units involved in the process by the time the ith failure occurs.

Assume that the cumulative number of failures versus the cumulative time is linear on logarithmic scales, as demonstrated by Equation (10.3.10). An effective corrective action may result in a change in the failure rate, i.e., the RG is not the same anymore after such a correction. Statistically speaking, there exist bi-modal Weibull distribution functions: one for the RG before and the other after the correction action is taken. Mathematically, they may be modeled separately using two different sets of data. However, the information collected during the first segment will be used when creating the model for the second segment.

The modeling of the first segment is

$$\beta_1 \approx N_1 / \{ \sum_{i=1}^{N_1} \ln (T_1 / t_i)\} \tag{10.3.19}$$

$$\text{and} \quad \lambda_1 \approx N_1 / (T_1)^{\beta 1} \tag{10.3.20}$$

Given that the MLE of the model parameters in the second segment can be estimated using the following two equations [Crow 2011]:

$$\beta_2 \approx N_2 / \{N \ln(T_2 / T_1) + [\sum_{i=N_1+1}^{N} \ln(T_2 / t_i)]\} \tag{10.3.21}$$

$$\text{and} \quad \lambda_2 \approx N / (T_2)^{\beta 2} \tag{10.3.22}$$

of which $N = N_1 + N_2$. Note that $\beta_2 > \beta_1$ is expected because of the corrective action. If $\beta_2 < \beta_1$, i.e., the RG is worsened because of the paradox "corrective action" made in-between them, if no other conditions have changed.

As an extension, multi-modal Weibull distributions will lead to identifying effectiveness of each corrective action. Each segment of data with "linear" RG can be modeled independently on the piecewise basis [Guo et al. 2010] when approaching a nonlinear RG.

10.3.6. RG Management

RG is to address factors for product improvement, such as the management strategy toward taking corrective actions, effectiveness of the fixed, reliability requirements, the initial reliability level, resource funding, and competitive factors, especially at the product design stage [Mohammadiam and Aït-Kadi 2010]. Testing steps include evaluation, qualification, RG modeling, and application of the FRACAS, as addressed by [National Academies of Sciences, Engineering, and Medicine 2002].

For example, RG curves can be used to chart the performance of repairable systems after being introduced into the market to provide the following information:

(a) Capturing and recording messages about unreliability issues.
(b) Corrective actions, e.g., periodic retrofits and redesign, during the product's life cycle.
(c) How often the manufacturing facility requires preventive maintenance.
(a) Assessing the number of replacement parts to have in inventory.
(d) Checking if the system runs at an acceptable level.
(b) Estimating costs incurred in connection with corrective actions.

It has been proven that considering designed-in reliability in the early design stage is more effective in terms of cost and time reductions in the lifetime management of a product.

10.4. Qualitative Accelerated Life Tests

In reality, automotive manufacturers do not have years to spend on burn-in at the DV stage or during the product break-in period after it is introduced into the market. RG of a product under development may rely on accelerated life tests and degradation tests. It means to test the product under conditions that are more extreme than normal operating conditions. Common acceleration factors for life tests include temperature cycling, thermal shock, vibration, corrosion, debris (e.g., dust), voltage, current, environment (humidity, O_2, H_2, etc.), clock variation, and loading path of the applied acceleration factors. Accelerated life tests can be classified into the following categories [McLean 2009]:

1. Failure mode verification test (FMVT)
2. Full system life test (FSLT)
3. ESS test
4. Highly accelerated life test (HALT)
5. Highly accelerated stress screening (HASS)
6. Step-stress accelerated life test (SSALT)

More severe conditions enable the product to degrade faster and fail sooner and thus make a reliability study in time. Nevertheless, results are often used to predict product reliability at normal use conditions. Comparison between constant-stress test and SSALT under time constraint is addressed in [Han and Ng 2013]. Detailed accelerated reliability and durability testing technology is presented in [Klyatis 2012]. DOEs can be incorporated into accelerated life tests to explore interactions between accelerating factors (variables) [Meeker et al. 2013].

An experimenter may take advantage of DOEs efficiencies by considering many factors in concert with the optimization objective while conducting these tests. Design for product RG via DOEs needs an overall proactive strategy for integration of product quality development and RG. It is recommended that one should also estimate product reliability under normal conditions for the final decision-making while being able to accelerate failure occurrences with stress variables at high levels. D-optimality provides good performance for acceleration model validation, especially when the testing feasibility region is asymmetric. Furthermore, it offers a significant reduction in sample size over legacy designs for a fixed precision [Monroe and Pan 2008].

10.4.1. FMVT

FMVT testifies to see whether a product meets its specifications by identifying failure modes and the related root causes. As the product is verified if it will function in its service environment by testing to failure, FMVT is determined for the life prediction and warranty need. It can be also carried out in a highly accelerated testing mode that uncovers product weaknesses quickly. Once successfully performed, the product is claimed to be free of failures. It is a "pass or fail" test for DV of parts or components.

10.4.2. FSLT

An FSLT lets the full system be tested under the loading conditions of the service environment with a significant number of sample size. It is in constant-stress testing; i.e., a unit is tested at a fixed stress level until failure or the termination time point of test. It is a "pass or fail" test for PV based on an entire system, e.g., automotive side doors, HVAC systems, and seating systems. It may reveal design defects or weaknesses that may not have been realized as designed and/or manufactured. The test will unveil the strength distribution and accurate reproduction of service conditions.

10.4.3. ESS Test

ESS is tested by exposing a newly manufactured product to environmental loads, often considered as stress factors, in order to identify and eliminate latent defects introduced during the manufacturing process, often involving many temperature and vibration cycles at extreme limits. Typical ESS guidance derived from the US Air Force (USAF) Rome Laboratory Reliability Engineer's Toolkit [MIL-HDBK-344A 1993] is summarized in Table 10.4.1. If no ESS is performed, the "break-in" period of a durable product like a passenger vehicle would take up to approximately one year for the population to naturally reach the expected steady-state failure rate observed during the "useful" life period. The inspection is therefore performed on 100% of the items manufactured.

It is extremely important to determine the correct limits for the stress screening process. If too little is undertaken, then some latent failures may remain in the unit under test. If the conditions that are applied to the unit are too severe, then additional failures may be introduced and there is a risk that the overall reliability may be impaired. One may refine the typical test conditions, as those provided in Table 10.4.1 for electronic equipment, in order to design a specific "screening strength" model [MIL-HDBK-344A 1993] for an individual product that has to undergo its own special screening conditions such as varying temperature levels and ranges, vibration levels and ranges, loading sequence, and their interactive impacts in the real world. Nevertheless, physical modeling (e.g., FEA and statistical tolerance design) with such complex loading functions shall be studied to understand the physics of potential failure modes [Chiang 2019].

TABLE 10.4.1 Typical ESS of electronic equipment [MIL-HDBK-344A 1993].

Parameter and conditions	Wiring assembly	Equipment system
Thermal cycling		
1. Temperature range	−50°C < T < 75°C	−40°C < T < 71°C
2. Temperature rate of change	20°C/min	15°C/min
3. Temperature dwell time	Until stabilized	Until stabilized
4. Temperature cycles	20–40 cycles	120–200 cycles
5. Power on or off	Not applicable	On heating; off cooling
6. Equipment monitoring	Not applicable	Go/no go
Random vibration		
7. Acceleration level	6 G_{rms}	6 G_{rms}
8. Frequency range	20–2000 Hz	20–2000 Hz
9. Axes excited	3	3
10. Vibration duration, minimum	10 min/axis	10 min/axis (serially)
	10 min	10 min (concurrently)
11. Power on or off	Not applicable	On heating; off cooling
12. Equipment monitoring	Not applicable	Go or no go
After screening test		
13. Electric testing after screening	Yes, at 23°C	Yes, at 23°C

10.4.4. HALT

HALT uses accelerated stress factors beyond product specifications and can be employed to lower product development costs and compress time to market, and it is a powerful tool if accompanied by DOEs. In HALT, by running life tests at higher stress levels than normal operating conditions, ALT quickly yields information on the statistical lifetime distribution of test units. A product is introduced to progressively higher loadings in order to quickly uncover design weaknesses, thereby increasing the operating margins of the product, translating to higher reliability.

Thus, HALT methodology is in need for RG, employing accelerated loading factors beyond the specification. The purpose of HALT is to find quickly design and construction weak points and to derive from its (by taking 80% of the "break levels") profiles for ESS. There are no pre-established limits, but the product determines the limits. The benefit of HALT can be summarized as follows: early product release, lower product development costs (manpower and timing), higher profits, no long-term product issues, and eventually great customer satisfaction.

Technically speaking, HALT is employed to determine functional and destructive limits. The target of HALT is not to determine the life of the product but to make the product reliable as possible, while samples are usually tested in a controlled environment. Stress factors, also called loads, applied in an accelerated test are typically

1. Temperature (mean), hot or cold
2. Temperature cycles
3. Mechanical mean stress/pressure
4. Stress/pressure cycles
5. Humidity
6. Random vibration
7. Voltage

8. Power margining
9. Power cycling
10. Foreign objects, e.g., sand, dust, and metal debris
11. Corrosive factors, e.g., brine

Test fixtures for automotive components or systems are enabled to transmit various vibration modes to the item under test at the determined frequencies and loading phase angles such as implemented in multiaxial fatigue test stands. The environmental chamber is capable of rapid changes in temperature. For example, 50°C per minute is considered a minimum rate of change. Usually, high-power resistive heating elements are used for heating and liquid nitrogen (N_2) is used for cooling. Besides selecting stress factors, a successful HALT requires the following:

(a) Understanding of failure modes and mechanisms to be accelerated
(b) Knowing how the applied stress factors affect the failure mechanisms
(c) Determining the intensity levels of these stress factors

While HALT is intended to quickly find failures and accelerate the improvement program, it is not a shake-and-bake method [Hobbs 2000]. When products are delivered to end users, there will be no potential failure modes. HALT would be normally incorporated into a RG program. RG of a product comes with a certain confidence level, or it is meaningless.

HALT is a good tool for quickly searching for what the weakest aspect of a product is, and then, the designer takes correlative action to enhance or improve it. Through HALT, the product reliability may be verified and enhanced in short time and thus speeding up introducing a robust product into the market. HALT can be used for a quantitative accelerated test, if used in combination with DOEs, FEA [or other computer-aided engineering (CAE) tools], and strain-life fatigue theory. Since both material and geometric nonlinearities are pretty strong in response to low cycle fatigue (LCF) tests when the product is subjected to high loads, three levels for each stress and design factor are recommended for setting up such a HALT-based DOEs.

Failure mechanisms that HALT did not find are long-term issues, such as thermal creep fatigue that results in barrel cracking in vias and high-cycle solder joint fatigue failures of resistors or capacitors on the printed circuit board assembly.

10.4.5. HASS

A HASS test uses accelerated loading factors as determined appropriate by earlier HALT testing at the design stage in order to detect product defects in manufacturing production screens [Charki et al. 2011]. After the design is ruggedized through HALT and the DVP&R has been satisfactorily completed, the production begins. The production process can render other failure modes that are not related to a faulty design, and the sustaining process can certainly introduce new design problems. HASS is intended to catch these new failure modes more effectively and faster than burn-in or other tests done in production. It is widely accepted to improve product reliability at the manufacturing site in a variety of industries, including electronics, computer, defense, automotive, medical, and others.

10.4.6. SSALT

SSALT is the major tool for quantifying the impact of various factors (design factors, manufacturing noises, and environmental noises) on the objective for the improvement in DFR. Algorithms for step-SSALT are detailed in Section 10.5.

10.4.7. Mechanical Components

When mechanical components are subjected to thermomechanical loads, there are two major "loading factors" that are used to accelerate the effective time of burn-in: temperature and stress. Increasing temperature (relative to the normal operating temperature) can accelerate the life damage by tenfold. An increase in stress level (relative to the normal operating level) can also provide higher acceleration factors on many durability tests.

HALT testing for mechanical components is largely applied to a thermomechanical fatigue process. Fatigue acts particularly on points of stress concentration and the strains induced due to either mechanical loadings, differential temperatures, and gradients create stress–strain cycles that progressively damage materials and the joints between them. As fatigue life is an exponential function of the load level, once the statistical distribution is identified (usually Weibull or lognormal), burn-in times of a few hundreds of hours can provide an effective estimate of warranty lifetime, significantly reducing the proportion of parts with infant mortality defects [Peiskammer et al. 2003].

10.4.8. Electronic Components

An electronic device is usually made of a suite of components that can be reasonably evaluated using HALT methods, since the thermal and mechanical environments are typical of those load types occurring in actual service. Generic screening tests, potential failure modes, and test conditions for on-ground vehicles, aircraft, medical devices, navigation systems, batteries, and semiconductors are described in Table 10.4.2.

TABLE 10.4.2 Typical screen tests for automotive electronics.

Screen tests	Failure modes detected	Test conditions
Temperature cycling	Die-substrate faults, sealing defects, fracture of solder balls and vias, and printed circuit board (PCB) delamination	Alternating cycles at temperature extremes: T_{max} = 125 or 140°C and T_{min} = −40°C; temperature change: $5 \le \Delta T/\Delta t \le 10$°C/min; ≤ dwell time ≤ 30 min, at temperature extremes; frequency ≥ 20 cycles/day
Burn-in test	Wire-bond defects and metallization faults	Pushing power through the device for a significant time duration ($48 \le \Delta t \le 168$ h), running at an elevated temperature (e.g., T = 70°C)
Parametric measurements (before burn-in)	Ionic contamination, surface defects, bond wire defects, metallization faults	Parametric characterization of devices (e.g., semiconductors and electrochemical cells) and electrical properties of materials
Parametric measurements (after burn-in)	Changes in electrical characteristics	Parametric characterization of devices (e.g., semiconductors and electrochemical cells) and electrical properties of materials
Soldering heat	Thermal degradation, change in electrical characteristics, and thermal stress because of heat	Components immersed in solder bath at a high temperature (e.g., T = 260°C for Δt = 10 sec)
Thermal shock	Thermomechanical stress-induced damage to seal and structural elements, leakage of filling materials, delamination, and change in electrical characteristics	Quick transfer between two temperature extremes (e.g., −40 to 125°C) for a certain number of times (e.g., 20 cycles)
Bias test	Simulating actual operating conditions in the field, to see if devices withstand the high voltage, permitting high-frequency response, great current density, and fast switching	Device samples are loaded slightly less than the maximum rated reverse breakdown voltage in a high-temperature environment (e.g., T = 125°C and H_R = 85%)

(Continued)

TABLE 10.4.2 (Continued) Typical screen tests for automotive electronics.

Screen tests	Failure modes detected	Test conditions
Salt-spray test	Resistance to corrosion (metallic parts and coatings)	Spray mist of 5% salt solution in water in an environmental chamber for 24 h
Moisture resistance	Effect of humidity on the resistance to corrosion of contacts and wires	Subject unit under test to alternate cycles of dampness and drying while cycling the temperature
Seal leak test	Defects in package seal	Fill package with a gas (e.g., helium) under pressure
Water immersion	Defects of sealing	Immerse units in water or salt solution at different temperatures (e.g., T = 0 and 35°C)
Humidity test	Corrosion of metallization, absorption of moisture, chemical action of pollutants, and electrolytic corrosion of joints	Subject units under test to H_r = 90–95% at 40°C for 96 h
Electrical overstress	Testing at vias levels with accelerated load factors to exposes infant mortality	Application of voltage in increasing steps from normal to 25 or 50% above ambient
Vibration test	Loose parts, wear and tear, solder joints, fatigue, noise, bonding defects, mechanical faults in devices	Vibrate units under test at the frequencies of expected operations, e.g., ranging from 20 Hz to 2 kHz
Drop test	Mechanical damage, change in electrical characteristics, leakage of electrolyte due to sealing defect, adhesion faults, bonding defects, and cracks in substrate	Evaluating a device's ability to withstand being dropped at a minimum defined number. Experimental factors including heights of drop, the surface to which the product is dropped, and orientation (e.g., corner, edge, and face drops).
High-temperature storage	Contact defects, oxidation, ingress of moisture metallization defects, thermal aging, softening, physical changes (degrading)	T = 50°C for plastic devices and T = 250°C for hermetically sealed devices, for a certain time duration (e.g., Δt = 24 h)
Normal storage	Contact defects, oxidation, ingress of moisture metallization defects, thermal aging, softening, physical changes (creep)	Simulating actual field storage conditions, e.g., T = 85°C and H_R = 85%
Signal transceiver	Automated driver assistance system (ADAS) used to identify the presence of any stationary infrastructure, pedestrian, or moving vehicle	Radio frequency (RF) transceivers of high bandwidths, using high-frequency waves (mmWave frequency)
Battery simulation	Manage the charge stored in the battery and regulate it during regenerative braking and charging	Testing battery management systems by battery simulators
EMC test by upsets (electromagnetic compatibility test)	Characterized by system performance upsets	The system is irradiated with an RF signal via an antenna and swept over a determined frequency range
EMC test by limits	RF signals characterized by different limits	RF signals emanating from the electronic system are measured with an antenna

Notes:
T and ΔT (°C) are temperature and temperature duration (e.g., dwell time), respectively
t and Δt (hours, minutes, or seconds) are time and time duration, respectively
H_r is the relative humidity

10.5. Cumulative Exposure Model for Stepwise Accelerated Life Test (SALT)

A SALT is a test-to-failure mechanism designed to quantify the shortened test life of specimens by subjecting them to higher levels of loadings (stresses, strains, strain energy, temperature, material degradation, environmental factors, etc.) than specified nominal test conditions. Loading strategies carried out such quantitative accelerated life tests can be divided into two groups:

1. Parallel Constant-Stress Accelerated Life Test (PCSALT): Test units are divided into several groups, and each group is tested under a different treatment (i.e., a set of stressing and environmental conditions) throughout the test. Algorithms presented in Volume 1, Chapters 2–6 are geared for PCSALTs.
2. SALT: The test begins by exposing all specimens to a prespecified treatment. For those test units that survive up to a prespecified time, the stress level is elevated and held constant for another prespecified treatment. The test continues until all units fail (uncensored test) or until a prespecified test termination time is attained (censored test).

Both can be employed to exploit the supplier–purchaser relationship by the contracted specification required.

Compared with a PCSALT, an SSALT is more practical in the sense that fewer specimens are required for product qualification tests [Tseng and Wen 2000]. Because specimens tested at higher load levels often incur higher costs, initial tests of all specimens at a lower load level under the SSALT strategy may lower the expected experimental time and cost. However, PCSALT reveals more information of the root causes of potential problems.

10.5.1. Material Transformations

Most modern on-ground vehicles are expected to operate without failure for at least ten years. However, only a few of vehicle components are really tested to failure or degrade to a significant amount in the real-world driving under normal service conditions. Estimating the time-to-failure distribution, as well as sustaining long-term performance of components of highly reliable products that involve material transformation, e.g., deformation and degradation, has never been an easy work task.

Traditionally, reliability assessment of new products has been based on accelerated life tests that account for test-to-failure modes and censoring times of products subjected to elevated load levels by design. However, this approach may offer little help for highly durable products, which are unlikely to fail in a reasonable time duration in the laboratory. An efficient way is to assess the reliability by the inferential mechanisms of material transformation based on physics of the concerned product characteristics [Elsayed and Liao 2004, Lee and Pan 2010, Nelson 2005] and then extrapolate performance degradation to do the product life prediction by statistics. In summary, the study on ALT serves for the following two purposes:

1. Inferential Mechanism: It is a trending model that derived from the acquired data from multi-factorial load cases, as being effective and adequate to represent how the physics of failure varies with respect to the applied loads. Inferential mechanism provides the means of characterizing a conversion (transformation) process in nature. Inferential mechanisms usually appear in an accelerated mode, such as the common degradation models for physics of failure of some electronic materials given in Table 10.5.1. The simplest inferential mechanism is a linear one, while the most complicated one is perhaps product fatigue life that consists of various evolving processes. Some inferential mechanisms based on natural behaviors of physics are discussed in the next chapter. Furthering one's understanding of more theories on inferential mechanisms of materials can

be found in [Chiang 2019]. It may include material transformations (e.g., curing [Chiang 2019] and degrading [Bae and Kvam 2004]), electronic conversions (e.g., photovoltaics), mechanical mechanisms (e.g., automobile ride), design for testability (e.g., with built-in sensors), and medical research [Altman 1991].

2. Statistical Model: It is utilized to obtain enough product life (failure) data at various load levels to accurately manifest which statistical distribution at use will present the failure mode with great effectiveness and adequacy. The statistical model that fits most must be applied. Nevertheless, it is interesting to know that the Weibull distribution (including exponential distribution) and lognormal distribution fit most failure modes.

On the DOEs, a traditional practice is to differentiate between-treatment variations by an acceleration model (inferential mechanism), while characterizing the within-treatment variation by a statistical model. The acceleration model may be approximated using a response surface based on DOEs and design factors can be transformed variables, such as factor $A = 1/(kT)$ instead of temperature T and factor $B = \ln(V)$ instead of electric voltage V, as a hint dropped by the acceleration model, such as the Arrhenius and Eyring models (detailed in Chapter 11), respectively.

Some researchers would like to mix the acceleration model and statistical model together into one single equation and call it mixed model. However, this cannot be done without exercising caution. For example, the shape factor of a Weibull distribution may vary from one treatment to another in a DOEs or even from a low-loaded step to a high-loaded step in stepwise accelerated testing. Furthermore, the additivity between two completely different statistical models may not exist.

TABLE 10.5.1 Degradation models of electronic materials.

Failure modes	Failure mechanisms	Influential factors	Models
Fractures*	Fatigue and creep	Temperature	Creep fatigue models
		Humidity	
		Voltage	
		Cyclic loading	
		Voids	
		Stress concentration	
		Stress intensity factor	
		Surface roughness	
Metallization	Corrosion	Chemical	Eyring
		Moisture	
		Temperature	
		Voltage	
	Electromigration	Temperature	Eyring
		Current intensity	
Metal traces	Diffusion and voiding	Stress	Eyring
		Temperature	
Metallic interfaces	Conductive filaments	Moisture	Power law
		Voltage	
Dielectrics	Breakdown	Voltage	Arrhenius
		Temperature	

* Fractures of electronic components include wire debonding, lessened solder leads, debonding pads, die detached, interfacial cracks (e.g., Vias/PTHS), and traces.

How to convert the lifetime (or cycles) at a higher load level to the normal operating load level in the service environment (as given in an engineering specification) is the key to a successful prediction of the lifetime in the real world. A generic nonlinear damage curve is usually better than a linear relationship in modeling the material degradation [Lee and Lu 2006, Nelson 2013]. Frequently used nonlinear product life models pertinent to material degradation mechanisms and their related statistical methodologies are given as follows:

1. Polynomial damage model
2. Arrhenius relationship
3. Eyring relationship inverse
4. Power-law relationship
5. Temperature–humidity relationship
6. Temperature non-thermal relationship
7. Multivariable relationships: general log-linear proportional hazards
8. Time-varying stress models

The challenge of conducting an accelerated life test lies in setting the amount by which various stress factors (e.g., temperature, moisture, vibration, and voltage) come with increasing levels. Because each stress factor of concern affects a specific failure mode differently, it is viable to set a good proportional increase individually. Two simple ways of thinking are given as follows:

(A) Even Load Steps: A simple method that has some merit is to first determine the destruct limit of the product for a given stress factor, e.g., 80°C in temperature for the automotive interior plastics, and then divide the stress levels between the service level (25°C) to the destruct level (80°C) into several even load steps. Here are the two imminent merits of this method: (1) simple to implement and (2) knowing that there is an approximately even increase in damage from each loading source.

(B) Even Damage Steps: An alternative is to determine the nonlinear (e.g., exponential) relationship between increasing each load and the effect on failure modes and determine a step level that keeps the stress damage accumulated from each stress factor in the same proportion, as observed under service conditions.

10.5.2. SALTs

SALT is a combination of traditional reliability testing and overstress testing. The purpose for conducting stepwise tests is to determine the product life and to find out its failure modes by overstressing the product in several incremental levels. Assume that the accelerated life tests are conducted with Q steps, of which the stress level is fixed at treatment q level, where $q = 1, 2, \ldots, Q$. Then, the cumulative distribution function F(t) can be written as the following stepped functions with Q steps:

$$F(t) = F_1(t) \qquad \text{when } 0 \le t < t_1 \tag{10.5.1}$$

$$F(t) = F_2[t_1 + (t - t_1)] \qquad \text{when } t_1 \le t < t_2 \tag{10.5.2}$$

$$\ldots$$

and

$$F(t) = F_Q[t_{Q-1} + (t - t_{Q-1})] \qquad \text{when } t_{Q-1} \le t < t_Q \tag{10.5.3}$$

The calculation process for implementing this methodology is also called cumulative exposure model that is designed for a stepwise plan [Nelson 1980, 2005], of which the effect of the change in stress levels on the remaining life time of a product is concerned. The additivity associated with different load levels for calculations may be based on the assumption that the damage is proportional only to the load level such as mechanical stresses, strains, strain energy, time, temperature, corrosion, creep, and travel distance. It may be a combination of some of these stress factors.

Assume that N_f specimens are failed and N_c specimens are censored within treatment q, i.e., during time interval (t_q, t_{q+1}), q = 1, 2, …, Q. Both T_{qf} and T_{qc} are herein used to denote the failure time and censoring time, respectively. When a test specimen fails at time T_{qf}, which falls between t_{q-1} and t_q, the actual equivalent product life at the normal operating stress level is

$$t_{0,eq} = \Delta t_0 + \Lambda_1 \Delta t_1 + \ldots + \Lambda_{q-1} \Delta t_{q-1} + \Lambda_q (T_{qf} - t_{q-1}) \quad (10.5.4)$$

and
$$\Delta t_q = t_q - t_{q-1} \quad (q = 1, 2, 3, \ldots, Q) \quad (10.5.5)$$

where

$t_{0,eq}$ is the equivalent time at stress level 0 in the service environment
$\Delta t_0, \Delta t_1, \Delta t_2, \ldots, \Delta t_{q-1}, \Delta t_q$ are the time durations of load levels 0, 1, 2, …, and q, respectively
$t_0, t_1, t_2, \ldots, t_{q-1}, t_q$ are the starting time of load levels 0, 1, 2, …, and q, respectively
$\Lambda_1, \Lambda_2, \ldots, \Lambda_{q-1}, \Lambda_q$ are the acceleration factors at stress levels 1, 2, …, q, respectively
T_{qf} is the time, at which the test specimen fails

Note that $t_{-1} = 0$. The transformation of each time duration by acceleration factors because of its individual applied stress level, into the accumulated equivalent product life at the operating stress level is demonstrated in Figure 10.5.1. This is called cumulative exposure model that works for most loading cases. Nevertheless, special attention must be paid to the following loading cases:

(a) Varying Shape Factor β: Equation (10.5.4) is made possible under the assumption that shape factor β in the Weibull distribution function does not change with different stress levels. This may not be true for every damage mechanism.

(b) Complex Failure Modes: The process involves more than two statistical distribution functions.

(c) Sequence of Applied Loads: The sequence of applied loads is not taken into consideration in Equation (10.5.4), but it does matter for some loading cases [Chiang 2019].

(d) Phase Angle of Applied Loads: The phase angle between normal and shear stresses is also a vital acceleration factor for multiaxial out-of-phase fatigue tests in addition to their stress magnitudes [Chiang 2019].

FIGURE 10.5.1 Contributions to equivalent life at operating stress level by two high stresses.

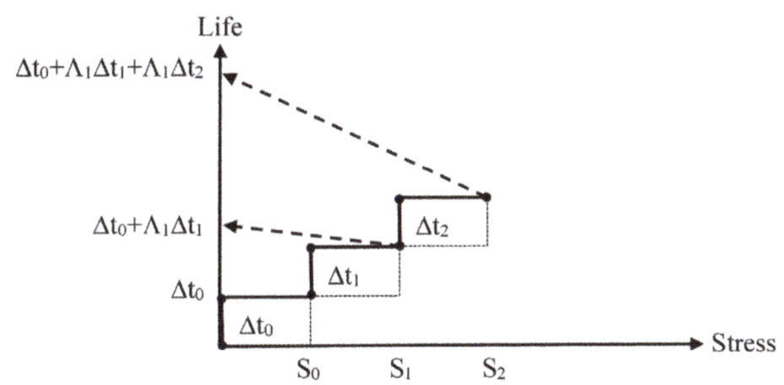

10.5.3. Cumulative Exposure Model for SALTs with One Stress Factor

Drop tests have been applied to a lot of consumer products such as cell phones and some automotive components. The example given here is a study on the drop test of automotive latches with a single stress factor, i.e., drop depth. Data resulting from the drop test of automobile latches are listed in Table 10.5.2, while the generic problem-solving procedure is described in Example 10.5.1. All the six identical latches pass ten drops under 1-m drop test. When the load step is elevated to 1.25-m drop (an accelerated life test), the second one survives four drops, the fourth one survives eight drops, the fifth one survives seven drops, and others pass ten drops. The experiment continued with 1.5-m drop test. All fail without reaching ten drops. This is a case study on test to failure with arbitrary censoring.

TABLE 10.5.2 Stepped-drop tests of six automotive latches by 3^1 factorial design [Schubert 2018].

Treatment	Drop test (h)	Y_1	Y_2	Y_3	Y_4	Y_5	Y_6
1	1 m	10	10	10	10	10	10
2	1.25 m	10	4	10	8	7	10
3	1.5 m	6	—	2	—	—	2
(A) $\Lambda_q \propto h$	Passed (m)	31.5	15	25.5	20	18.75	25.5
	Failed (m)	33	16.25	27	21.25	20	27
(B) $\Lambda_q \propto h^2$	Passed (m)	39.125	16.25	30.125	22.5	20.9375	30.125
	Failed (m)	41.375	17.8125	32.375	24.0625	22.25	32.375

Notes:
Y_n is the number of drops; once surviving 10 drops, it moves to the next stress level.
Subscript n is the latch number, ranging from 1 to 6.

EXAMPLE 10.5.1

A study on the stepped-drop test of automotive latches is conducted for meeting the 1-m drop test requirement (FMVSS). Each specimen is assigned to go through the same set of run conditions, i.e., the same treatment orientation, drop height, until it fails. Once a specimen survives ten drops, it moves to the next stress level, i.e., height. The test data resulting from six specimens are postulated in Table 10.5.2. Two different assumptions are made as follows:

(A) Acceleration factor Λ_q is proportional to the height.

(B) Acceleration factor Λ_q is proportional to the square of height.

How many meters a sample latch is expected to pass without failure at the operating stress level (i.e., 1-m drop test)?

Solution:

(A) Assume that the acceleration factor Λ_q is proportional to the stress level, i.e., the height. The total equivalent meters at the operating stress level passed without failure and with failure are, respectively,

Latch 1: $L_{no\text{-}failure} = 1 \times 10 + 1.25 \times 10 + 1.5 \times 6 = 31.5$ meters
$L_{failure} = L_{no\text{-}failure} + 1.5 = 33$ meters

Latch 2: $L_{no\text{-}failure} = 1 \times 10 + 1.25 \times 4 = 15$ meters
$L_{failure} = L_{no\text{-}failure} + 1.25 = 16.25$ meters

Latch 3: $L_{no\text{-}failure} = 1 \times 10 + 1.25 \times 10 + 1.5 \times 2 = 25.5$ meters
$L_{failure} = L_{no\text{-}failure} + 1.5 = 27$ meters

Latch 4: $L_{no\text{-}failure} = 1 \times 10 + 1.25 \times 8 = 20$ meters
$L_{failure} = L_{no\text{-}failure} + 1.25 = 21.5$ meters

Latch 5: $L_{no\text{-}failure} = 1 \times 10 + 1.25 \times 7 = 18.75$ meters
$L_{failure} = L_{no\text{-}failure} + 1.25 = 20$ meters

Latch 6: $L_{no\text{-}failure} = 1 \times 10 + 1.25 \times 10 + 1.5 \times 2 = 25.5$ meters
$L_{failure} = L_{no\text{-}failure} + 1.5 = 27$ meters

This is a case study on test to failure with arbitrary censoring. The statistical significance can be figured out using Minitab as

Stat → Reliability/survival

→ Distribution analysis (arbitrary censoring)

→ Parametric distribution analysis

The two (passed and failed) rows in Table 10.5.2(A) will be the input data to Minitab's "start variables" and "end variables," respectively. The Q-Q plot for the total meters passed is shown in Figure 10.5.2(a).

(B) Assume that the acceleration factor Λ_q is proportional to the square of height. The total meters passed without failure and with failure are, respectively,

Latch 1: $L_{no\text{-}failure} = 1 \times 10 + (1.25 / 1)^2 \times 10 + (1.5/1)^2 \times 6 = 39.125$ meters
$L_{failure} = L_{no\text{-}failure} + (1.5 / 1)^2 = 41.375$ meters

Latch 2: $L_{no\text{-}failure} = 1 \times 10 + (1.25 / 1)^2 \times 4 = 16.25$ meters
$L_{failure} = L_{no\text{-}failure} + (1.25 / 1)^2 = 17.8125$ meters

Latch 3: $L_{no\text{-}failure} = 1 \times 10 + (1.25 / 1)^2 \times 10 + (1.5/1)^2 \times 2 = 30.125$ meters
$L_{failure} = L_{no\text{-}failure} + (1.5 / 1)^2 = 32.375$ meters

Latch 4: $L_{no\text{-}failure} = 1 \times 10 + (1.25 / 1)^2 \times 8 = 22.5$ meters
$L_{failure} = L_{no\text{-}failure} + (1.25 / 1)^2 = 24.0625$ meters

Latch 5: $L_{no\text{-}failure} = 1 \times 10 + (1.25 / 1)^2 \times 7 = 20.9375$ meters
$L_{failure} = L_{no\text{-}failure} + (1.25 / 1)^2 = 22.5$ meters

Latch 6: $L_{no\text{-}failure} = 1 \times 10 + (1.25 / 1)^2 \times 10 + (1.5 / 1)^2 \times 2 = 30.125$ meters
$L_{failure} = L_{no\text{-}failure} + (1.5 / 1)^2 = 32.375$ meters

Again, the statistical significance can be carried out using Minitab. The Q-Q plot for the total meters passed is shown in Figure 10.5.2(b).

(C) Design contrasts because of two different damage models are demonstrated as follows:

Load-life model	Shape	Scale	Mean	Life (R90/C95)	Life (R50/C95)
Linear	4.77	25.57	23.41	10.79 times	23.41 times
Quadratic	3.966	30.57	27.70	10.82 times	27.70 times

FIGURE 10.5.2 Probability plot for the total meters passed.

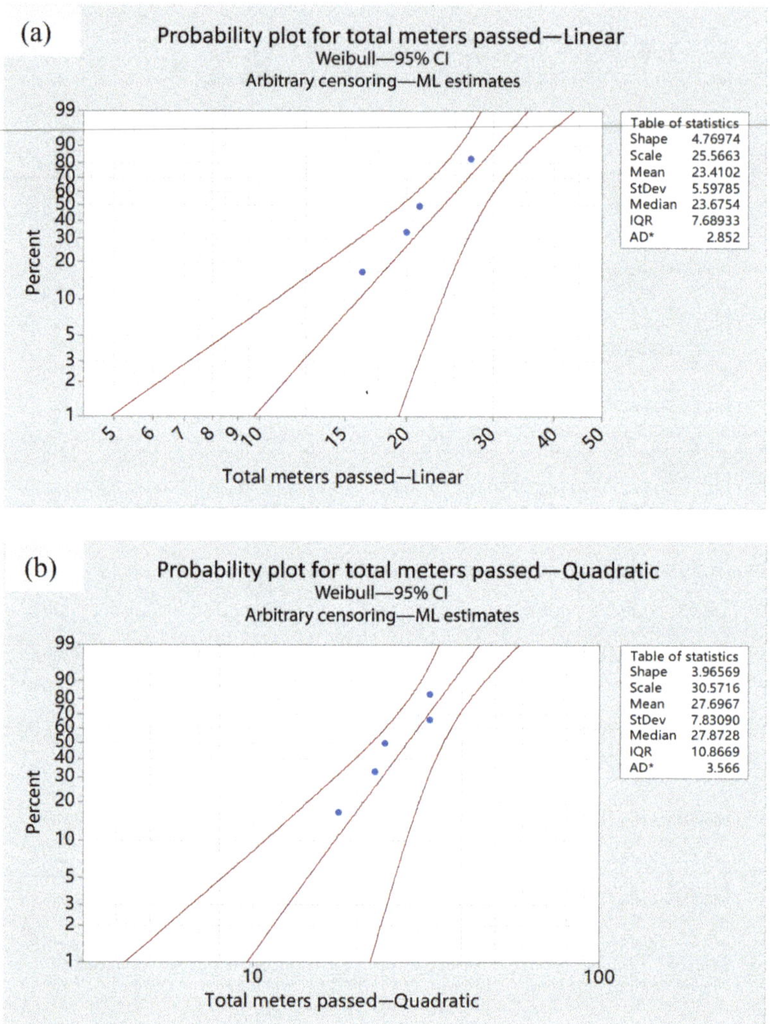

10.6. SSALTs by Weibull Statistics

A SSALT (or SALT) means to subject specimens to stepwise loads at higher levels than operating loads with the objective of increasing the likelihood of observing failures, from which statistical inferences can be drawn for an unbiasedly estimate of the true lifetime distribution under normal operating conditions [Lu et al. 2003]. Here, "stress" stands for "load," although originating from mechanical stress. The lifetime distribution of a product is often used to assess the likelihood of cumulative distribution function of the product life. It can be utilized to establish a warranty period or when pricing extended warranties with regard to such a failure mode.

SSALTs can be also utilized to estimate the life expectancy in a timely manner and statistical distribution related to the potential failure mode before the product is introduced into the market. The additivity associated with different stress levels for the calculations may be based on the assumption that the damage is proportional only to the stress level (travel distance). This is usually not true for SALTs that generally follow a statistical distribution such as the Weibull distribution and lognormal distribution. Thus, a corresponding adjustment according to the nonlinearity of the inferential mechanism has to be incorporated, as shown in Figure 10.6.1. For example, the strain–life curve can be used as perspective for "life-cycle" adjustment at different load levels when dealing with fatigue life of materials. The strain is therefore identified as the acceleration factor. In fact, the so-called load level means a working condition that is a combination of various loading functions such as temperature, moisture, mechanical forces, electromagnetic forces, corrosion, creep, and environmental factors.

FIGURE 10.6.1 Nonlinear impact of stress level on product life relative to operating load S_0.

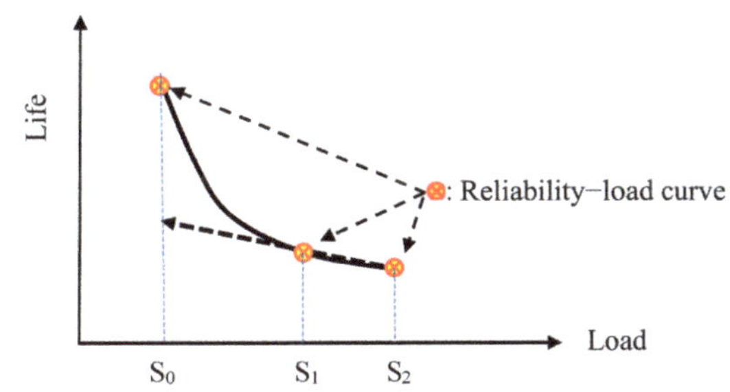

10.6.1. Stepwise Loading

SSALT methodology is often employed to determine the functional and destructive limits of products [Xiong 2003]. As being a quantitative accelerated test, it is suitable for examining a highly reliable product with a suitable statistical distribution, such as a Weibull (or lognormal) distribution, whose characteristic life is a log function of the load factor (also called stress factor). Under the assumption that the failure mechanism is unchanged under different applied load levels and the failure times at each load level are often assumed to be nonlinearly distributed, the statistical likelihood function is constructed from the observed failure time data, and a conjugated joint prior distribution of model parameters is derived. See [Li 2009] for detailed quantitative statistics on SSALT.

Lifetime data of a product may constitute a distinct statistical distribution of its own. Identification of statistical distributions relies extensively on time-to-failure data (lifetime data) at different load levels. For the purpose of illustration, consider a simple test case that consists of Q load levels and the corresponding load levels are in the ascending order as

$$S_0 < S_1 < S_2 < \ldots < S_Q \quad (10.6.1)$$

where

S_0 is the operating load level in service environment
S_1 is accelerated load level 1
S_2 is accelerated load level 2
S_Q is accelerated load level Q

Then, the dimensionless load level (s_q) can be defined as

$$s_q \equiv (S_q - S_0) / (S_Q - S_0) \quad (10.6.2)$$

Assume that there are only two load levels, Q = 2 for a simply linear case. This turns into a study on the linear scale interpolated between 0 and 1, and the following relationships can be derived

$$s_0 = 0 \tag{10.6.3}$$

$$s_1 = (S_1 - S_0) / (S_2 - S_0) \tag{10.6.4}$$

and $$s_2 = 1 \tag{10.6.5}$$

The above equations can be generalized to include arbitrary Q load levels

$$s_0 = 0$$

$$s_q = (S_q - S_0) / (S_Q - S_0) \tag{10.6.6}$$

and $$s_Q = 1$$

where q = 1, 2, 3, …, and Q.

The optimal allocation for each stress level can be found according to the test-plan efficiency, which may be measured in terms of the variance of the parameters that are being sought [Fard and Li 2009, Li 2009, Zhao and Elsayed 2005]. Optimal test plans for simple SSALTs under type I censoring developed for two-parameter Weibull statistics, including the exponential distribution, are explored by [Tang et al. 2012] based on the criterion of the minimization of the asymptotic variance of the MLE of the pth percentile of the lifetime under the normal operating condition with a constant shape factor. Nevertheless, this usually does not make engineering sense. Instead, one may consider the DOEs with the objective function expressed in terms of "time to failure" based on Equation (9.2.9), combined with inferential mechanisms at different loads, to figure out the load steps. For example, when a plastic part works at a temperature higher than its glass transition point, the associated material properties change dramatically. For another example, a failure mode may be process (path)-dependent such as cracks emanating from a hole.

The proportionate allocation of units at one test stress should not be too small relative to the others [Bai et al. 1989]. Nevertheless, the accuracy for generating a robust test plan depends on having enough failures at each stress level. Most commercial statistical codes, such as Minitab and ReliaSoft, provide this capacity. Assume that there are k stress levels. The ranges for the proportions at the different stress levels suggested by Minitab are given as follows:

1. Range for the Lowest Test Stress: (1/K, 1/K + 0.35)
2. Ranges for Other Test Stresses: $((1 - 1/K - 0.35)/2^K, 1/K)$
3. Proportion for the Highest Stress: Chosen to assure that all proportions sum up to 100%.

For example, for a two-step stress test (K = 2), the range of the proportions at the lowest stress falls in the range of (50%, 85%) and the range of the proportions at the highest stress falls in the range of (7.5%, 50%).

For a three-step stress test (K = 3), the range of the proportion for the lowest test stress level falls in the range of (33.3%, 69.3%), the range of the proportion for the second stress level falls in the range of (4%, 33.3%), and the balance will be used for the third stress level.

10.6.2. Weibull Statistics for Accelerated Life Tests

The information derived from accelerated life tests is to be utilized to predict the product life or long-term product performance at the nominal load level through a reasonable statistical method based on physical laws [Escobar and Meeker 2006]. It has been pointed out that both the scale parameter and shape parameter of the Weibull distribution may vary with respect to the arbitrarily assigned load levels [Dharmadhikari and

Rahman 2002, Khamis and Higgins 1998]. Let t_{qn} denote the life of each individual data collected from the experiment. For a general case,

$$t_{qn} \sim \text{Weibull}\,(\beta_q, \eta_q) \tag{10.6.7}$$

where

t_{qn} is the product life
q is the treatment q
β_q is the shape factor with treatment q
η_q is the scale parameter with treatment q
n is the observation within treatment q

Note that n is the numbering of data, acquired at load level q (treatment q in a DOE design matrix). As enlightened by Equations (10.6.1)–(10.6.3), the cumulative distribution functions at individual load steps can be written as [Zhao and Elsayed 2005]

$$F(t) = 1 - \exp[-(t / \eta_1)^\beta] \qquad \text{when } 0 < t < t_1 \tag{10.6.8}$$

$$F(t) = 1 - \exp\{-[(t - t_1) / \eta_2]^\beta - (t / \eta_1)^\beta\} \qquad \text{when } t_1 < t < t_2 \tag{10.6.9}$$

...

and $$F(t) = 1 - \exp\{-[(t - t_{Q-1}) / \eta_Q]^\beta - (t / \eta_{Q-1})^\beta\} \qquad \text{when } t_{Q-1} < t < T_Q \tag{10.6.10}$$

As long as the shape factor is constant, the hazard rate function based on the two-parameter Weibull statistics with shape factor β and scale parameter η can be written as follows [Khamis and Higgins 1998]:

$$h(t) = h(t \mid \eta_1) \qquad \text{when } 0 < t < t_1 \tag{10.6.11}$$

$$h(t) = (\eta_1 / \eta_2)\, h(t \mid \eta_2) \qquad \text{when } t_1 < t < t_2 \tag{10.6.12}$$

...

and $$h(t) = (\eta_{Q-1} / \eta_Q)\, h(t \mid \eta_Q) \qquad \text{when } t_{Q-1} < t < T_Q \tag{10.6.13}$$

The above equations render a group of hazard rates resulting from the cumulative damage model as proposed by [Nelson 1980, 2005], which can be regarded as a composite of mixed Weibull distributions with Q scale factors ($\eta_1, \eta_2, \ldots, \eta_Q$). The hazard rate at a higher load level can be related to the hazard rate at the normal operating load level by an acceleration factor as

$$h(t_q) = \Lambda_q\, h(t_0) \tag{10.6.14}$$

of which Λ_q is the acceleration factor that depends on the inferential mechanism. Therefore, the accumulated degradation at the higher load level can be related to the accumulated degradation at the normal operating load level by

$$h(t_q)\, \Delta t_q = \Lambda_q\, h(t_0)\, \Delta t_q \tag{10.6.15}$$

i.e., $$\Delta t_q\, h(t_q) = (\Lambda_q\, \Delta t_q)\, h(t_0) \tag{10.6.16}$$

In other words, the time duration of the damage experienced by a test specimen at a higher load level can be converted into the time duration at the operating load level as enlarged (or reduced) by the acceleration factor.

10.6.3. Load–Life Relationship

Assume that the variance alters with respect to the applied stress level and the variance at each accelerated load level can be obtained by the interpolation between those at the operating load level and the highest load level. In the domain of Weibull statistics, here is a simple equation for describing the fundamental relationship between the variances of load level q (treatment q) and the normal operating level:

$$\sigma_q^2 = \sigma_0^2 + \gamma\, s_q,\ q = 1, 2, \ldots, Q \tag{10.6.17}$$

where γ is the proportionality.

It means that the variance at load level q can be assessed using dimensionless load level s_q, defined by Equation (10.6.2), and a constant (γ). For the special case that $\gamma = 0$, it creates a case study on a constant shape parameter. Assume that the fatigue life of the test specimens is an independently and identically distributed random variable and the remaining life of a test specimen depends on the cumulative exposure to the experimental conditions. The characteristic of a product life is to be modeled using an inferential mechanism, e.g., strain–life curve for fatigue cycling [Chiang 2019]. For simple cases, scale parameter η_q of treatment q is modeled using the linear or quadratic function on the logarithmic scale as [Nelson 1980, 2005]

$$\ln(\eta_q) = C_{0q} + C_{1q}\, S_q \tag{10.6.18}$$

or

$$\ln(\eta_q) = C_{0q} + C_{1q}\, S_q + C_{2q}\, S_q^2 \tag{10.6.19}$$

of which constant C_{0q} and coefficients C_{1q} and C_{2q} are unknowns yet to seek out through the logarithmic-transformed experimental data. Note that scale parameter η is also called characteristic parameter, as it is used for characterizing a Weibull distribution. If each of the collected sample observation, t_{qn}, fits in Weibull statistics, $\ln(t_{qn})$ will follow the SEV distribution [Ma and Meeker 2008],

$$Y_{qn} = \ln(t_{qn}) \sim \mathrm{SEV}(\mu_q, \sigma_q) \tag{10.6.20}$$

where $$\mu_q = \ln(\eta_q) = C_{0q} + C_{1q}\, S_q \quad \text{(Linear case)} \tag{10.6.21}$$

or $$\mu_q = \ln(\eta_q) = C_{0q} + C_{1q}\, S_q + C_{2q}\, S_q^2 \quad \text{(Quadratic case)} \tag{10.6.22}$$

and $$\sigma_q = \beta_q^{-1} \tag{10.6.23}$$

Note that load level q is considered as treatment q as defined for the applied DOEs.

10.6.4. MLE of Parameters

Assume that the variation in load level changes only the scale parameter in the Weibull statistics. The probability density function in response to the SALT at treatment q can be written as [Alhadeed and Yang 2002, Shen et al. 2011, Tang et al. 2012]

$$f(t_{qn}) = \frac{\beta}{\eta_q}\left(\frac{t_{qn}}{\eta_q}\right)^{\beta-1} \exp\left[-\left(\frac{t_{qn}}{\eta_q}\right)^{\beta}\right] \tag{10.6.24}$$

$$\text{or} \quad f(t_{qn}) = \frac{\beta}{\eta(S_q)} \left[\frac{t_{qn}}{\eta(S_q)}\right]^{\beta-1} \exp\left\{-\left[\frac{t_{qn}}{\eta(S_q)}\right]^{\beta}\right\} \tag{10.6.25}$$

After the load–life relationship, e.g., Equation (10.6.21), Equation (10.6.22), or a more complicated model of inferential mechanism, is reckoned, the above probability density function at stress level q can be rewritten in reference to the operating stress level, i.e., stress level 0, as [Zhao and Elsayed 2005]

$$f(t_{qn}) \rightarrow f(t_{0,eq}) = \frac{\beta}{\eta(S_0)} \left[\frac{\Delta t_0 + \Lambda_1 \Delta t_1 + \ldots + \Lambda_{q-1} \Delta t_{q-1} + \Lambda_q (T_{qn} - t_q)}{\eta(S_0)}\right]^{\beta-1}$$

$$\exp\left\{-\left[\frac{\Delta t_0 + \Lambda_1 \Delta t_1 + \ldots + \Lambda_{q-1} \Delta t_{q-1} + \Lambda_q (T_{qn} - t_q)}{\eta(S_0)}\right]^{\beta}\right\} \tag{10.6.26}$$

With a piece of data censored at T_{qc}, the reliability function corresponding to the above probability density function is

$$R(T_{qc}) = \exp\left\{-\left[\frac{\Delta t_0 + \Lambda_1 \Delta t_1 + \ldots + \Lambda_{q-1} \Delta t_{q-1} + \Lambda_q (T_{qc} - t_q)}{\eta(S_0)}\right]^{\beta}\right\} \tag{10.6.27}$$

The likelihood function for a general SALT described by Weibull statistics with a constant shape factor can be written as

$$L(\mu_1, \mu_2, \sigma_0) = \left\{\prod_{q=1}^{Q} \left[\prod_{n=1}^{N_q} f(t_{qn})\right]\right\} \left\{\prod_{c=1}^{Q} \left[\prod_{n=1}^{N_c} R(T_{qc})\right]\right\} \tag{10.6.28}$$

where

N_q is the number of surviving data within treatment q, up to moving to the next level or failure

N_c is the number of censored data within treatment q, up to censoring

Substituting Equations (10.6.26) and (10.6.27) into the above equation and taking a logarithmic transformation yield

$$\ln[L(\mu_1, \mu_2, \sigma_0)] = \sum_{q=1}^{Q} \sum_{n=1}^{N_q} \ln\left\{\frac{\beta}{\eta(S_0)} \left[\frac{\Delta t_0 + \Lambda_1 \Delta t_1 + \ldots + \Lambda_{q-1} \Delta t_{q-1} + \Lambda_q (T_{qn} - t_q)}{\eta(S_0)}\right]^{\beta-1}\right.$$

$$\left.\exp\left[-\left[\frac{\Delta t_0 + \Lambda_1 \Delta t_1 + \ldots + \Lambda_{q-1} \Delta t_{q-1} + \Lambda_q (T_{qn} - t_q)}{\eta(S_0)}\right]^{\beta}\right]\right\}$$

$$-\sum_{q=1}^{Q} \sum_{n=1}^{N_c} \left[\frac{\Delta t_0 + \Lambda_1 \Delta t_1 + \ldots + \Lambda_{q-1} \Delta t_{q-1} + \Lambda_q (T_{qc} - t_q)}{\eta(S_0)}\right]^{\beta}\Bigg\} \tag{10.6.29}$$

10.6.5. Estimating Reliability and Associated Confidence Bounds: Weibull Statistics

After Weibull parameters are calculated, the product reliability at the operating stress level and its associated confidence bounds can be evaluated. The two-sided confidence bounds of product reliability can be obtained from the following inequality:

$$\hat{R}(t) + [1 - \hat{R}(t)] \exp\{-z_{\alpha/2}\, \sigma_{R(t)} / [\hat{R}(t)\, (1 - \hat{R}(t))]\} \leq R(t) \leq$$
$$\hat{R}(t) + [1 - \hat{R}(t)] \exp\{ z_{\alpha/2}\, \sigma_{R(t)} / [\hat{R}(t)\, (1 - \hat{R}(t))]\} \tag{10.6.30}$$

where

α is the significance level based on the Weibull distribution, e.g., $\alpha = 10\%$
$\hat{R}(t)$ is the estimated reliability by the model
$R(t)$ is the sample reliability

10.6.6. Conditional Reliability: Weibull Statistics

When a product unit has been in use at load level q for a time period of Δt_q, the remaining product life of the product to be used at the operating stress level can be predicted using the following conditional Weibull statistics as

$$F(t \mid S_0, \Delta t_q) = \exp\{-[(t + \Lambda_q\, \Delta t_q) / \eta(S_0)]^{\beta}\} / \exp\{-[(\Lambda_q\, \Delta t_q) / \eta(S_0)]^{\beta}\}$$

$$\text{i.e.,} \quad F(t \mid S_0, \Delta t_q) = \exp\{-[(t + \Lambda_q\, \Delta t_q) / \eta(S_0)]^{\beta} + (\Lambda_q\, \Delta t_q) / \eta(S_0)]^{\beta}\} \tag{10.6.31}$$

Note that for the special case at the operating load level, $\Lambda_0 = 1$ at $q = 0$.

10.7. SALT in Lognormal Distribution

SALT is a commonly used test methodology for identifying failure mechanisms related to product reliability, such as tire wear, cracks, lasers, or light bulbs. It can provide measured data with regard to the product's degradation that would eventually lead to weakness or hard failures. The first possible statistical model aiming at this purpose in the academic study is the lognormal distribution. Variable t is lognormally distributed, if

$$y = \ln(t) \tag{10.7.1}$$

is normally distributed with "ln" denoting the natural logarithm. The probability density function of lognormal-distributed variable t is

$$f(t) = \frac{1}{(t - t_o)\, \sigma_y\, (2\pi)^{1/2}} \exp\{-½ [\frac{\ln[(t - t_o) / \eta]}{\sigma_y}]^2\} \tag{10.7.2}$$

$$\text{or} \quad f(t) = \Phi\left[\frac{\ln(t)}{\sigma_y}\right] \tag{10.7.3}$$

where

σ_y is the standard deviation of y and it is the shape factor of a lognormal distribution
η is the location parameter or scale factor
Φ is the notation to denote the probability density function of a normal distribution

When $t_o = 0$ and $\eta = 1$, the above equation reduces to the standard lognormal distribution function

$$f(t) = \frac{1}{t - \sigma_y (2\pi)^{1/2}} \exp\left\{-\tfrac{1}{2}\left[\frac{\ln(t)}{\sigma_y}\right]^2\right\}, \text{ where } t > 0 \text{ and } \sigma_y > 0 \tag{10.7.4}$$

Note that $f(t) \equiv 0$, when $t \leq 0$. The mean of the lognormal probability density function can be expressed in terms of μ_y and σ_y.

$$\text{Mean} = E[t] = \exp(\mu_y + \tfrac{1}{2}\sigma_y^2) \tag{10.7.5}$$

The equation for the cumulative distribution function of a lognormal distribution function can be written as

$$F(t) = = \Phi\left[\ln\left(\frac{y}{\sigma_y}\right)\right] = \Phi\left[\ln\left(\frac{\ln(t)}{\sigma_y}\right)\right], \text{ where } t > 0 \text{ and } \sigma_y > 0 \tag{10.7.6}$$

where Φ is the notation that denotes the cumulative normal distribution function.

10.7.1. MLE for Lognormal Distribution

The lognormal distribution is useful in modeling continuous random variables, e.g., the product life of an electronic components and the size of a dimension, which are greater than or equal to zero. Subject to stepwise loads, the probability density function in response to each piece of acquired data, t_{qn}, up to failure can be derived from Equation (10.7.3) and expressed as

$$f(t_{qn}) = \Phi\left\{\frac{1}{\sigma} \ln\left[\frac{\Delta t_0 + \Lambda_1 \Delta t_1 + \ldots + \Lambda_{q-1} \Delta t_{q-1} + \Lambda_q (T_{qn} - t_q)}{\eta(S_0)}\right]\right\}$$
$$/ \{[\Delta t_0 + \Lambda_1 \Delta t_1 + \ldots + \Lambda_{q-1} \Delta t_{q-1} + \Lambda_q (T_{qn} - t_q)]\, \sigma\} \tag{10.7.7}$$

where

T_{qn} is the nth data within treatment q in a DOE design matrix
$\eta(S_0)$ is the scale factor of the logarithmic distribution function
S_0 is the load level at the normal operation
σ is used for short, instead of σ_y that is the shape factor of a logarithmic distribution

The reliability function corresponding to the above cumulative distribution function with a piece of data censored at T_{qc} can be derived from Equation (10.7.5) and expressed as

$$R(T_{qc}) = 1 - F(t) = 1 - \Phi\{\frac{1}{\sigma}[\frac{\Delta t_0 + \Lambda_1 \Delta t_1 + \ldots + \Lambda_{q-1} \Delta t_{q-1} + \Lambda_q (T_{qc} - t_q)}{\eta(S_0)}]^{\beta}\} \quad (10.7.8)$$

The overall logarithmic likelihood, combining Equations (10.7.7) and (10.7.8) together, becomes

$$\ln[L(\eta, \sigma)] = \sum_{q=1}^{Q} \sum_{n=1}^{N_q} \ln\{\frac{1}{[\Delta t_0 + \Lambda_1 \Delta t_1 + \ldots + \Lambda_{q-1} \Delta t_{q-1} + \Lambda_q (T_{qn} - t_q)]\sigma}$$

$$\Phi[\frac{1}{\sigma} \ln(\frac{\Delta t_0 + \Lambda_1 \Delta t_1 + \ldots + \Lambda_{q-1} \Delta t_{q-1} + \Lambda_q (T_{qn} - t_q)}{\eta(S_0)})^{\beta}]\}$$

$$- \sum_{q=1}^{Q} \sum_{n=1}^{N_c} \ln\{1 - \Phi[\frac{1}{\sigma} \ln(\frac{\Delta t_0 + \Lambda_1 \Delta t_1 + \ldots + \Lambda_{q-1} \Delta t_{q-1} + \Lambda_q (T_{qc} - t_q)}{\eta(S_0)})]\} \quad (10.7.9)$$

10.7.2. Estimating Reliability and Associated Confidence Bounds: Lognormal Statistics

After lognormal statistics are identified, the product reliability at the operating load level and its associated confidence bounds are to be evaluated. The confidence bounds of product reliability is given here without a derivation as

$$\hat{R}(t) + [1 - \hat{R}(t)] \exp\{-z_{\alpha/2}\, \sigma_{R(t)} / [\hat{R}(t)\,(1 - \hat{R}(t))]\} \le R(t) \le$$

$$\hat{R}(t) + [1 - \hat{R}(t)] \exp\{z_{\alpha/2}\, \sigma_{R(t)} / [\hat{R}(t)\,(1 - \hat{R}(t))]\} \quad (10.7.10)$$

where

α is the significance level based on lognormal distribution, e.g., $\alpha = 10\%$
$\hat{R}(t)$ is the estimated reliability by the model
$R(t)$ is the sample reliability

10.7.3. Conditional Reliability: Lognormal Statistics

When a product unit has been in use at load level q for a time period of ΔT_q, the remaining product life of the product to be used at the operating load level can be predicted using the following conditional lognormal statistics:

$$R(t \mid S_0, \Delta T_q) = \frac{1 - \Phi\{\sigma^{-1} \ln[(t + \Lambda_q \Delta T_q) / \eta(S_0)]\}}{1 - \Phi\{\sigma^{-1} \ln[(\Lambda_q \Delta T_q) / \eta(S_0)]\}} \quad (10.7.11)$$

Note that $\Lambda_0 = 1$ for the special case that $q = 0$, i.e., at the normal operating level.

10.8. Fatigue Life Prediction Subject to Dynamic Load Steps

The fatigue life of a part, mechanical or electronic, is a complex material-dependent thermomechanical degradation under various loading functions including forces, working temperature, and environmental factors. Researchers attempted to identify statistical distribution functions for fatigue of structures [Schijve 2005, Wu et al. 2017]. Nevertheless, material properties are nonlinearly degraded during fatigue life cycling with an increasing mechanical stress–strain level (dislocations), temperature, crack length and branching, creep, oxidation (e.g., corrosion, rusting, and erosion), etc. In other words, it is a systemic problem across different statistical distributions holistically.

In general, a load–life relationship of material at a given temperature can be characterized by ε-N curves and/or a strain energy-based fatigue theory, with at least the following eight material parameters [Sun et al. 2014]:

1. Fatigue strength coefficient σ_f'
2. Fatigue strength exponent b
3. Fatigue strain coefficient ε_f'
4. Fatigue strain exponent c
5. Cyclic hardening coefficient K′
6. Cyclic hardening exponent n′
7. Loading sequence
8. Nonproportionality of multiaxial loads (phase angle of loads)

Fatigue damage of materials and cumulative damage laws are of major interest. It would be appropriate for a multiaxial stress, strain, or strain energy-based fatigue theory to work with the law of statistical distribution because of random material properties. It is better characterized by a strain-based fatigue theory than a stress-based fatigue theory (S-N curves) [Nagode and Hack 2004]. Step-strain or step-strain energy ALT based on DOEs can be used as a replacement of simply SALT [Chiang 2019].

The complete definition of an applied loading function is representative of customer's uses of a product, and it includes dynamic cyclic loading amplitude and phase angles between strain components, as well as the counting method.

10.8.1. Fatigue Damage of Materials

When it is a loading order-oriented fatigue process, or saying that the loading sequence is of concern, the relationship between the strain amplitude, mean stress, damage parameter, and material properties may be written as

$$\varepsilon_a = \frac{\sigma_f' - \sigma_m}{E\,(1 - D)} (2N_f)^b + \varepsilon_f' (2N_f)^c \tag{10.8.1}$$

where

ε_a is the strain amplitude
σ_m is the mean stress
$2N_f$ is cycles, i.e., reversals to failure
E is Young's modulus
D is the damage parameter, used before applying the next load step

The above equation is extended from Miner's equation [Miner 1945], and it is called the Manson–Coffin–Basquin–Morrow equation [Morrow 1965], in which D = 0. Damage parameter D is updated before applying the next load step, as it accounts for the loading sequence. A schematic strain–life curve described by the equation (with D = 0) is depicted in Figure 10.8.1. This is one of the load–life relationships that work for generic steels [Schlitz et al. 1990]. More proposals for fitting fatigue life cycles of various materials can be found in [Chiang 2019].

FIGURE 10.8.1 Influence of strain amplitude and mean stress on the fatigue life.

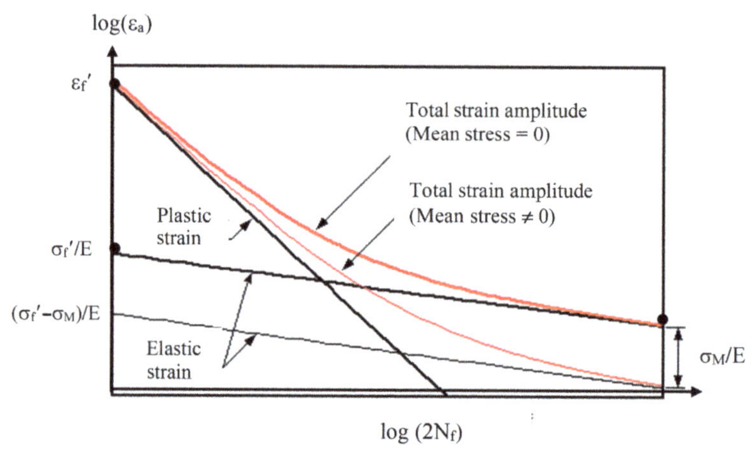

10.8.2. Cumulative Damage

Damage parameter D is also called diffusion damage, under the assumption that D < 1 before the structure loses the entire stiffness. As D is cumulative, a simple form of accumulation is

$$D = \sum_{i=1} \Delta D_i = \sum_{i=1} \frac{2N_i}{2N_f} \tag{10.8.2}$$

where

Subscript i is the load step

ΔD_i is the partial damage at load step i

$2N_i$ is the number of cycles experienced at load step i

$2N_f$ is the number of cycles at fatigue failure due to the same damage capacity at load step i

Initially, $\Delta D_0 = 0$ at time = 0, unless the part is partially damaged before the test starts. When D = 1, the material fails. Some other damage equations can be found in [Chiang 2019]. One counting algorithm of $2N_f$ for power semiconductor is exhibited in [GopiReddy et al. 2015].

10.8.3. Dynamic Cyclic Loading

Dynamic cyclic loadings are what is actually applied during fatigue cycling and they are more stressful than static stress–strain (σ-ε) curves, as shown in Figure 10.8.2. The following cyclic relationship between the stress and strain with two more fatigue material parameters (K′ and n′) has to be used for describing hysteresis cycles under proportional loadings [Ramberg and Osgood 1943]:

$$\varepsilon_a = \frac{\sigma_a}{E} + \left(\frac{\sigma_a}{K'}\right)^{1/n'} + \varepsilon_{creep} \tag{10.8.3}$$

where

ε_a is the stain amplitude
σ_a is the stress amplitude
K′ (MPa) is the cyclic hardening coefficient
n′ is the cyclic hardening exponent
ε_{creep} is the creep strain

The above equation, also called the Ramberg–Osgood relation, is adequate to model most metals, including bi-modulus materials such as gray cast iron, with different compressive and tensile cyclic loadings.

FIGURE 10.8.2 Cyclic σ-ε curves versus quasi-static σ-ε curves.

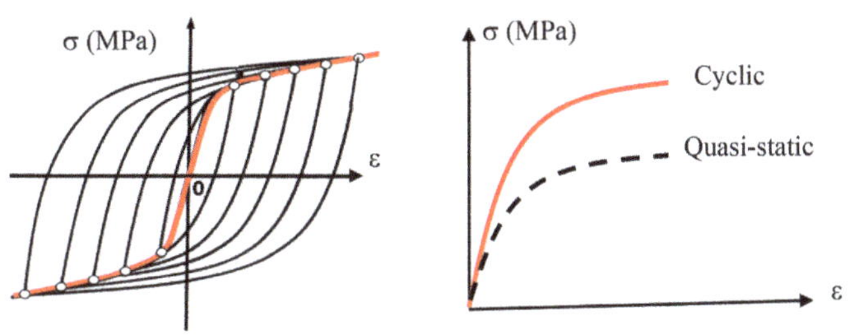

(a) Stabilized cyclic σ-ε curves (b) Typical σ-ε curves for Al alloys

As argued by some researchers [Dowling 2007] that theoretical relationships among the six fitting constants in the above three "dissimilar" equations, i.e., Equations (10.8.1)–(10.8.3), are not invoked. It is recommended that Equations (10.8.1)–(10.8.3) should be fitted separately to stress–strain–life test data, for the purpose of accurately representing each curve.

EXAMPLE 10.8.1

Given that the material properties of the lower control arm (AISI 1513) of an automotive suspension are as follows: Young's modulus E = 210 GPa, fatigue strength coefficient σ_f' = 1089 MPa, fatigue strength exponent b = −0.07, fatigue strain coefficient ε_f' = 0.41, and fatigue strain exponent c = −0.54, cyclic hardening coefficient, K′ = 978 MPa, and cyclic hardening exponent n′ = 0.11. How much is the damage done with a uniaxial loading strain that fluctuates from 0.016 to −0.012, referring to Figure 10.8.2(b): load block 1. Assume that the initial damage is zero. How much is the damage done in load block 1 (five cycles)?

Solution:

(a) Cyclic stresses at the lowest strain at −0.012 and the highest strain at 0.016 can be assessed using Equation (10.8.3), without creep strain.

$$\varepsilon_a = \frac{\sigma_a}{E} + \left(\frac{\sigma_a}{K'}\right)^{1/n'} = \frac{\sigma_a}{210000} + \left(\frac{\sigma_a}{978}\right)^{1/0.11}$$

When $\varepsilon_a = 0.016$, $\sigma_a = 326.5 \rightarrow \sigma_{max} = 326.5$ MPa

When $\varepsilon_a = 0.012$, $\sigma_a = 251.0 \rightarrow \sigma_{min} = -251.0$ MPa

$$\sigma_m = (\sigma_{max} - \sigma_{min}) / 2 = 288.75 \text{ MPa}$$

(b) The actual strain amplitude applied in load block 1 is $\varepsilon_a = [0.016 - (-0.012)]/2 = 0.014$

(c) According to Equation (10.8.1) with D = 0, one has

$$\varepsilon_a = \frac{\sigma_f' - \sigma_m}{E} (2N_f)^b + \varepsilon_f' (2N_f)^c$$

$$\text{i.e.,} \quad 0.014 = \frac{1089 - 288.75}{210000} (2N_f)^{-0.07} + 0.41\,(2N_f)^{-0.54}$$

which can be solved for $2N_f \approx 736.7$ cycles.

Thus, the damage done for the five cycles ($N_i = 5$) in load block 1, referring to Figure 10.9.2(b), can be assessed using Equation (10.8.2) as

$$\Delta D_i = 2N_i / 2N_f = 5 / 736.7 = 0.006787$$

10.8.4. Nonproportionality of Loads

The phase angle between any two strain components of a dynamic load would generate the nonproportionality between them. An applied load is not proportional when its related six strain components (ε_{xx}, ε_{yy}, ε_{zz}, ε_{xy}, ε_{yz}, and ε_{zx}) are not in phase with each other and their resulting principal strain (stress) angle is thus not constant [Sun et al. 2014]. Note that each engineering strain is defined to double the size of its counter tensor strain: $\gamma_{xy} = 2\varepsilon_{xy}$, $\gamma_{yz} = 2\varepsilon_{yz}$, and $\gamma_{zx} = 2\varepsilon_{zx}$. For example, normal strain

$$\varepsilon = \varepsilon_o \sin(\omega t) \tag{10.8.4}$$

in combination with shear strain:

$$\gamma = \gamma_o \sin(\omega t + \psi) \tag{10.8.5}$$

of which $\psi \neq$ constant. The influence of nonproportionality on fatigue is to cause more damage than proportional loading for most cases. Its impact on the mechanism of dislocation evolution is the main cause of the decrease in fatigue strength and the reduction in the fatigue limit [Skibicki 2014]. Nevertheless, there is also the influence of nonproportionality on the course of structural changes (e.g., phase transformations in steel) induced by plastic strain [Skibicki and Dymski 2010]. The transformation depended on intensity and the type of applied loading.

Loadings with sine and cosine waves are proposed by [Chattopadhyay and Nathan 2013] to explain the concepts of proportionality and nonproportionality based on rotating shafts that experience time-varying loads such as transmission shafts, connecting rods, and crankshafts under multiaxial loads. Assume that a rotating shaft experiences significant torsional and bending strains that can be represented by the following three strain components: ε_{xx}, ε_{yy}, and γ_{xy}. The concept of rotating phase angle can be illustrated using Mohr's circles based on principal strains, as demonstrated using Example 10.8.2 [Chiang 2019].

EXAMPLE 10.8.2

Given that ε_{xx} = 4000 μm/m, ε_{yy} = 0 μm/m, and ε_{xy} = 2000 sin(ωt) μm/m [i.e., γ_{xy} = 4000 sin(ωt) μm/m]. How does the phase angle vary in the space?

Solution:

According to Equations (10.8.4) and (10.8.5), the principal strain angle and principal strain are, respectively,

$$\theta = \tfrac{1}{2}\tan^{-1}\left[\frac{\varepsilon_{xy}}{\tfrac{1}{2}(\varepsilon_{xx}-\varepsilon_{yy})}\right] = \tfrac{1}{2}\tan^{-1}\left[\frac{2000\sin(\omega t)}{2000}\right] = \tfrac{1}{2}\tan^{-1}[\sin(\omega t)]$$

and

$$\varepsilon = \tfrac{1}{2}(\varepsilon_{xx}+\varepsilon_{yy}) \pm \{[\tfrac{1}{2}(\varepsilon_{xx}-\varepsilon_{yy})]^2 + \varepsilon_{xy}^2\}^{1/2}$$

$$= 2000 \pm [2000^2 + 2000^2 \sin^2(\omega t)]^{1/2}$$

$$= 2000 \pm 2000\,[1+\sin^2(\omega t)]^{1/2}$$

Thus,

$$\varepsilon_1 = 2000\,\{1 + [1+\sin^2(\omega t)]^{1/2}\}\ \mu\text{m/m} \quad \text{(1st Principal Strain)}$$

$$\varepsilon_3 = 2000\,\{1 - [1+\sin^2(\omega t)]^{1/2}\}\ \mu\text{m/m} \quad \text{(3rd Principal Strain)}$$

$$\Delta\varepsilon = (\varepsilon_1 - \varepsilon_3) = 4000\,[1+\sin^2(\omega t)]^{1/2}\}\ \mu\text{m/m} \quad \text{(Strain Range)}$$

$$\varepsilon_a = \tfrac{1}{2}(\varepsilon_1 - \varepsilon_3) = 2000\,[1+\sin^2(\omega t)]^{1/2}\}\ \mu\text{m/m} \quad \text{(Strain Amplitude)}$$

$$\varepsilon_m = \tfrac{1}{2}(\varepsilon_1 + \varepsilon_3) = 2000\ \mu\text{m/m} \quad \text{(Mean Strain)}$$

and

$$\varepsilon_{13max} = \tfrac{1}{2}(\varepsilon_1 - \varepsilon_3) = 2000\,[1+\sin^2(\omega t)]^{1/2}\}\ \mu\text{m/m} \quad \text{(Max. Shear Strain)}$$

The phase angle is not constant. This is a nonproportional load case. Note that the second principal strain is zero (ε_2 = 0) and the maximum shear strain (ε_{13max}) is of the same magnitude as the fluctuating amplitude of normal strain (ε_a). Assume that ω = 2π/sec and 4π/sec (i.e., 1 and 2 Hz). The variations of the phase angle and strain amplitude with respect to time for both angular speeds can be plotted in Figures 10.8.3 and 10.8.4, respectively.

FIGURE 10.8.3 Plots of phase angle versus time at two angular speeds (1 and 2 Hz).

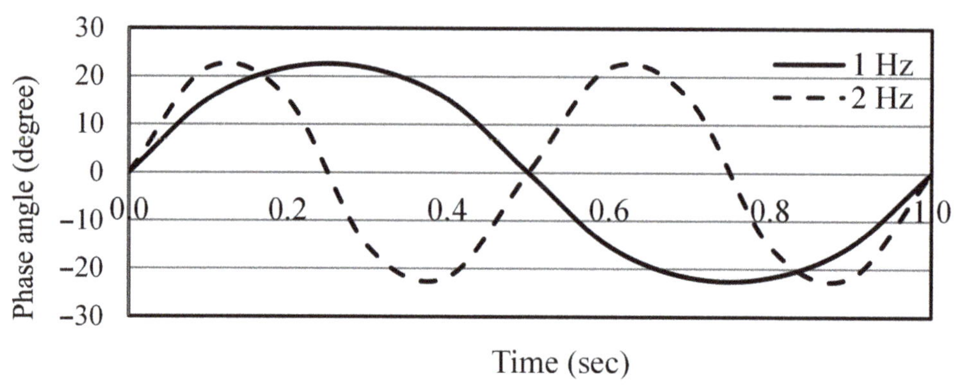

FIGURE 10.8.4 Plots of normal strain versus mean strain as functions of time at two angular speeds (1 and 2 Hz).

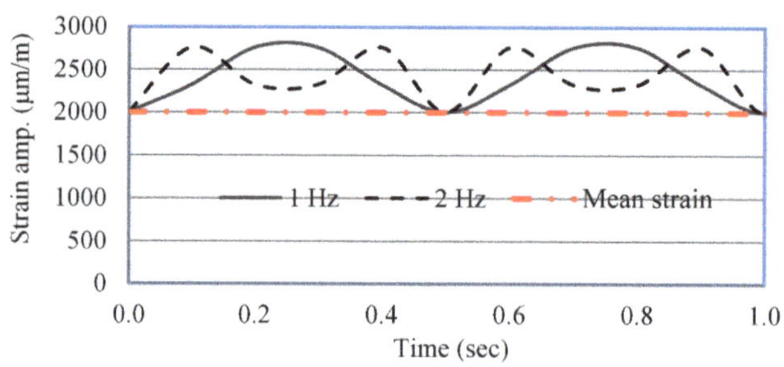

10.8.5. Acceleration Factor for Fatigue Life

The fatigue life of a product is a function of a good number of influential variables including material properties (σ_f', b, ε_f', c), dynamic cyclic loads (K′, n′), damage (D), nonproportionality (ψ) of loads because of time-varying phase angle of strain (ε_a), and the working environment (e.g., temperature and humidity). The life cycle is generally delineated as

$$2N_f = F(\sigma_f', b, \varepsilon_f', c, K', n', D, \psi, T, H_R, \ldots) \tag{10.8.6}$$

where

ψ is the time-varying phase angle of strain amplitude ε_a in Equation (10.8.5)
T is the temperature
H_R is the relative humidity
… is more environmental conditions such as H_2 and salt

There is no apparent single equation to relate the fatigue life to these indirectly related groups of variables in Equations (10.8.1)–(10.8.3). When stepwise load levels are applied, an experimenter has to resolve all these equations sequentially and take a procedure to estimate the damage done. The acceleration factor can be figured out using the following equation:

$$\Lambda_q = \frac{\Delta D_q}{\Delta D_0} \quad (10.8.7)$$

where

ΔD_0 is the damage done at the normal load, e.g., a section of road with typical roughness

ΔD_q is the damage done at the qth load step, e.g., a section of rough road

Acceleration factor Λ_q is assessed for reference only, since the damage is usually calculated directly with Equation (10.8.1) or similar equations. This may involve risk analysis that accounts for manufacturing scatters, including materials and fabrication process, with regard to the statistical distribution of the damage incurred. This is to be illustrated using an example of automotive suspension components: coil springs and lower arms, subject to road roughness.

10.8.6. Fatigue Damage of Automotive Suspension Components Subjected to Road Roughness

Fatigue of each automotive suspension component is described as the weakening of its material as a result of cyclic loading. Coil springs and lower control arms are two weak links for automotive suspensions. This loading process may cause localized structural damage, initiating cracks, and the growth of progressive cracks. Their materials and related material properties for a specific vehicle are listed in Table 10.8.1. As a whole, these two parts in the frame are first analyzed using FEMs to identify the high-stress spots, where strain gauges are applied as aligned with the directions of principal strains. The measured amplitudes of fluctuating strains (μm/m) expressed as functions of time are displayed in Figure 10.8.5. The von Mises strains are then calculated [Putra and Ikibal 2021]. The cumulative damage D is estimated using Equation (10.8.2) accordingly and the results are exhibited in Figure 10.8.6.

TABLE 10.8.1 Material properties of the coil spring and lower control arm of an automotive suspension of interest.

Material property	Coil spring	Lower control
Material	SAE 5160	AISI 1513
Modulus of elasticity (GPa)	201	210
Yield strength (MPa)	1487	450
Tensile strength (MPa)	1584	585
Fatigue strength coefficient σ_f' (MPa)	2063	1089
Fatigue strength exponent b	−0.08	−0.07
Fatigue strain coefficient ε_f'	9.56	0.41
Fatigue strain exponent c	−1.05	−0.54
Cyclic hardening coefficient K′ (MPa)	1940	978
Cyclic hardening exponent n′	0.05	0.11
Loading sequence	Road profile	Road profile
Nonproportionality of multiaxial loads	Proportional	Proportional

FIGURE 10.8.5 Strains (μm/m) measured at the coil spring and lower control arms [Putra and Ikibal 2021].

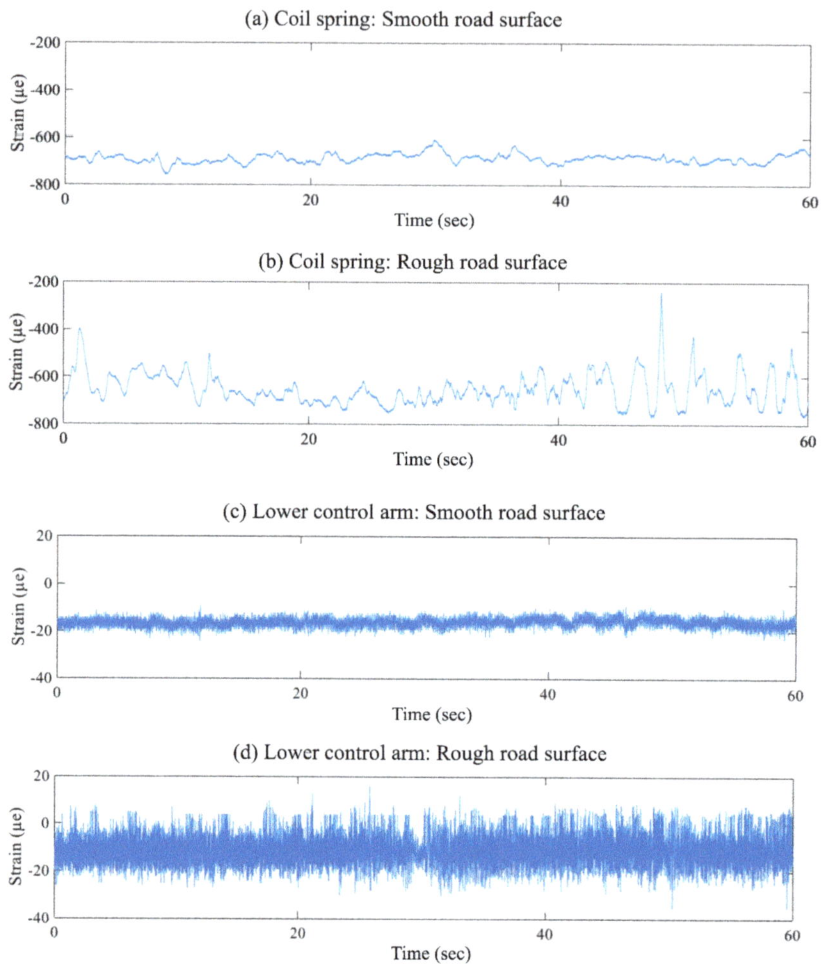

FIGURE 10.8.6 Damage as a function of mean strain and strain range, according to Equation (10.8.1) with D = 0, at the coil spring and lower control arms [Putra and Ikibal 2021].

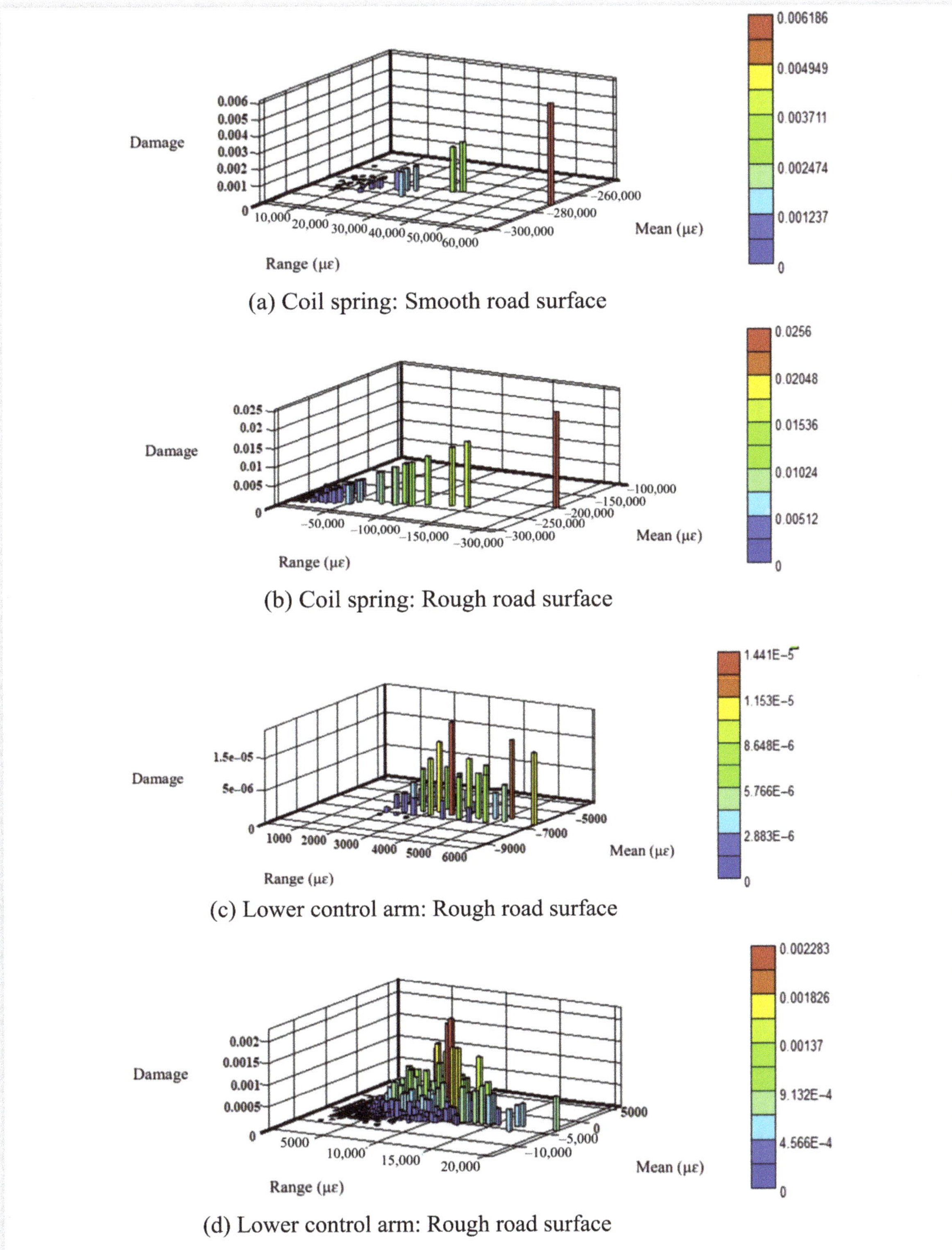

10.8.6.1. Counting Damaging Cycles

It is well known that fatigue damage depends more on the effective strain amplitude and mean strain than on the applied apparent stress levels. Every effective strain (stress) cycle accountable for material fatigue in the load history must be included in the damage accumulation model. For example, an actually measured strain history at a frame joint of a mining truck is depicted in Figure 10.8.7: (1) The truck is loaded in the park gear, (2) the truck runs at the full load, and (3) the truck returns with no load.

FIGURE 10.8.7 Measured real-time strain history (rosette) of the truck frame welding point on a highly irregular mining route (vertical axis—strain, μm/m; horizontal axis—time, sec).

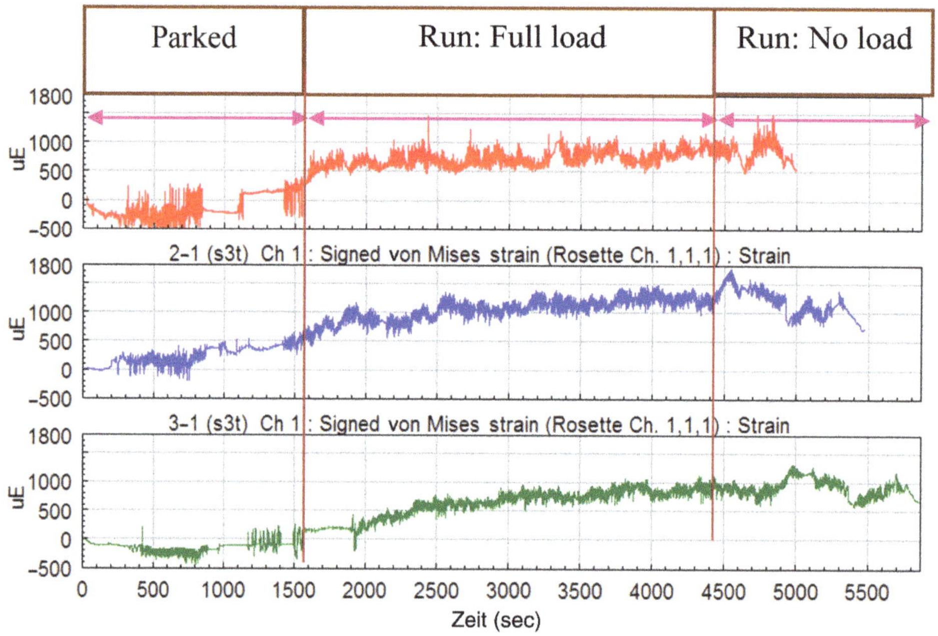

10.9. Counting Dynamic Load Cycles

It is well known that fatigue damage depends on the strain amplitude, mean strain, phase angle, and loading sequence rather than just on the applied stress. The four damaging factors are explained here as follows:

1. Strain Amplitude and Mean Strain: A uniaxial loading sequence of varying strain can be broken into an equivalent set of constant amplitude strain reversals by the rainflow counting method. In practice, the total data in each section of sequential time series are divided into $C_1 \times C_2$ classes of constant-width interval, i.e., steps in a 2D array, and all the extreme values (peaks and valleys) located in a given load block are replaced by the representative values in terms of mean value and fluctuating amplitude, as shown in Figure 10.9.2. Standard practice in the industry is to take $C_1 = 8$ and $C_2 = 8$, and thus, $C_1 \times C_2 = 64$ [Schlitz et al. 1990].
2. Phase Angle: Another concern for calculating fatigue life cycles is the phase angle. When there are multiple loading sequences, phase angles between individual loading sequences, including both normal and shear components, have to be also considered for calculating the damage.

3. Load History: Damage done by fatigue is usually interpreted as micro-crack growth in opening and/or sliding modes, while the effective strain (stress) value results from the crack closure phenomenon due to which the applied stress range is reduced; such a mechanism strongly depends on the load history. The rainflow counting method emerges as an examination procedure that can be used to store service measurements in a form suitable for fatigue life prediction of materials and structures, as well as for reconstructing a sequence from the data in order to provide prominent loading data for fatigue life testing.

Counting algorithms have initially been developed for the study of fatigue damage generated in aeronautical structures. Among them, the rainflow counting method is widely applied to the analysis of fatigue data for transforming the time history of a fluctuating load (e.g., strain, stress, strain energy, and temperature) into a set of reversals of simple loading components (two reversals = one cycle) in terms of the load amplitude and mean load. Detailed algorithms for counting load cycles can be found in [Amzallag et al. 1994, ASTM E1049-85 2017, Glinka and Kam 1987, McInners and Meehan 2008, Rychlik 1987].

For example, an actually measured strain history at a frame joint of a mining truck is depicted in Figure 10.9.1, following its three distinct load histories:

1. The truck is loaded while in the park gear.
2. The truck runs at the full load from the mining location to unload the mine.
3. The truck comes back at no load.

Every effective load reversal (cycle) that is accountable for material fatigue is included in the damage accumulation model. Reversals (cycles) with small amounts of mean value and fluctuating amplitude are to be truncated, but the threshold of "being small" depends on the material fatigue properties and working environment. Note that phase transformations of metals and glass points of plastics may change the material properties dramatically.

FIGURE 10.9.1 Measured real-time strain history (rosette) of the truck frame welding point on a highly irregular mining route (vertical axis—strain, μm/m; horizontal axis—time, sec).

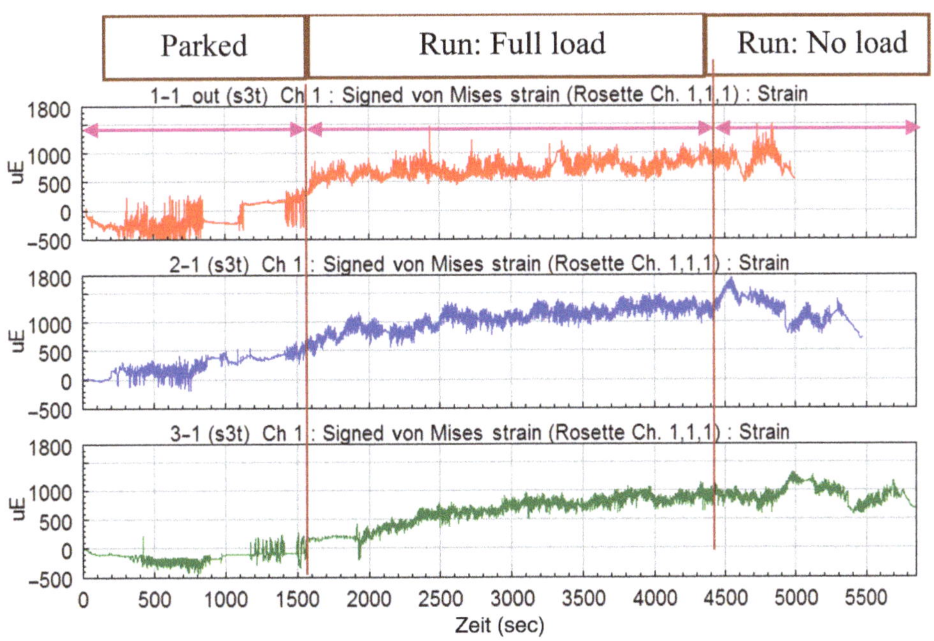

10.9.1. Load Bock

When a test specimen is loaded cyclically until failure, the rainflow counting method in the time domain can be applied to converting an irregular fluctuating load history into a histogram of loads [ASTM E1049-85 2017] in terms of mean load and load amplitude. Similar loads can be grouped together to form a load block. In many load cases, such as the one shown in Figure 10.9.1, a vehicle repeatedly undergoes a good number of load blocks.

How to convert a time series of mechanical loads into load blocks is demonstrated here using a simple strain history as shown in Figure 10.9.2, which can be schematically allocated into five blocks according to the mean strain and strain range. These load blocks can be loaded into Equations (10.8.1) and (10.8.2) to predict the accumulated damage D. When damage D reaches 100%, the material fails.

FIGURE 10.9.2 Allocating historical data into load blocks according to strain range and mean strain.

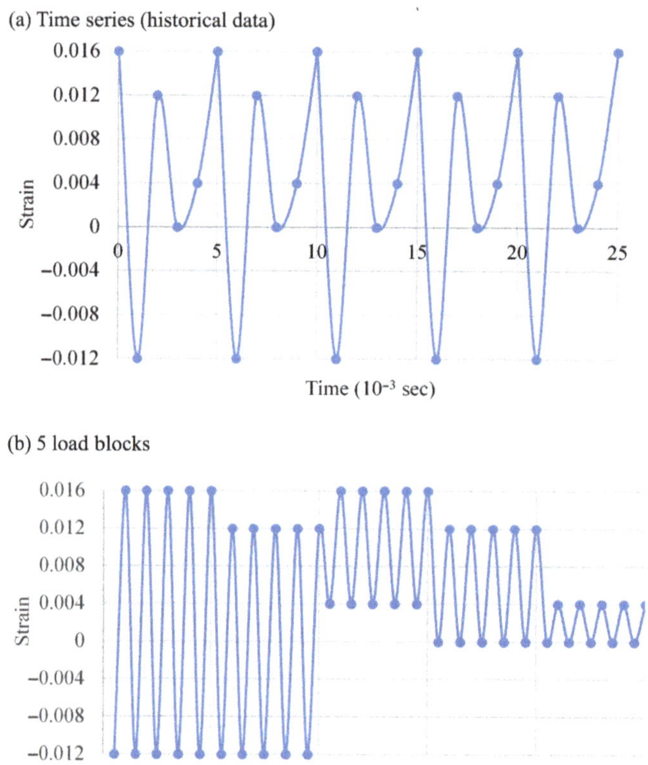

10.9.2. Rainflow Counting Method of Proportional Stress/Strain Fluctuations in Time Domain

The rainflow counting method, also known as the rainflow method, is used in the analysis of fatigue data in order to transform the time history of a fluctuating load (e.g., strain, stress, energy, and temperature) into a set of reversals of simple loading components (two reversals = one cycle). Before conducting the rainflow counting method, one has to orderly convert the loading cycles of a spectrum of the fluctuating load of interest into a set of simple reversals [ASTM E1049-85 2017, Matsuishi and Endo 1968].

A rainflow counting may begin with hysteresis filtering, which can be used to set a threshold and remove cycles (or load blocks) whose contribution to the total damage is small. For example, load block 3 in Figure 10.9.2(b) may be neglected according to its strain amplitude and mean strain. Nevertheless, it is never too cautious to look into other factors before jumping to any conclusion because the threshold of "being small" depends on the material fatigue properties and working environment.

The time history is transformed to a time series having various extrema peaks and valleys after hysteresis filtering. The peaks are in either tension or compression, and so are the valleys. Peak–valley filtering is then applied to preserving only the maximum and minimum value of the cycles and remove the points in between. The rainflow method, as an algorithm for counting fatigue cycles from a response time history that comprises loading factors, such as strain, stress, equivalent strain, equivalent stress, strain energy, and/or temperature reversals, is then carried out to count the number of reversals (or cycles) for each individual "similar" damaging capacity.

The rainflow method based on peak–valley filtering begins successively at the inside of each peak or valley, and every part of the history is counted once and only once [Glinka and Kam 1987]. One may turn the strain-time history from left to right, clockwise 90°, as shown in Figure 10.9.3. It then resembles a pagoda roof with the strain at the earliest time being on top. One may think of how the rain falls along a 2D pagoda roof. The number of total cyclic reversals, i.e., half-cycles, can be counted as follows:

1. Start with peaks. Each peak is perceived as a source of water that "drips" down the pagoda. Count the number of reversals by looking for terminations in the flow if one of the following conditions is met:
   (a) It reaches the end of the time history.
   (b) It merges with a flow that started at an earlier tensile peak.
   (c) It flows until (when) an opposite tensile peak has a greater magnitude.
2. Repeat the above step for valleys.
3. The strain difference between its start and termination is identified as a reversal.
4. Pair up reversals of identical magnitude, but with an opposite sense, to count the number of complete cycles. There are residual reversals usually.

The rainflow method allows most damage accumulation models, e.g., Equations (10.8.1) and (10.8.2), to assess the fatigue life of a mechanical structure subject to complex loadings in either stationary vibrational state or nonstationary vibrational state.

The rainflow method has been given a mathematical definition by [Rychlik 1987], and thus, the analytic closed-form calculation based on the statistical properties of the designated design variable can be carried out. There are two fundamental assumptions made to rearrange the loads into blocks using the rainflow counting method. Applied loads are independent of each other, and the loading order is not rigorously taken into consideration.

Two different approaches to the rainflow counting based on peak–valley filtering are in use. They are four-point algorithm and three-point algorithm:

1. Four-Point Algorithm: Comparing design ranges.
2. Three-Point Algorithm: Comparing the relative levels.

The four-point algorithm is simpler to understand than the three-point algorithm. Nevertheless, both would lead to the same result [McInners and Meehan 2008].

FIGURE 10.9.3 Conceptual pagoda roof for counting fluctuating reversals for fatigue analysis [Fatigue Life LLC 2023].

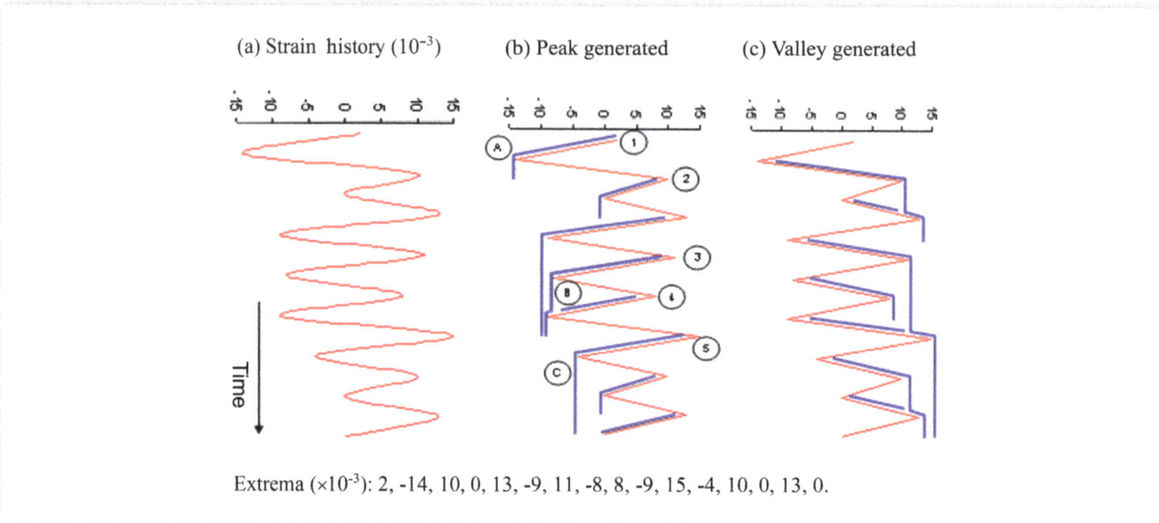

10.9.3. Four-Point Algorithm

The four-point algorithm is a widely used rainflow counting method. Both the amplitude and mean value of the design variable of interest can be counted by the four-point algorithm. Let ε represent a strain component, and the recorded strain data are the consecutive values ε_i $(1 \le i \le N)$ of the peaks and valleys of the sequence. Cycle extractions are demonstrated using the load in Figure 10.9.3. Assume that the cycles made of the first four points appear as a close-looped stress–strain curve as shown in the figure, which shows that there are two possible types of cycles that can occur. The principle of the extraction of a cycle based on four successive points, namely ε_1, ε_2, ε_3, and ε_4, is illustrated as follows [Amzallag et al. 1994]:

(a) Three consecutive ranges are determined using four points: $\Delta\varepsilon_a = |\varepsilon_2 - \varepsilon_1|$, $\Delta\varepsilon_b = |\varepsilon_3 - \varepsilon_2|$, and $\Delta\varepsilon_c = |\varepsilon_4 - \varepsilon_3|$. If $\Delta\varepsilon_b \le \Delta\varepsilon_a$ and $\Delta\varepsilon_b \le \Delta\varepsilon_c$, the cycle represented by strain range $(\varepsilon_2, \varepsilon_3)$ is extracted. Then, both points ε_2 and ε_3 are discarded. Point ε_1 is then connected directly to point ε_4.

(b) If "$\Delta\varepsilon_b \le \Delta\varepsilon_a$ and $\Delta\varepsilon_b \le \Delta\varepsilon_c$" is not true, read the next point (ε_5). Reversal ε_1-ε_2 is a residue (reversal). Then, use $\Delta\varepsilon_a = |\varepsilon_3 - \varepsilon_2|$, $\Delta\varepsilon_b = |\varepsilon_4 - \varepsilon_3|$, and $\Delta\varepsilon_c = |\varepsilon_5 - \varepsilon_4|$ to detect the next feasible cycle.

(c) The procedure, given in steps (a) and (b), is repeated until the last point of the sequence is reached. Once all the comparisons are completed, the remaining points constitute the so-called residue. Note that the maximum number of points cannot exceed 2L – 1, where L is the number of possible levels.

(d) The resulting counted cycles are then input into a matrix that stores the following three pieces of information: mean value, amplitude (or range), and number of reversals (cycles).

Retraction of load cycles (reversals) from a time series of data to stress–strain curves is demonstrated in Figure 10.9.4.

FIGURE 10.9.4 Cycle extraction from data in time series to stress-strain curves using four-point method [GopiReddy et al. 2015].

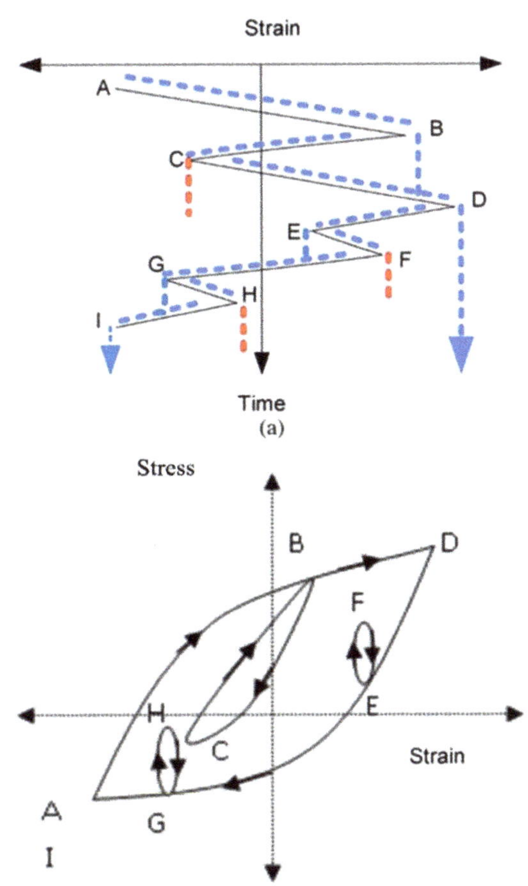

EXAMPLE 10.9.1

A series of 24 strain extrema (including peaks and valleys) measured at the critical location of a steel part are listed as follows: 0.0004, 0.0007, 0.0002, 0.0010, 0.0005, 0.0009, 0.0003, 0.0004, 0.0002, 0.0012, 0.0005, 0.0011, 0.0001, 0.0004, 0.0003, 0.0010, 0.0006, 0.0012, 0.0004, 0.0008, 0.0001, 0.0009, 0.0004, and 0.0006. They are plotted as follows:

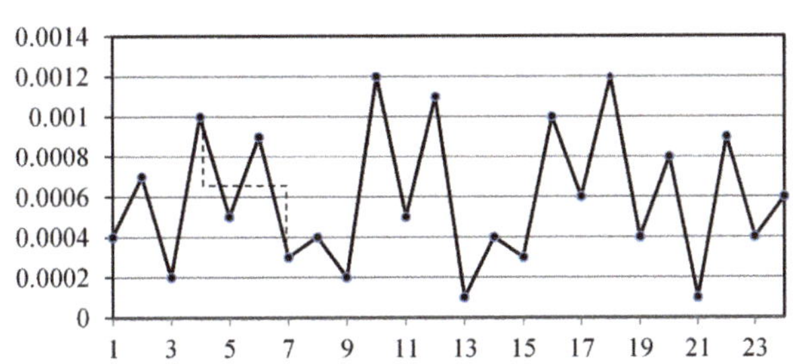

Note that the yield strain of steel is 0.002 or 0.2%. Please extract strain cycles and residues using the four-point algorithm.

Solution:

(a) Consider the first four points. Since $\Delta\varepsilon_a = |\varepsilon_2 - \varepsilon_1| = 0.03\%$, $\Delta\varepsilon_b = |\varepsilon_3 - \varepsilon_2| = 0.05\%$, and $\Delta\varepsilon_c = |\varepsilon_4 - \varepsilon_3| = 0.08\%$, "$\Delta\varepsilon_b \le \Delta\varepsilon_a$ and $\Delta\varepsilon_b \le \Delta\varepsilon_c$" is not true. Thus, ε_1-ε_2, going from 0.0002 to 0.0010, is a residue (reversal) and no cycle is extracted.

(b) Read the next piece of data, i.e., ε_5; then, the new four points are ε_2, ε_3, ε_4, and ε_5. Since $\Delta\varepsilon_a = |\varepsilon_3 - \varepsilon_2| = 0.05\%$, $\Delta\varepsilon_b = |\varepsilon_4 - \varepsilon_3| = 0.08\%$, and $\Delta\varepsilon_c = |\varepsilon_5 - \varepsilon_4| = 0.05\%$, "$\Delta\varepsilon_b \le \Delta\varepsilon_a$ and $\Delta\varepsilon_b \le \Delta\varepsilon_a$" is not true. Thus, ε_2-ε_3 is a residue (reversal), going from 0.0007 to 0.0002; no cycle is extracted.

(c) Read the next piece of data, i.e., ε_6; then, the new four points are ε_3, ε_4, ε_5, and ε_6. Since $\Delta\varepsilon_a = |\varepsilon_4 - \varepsilon_3| = 0.08\%$, $\Delta\varepsilon_b = |\varepsilon_5 - \varepsilon_4| = 0.05\%$, and $\Delta\varepsilon_c = |\varepsilon_6 - \varepsilon_5| = 0.04\%$, "$\Delta\varepsilon_b \le \Delta\varepsilon_a$ and $\Delta\varepsilon_b \le \Delta\varepsilon_c$" is not true. Thus, ε_3-ε_4 is a residue (reversal), going from 0.0002 to 0.0010; no cycle is extracted.

(d) Read the next piece of data; i.e., ε_7; then, the new four points are ε_4, ε_5, ε_6, and ε_7. Since $\Delta\varepsilon_a = |\varepsilon_5 - \varepsilon_4| = 0.05\%$, $\Delta\varepsilon_b = |\varepsilon_6 - \varepsilon_5| = 0.04\%$, and $\Delta\varepsilon_c = |\varepsilon_7 - \varepsilon_6| = 0.06\%$, "$\Delta\varepsilon_b \le \Delta\varepsilon_a$ and $\Delta\varepsilon_b \le \Delta\varepsilon_c$" is true. The cycle represented by strain range $(\varepsilon_5, \varepsilon_6)$ = (0.0005, 0.0009) that has mean = 0.0007 and amplitude = 0.0002 is extracted. Then, both points ε_5 and ε_6 are discarded from the picture. Point ε_4 is then connected directly to point ε_7.

(e) Read the next two pieces of data, and the new four points are ε_4, ε_7, ε_8, and ε_9. Since $\Delta\varepsilon_a = |\varepsilon_7 - \varepsilon_4| = 0.07\%$, $\Delta\varepsilon_b = |\varepsilon_8 - \varepsilon_7| = 0.01\%$, and $\Delta\varepsilon_c = |\varepsilon_9 - \varepsilon_8| = 0.02\%$, "$\Delta\varepsilon_b \le \Delta\varepsilon_a$ and $\Delta\varepsilon_b \le \Delta\varepsilon_c$" is true. The cycle represented by strain range $(\varepsilon_7, \varepsilon_8)$ = (0.0003, 0.0004) that has mean = 0.00035 and amplitude = 0.00005 is extracted. Then, both points ε_7 and ε_8 are discarded from the picture. Point ε_4 is then connected directly to point ε_9.

(f) Read the next two pieces of data, and the new four points are ε_4, ε_9, ε_{10}, and ε_{11}. Since $\Delta\varepsilon_a = |\varepsilon_9 - \varepsilon_4| = 0.08\%$, $\Delta\varepsilon_b = |\varepsilon_{10} - \varepsilon_9| = 0.10\%$, and $\Delta\varepsilon_c = |\varepsilon_{11} - \varepsilon_{10}| = 0.07\%$, "$\Delta\varepsilon_b \le \Delta\varepsilon_a$ and $\Delta\varepsilon_b \le \Delta\varepsilon_c$" is not true. Thus, ε_9-ε_{10} is a residue (reversal), going from 0.0002 to 0.0012; no cycle is extracted.

(g) Read the next piece of data; i.e., ε_{12} and the new four points are ε_4, ε_{10}, ε_{11}, and ε_{12}. Since $\Delta\varepsilon_a = |\varepsilon_{10} - \varepsilon_4| = 0.10\%$, $\Delta\varepsilon_b = |\varepsilon_{11} - \varepsilon_{10}| = 0.07\%$, and $\Delta\varepsilon_c = |\varepsilon_{12} - \varepsilon_{11}| = 0.06\%$, "$\Delta\varepsilon_b \le \Delta\varepsilon_a$ and $\Delta\varepsilon_b \le \Delta\varepsilon_c$" is not true. Thus, ε_9-ε_{10} is a residue (reversal), going from 0.0002 to 0.0012; no cycle is extracted.

(h) Read the next piece of data; i.e., ε_{13}; then, the new four points are ε_{10}, ε_{11}, ε_{12}, and ε_{13}. Since $\Delta\varepsilon_a = |\varepsilon_{11} - \varepsilon_{10}| = 0.07\%$, $\Delta\varepsilon_b = |\varepsilon_{12} - \varepsilon_{11}| = 0.06\%$, and $\Delta\varepsilon_c = |\varepsilon_{13} - \varepsilon_{12}| = 0.10\%$, "$\Delta\varepsilon_b \le \Delta\varepsilon_a$ and $\Delta\varepsilon_b \le \Delta\varepsilon_c$" is true. The cycle represented by strain range $(\varepsilon_{11}, \varepsilon_{12})$ = (0.0005, 0.0011) that has mean = 0.0008 and amplitude = 0.0003 is extracted. Then, both point ε_{11} and point ε_{12} are discarded from the picture. Point ε_{10} is then connected directly to point ε_{13}.

(i) Read the next two pieces of data, i.e., ε_{14} and ε_{15}; then, the new four points are ε_{10}, ε_{13}, ε_{14}, and ε_{15}. Since $\Delta\varepsilon_a = |\varepsilon_{13} - \varepsilon_{10}| = 0.11\%$, $\Delta\varepsilon_b = |\varepsilon_{14} - \varepsilon_{13}| = 0.03\%$, and $\Delta\varepsilon_c = |\varepsilon_{15} - \varepsilon_{14}| = 0.01\%$, "$\Delta\varepsilon_b \le \Delta\varepsilon_a$ and $\Delta\varepsilon_b \le \Delta\varepsilon_c$" is not true. Thus, ε_{10}-ε_{13} is a residue (reversal), going from 0.0012 to 0.0001; no cycle is extracted.

(j) Read the next piece of data, i.e., ε_{16}; then, the new four points are ε_{13}, ε_{14}, ε_{15}, and ε_{16}. Since $\Delta\varepsilon_a = |\varepsilon_{14} - \varepsilon_{13}| = 0.03\%$, $\Delta\varepsilon_b = |\varepsilon_{15} - \varepsilon_{14}| = 0.01\%$, and $\Delta\varepsilon_c = |\varepsilon_{16} - \varepsilon_{15}| = 0.07\%$, "$\Delta\varepsilon_b \le \Delta\varepsilon_a$ and $\Delta\varepsilon_b \le \Delta\varepsilon_c$" is true. The cycle represented by strain range $(\varepsilon_{14}, \varepsilon_{15})$ = (0.0004, 0.0003) that has mean = 0.00035 and amplitude = 0.00005 is extracted. Then, both points ε_{14} and ε_{15} are discarded from the picture. Point ε_{13} is then connected directly to point ε_{16}.

(k) Read the next two pieces of data, i.e., ε_{17} and ε_{18}; then, the new four points are ε_{13}, ε_{16}, ε_{17}, and ε_{18}. Since $\Delta\varepsilon_a = |\varepsilon_{16} - \varepsilon_{13}| = 0.09\%$, $\Delta\varepsilon_b = |\varepsilon_{17} - \varepsilon_{16}| = 0.04\%$, and $\Delta\varepsilon_c = |\varepsilon_{18} - \varepsilon_{17}| = 0.06\%$, "$\Delta\varepsilon_b \leq \Delta\varepsilon_a$ and $\Delta\varepsilon_b \leq \Delta\varepsilon_c$" is true. The cycle represented by strain range (ε_{16}, ε_{17}) = (0.0010, 0.0006) that has mean = 0.0008 and amplitude = 0.0002 is extracted. Then, both points ε_{16} and ε_{17} are discarded from the picture. Point ε_{13} is then connected directly to point ε_{18}.

(l) Read the next two pieces of data, i.e., ε_{19} and ε_{20}; then, the new four points are ε_{13}, ε_{18}, ε_{19}, and ε_{20}. Since $\Delta\varepsilon_a = |\varepsilon_{18} - \varepsilon_{13}| = 0.11\%$, $\Delta\varepsilon_b = |\varepsilon_{19} - \varepsilon_{18}| = 0.08\%$, and $\Delta\varepsilon_c = |\varepsilon_{20} - \varepsilon_{19}| = 0.04\%$, "$\Delta\varepsilon_b \leq \Delta\varepsilon_a$ and $\Delta\varepsilon_b \leq \Delta\varepsilon_c$" is not true. Thus, ε_{13}-ε_{18} is a residue (reversal), going from 0.0001 to 0.0012; no cycle is extracted.

(m) Read the next piece of data, i.e., ε_{21}; then, the new four points are ε_{18}, ε_{19}, ε_{20}, and ε_{21}. Since $\Delta\varepsilon_a = |\varepsilon_{19} - \varepsilon_{18}| = 0.08\%$, $\Delta\varepsilon_b = |\varepsilon_{20} - \varepsilon_{19}| = 0.04\%$, and $\Delta\varepsilon_c = |\varepsilon_{21} - \varepsilon_{20}| = 0.07\%$, "$\Delta\varepsilon_b \leq \Delta\varepsilon_a$ and $\Delta\varepsilon_b \leq \Delta\varepsilon_c$" is true. The cycle represented by strain range (ε_{19}, ε_{20}) = (0.0004, 0.0008) that has mean = 0.0006 and amplitude = 0.0002 is extracted. Then, both points ε_{19} and ε_{20} are discarded from the picture. Point ε_{18} is then connected directly to point ε_{21}.

(n) Read the next two pieces of data, i.e., ε_{22} and ε_{23}; then, the new four points are ε_{18}, ε_{21}, ε_{22}, and ε_{23}. Since $\Delta\varepsilon_a = |\varepsilon_{21} - \varepsilon_{18}| = 0.11\%$, $\Delta\varepsilon_b = |\varepsilon_{22} - \varepsilon_{21}| = 0.08\%$, and $\Delta\varepsilon_c = |\varepsilon_{23} - \varepsilon_{22}| = 0.05\%$, "$\Delta\varepsilon_b \leq \Delta\varepsilon_a$ and $\Delta\varepsilon_b \leq \Delta\varepsilon_c$" is not true. Thus, ε_{18}-ε_{21} is a residue (reversal), going from 0.0012 to 0.0001; no cycle is extracted.

(o) Read the next piece of data, i.e., ε_{24}; then, the new four points are ε_{21}, ε_{22}, ε_{23}, and ε_{24}. Since $\Delta\varepsilon_a = |\varepsilon_{22} - \varepsilon_{21}| = 0.08\%$, $\Delta\varepsilon_b = |\varepsilon_{23} - \varepsilon_{22}| = 0.05\%$, and $\Delta\varepsilon_c = |\varepsilon_{24} - \varepsilon_{23}| = 0.02\%$, "$\Delta\varepsilon_b \leq \Delta\varepsilon_a$ and $\Delta\varepsilon_b \leq \Delta\varepsilon_c$" is not true. Thus, ε_{21}-ε_{22} is a residue (reversal), going from 0.0001 to 0.0009; no cycle is extracted.

(p) Read the next piece of data, i.e., ε_1; then, the new four points are ε_{22}, ε_{23}, ε_{24}, and ε_1. Since $\Delta\varepsilon_a = |\varepsilon_{23} - \varepsilon_{22}| = 0.05\%$, $\Delta\varepsilon_b = |\varepsilon_{24} - \varepsilon_{23}| = 0.02\%$, and $\Delta\varepsilon_c = |\varepsilon_1 - \varepsilon_{24}| = 0.02\%$, "$\Delta\varepsilon_b \leq \Delta\varepsilon_a$ and $\Delta\varepsilon_b \leq \Delta\varepsilon_c$" is true. The cycle represented by strain range (ε_{23}, ε_{24}) = (0.0004, 0.0006) that has mean = 0.0005 and amplitude = 0.0001 is extracted. Then, both points ε_{19} and ε_{20} are discarded from the picture. Point ε_{18} is then connected directly to point ε_{21}.

In summary,

Reversal	Range	Mean	Amplitude
2	(ε_5, ε_6) = (0.0005, 0.0009)	0.0007	0.0002
2	(ε_7, ε_8) = (0.0003, 0.0004)	0.00035	0.00005
2	(ε_{11}, ε_{12}) = (0.0005, 0.0011)	0.0008	0.0003
2	(ε_{14}, ε_{15}) = (0.0004, 0.0003)	0.00035	0.00005
2	(ε_{16}, ε_{17}) = (0.0010, 0.0006)	0.0008	0.0002
2	(ε_{19}, ε_{20}) = (0.0004, 0.0008)	0.0006	0.0002
1	$\varepsilon_1 \rightarrow \varepsilon_2$ (0.0002, 0.0010)	0.0006	0.0004
1	$\varepsilon_2 \rightarrow \varepsilon_3$ (0.0007, 0.0002)	0.00045	0.00025
1	$\varepsilon_3 \rightarrow \varepsilon_4$ (0.0002, 0.0010)	0.0006	0.0004
1	$\varepsilon_9 \rightarrow \varepsilon_{10}$ (0.0002, 0.0012)	0.0007	0.0005
1	$\varepsilon_{10} \rightarrow \varepsilon_{13}$ (0.0012, 0.0001)	0.00065	0.00055
1	$\varepsilon_{13} \rightarrow \varepsilon_{18}$ (0.0001, 0.0012)	0.00065	0.00055
1	$\varepsilon_{18} \rightarrow \varepsilon_{21}$ (0.0012, 0.0001)	0.00065	0.00055
1	$\varepsilon_{21} \rightarrow \varepsilon_{22}$ (0.0001, 0.0009)	0.0005	0.0004
1	$\varepsilon_{22} \rightarrow \varepsilon_{23}$ (0.0009, 0.0004)	0.00065	0.00025
1	$\varepsilon_{23} \rightarrow \varepsilon_{24}$ (0.0004, 0.0006)	0.0005	0.0001

Assume that mean strain and strain amplitude are the only two damaging factors of concern without considering the influences of loading sequence and others. The data shown above exhibit two additional cycles, i.e., ($\varepsilon_{13} \rightarrow \varepsilon_{18}$, $\varepsilon_{18} \rightarrow \varepsilon_{21}$) and ($\varepsilon_1 \rightarrow \varepsilon_2$, $\varepsilon_3 \rightarrow \varepsilon_4$), which can be regrouped into the following patterns:

Reversal	Range	Mean	Amplitude
2	($\varepsilon_1 \rightarrow \varepsilon_2$, $\varepsilon_3 \rightarrow \varepsilon_4$)	0.0006	0.0004
2	(ε_5, ε_6) = (0.0005, 0.0009)	0.0007	0.0002
2	(ε_7, ε_8) = (0.0003, 0.0004)	0.00035	0.00005
2	(ε_{11}, ε_{12}) = (0.0005, 0.0011)	0.0008	0.0003
2	($\varepsilon_{13} \rightarrow \varepsilon_{18}$, $\varepsilon_{18} \rightarrow \varepsilon_{21}$)	0.00065	0.00055
2	(ε_{14}, ε_{15}) = (0.0004, 0.0003)	0.00035	0.00005
2	(ε_{16}, ε_{17}) = (0.0010, 0.0006)	0.0008	0.0002
2	(ε_{19}, ε_{20}) = (0.0004, 0.0008)	0.0006	0.0002
1	$\varepsilon_2 \rightarrow \varepsilon_3$ (0.0007, 0.0002)	0.00045	0.00025
1	$\varepsilon_9 \rightarrow \varepsilon_{10}$ (0.0002, 0.0012)	0.0007	0.0005
1	$\varepsilon_{10} \rightarrow \varepsilon_{13}$ (0.0012, 0.0001)	0.00065	0.00055
1	$\varepsilon_{21} \rightarrow \varepsilon_{22}$ (0.0001, 0.0009)	0.0005	0.0004
1	$\varepsilon_{22} \rightarrow \varepsilon_{23}$ (0.0009, 0.0004)	0.00065	0.00025
1	$\varepsilon_{23} \rightarrow \varepsilon_{24}$ (0.0004, 0.0006)	0.0005	0.0001

An example of allocation of load blocks of temperature exerted on IGBTs in the real world is depicted in Figure 10.9.5(b) based on mean temperature and temperature range, which are obtained from the repeating time history of temperature observed in a load section (30 days) as shown in Figure 10.9.5(a). Because the breakdown temperature of an IGBT is relatively high (>250°C) and it is usually operated within its maximum rated temperature (≈150°C) as designed, the time-dependent equation can be used for calculations toward its fatigue failure [Nagode and Hack 2004]. Each load block that exceeds a predefined threshold will be taken into consideration for fatigue analysis in terms of its mean value, range (amplitude), and loading sequence [GopiReddy et al. 2015].

FIGURE 10.9.5 Allocating daily temperature variation of the hot spot of an IGBT as a function of temperature range and mean [GopiReddy et al. 2015].

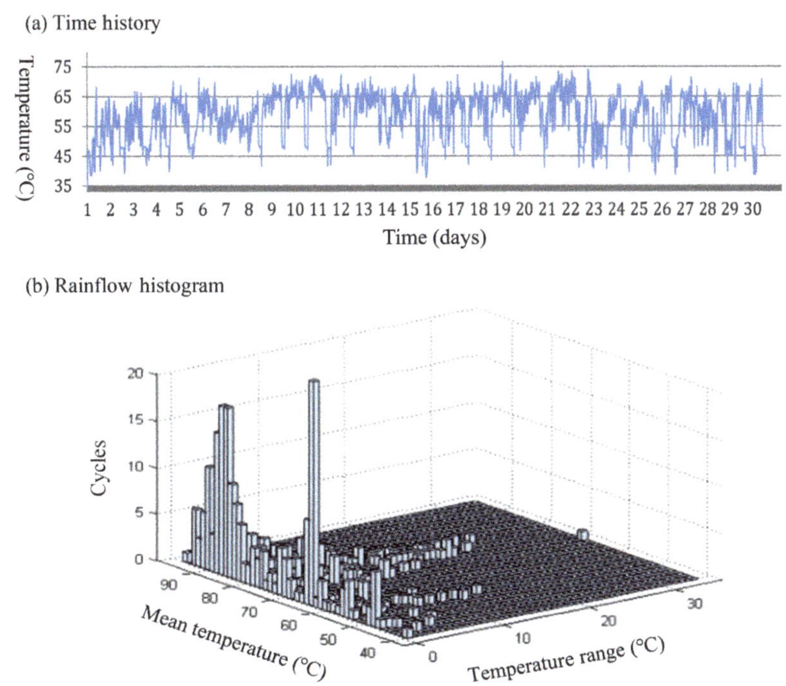

10.9.4. Three-Point Algorithm

The three-point algorithm is another way to carry out rainflow counting. It follows the guideline given as follows:

If $|\varepsilon_{i+1} - \varepsilon_i| > |\varepsilon_i - \varepsilon_{i-1}|$, one cycle is identified.

of which i is one of the strain extrema. The algorithm is based on the assumption that the mean and amplitude ($\equiv 1/2$ range) are the only two damaging factors of concern without considering the influences of loading sequence and others. It is here demonstrated how to allocate the temperature distribution according to its amplitude and mean on IGBT connections for an electronic power module, similar to the case study presented by [Nicola 2013]. The temperature profile as a function of time is exhibited in Figure 10.9.6(a).

EXAMPLE 10.9.2

IGBT connections are often identified as the weak link in an electronic power module. Please allocate the temperature distribution of the time history given in Figure 10.9.6(a) using the three-point algorithm.

Solution:

Temperature amplitudes and mean values counted using the three-point algorithm are given as follows:

(a) A history of temperature fluctuating between 30 and 70°C has been recorded as shown in Figure 10.9.6(a), including 17 peaks and valleys.

(b) Values of the 17 extrema, connected as shown in Figure 10.9.6(b), are given as follows:

$P_1 = 44.8°C$ $P_2 = 55.8°C$ $P_3 = 32.3°C$ $P_4 = 55.2°C$ $P_5 = 33.7°C$

$P_6 = 58.2°C$ $P_7 = 36.2°C$ $P_8 = 59°C$ $P_9 = 66.7°C$ $P_{10} = 61.2°C$

$P_{11} = 38.7°C$ $P_{12} = 61.8°C$ $P_{13} = 39°C$ $P_{14} = 63.4°C$ $P_{15} = 40.5°C$

$P_{16} = 63.8°C$ $P_{17} = 37.4°C$

(c) Check in the first three points, i.e., P_1, P_2, and P_3. Since $|P_3 - P_2| > |P_2 - P_1|$, P_1 is discarded and a reversal (half cycle) is designated to segment P_1-P_2, as shown in Figure 10.9.6(c), of which

$$½\Delta T = (P_2 - P_1)/2 = (55.8 - 44.8)/2 = 5.5^\circ C \quad \text{(Amplitude)}$$

$$T_{mean} = (P_2 + P_1)/2 = (55.8 + 44.8)/2 = 50.3^\circ C \quad \text{(Mean)}$$

Discard P_1, and check in the next point (P_4). P_2 is the new starting point. Since $|P_4 - P_3| < |P_3 - P_2|$, no cycle or reversal is identified. Check in the next point (P_5). Since $|P_5 - P_4| < |P_4 - P_3|$, no cycle or reversal is identified. Check in the next point (P_6). Since $|P_6 - P_5| > |P_5 - P_4|$, a cycle at "$P_4$-$P_5$" level is formed as

$$½\Delta T = |P_5 - P_4|/2 = |33.7 - 55.2|/2 = 10.75^\circ C \quad \text{(Amplitude)}$$

$$T_{mean} = (P_5 + P_4)/2 = (33.7 + 55.2)/2 = 44.45^\circ C \quad \text{(Mean)}$$

(d) P_4 and P_5 are discarded, and P_3 is connected to P_6. Now the new three points of concern are P_2-P_3-P_6, as shown in Figure 10.9.6(d). Since $|P_6 - P_3| > |P_3 - P_2|$, a reversal is identified at P_2-P_3 level.

$$½\Delta T = |P_3 - P_2|/2 = |32.3 - 55.8|/2 = 11.75^\circ C \quad \text{(Amplitude)}$$

$$T_{mean} = (P_3 + P_2)/2 = (32.3 + 55.8)/2 = 44.05^\circ C \quad \text{(Mean)}$$

(e) Check in the next point (P_7), discard P_2, and P_3 is the new starting point. The new three points of concern are P_3-P_6-P_7, as shown in **Figure 10.9.6(e)**. Since $|P_7 - P_6| < |P_6 - P_3|$, no cycle or reversal is identified. Discard P_3, and check in the next point (P_8). Since $|P_7 - P_6| < |P_8 - P_7|$, a cycle at "$P_6$-$P_7$" level is formed as

$$½\Delta T = |P_7 - P_6|/2 = |36.2 - 58.2|/2 = 11^{\circ}C \quad \text{(Amplitude)}$$

$$T_{mean} = (P_7 + P_6)/2 = (36.2 + 58.2)/2 = 47.2^{\circ}C \quad \text{(Mean)}$$

(f) P_6 and P_7 are discarded, and P_3 is connected to P_8. Now the three points of concern are P_3-P_8-P_9, as shown in **Figure 10.9.6(f)**. Since $|P_9 - P_8| < |P_8 - P_3|$, no cycle or reversal is identified. Discard P_3 and check in the next point (P_{10}). Since $|P_{10} - P_9| > |P_9 - P_8|$, a cycle at "$P_8$-$P_9$" level is formed as

$$½\Delta T = |P_9 - P_8|/2 = |36.7 - 59|/2 = 11.15^{\circ}C \quad \text{(Amplitude)}$$

$$T_{mean} = (P_9 + P_8)/2 = (36.7 + 59)/2 = 47.85^{\circ}C \quad \text{(Mean)}$$

(g) P_8 and P_9 are discarded. P_{10} and P_{11} are checked in, and P_3 is connected to P_{10}. Now the three points of concern are formed, i.e., P_3-P_{10}-P_{11}, as shown in **Figure 10.9.6(g)**. Since $|P_{11} - P_{10}| < |P_{10} - P_3|$, no cycle or reversal is identified. P_3 is discarded, P_{12} is checked in, and the new three points of concern are P_{10}-P_{11}-P_{12}. Since $|P_{12} - P_{11}| > |P_{11} - P_{10}|$, a cycle at "$P_{10}$-$P_{11}$" level is formed as

$$½\Delta T = |P_{11} - P_{10}|/2 = |38.7 - 61.2|/2 = 11.25^{\circ}C \quad \text{(Amplitude)}$$

$$T_{mean} = (P_{11} + P_{10})/2 = (38.7 + 61.2)/2 = 49.95^{\circ}C \quad \text{(Mean)}$$

(h) P_{10} and P_{11} are discarded. P_3 is connected to P_{12}. Now the three points of concern are formed, i.e., P_3-P_{12}-P_{13}, as shown in **Figure 10.9.6(h)**. Since $|P_{13} - P_{12}| < |P_{12} - P_3|$, no cycle or reversal is identified. P_3 is neglected, P_{14} is checked in, and the new three points of concern are P_{12}-P_{13}-P_{14}. Since $|P_{14} - P_{13}| > |P_{13} - P_{12}|$, a cycle at "$P_{12}$-$P_{13}$" level is formed as

$$½\Delta T = |P_{13} - P_{12}|/2 = |39 - 61.8|/2 = 11.4^{\circ}C \quad \text{(Amplitude)}$$

$$T_{mean} = (P_{13} + P_{12})/2 = (39 + 61.8)/2 = 50.4^{\circ}C \quad \text{(Mean)}$$

(i) P_{12} and P_{13} are discarded. P_3 is connected to P_{14}. Now the three points of concern are formed, i.e., P_3-P_{14}-P_{15}, as shown in **Figure 10.9.6(i)**. Since $|P_{15} - P_{14}| < |P_{14} - P_3|$, no cycle or reversal is identified. P_3 is neglected, P_{16} is checked in, and the new three points of concern are P_{14}-P_{15}-P_{16}. Since $|P_{16} - P_{15}| > |P_{15} - P_{14}|$, a cycle at "$P_{15}$-$P_{14}$" level is formed as

$$\tfrac{1}{2}\Delta I = |P_{15} - P_{14}|/2 = |40.5 - 63.4|/2 = 11.45^\circ C \quad \text{(Amplitude)}$$

$$T_{mean} = (P_{15} + P_{14})/2 = (40.5 + 63.4)/2 = 51.95^\circ C \quad \text{(Mean)}$$

(j) P_{14} and P_{15} are discarded. P_3 is connected to P_{14}. Now the three points of concern are P_3-P_{16}-P_{17}, as shown in **Figure 10.9.6(j)**. Since $|P_{17} - P_{16}| < |P_{16} - P_3|$, no cycle or reversal is identified. P_3 is neglected, but no more points to be checked in. Here, one may assume there are two reversals: one at P_3-P_{16} level and the other at P_{16}-P_{17} level, respectively, having

$$\tfrac{1}{2}\Delta T = |P_3 - P_{16}|/2 = |32.3 - 63.8|/2 = 15.75^\circ C \quad \text{(Amplitude)}$$

$$T_{mean} = (P_{15} + P_{14})/2 = (32.3 + 63.8)/2 = 48.05^\circ C \quad \text{(Mean)}$$

and $$\tfrac{1}{2}\Delta T = |P_{16} - P_{17}|/2 = |63.8 - 37.4|/2 = 13.2^\circ C \quad \text{(Amplitude)}$$

$$T_{mean} = (P_{16} + P_{17})/2 = (63.8 + 37.4)/2 = 50.6^\circ C \quad \text{(Mean)}$$

In summary, the rainflow counting in terms of temperature amplitude and mean temperature is given as follows:

Reversal	Amplitude	Mean
1	5.5	40.3
2	10.75	34.45
2	11	37.2
2	11.15	37.85
2	11.25	39.95
2	11.4	40.4
2	11.45	41.95
1	11.75	34.05
1	13.2	40.6
1	15.75	38.05

Accordingly, the number of reversals (1/2 cycles) allocated in terms of temperature amplitude and mean temperature is presented in **Figure 10.9.7**.

FIGURE 10.9.6 Rainflow counting of temperature fluctuations at an IGBT junction.

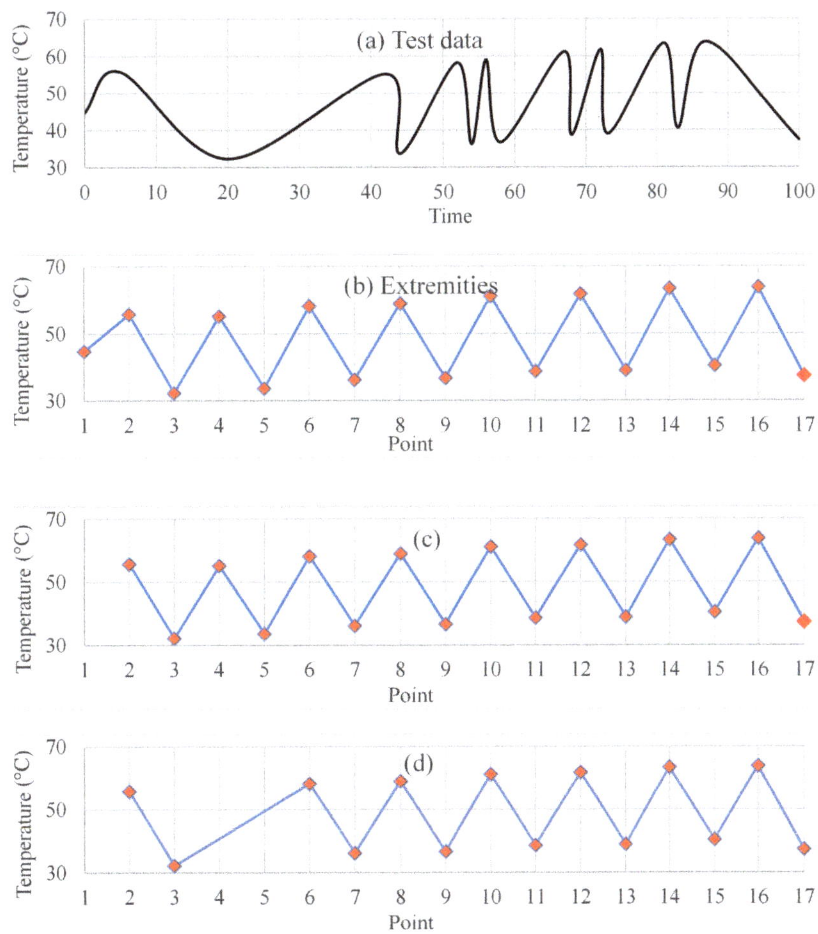

(Continued)

FIGURE 10.9.6 (Continued) Rainflow counting of temperature fluctuations at an IGBT junction.

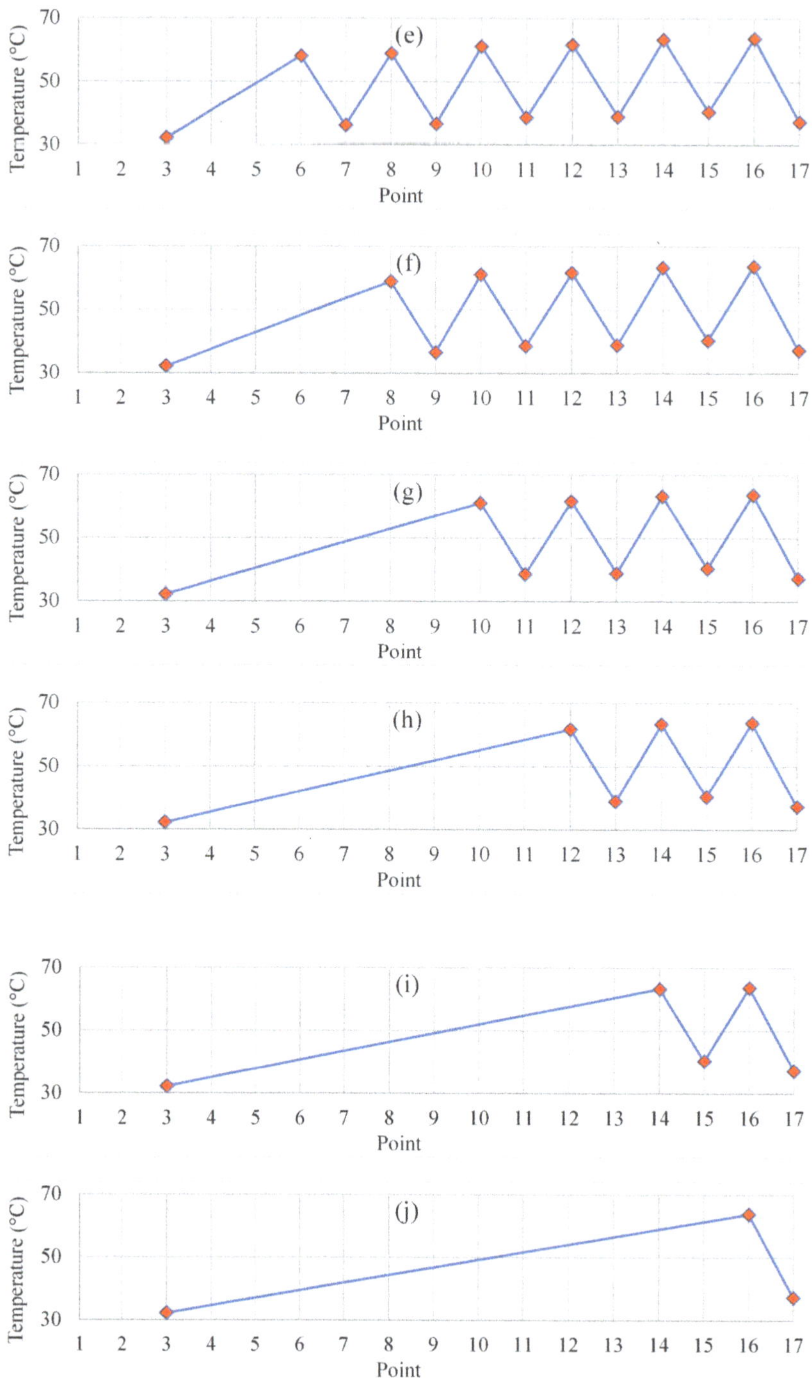

FIGURE 10.9.7 Number of reversals (1/2 cycles) as a function of temperature amplitude and mean temperature.

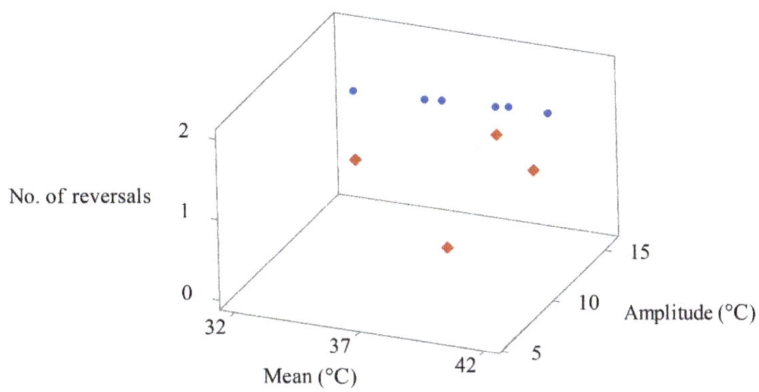

References

Alhadeed, A. A. and Yang, S. S. (2002), "Optimal Simple Step-Stress Plan for Khamis-Higgins Model," *IEEE Transactions on Reliability*, 51, pp. 212-215.

Altman, D. G. (1991), *Practical Statistics for Medical Research*, Chapman & Hall, London, UK.

AMSAA (2010), "The U.S. Army Material Systems Analysis Activity (AMSAA)," Aberdeen Proving Ground, MD.

Amzallag, C., Gerey, J., Robert, J., and Bahuaud, J. (1994), "Standardization of the Rainflow Counting Method for Fatigue Analysis," *Int'l Journal of Fatigue*, 16, pp. 287-293.

ASTM E1049-85 (Reapproved 2017), "Standard Practices for Cycle Counting in Fatigue Analysis," ASTM Int'l, West Conshohocken, PA.

Bae, S. J. and Kvam, P. H. (2004), "A Nonlinear Random Coefficients Model for Degradation Testing," *Technometrics*, 46(4), 460-469.

Bai, D., Kim, M., and Lee, S. (1989), "Optimum Simple Step-Stress Accelerated Life Tests with Censoring," *IEEE Transactions on Reliability*, 38(5), pp. 528-532.

Barringer, H. P. (2006), "Use Crow-AMSAA Reliability Growth Plots to Forecast Future System Failures," Humble, TX.

Charki, A. et al. (2011), "Robustness Evaluation Using Highly Accelerated Life Testing," *International Journal of Materials and Product Technology*, 56, pp. 1253-1261.

Chattopadhyay, S. and Nathan, R. (2013), "Illustrating Rotating Principal Stresses in a Materials Science Course," *120th ASEE Annual Conference and Exhibition*, June 23–26, 2013, Paper ID #7026.

Chiang, Y. J. (2019), *Mechanics and Design for Product Life Prediction*, Chongqing University Press, Chongqing, China; ISBN: 978-7-5689-1917-6.

Crow, L. H. (2011), "Planning a Reliability Growth Program Utilizing Historical Data," *2011 Proceedings of Annual Reliability and Maintainability Symposium*, Lake Buena Vista, FL, 2011, pp. 1-6.

Dharmadhikari, A. D. and Rahman, M. (2002), "Step-Stress Accelerated Life Testing for Weibull Distribution with Non-constant Shape Parameter," *Calcutta Statistical Association Bulletin*, 53, pp. 33-51.

Dowling, N. E. (2007), *Mechanical Behavior of Materials: Engineering Methods for Deformation, Fracture and Fatigue*, Pearson Prentice Hall, Upper Saddle River.

Duane, J. T. (1964), "Learning Curve Approach to Reliability Monitoring," *IEEE Transactions on Aerospace*, 2, pp. 563-566.

Elsayed, E. A. and Liao, H. T. (2004), “Reliability Prediction and Testing Plan Based on an Accelerated Degradation Rate Model,” *Int’l Journal of Material and Product Technology*, 21(5), pp. 402-422.

Escobar, L. A. and Meeker, W. Q. (2006), “A Review of Accelerated Test Models,” *Statistical Science*, 21, pp. 552-577.

Fard, N. and Li, C. (2009), “Optimal Simple Step Stress Accelerated Life Test Design for Reliability Prediction,” *Journal of Statistical Planning and Inference*, 239, pp. 1799-1808.

Fatigue Life LLC (Retrieved March 21, 2023), “Metal Fatigue Life Prediction,” fatigue-life.com/rainflow counting, Cheshire, CT.

Glinka, G. and Kam, J. (1987), “Rainflow Counting Algorithm for Very Long Stress Histories,” *Int’l Journal of Fatigue*, 9(4), pp. 223-228.

GopiReddy, L. et al. (2015), “Rainflow Algorithm Based Lifetime Estimation of Power Semiconductors in Utility Applications,” *IEEE Transactions on Industry Applications*, 51(4), pp. 3368-3375.

Guo, H., Mettas, A., Sarakakis, G., and Niu P. (2010), “Piecewise NHPP Models with Maximum Likelihood Estimation for Repairable Systems,” *2010 Proceedings of the Annual Reliability and Maintainability Symposium*, San Jose, April 15, 2010.

Han, D. and Ng, H. (2013), “Comparison between Constant-Stress and Step-Stress Accelerated Life Tests under Time Constraint,” *Naval Research Logistics*, 60(7), pp. 541-556.

Hauser, J. R. and Clausing, D. (1988), “The House of Quality,” *Harvard Business Review*, May–June, 1988, pp. 63-73.

Hobbs, G. K. (2000), *Accelerated Reliability Engineering: Halt and Hass*, Wiley Series in Quality and Reliability Engineering, Wiley, Hoboken.

Kaplan, E. L. and Meier, P. (1958), “Nonparametric Estimation from Incomplete Observations,” *Journal of the American Statistical Association*, 53, pp. 457-481.

Khamis, I. H. and Higgins, S. H. (1998), “A New Model for Step-Stress Testing,” *IEEE Transactions on Reliability*, 47, pp. 131-134.

Klyatis, L. M. (2012), *Accelerated Reliability and Durability Testing Technology*, John Wiley & Sons, New York.

Lawless, J. F. (2003), *Statistical Models and Methods for Lifetime Data*, 3rd Edition, John Wiley & Sons, New York.

Lee, Y. and Lu, M. (2006), “Damage-Based Models for Step-Stress Accelerated Life Testing,” *Journal of Testing and Evaluation*, 34(6), pp. 494-503.

Lee, J. and Pan, R. (2010), “Analyzing Step-Stress Accelerated Life Testing Data Using Generalized Linear Models,” *IIE Transactions*, 42, pp. 589-598.

Li, C. (2009), “Optimal Step-Stress Plans for Accelerated Life Testing Considering Reliability/Life Prediction,” PhD dissertation, Northeastern University, Boston.

Lu, M., Leiman, J. E., Rudy, R. J., and Lee, Y. (2003), “Step-Stress Accelerated Life Test Method - A Variational Study,” *SAE Technical Paper 2003-01-0470*, https://doi.org/10.4271/2003-01-0470.

Ma, H. and Meeker, W. Q. (2008), “Optimal Step-Stress Accelerated Life Test Plans for Log-Location-Scale Distributions,” *Naval Research Logistics*, 55, pp. 551-562.

Matsuishi, M. and Endo, T. (1968), “Fatigue of Metals Subjected to Varying Stress,” Paper Presented to *Japanese Society of Mechanical Engineering*, Fukuoka, Japan.

McInners, C. and Meehan, P. (2008), “Equivalence of Four-Point and Three-Point Rainflow Cycle Counting Algorithm,” *Int’l Journal of Fatigue*, 30(3), pp. 547-559.

McLean, H. (2009), *HALT, HASS, and HASA Explained*, ASQ Quality Press, Milwaukee, WI.

Meeker, W. Q., Sarakakis, G., and Gerokostopoulos, A. (2013), “More Pitfalls of Accelerated Tests,” *Journal of Quality Technology*, 45, pp. 1-11.

MIL-HDBK-344A (August 16, 1993), “Environmental Stress Screening of Electronic Equipment,” Published by the Institute of Environmental Sciences and Technology, USAF Rome Laboratory.

Miner, M. A. (1945), "Cumulative Damage in Fatigue," *Journal of Applied Mechanics*, 12, pp. A159-A164.

Mohammadiam, H. S. and Aït-Kadi, D. (2010), "Design Stage Confirmation of Lifetime Improvement for Newly Modified Products through Accelerated Life Testing," *Reliability Engineering and System Safety*, 95, pp. 897-905.

Monroe, E. and Pan, R. (2008), "Experimental Design Considerations for Accelerated Life Tests for Nonlinear Constraints and Censoring," *Journal of Quality Technology*, 40(4), pp. 355-367.

Morrow, J. (1965), "Cyclic Plastic Strain Energy and Fatigue of Metals," Internal Friction, Damping, and Cyclic Plasticity, ASTM STP 378.

Nagode, M. and Hack, M. (2004), "An Online Algorithm for Temperature Influenced Fatigue Life Estimation: Stress-Life Approach," *International Journal of Fatigue*, 26, pp. 163-171.

National Academies of Sciences, Engineering, and Medicine (2002), *Reliability Issues for DOD Systems: Report of a Workshop*, The National Academies Press, Washington, DC, https://doi.org/10.17226/10561.

Nelson, B. L. (2013), "Experiment Design and Analysis," *Foundations and Methods of Stochastic Simulation*, Springer, Boston, MA; On-line ISBN: 978-1-4614-6160-9.

Nelson, W. B. (1980), "Accelerated Life Testing-Step-Stress Models and Data Analysis," *IEEE Transactions on Reliability*, R-29, pp. 103-108.

Nelson, W. B. (2005), "A Bibliography of Accelerated Test Plans," *IEEE Transactions on Reliability*, 54, pp. 194-197.

Nicola, L. (2013), "Lifetime Estimation of IGBT Power Modules," Master thesis, Aalborg University, Aalborg.

Peiskammer, D., Dannbauer, H., and Dutzler, E. (2003), "Using Fatigue Life Prediction to Increase Reliability of Automotive Structures," *SAE Technical Paper 2003-01-0471*, https://doi.org/10.4271/2003-01-0471.

Peterson, A. V., Jr. (1977), "Expressing the Kaplan-Meier Estimator as a Function of Empirical Subsurvival Functions," *Journal of the American Statistical Association*, 72, pp. 854-858.

Peto, J. (1984), "The Calculation and Interpretation of Survival Curves," in *Cancer Clinical Trials. Methods and Practice*, Edited by Buyse, M.E., Staquet, M.J., and Sylvester, R.J., Oxford Medical Publications, Oxford University Press, Oxford, pp. 361-380.

Putra, T. E. and Ikibal, H. M. (2021), "Automotive Suspension Component Behaviors Driven on Flat and Rough Road Surfaces," *Heliyon*, 7(7), e07528.

Raheja, D. and Allocco, M. (2006), *Assurance Technologies, Principles and Practices*, 2nd Edition, Wiley, Hoboken.

Ramberg, W. and Osgood, W. R. (1943), "Description of Stress-Strain Curves by Three Parameters," Technical Note No. 902, National Advisory Committee for Aeronautics, Washington, DC.

Rigdon, S. E., Pan, R., Montgomery, D. C., and Borror, C. M. (2022), *Experiments for Reliability Achievement*, Wiley, Hoboken, 416 pages; ISBN: 978-1119237693.

Rychlik, I. (1987), "A New Definition of the Rainflow Cycle Counting Method," *Int'l Journal of Fatigue*, 9(2), pp. 119-121.

Schijve, J. (2005), "Statistical Distribution Functions and Fatigue of Structures," *Int'l Journal of Fatigue*, 27, pp. 1031-1039.

Schlitz, D., Klitschke, H., Steinhilber, H., Heuler, P., and Schfitz, W. (1990), "Standardized Load Sequences for Car Wheel Suspension Components - Car Loading Standard CARLOS," LBF Report No. FB-191, 1990/IABG Report No. TF-2695, LBF/IABG, Germany.

Schubert, R. (Retrieved December 31, 2018), "Overview of Reliability Functions for Life Testing in Minitab," Legendary Performance, Sure, Inc.

Shen, K., Shen, Y., and Leu, L. (2011), "Design of Optimal Step-Stress Accelerated Life Tests under Progressive Type I Censoring with Random Removals," *Quality & Quantity*, 45(3), pp. 587-597.

Skibicki, D (2014), *Phenomena and Computational Models for Nonproportional Fatigue of Materials*, Springer, Cham; ISBN: 978-3-319-01564-4.

Skibicki, D. and Dymski, S. (2010), "The Influence of Fatigue Loading on the Microstructure of an Austenitic Stainless Steel," *Materials Testing*, 52(11), pp. 787-794.

Sun, N., Chiang, Y. J., Li, G., and Wang, Y. (2014), "A Comparative Study of Fatigue Criteria for Engineering Application," *Applied Mechanics & Materials*, 538, pp. 356-359.

Tang, Y., Guan, Q., Xu, P., and Xu, H. (2012), "Optimum Design for Type-I Step-Stress Accelerated Life Tests of Two-Parameter Weibull Distributions," *Communications in Statistics - Theory and Methods*, 41(21), pp. 3863-3877.

Tseng, S. T. and Wen, Z. C. (2000), "Step-Stress Accelerated Degradation Analysis for Highly Reliable Products," *Journal of Quality Technology*, 32, pp. 209-216.

Wu, Z., Li, J., and Bai, C. (2017), "Scaling Relations of Lognormal Type Growth Process with an Extremal Principle of Entropy," *Entropy*, 19(56), pp. 1-14.

Xiong, C. (2003), "Step-Stress Accelerated Life Testing," *Handbook of Reliability Engineering*, Edited by Pham, H., Springer, Berlin, pp. 457-469; ISBN: 1-85233-453-3.

Zhao, W. and Elsayed, E. A. (2005), "A General Acceleration Life Model for Step-Stress Testing," *IIE Transactions*, 37(11), pp. 1059-1069.

Problems

P10.1: Please use the four-point algorithm to allocate the time series in Figure 10.9.2(a) into five load blocks [as shown in Figure 10.9.2(b)], according to the strain range and mean strain.

P10.2: Please use the four-point algorithm to allocate the strain data in time series in Figure 10.9.3 in terms of the strain range and mean strain.

11

Inferential Mechanisms

Various natural laws have sketched out work tasks for identifying the meaning or guiding the statistical inferences with the goal of using the information expressed by experimental design. Different approaches to the problems of building a DOEs to perform such a tactic constitute the central theme of problem-solving. Many applications in science, engineering, and business operation make extensive use of complex simulations describing the dynamics of the physical transformation being investigated. Inferential mechanisms, whether they are formalized or not, possess various methodological virtues.

11.1. Drawing Inferences from Natural Instances

Each test plan in the laboratory or on the factory floor is to be implemented in accordance with the related inferential mechanisms by nature. Inferential mechanisms characterize how the material transforms and its changing process in response to the working condition. Such models are derived from knowledge of the mechanisms and principles underlying the data-generating process and are of critical importance for scientific hypothesis building and testing. For example, fatigue life cycles of automotive components in operation follow the physics of transformation that are often not normally distributed with respect to the related design parameters. Furthermore, the curing of automotive plastics and rubber obeys their respective kinematic laws. Linking complex physical models to empirical measurements can be challenging, since many models are defined implicitly through stochastic simulators. If perception is to be thought of as some sort of thinking, its processes must necessarily first include transformations of states that are expressed in symbolic or propositional forms, such as polynomial conversion, Arrhenius model, Eyring model, photovoltaics (PVs), crack propagation, oxidation, creep of materials, mechanical fatigue, and even automobile ride.

11.2. Polynomial Conversion

This is the most straightforward inferential mechanism. The transformation is generally represented by a polynomial in terms of the applied load level as [Lee and Pan 2010]

$$C(S) = C_0 + C_1 S + C_2 S^2 + C_3 S^3 + \ldots \quad (11.2.1)$$

where

C(S) is the conversion characteristics, e.g., fatigue damage and material degradation
S is the load factor of interest, e.g., stress, strain, strain energy, temperature, and humidity
C_0 is the initial conversion (damage) at S = 0
C_1, C_2, and C_3 are the coefficients

Coefficients given above are deterministically constant for each identifiable unit by the law of physics, but they may vary from one unit to another randomly as they are manufactured. The first-order and second-order damage models are more practical for most engineering applications, respectively,

$$C(S) = C + C_1 S \quad (11.2.2)$$

$$\text{and} \quad C(S) = C_0 + C_1 S + C_2 S^2 \quad (11.2.3)$$

The corresponding conversion rates, also called hazard rate, can be obtained from Equation (11.2.1) as

$$h(S_0) = d[C(S_0)] / dS_0 = C_1 + 2 C_2 S + 3 C_3 S^2 + \ldots \quad (11.2.4)$$

The acceleration factor at the qth load step relative to the state at the very beginning is

$$A_f = \frac{h_q}{h_0} = \frac{C_1 + 2 C_2 S_q + 3 C_3 S_q^2 + \ldots}{C_1 + 2 C_2 S_0 + 3 C_3 S_0^2 + \ldots} \quad (11.2.5)$$

The typical design methodology with a polynomial conversion of damage subject to one load factor can be demonstrated using the drop test, as given in Example 10.5.1.

11.3. Arrhenius Model

The Arrhenius equation is employed to characterize the relationship between the rate constant (of a chemical reaction) and the absolute temperature since 1895 [Arrhenius 1895]. Temperature and humidity are two important environmental factors that have a great impact on material degradation, especially for electronic components.

11.3.1. Arrhenius Model for Temperature

The Arrhenius model is originally developed to describe the behavior of conversion as a function of the applied temperature in chemistry [Arrhenius 1895]. The conversion rate (hazard rate), which is defined as the transformation rate divided by the balance of the survivor portion as described by Equation (8.1.1), follows the Arrhenius model in chemical reaction as a function of temperature. It is written as

$$h(T) = A \exp\left(\frac{-E_a}{KT}\right) = A \exp\left(\frac{B}{T}\right) \tag{11.3.1}$$

$$K = 8.617385 \times 10^{-5}\ \mathrm{eV/^{\circ}K} \tag{11.3.2}$$

and

$$B = -E_a / K \tag{11.3.3}$$

where

- T (°K) is the working temperature (Kelvin scale), and 273.16°K = 0°C
- A is the pre-exponential factor, also called frequency factor
- E_a is the activation energy in eV (electron volts), which is material-dependent; $E_a > 0$
- K is Boltzmann's constant
- B is the material-dependent constant, e.g., which is a negative number

Pre-exponential factor A is often taken as constant across a small temperature range. Note that the inverse of hazard rate is the MTTF, where the failure refers to the state of full conversion. When logarithms are taken on both sides of Equation (11.3.1), the Arrhenius equation turns into

$$\ln[h(T)] = \ln(A) - (E_a / K) / T \tag{11.3.4}$$

Assume that ln(A) and E_a/K are constants for a given material. Then, the plot of ln[h(T)] versus 1/T yields a straight line. It is called an Arrhenius plot.

The acceleration factor at the qth load step relative to the state at the very beginning (assumed to be at normal operating temperature, i.e., the reference temperature) can be calculated as

$$\Lambda_q = \frac{h_q}{h_0} = \exp\left[\left(\frac{-E_a}{K}\right)\left(\frac{1}{T_q} - \frac{1}{T_0}\right)\right] \tag{11.3.5}$$

where

- Λ_q is the hazard rate ratio, i.e., acceleration factor
- T_0 is the reference temperature, e.g., the normal operating temperature
- T_q is the qth temperature level (load level)

The Arrhenius equation is simple but profoundly accurate for most temperature-dependent inferential mechanisms such as chemical reactions and material degradations. Empirical values of activation energy (E_A) in different failure modes (inferential mechanisms) of electronic components and the corresponding acceleration factors (Λ_f) are listed in Table 11.3.1.

TABLE 11.3.1 Potential acceleration factors of electronics degradation at elevated temperatures (85–75°C) in reference to the state at 25°C [Lakshminarayanan 2001].

Mechanism	E_A (eV)	Λ_f-85	Λ_f-105	Λ_f-125	Λ_f-150	Λ_f-175
Intermetallic	0.3	7	12	19	32	50
Corrosion	0.45	19	41	82	178	355
Metallization	0.5	26	62	134	317	682
Silicon (bulk)	Up to 0.5	26	62	134	317	682
Gate-oxide defect	Up to 0.5	26	62	134	317	682
Electromigration	0.6 (A1)	50	141	356	1004	2515
Electromigration	0.9 (Contact)	357	1676	6723	31,825	126,121
Photoresist (mask)	0.7	97	322	948	3178	9274
Ionic contamination	1.0	687	3824	17,901	10×10^4	47×10^4
Charge injection	1.3	4874	45,420	34×10^4	32×10^5	23×10^6
Al-Si dissolution	1.7	66,463	12×10^5	17×10^6	32×10^7	43×10^8

EXAMPLE 11.3.1

Accelerated life tests were conducted on an electronic component. The MTTF of an electronic component is 1000 h at 125°C and 600 h at 175°C, respectively. What are the MTTFs of this component at 23°C (normal operating room temperature) and at 40°C (factory floor with no air conditioning), respectively?

Solution:

By Equation (11.3.1), one has the hazard rate ratio at both elevated temperatures as

$$h_1 / h_2 = \exp\left[\frac{-E_a}{K\,(125+273)}\right] / \exp\left[\frac{-E_a}{K\,(175+273)}\right] = \exp\left[\frac{-E_a}{K}\left(\frac{1}{448} - \frac{1}{398}\right)\right]$$

Since $MTTF_1/MTTF_2 = h_1/h_2$, one has $\ln(MTTF_1/MTTF_2) = \ln(h_1/h_2)$, which leads to

$$\ln(1000/600) = (-E_a/K)\left[448^{-1} - 398^{-1}\right]$$

i.e., $E_a/K = 1821.645$

Thus, $E_a = 1821.645\,K = 1821.645\,(8.617 \times 10^{-5}) = 0.157$ eV.

The MTTF at the room temperature (23°C), referring to $MTTF_0$, can be calculated as

$$MTTF_0 / MTTF_1 = h_0 / h_1 = \exp\left[\frac{-E_a}{K}\left(\frac{1}{398} - \frac{1}{296}\right)\right] = 4.8414$$

$$MTTF_0 = 4.8414\ MTTF_1 = 4814.4 \text{ hours}$$

The MTTF at the elevated temperature (50°C), i.e., $MTTF_{50}$, can be then calculated as

$$MTTF_{50} / MTTF_1 = h_{50} / h_1 = \exp\left[\frac{-E_a}{K}\left(\frac{1}{398} - \frac{1}{323}\right)\right] = 2.8944$$

$$MTTF_{50} = 2.8944\ MTTF_1 = 2894.4 \text{ hours}$$

Here is an example referring to the stepwise temperature test of 32 light-emitting diodes (LEDs) with the same inferential mechanisms given in Example 11.3.1. The reliability of LEDs that work as an optical source is dependent on the degradation mode and device characteristics such as the current versus optical output power characteristic and its operating temperature. At higher current densities, voltages, or temperatures, rapid power reduction is due to dark spot/line defect generation [Chiao and Hamada 1996, Fukuda et al. 1983]. Assume that the underlying life distribution follows the Weibull statistics and that the Arrhenius life–load relationship holds as the temperature varies.

EXAMPLE 11.3.2

Accelerated stepwise temperature tests of 32 LEDs are done by 4^1 factorial design [Elsayed and Liao 2003, 2004, Zhao and Elsayed 2005], referring to Example 11.3.1. These 32 LEDs are given as follows:

Run	Time (h)	Temperature (°K)	Y_1	Y_2	Y_3	Y_4	Y_5	Y_6	Y_7	Y_8
1	300	363	300^+	300	300	300	300	300	300	300
2	200	413	—	47	97	132	191	200^+	200^+	200
3	100	433	—	—	—	—	—	—	—	12
4	120	448	—	—	—	—	—	—	—	—

Run	Time (h)	Temperature (°K)	Y_9	Y_{10}	Y_{11}	Y_{12}	Y_{13}	Y_{14}	Y_{15}	Y_{16}
1	300	363	300	300	300	300	300	300	300	300
2	200	413	200	200	200	200	200	200	200	200
3	100	433	67	74	88	97	100^+	100^+	100	100
4	120	448	—	—	—	—	—	—	3	5

Run	Time (h)	Temperature (°K)	Y_{17}	Y_{18}	Y_{19}	Y_{20}	Y_{21}	Y_{22}	Y_{23}	Y_{24}
1	300	363	300	300	300	300	300	300	300	300
2	200	413	200	200	200	200	200	200	200	200
3	100	433	100	100	100	100	100	100	100	100
4	120	448	15	33	34	37	44^+	53^+	75	84

Run	Time (h)	Temperature (°K)	Y_{25}	Y_{26}	Y_{27}	Y_{28}	Y_{29}	Y_{30}	Y_{31}	Y_{32}
1	300	363	300	300	300	300	300	300	300	300
2	200	413	200	200	200	200	200	200	200	200
3	100	433	100	100	100	100	100	100	100	100
4	120	448	99	106	118	120	120^+	120^+	120^+	120^+

Note that "superscript +" means "censored." Please identify the past hours (equivalent hours) at 23°C.

Solution:

The acceleration factor pertaining to arbitrary temperature T_q (°K) (where q = 1, 2, 3, 4) is

$$\Lambda_q = \frac{h_q}{h_0} = \exp\left[\left(\frac{-E_a}{K}\right)\left(\frac{1}{T_q} - \frac{1}{T_0}\right)\right] = \exp[-1821.645\,(T_q^{-1} - 298^{-1})]$$

Thus, $\Lambda_1 = 2.988$ at $T_1 = 363°K$ (90°C)
$\Lambda_3 = 5.486$ at $T_2 = 413°K$ (140°C)
$\Lambda_3 = 6.725$ at $T_3 = 433°K$ (160°C)
and $\Lambda_4 = 7.743$ at $T_4 = 448°K$ (175°C)

The equivalent hours for testing at 296°K (23°C) can be calculated as follows:

$$Y_{1,\ equivalent} = 300^{+}\ \Lambda_1 = 300^{+} * 2.988 = 896.4^{+} \text{ hours (Censored)}$$

$$Y_{2,\ equivalent} = 300\ \Lambda_1 + 47\ \Lambda_2 = 300 * 2.988 + 47 * 5.486 = 1154.242 \text{ hours (Failure)}$$

...

$$Y_{28,\ equivalent} = 300\ \Lambda_1 + 200\ \Lambda_2 + 100\ \Lambda_3 + 120\ \Lambda_4$$
$$= 300 * 2.988 + 200 * 5.486 + 100 * 6.725 + 120 * 7.743$$
$$= 3595.26 \text{ hours (Failure)}$$

...

$$Y_{32,\ equivalent} = 300\ \Lambda_1 + 200\ \Lambda_2 + 100\ \Lambda_3 + 120\ \Lambda_4$$
$$= 300 * 2.988 + 200 * 5.486 + 100 * 6.725 + 120^{+} * 7.743$$
$$= 3595.26^{+} \text{ hours (Censored)}$$

In summary,

$$(Y_1, Y_2, \ldots, Y_{32}) = \begin{pmatrix} 896.4^{+}, 1154.2, 1428.5, 1620.6, 1944.2, 1993.6^{+}, \\ 1993.6^{+}, 2074.3, 2444.2, 2491.3, 2585.4, 2645.9, 2666.1^{+}, 2666.1^{+}, 2689.3, 2704.8, \\ 2782.2, 2921.6, 2929.4, 2952.6, 3006.8^{+}, 3076.5^{+}, 3246.8, 3316.5, \\ 3432.7, 3486.9, 3579.8, 3595.3, 3595.3^{+}, 3595.3^{+}, 3595.3^{+}, 3595.3^{+} \end{pmatrix}$$

This is a case study on test to failure with arbitrary censoring. The statistical significance can be figured out using Minitab as

Stat ➔ Reliability/Survival.
➔ Distribution Analysis (Right Censoring).
➔ Parametric Distribution Analysis.
➔ In Variables, enter C1 (*test data, failure and right censored*).
In Assumed distribution, choose Weibull.
Click Censor.
Choose Use censoring columns, and enter C2 (*e.g., value 1 or 0*).
Enter the value from C2 that indicates a censored observation (*e.g., 0*).
Click OK. The plot is shown in **Figure 11.3.1**.

It is shown in **Figure 11.3.1** that all 21 data (failure) fit the two-parameter Weibull distribution (shape factor $\beta = 4.38922$ and scale factor $\eta = 3271.41$ h) well with a CI of 95%. Note that only the failure data can be plotted.

FIGURE 11.3.1 Probability plot for accelerated stepwise temperature tests of LEDs.

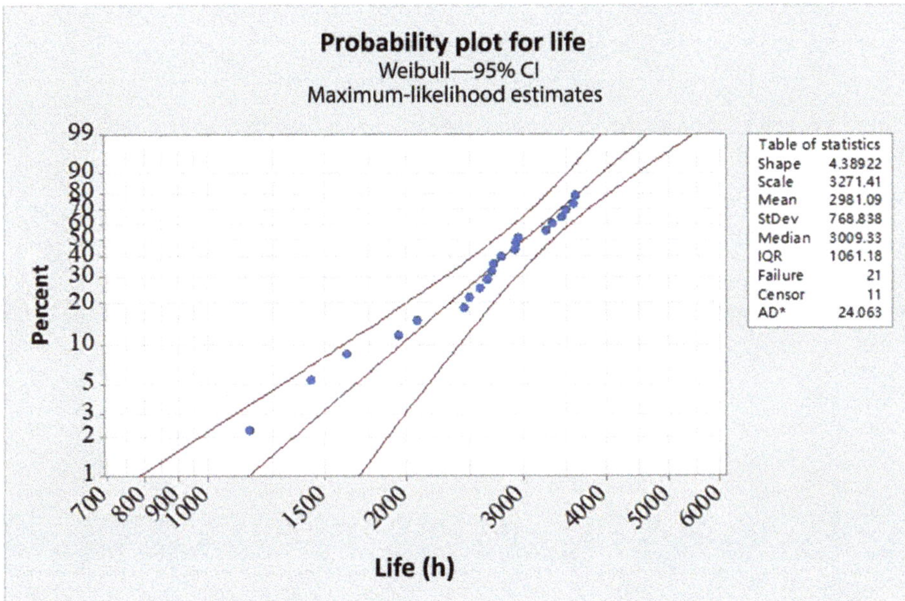

11.3.2. Arrhenius Model for Electric Voltage

When the temperature is generalized to electric voltage, namely, V, the Arrhenius model for the hazard rate, or called conversion rate, becomes

$$h(V) = A \exp\left(\frac{B}{V}\right) \tag{11.3.6}$$

$$\text{Thus,} \quad \ln[h(V)] = \ln(A) + B / V \tag{11.3.7}$$

If an equation expressed as Equation (11.3.7) fitted with test data does not yield a straight line, the modeling using the Arrhenius relationship is not valid. For a stepwise acceleration test, acceleration factor Λ_q can be represented, respectively, as

$$\Lambda_q = \frac{h_q}{h_0} = \exp\left[B\left(\frac{1}{V_q} - \frac{1}{V_0}\right)\right] \tag{11.3.8}$$

where

Λ_q is the acceleration factor corresponding to voltage level q

h_q is the hazard rate at voltage level q

h_0 is the hazard rate at voltage level 0, i.e., normal operating load (humidity) level

V_q (V) is the applied voltage level

V_0 (V) is the reference voltage level, usually taken at the operating temperature, i.e., level 0

The hazard rate defined above can be used for calculating the MTTF. If the applied voltage is a smoothly continuous function, instead of a step function, the acceleration factor becomes

$$\Lambda_q = \frac{1}{t_{q2} - t_{q1}} \int^{t_{q2}} \exp\left[B\left(\frac{1}{V_0} - \frac{1}{V_q}\right)\right] dt \qquad (11.3.9)$$

of which t_{q1} and t_{q2} are the starting time and end time of the continuous-varying "voltage level q" of interest, respectively.

EXAMPLE 11.3.3

Accelerated stepwise voltage testing is planned for quantifying the reliability of the newly proposed insulating fluid for transformers, based on factorial design 3^1 of which the load factor is voltage (25, 30, and 40 kV). Time to breakdown (seconds) of the insulating fluid recorded at three different voltages on 12 transformers is given as follows [DeSourza and Rocha 2012]:

V (kV)	Y_1	Y_2	Y_3	Y_4	Y_5	Y_6	Y_7	Y_8	Y_9	Y_{10}	Y_{11}	Y_{12}
25	2500	4056	12,553	40,290	$40{,}290^+$	$40{,}290^+$	$40{,}290^+$	$40{,}290^+$	$40{,}290^+$	$40{,}290^+$	$40{,}290^+$	$40{,}290^+$
30	50	134	187	882	1448	1468	2290	4138	15,750	$15{,}750^+$	$15{,}750^+$	$15{,}750^+$
40	1	1	1	3	12	25	46	56	68	109	323	417

Note that "superscript +" means "censored." Please identify the past hours (equivalent hours) at 25 kV, i.e., the normal operating voltage.

Solution:

Assume that the two-parameter Weibull distribution is a good fit for the entire process. The MTTF for each applied voltage can be obtained from Equation (9.3.3) as

$$E(t_{qn}) = \eta_q \Gamma(1 + 1/\beta_q) \quad \text{where } q = 1,2,3; n = 1,2,\ldots,12 \qquad (a)$$

Fitting each data set of individual voltages (treatment) using Minitab (→ Reliability/Survival → Distribution Analysis (Right Censoring → Parametric Distribution Analysis), one can obtain two parameters of its corresponding Weibull distribution function with 95% CI as

$$25\,\text{kV}: \quad \beta_0 = 0.703668 \quad \text{and} \quad \eta_0 = 143{,}554 \qquad (b)$$

$$30\,\text{kV}: \quad \beta_1 = 0.487358 \quad \text{and} \quad \eta_1 = 6991.76 \qquad (c)$$

$$40\,\text{kV}: \quad \beta_2 = 0.551509 \quad \text{and} \quad \eta_2 = 53.5049 \qquad (d)$$

Since the hazard rate is the inverse of MTTF, represented by Equation (a), the acceleration factors for 30 kV and 40 kV are, respectively,

$$\Lambda_1 = \frac{\eta_0\,\Gamma(1 + 1/\beta_0)}{\eta_1\,\Gamma(1 + 1/\beta_1)} = \frac{143554\ \Gamma(1 + 1/0.703668)}{6991.76\,\Gamma(1 + 1/0.487358)} = \frac{180815.6}{14676.994} = 12.32$$

and

$$\Lambda_2 = \frac{\eta_0\,\Gamma(1 + 1/\beta_0)}{\eta_2\,\Gamma(1 + 1/\beta_2)} = \frac{143554\ \Gamma(1 + 1/0.703668)}{53.5049\,\Gamma(1 + 1/0.551509)} = \frac{180815.6}{90.70518} = 1993.4$$

The equivalent hours for testing at the load step of 25 kV (normal operating load) can be calculated as follows:

$$Y_{1,\text{equivalent}} = 2500 + 50(12.32) + 1(1993.4) = 5109\ \text{sec} \quad (\text{Failure})$$

$$Y_{2,\text{equivalent}} = 4056 + 134(12.32) + 1(1993.4) = 7700\ \text{sec} \quad (\text{Failure})$$

$$Y_{3,\text{equivalent}} = 12{,}553 + 187(12.32) + 1(1993.4) = 16{,}850\ \text{sec} \quad (\text{Failure})$$

$$Y_{4,\text{equivalent}} = 40{,}290 + 882(12.32) + 3(1993.4) = 57{,}137\ \text{sec} \quad (\text{Failure})$$

$$Y_{5,\text{equivalent}} = 40{,}290^{+} + 1448(12.32) + 12(1993.4) = 82{,}051^{+}\ \text{sec} \quad (\text{Censored})$$

$$Y_{6,\text{equivalent}} = 40{,}290^{+} + 1468(12.32) + 25(1993.4) = 108{,}212^{+}\ \text{sec} \quad (\text{Censored})$$

$$Y_{7,\text{equivalent}} = 40{,}290^{+} + 2290(12.32) + 46(1993.4) = 160{,}201^{+}\ \text{sec} \quad (\text{Censored})$$

$$Y_{8,\text{equivalent}} = 40{,}290^{+} + 4138(12.32) + 56(1993.4) = 202{,}903^{+}\ \text{sec} \quad (\text{Censored})$$

$$Y_{9,\text{equivalent}} = 40{,}290^{+} + 15{,}750(12.32) + 68(1993.4) = 369{,}884^{+}\ \text{sec} \quad (\text{Censored})$$

$$Y_{10,\text{equivalent}} = 40{,}290^{+} + 15{,}750(12.32) + 109(1993.4) = 451{,}615^{+}\ \text{sec} \quad (\text{Censored})$$

$$Y_{11,\text{equivalent}} = 40,290^{+} + 15,750(12.32) + 323(1993.4) = 878,212^{+} \text{ sec} \quad (\text{Censored})$$

and $$Y_{12,\text{equivalent}} = 40,290^{+} + 15,750(12.32) + 417(1993.4) = 1065,595^{+} \text{ sec} \quad (\text{Censored})$$

This is a case study on test to failure with arbitrary censoring. The statistical significance can be figured out using Minitab as

Stat → Reliability/Survival.
→ Distribution Analysis (Right Censoring).
→ Parametric Distribution Analysis.
→ In Variables, enter C1 (*test data, failure and right censored*).
In Assumed distribution, choose Weibull.
Click Censor.
Choose Use censoring columns, and enter C2 (*e.g., value 1 or 0*).
Enter the value from C2 that indicates a censored observation (*e.g., 0*).
Click OK. The plot is shown in Figure 11.3.2.

It is shown in Figure 11.3.2 that all 21 data (failure) fit the two-parameter Weibull distribution (shape factor β = 0.350745 and scale factor η = 3,744,709 sec) well with a CI of 95%. It will run 6123.51 sec with R = 90% and C = 95%. Note that only the failure data can be plotted. Since the shape parameters are not close to one, exponential distribution cannot represent well the MTTF (time to breakdown of insulating fluid) between electrodes.

FIGURE 11.3.2 Probability plot for accelerated stepwise voltage tests of insulating fluid of transformer.

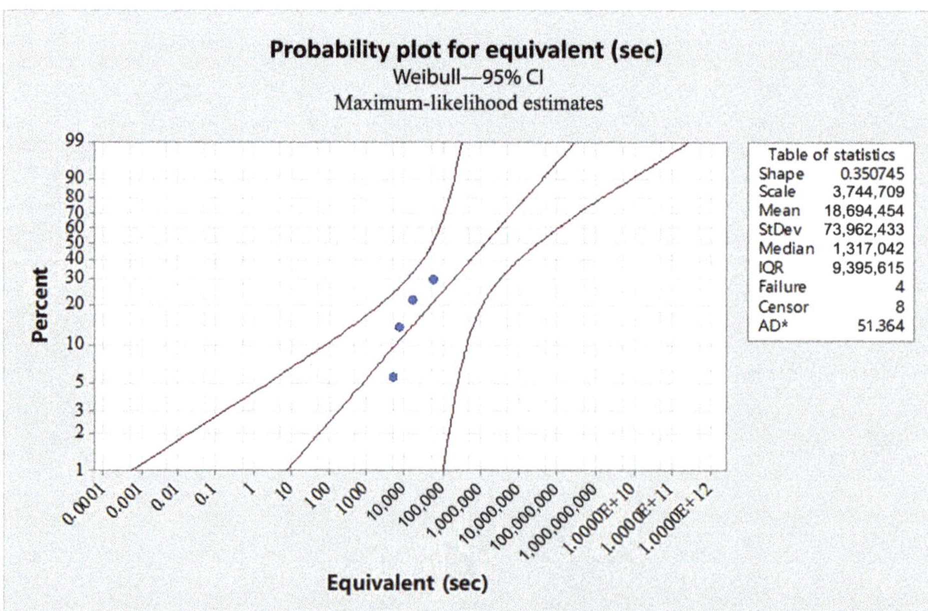

11.3.3. Arrhenius Model for Temperature and Humidity

Humidity is often a viable acceleration factor, by which the conversion rate varies with respect to the humidity change inversely like "temperature" by nature.

$$h(H_R) \propto \exp(H_R^{-1}) \tag{11.3.6b}$$

of which H_R is the relative humidity under the nominal working condition. The relative humidity (H_R) usually work together interactively with the temperature variation. Combining humidity with temperature accelerates material transformation (e.g., degradation), and the Arrhenius equation is used to characterize the effect in the long run. Combining Equations (11.3.1) and (11.3.6) yields

$$h(T, H_R) = A \exp\left(\frac{-E_a}{K\,T}\right) \exp\left(\frac{C}{H_R}\right) = A \exp\left(\frac{-E_a}{K\,T} + \frac{C}{H_R}\right) = A \exp\left(\frac{B}{T} + \frac{C}{H_R}\right) \tag{11.3.7b}$$

The acceleration factor at the qth load step relative to the state at the normal operating load step is

$$\Lambda_f = \frac{h_q(T, H_R)}{h_0(T_0, H_{R0})} = \exp\left[B\left(\frac{1}{T_q} - \frac{1}{T_0}\right) + C\left(\frac{1}{H_{R0}} - \frac{1}{H_{Rq}}\right)\right] \tag{11.3.8b}$$

where

H_{R0} is the reference relative humidity, e.g., the operating relative humidity
H_{Rq} is the relative humidity at the qth load step
A is the a function of time, e.g., Weibull or lognormal distribution or just a constant
B is the coefficient of temperature; note that $B = -E_a/K$, which is a negative number
C is the coefficient of humidity; C is a positive number

For an environment with continuous-varying humidity and temperature [Benavides 2011], the above equation can be rewritten as

$$\Lambda_f = \frac{1}{t_{q2} - t_{q1}} \int^{t_{q2}} A \exp\left[B\left(\frac{1}{T_0} - \frac{1}{T(t)}\right) + C\left(\frac{1}{H_{R0}} - \frac{1}{H_R(t)}\right)\right] dt \tag{11.3.9b}$$

The above equation is valid as long as the applied conditions are kept the same or approximately unchanged over the time duration ($t_{q1} \le t \le t_{q2}$).

When selecting temperature and humidity levels of the environmental air, it must be realized that humidity can only be controlled up to approximately 98°C as water will boil around 100°C or below 5°C as water will freeze. As a variation to Equation (11.3.8), the acceleration factor at the qth load step relative to the state at the normal operating load step is [Hallberg and Peck 1991, Hardikar et al. 2013]

$$\Lambda_f = \frac{h_q(T, H_R)}{h_0(T_0, H_{R0})} = \left(\frac{H_{R0}}{H_{Rq}}\right)^{-m} \exp\left[\frac{-E_a}{K}\left(\frac{1}{T_q} - \frac{1}{T_0}\right)\right] \tag{11.3.10}$$

Exponent $-m$ is a negative number and usually $m \ge 1$, although the exact value depends on the applied material. Equation (11.3.10) is also called the Arrhenius–Peck model [Peck 1986].

11.3.4. Arrhenius–Coffin–Manson's Lifetime Model

An extended application of the Arrhenius equation to fatigue life prediction subjected to temperature variations is given as follows [Held et al. 1997]:

$$2N_f(\Delta T, T_m) = A\,(\Delta T)^{-n} \exp\left[\left(\frac{-E_a}{K\,T_m}\right)\right] \tag{11.3.11}$$

where

$2N_f$ is the reversal to cyclic fatigue failure
ΔT (°K) is the temperature range as the load factor
T_m (°K) is the mean temperature
n is the exponent, usually positive

The above equation is called the Arrhenius–Coffin–Manson's lifetime model, by which the fatigue life is a function of temperature range and mean temperature. The dependence on the mean temperature follows the Arrhenius model.

The Arrhenius–Coffin–Manson's lifetime model has been extensively applied for power cycling tests of IGBT modules in electronic devices [Bouarroudj et al. 2007, Smet et al. 2011]. The related total fatigue damage (namely D) of an IGBT of interest because of the linear accumulation of the fatigue damage subjected to cyclic temperature variation in a load block (ΔD_i) is given as [Bannantine et al. 1990]

$$D = \sum_{i=1} \Delta D_i = \sum_{i=1} \frac{2N_i(\Delta T, T_m)}{2N_f(\Delta T, T_m)} \tag{11.3.12}$$

where

ΔD_i is the damage done by temperature variation in load block i
$2N_i$ is the reversals applied for daily applications in load block i

11.4. Eyring Model

The reaction rate because of the Eyring model originates from chemical reaction theory, and it has led to a general and versatile model as an inferential mechanism involving more load factors. By Eyring law, each additional factor can be incorporated into the exponential function in addition to thermal load $-E_a/(KT)$. There are several versions because of the modifications after the model is disclosed [Eyring et al. 1941]. One general form of the frequently referred model, involving temperature and another load factor, is given as follows:

$$h(T, S) = A\,S^{-m} \exp\left[\left(\frac{-E_a}{K\,T}\right) + \left(B + \frac{C}{K\,T}\right) S\right] \tag{11.4.1}$$

where

S is the load factor
S^{-m} is the pre-exponential load factor S with exponent $-m$
B, C are the exponential constants

11.4.1. Characterization of Eyring Model

Suppose that $h(T_0, S_0)$ is the conversion rate at the normal operating temperature and normal operating load (variable S). Conversion rate $h(T_q, S_q)$ is the conversion rate at treatment q, i.e., the qth load level. Equations (11.4.1) and (11.4.2) are valid only if h(T, S) has the same log-location-scale distribution as that of load factors T and S. Equation (11.4.1) looks like an extension of the Arrhenius equation. The acceleration factor is formulated as

$$\Lambda_q = \frac{h(T_q, S_q)}{h(T_0, S_0)} \tag{11.4.2}$$

Load factor S can be mechanical stress, pressure, voltage, current, humidity, etc. For most practical applications where the temperature range of variation (on °K scale) is small, $C \approx 0$, then Equation (11.4.1) becomes

$$h(T, S) = A\ S^{-m} \exp[(\frac{-E_a}{K\ T}) + B\ S] \tag{11.4.3}$$

For some simple cases, B = 0. The above equation reduces to

$$h(T, S) = A\ S^{-m} \exp(\frac{-E_a}{K\ T}) \tag{11.4.4}$$

$$\text{Thus,}\quad \ln[h(T, S)] = \ln(A) - m\ln(S) + (\frac{-E_a}{K\ T}) \tag{11.4.5}$$

Since the inverse of hazard rate is the MTTF (denoted by L referring to life expectancy), the above equation can be translated into

$$\text{Thus,}\quad -\ln(L) = \ln(A) - m\ln(S) + (\frac{-E_a}{K\ T}) \tag{11.4.6}$$

11.4.2. SALT with Multi-Factorial Design

Modeling feasibility of SALTs discussed so far (examples given in Chapter 10, Sections 10.7 and 10.8) has been based on a single acceleration factor. For coping with a complex situation with multiple acceleration factors, in which data are collected from different experimental treatments according to a suitable DOE design matrix that has come in quest of resolution. All the test samples within the same treatment are exposed to the same specified acceleration factors, but with possibly varying load levels for different treatments.

When a part is subjected to two different types of loads or more, the life cycles are more susceptible to a varying PDF in the statistical model with respect to the extent of the applied load at each load step. For example, both scale and shape factors in Weibull statistics may change with respect to an increasing load level, as depicted schematically by the varying PDF, as demonstrated in Figure 11.4.1.

FIGURE 11.4.1 Shape and scale factors in Weibull statistics varying with an increasing load step.

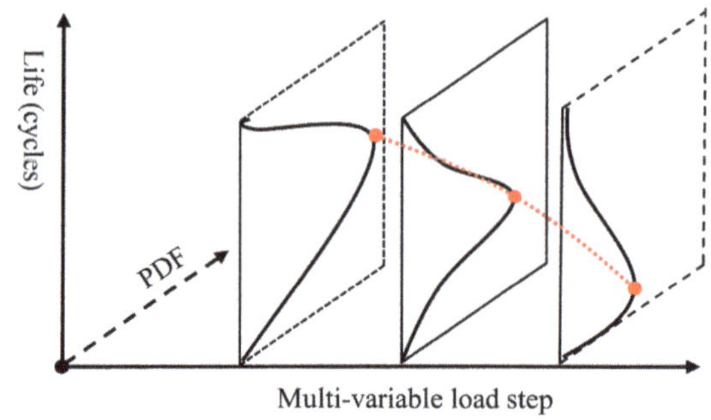

This is to be further illustrated using a published example, in which the life prediction of glass capacitors subjected to both temperature (T) and voltage (V) variations was conducted [Zelen 1959]. Assume that Equation (11.4.4) is true for this case study. In light of the concept of Arrhenius equation, the transformation of temperature is taken as follows:

$$X = 1 / (k\,T) \tag{11.4.7}$$

of which T is the applied temperature. The natural logarithmic transformation of voltage is taken as

$$Y = \ln(V) \tag{11.4.8}$$

of which V is the applied voltage that is the second variable S in Equations (11.4.4) and (11.4.5). By means of the above two equations, the load variables used for the original data (temperature in °C and voltage in V) are transformed into such load variables that Equation (11.4.6) turns into

$$-\ln(L) = \ln(A) - m\,Y - E_a\,X \tag{11.4.9}$$

It means the logarithmic life expectancy is a linear function of variables X and Y.

In a completely randomized design, each glass capacitor should be randomly assigned to a test stand and tested at the same time. Instead, data in [Zelen 1959] were generated from eight test stands (as a block factor) with four glass capacitors (i.e., subsamples) per test stand [Seo 2017]. It is here regarded as a randomized complete block design (RCBD) model. In light of Equation (11.4.8), each outcome Z can then be represented by the following linear statistical equation:

$$Z_{qij} = \mu + W_q + [X_{i(q)} + Y_{j(q)} + X_{i(q)}\,Y_{j(q)} + \varepsilon_{ij(q)}] + \text{higher-ordered terms} + \varepsilon_{qij} \tag{11.4.10}$$

where

- μ is the grand average accounting for the mean
- W_q is the whole plot block (test stand) at the qth load level, as a block factor statistically
- q is the block number, $1 \le q \le 8$; the load level, too
- $X_{i(q)}$ is the factor A at the ith level, as randomized over the whole plot
- $Y_{j(q)}$ is the factor B at the jth level, as randomized over the whole plot
- $X_{i(q)}\,Y_{j(q)}$ is the interaction between factors A and B, as randomized over the whole plot
- $\varepsilon_{ij(q)}$ is the sample error because of random noises by factors A, B, and their interactions
- ε_{qij} is the noise, $\varepsilon_{qij} - N(0, \sigma_\varepsilon^2)$

An experiment test to characterize the life expectancy of glass capacitors was conducted with two distinct loads (variables)—voltage and temperature. The acquired data set consists of 32 failure/censored observations under eight different treatments (factorial design $8^1 \times 2^1 \times 4^1$). The test stand is suspended as soon as the fourth test data available significantly exceed the third data. Life data of glass capacitors obtained from $2^1 \times 4^1$ design with two load factors are given Table 11.4.1. How do these two applied loads affect the life expectancy?

TABLE 11.4.1 Life testing of glass capacitors subject to stepwise load levels of temperature and voltage [Zelen 1959].

Run	T (°C)	V (V)	A	B	Failure time L (h)				$L \approx E(t_q)$
1	170	200	26.19	5.300	439	904	1092	1105	890.7
2	170	250	26.19	5.520	572	690	904	1090	815.9
3	170	310	26.19	5.737	315	315	439	628	424.9
4	170	350	26.19	5.858	258	258	347	588	363.9
5	180	200	25.61	5.300	959	1065	1065	1087	1047.3
6	180	250	25.61	5.520	216	315	455	473	366.9
7	180	310	25.61	5.737	241	315	332	380	317.7
8	180	350	25.61	5.858	241	241	435	455	345.0

Notes:
A = 1/(K T), where T is the temperature in °K; 0°C = 273.16°K
B = ln V, where V is the voltage in Volts
L (h) = life, failure time; data obtained from [Zelen 1959]

Assume that the two-parameter Weibull distribution is a good fit for the entire process. The MTTF for each applied voltage can be obtained from Equation (9.3.3) as

$$E(t_{qn}) = \eta_q \Gamma(1 + 1 / \beta_q), \text{ where } q = 1, 2, \ldots, 8; n = 1, 2, 3, 4 \tag{11.4.11}$$

Fitting each data set within individual treatment (q = 1, 2, 3, 4, 5, 6, 7, 8) using Minitab (➔ Reliability/Survival ➔ Distribution Analysis (Right Censoring ➔ Parametric Distribution Analysis), one can obtain two parameters of its corresponding Weibull distribution function with 95% CI as

$$q=1: \quad \beta_1 = 4.24062 \quad \text{and} \quad \eta_1 = 979.381 \tag{a}$$

$$q=2: \quad \beta_2 = 4.60573 \quad \text{and} \quad \eta_2 = 892.808 \tag{b}$$

$$q=3: \quad \beta_3 = 3.53774 \quad \text{and} \quad \eta_3 = 471.946 \tag{c}$$

$$q=4: \quad \beta_4 = 2.85697 \quad \text{and} \quad \eta_4 = 408.373 \tag{d}$$

$$q=5: \quad \beta_5 = 32.9826 \quad \text{and} \quad \eta_5 = 1064.95 \tag{e}$$

$$q=6: \quad \beta_6 = 4.15570 \quad \text{and} \quad \eta_6 = 403.860 \tag{f}$$

$$q=7: \quad \beta_7 = 7.72387 \quad \text{and} \quad \eta_7 = 337.950 \tag{g}$$

$$q=8: \quad \beta_8 = 3.89303 \quad \text{and} \quad \eta_8 = 381.179 \tag{h}$$

Substituting Equations (a)–(h) into Equation (11.4.11) yields

$$E(t_1) = \eta_1\,\Gamma(1+1/\beta_1) = 890.7$$

$$E(t_2) = \eta_2\,\Gamma(1+1/\beta_2) = 815.9$$

$$E(t_3) = \eta_3\,\Gamma(1+1/\beta_3) = 424.9$$

$$E(t_4) = \eta_4\,\Gamma(1+1/\beta_4) = 363.9$$

$$E(t_5) = \eta_5\,\Gamma(1+1/\beta_5) = 1047.3$$

$$E(t_6) = \eta_6\,\Gamma(1+1/\beta_6) = 366.9$$

$$E(t_7) = \eta_7\,\Gamma(1+1/\beta_7) = 317.7$$

and $$E(t_8) = \eta_8\,\Gamma(1+1/\beta_8) = 345.0$$

The eight data of life expectancy given above are registered in column $L = E(T_q)$ in Table 11.4.1. The data analysis is done with Minitab (Stat ➔ Regression ➔ Regression ➔ Fit Regression Model). The "most insignificant" factor or interaction that comes with the largest p-value is weeded out one by one until all p-values are less than 10% that is here defined as the parting significance level based on F-distribution. In light of the ANOVA presented in Table 11.4.2, the final predictive equation for the life expectancy based on the factorial design is summarized as follows:

$$L_P = 8397125 - 321659\,A - 3008592\,B + 268958\,B^2 + 115296\,A\,B - 10311\,A\,B^2$$

$$(R = 98.5\%,\ R_{adj} = 94.7\%,\ \text{and}\ R_{pred} = 0\%) \tag{11.4.12}$$

The above predictive equation is effective with high correlation and adjusted correlation. Unfortunately, the predictive correlation is so low. This can be improved with more data for each treatment (run), e.g., 22 specimens. It can be seen that the impacts of temperature and voltage on the life expectancy of the tested glass capacitors are more complicated than what can be characterized using a log-linear model like Equation (11.4.9).

TABLE 11.4.2 ANOVA on life testing of glass capacitors subject to stepwise load levels of temperature and voltage.

Source	SS_{adj}	DOF	MS_{adj}	F-value	P-value
Regression	591,985	5	118,397	13.06	7.3%
A	77,643	1	77,643	8.56	10.0%
B	78,562	1	78,562	8.66	9.9%
B^2	78,152	1	78,152	8.62	9.9%
AB	77,405	1	77,405	8.54	10.0%
A^2	77,056	1	77,056	8.50	10.0%
Error	18,136	2	9068		
Subtotal	610,121	7			
Grand ave.	—	1			
Total	—	8			

Substituting Equations (11.4.7) and (11.4.8) into Equation (11.4.12) leads to

$$L_P = 8397125 - 321659/(K\,T) - 3008592\ln(V) + 268958[\ln(V)]^2 + 115296\ln(V)/(K\,T) - 10311\ln(V)^2/(K\,T) \tag{11.4.13}$$

According to the above equation, the response surface of MTTF expressed in terms of temperature and voltage is depicted in Figure 11.4.2. It can be seen that both the influences of temperature and voltage on the life expectancy of the glass capacitors tested are highly nonlinear.

FIGURE 11.4.2 Response surface of life expectancy of glass capacitors subject to temperature and voltage.

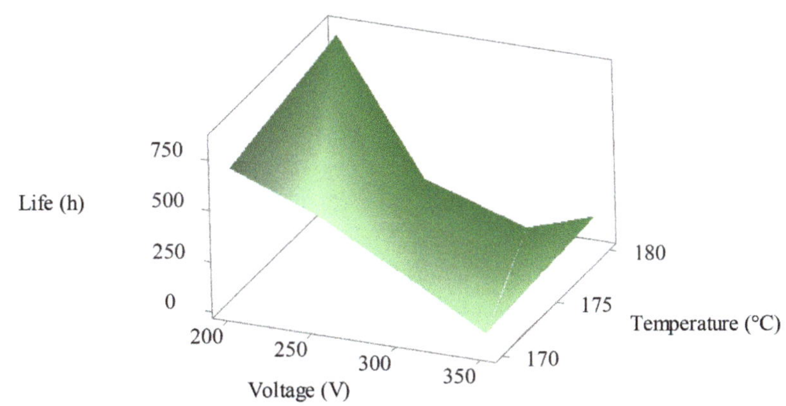

11.5. Oxidation, Corrosion, Rusting, and Erosion

When a part is exposed to the external environment, it can be destroyed because of oxidation taking place on its surface [Chen and Li 2011]. As a type of oxidation, corrosion happens when the metal atoms on a metal surface get oxidized in the presence of oxygen and water. The corrosion in iron or steel is recognized as rusting. Outer surface oxide scales result from the oxidation nucleate micro-voids and oxide particles right beneath oxide scales. It is atmospheric corrosion. Thus, applying a cyclic mechanical load may not change the nature of oxidation chemically but tends to create stress (strain) singularity and/or stress (strain) concentration that may lead to a product failure mechanically. As one type of oxidation, corrosion is the process of material deterioration resulting from chemical, electrochemical, or some other reactions.

For example, the influence of oxidation on the fatigue crack growth in a traditional polycrystalline nickel-based superalloy, namely low solvus high refractory (LSHR), is depicted in Figure 11.5.1 [Jiang et al. 2017]. This type of oxidation damage is profoundly manifested by nickel-based superalloys that are used for disc rotors in high-pressure compressor and turbine of aero-engines in the air [Karabela et al. 2011]. When the working temperature rises higher than 700°C, the interlaminar stress intensity because of the combined mechanical and thermal loads becomes significantly damaging.

FIGURE 11.5.1 Influence of oxidation on the fatigue crack growth in nickel-based superalloy LSHR [Jiang et al. 2017].

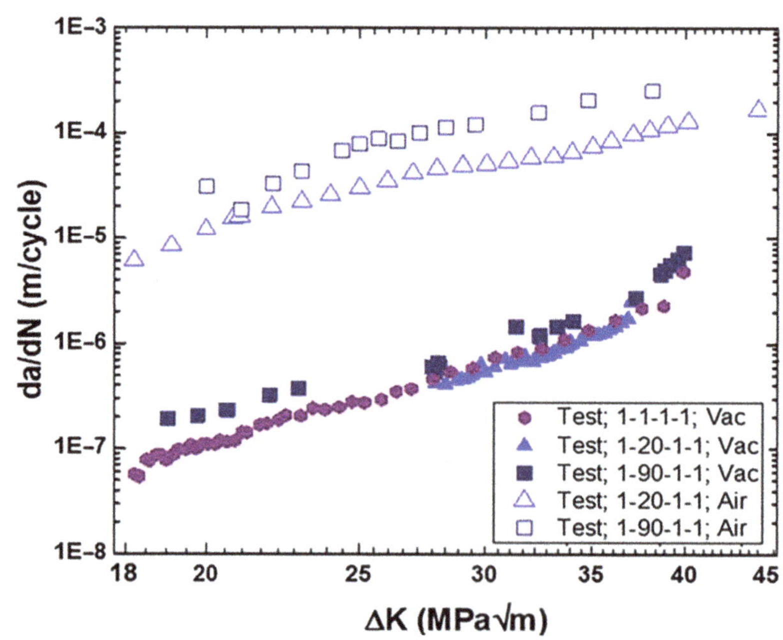

11.5.1. Kinetic Mechanism of Oxidation

An oxidation kinetic mechanism of metals subject to oxygen diffusion comes with two different reaction periods: (a) initial transient period at a faster pace and (b) steady-state period afterward. A general description of mass gain per unit area during an oxidation process can be described by a modified Arrhenius equation as

$$(\Delta m - \Delta m_{in})^n = k_p\,(t - t_{in})^m = k_0\,(t - t_{in})^m \exp\left(\frac{-Q_o}{R\,T_k}\right) \tag{11.5.1}$$

and
$$k_p = k_0 \exp\left(\frac{-Q_o}{R\,T_k}\right) \tag{11.5.2}$$

where

Δm (mg/cm^2 or kg/m^2) is the steady-state mass change per unit area; $\Delta m \geq \Delta m_{in}$
Δm_{in} is the initial mass change in the transient reaction period at a faster pace
t (h) is the time
t_{in} is the initial transient reaction period
n is the exponent
m is the exponent
k_p ($kg^2/m^4/h$ or $mg^2/cm^4/h$) is the constant at a fixed temperature
Q_o (J/mol) is the activation energy for oxide growth
R (=8.314 J/mol/°C) is the universal gas constant
T_k (°K) is the exposure temperature
k_0 is the constant of material oxidation, a material-dependent property

Alloying elements chromium and silicon improve the resistance to heavy scaling by forming a light surface oxide (e.g., CrO) that is impervious to oxidizing atmospheres. However, both elements reduce the toughness and thermal shock resistance of the metal. Equation (11.5.1) can be rewritten as

$$\Delta m = \Delta m_{in} + k_0\,(t - t_{in})^{m/n} \left[\exp\left(\frac{-Q_o}{R\,T_k}\right)\right]^{1/n} \tag{11.5.3}$$

11.5.2. Rust of Iron and Steel

Iron oxidizes to form three well-known oxides, namely, wurtzite (FeO), magnetite (Fe_3O_4), and hematite (Fe_2O_3) in proportions determined by reaction kinetics, where the predominant oxide is magnetite [Konishi et al. 2005]. Corrosion is a type of oxidation, whereas rusting is partially caused by corrosion. The main difference though between corrosion and rust is that corrosion occurs as a result of the chemical influence and it affects a lot of materials, whereas rusting is only accelerated by certain chemicals and usually affects iron substances. One of the main types of corrosion that affects the esthetic look is edge rust on sheet metal [Balasubramanian et al. 2019], as edges act as starting points of rusting.

Intermediate or mixed control oxidation may put exponent n in the range between 1 and 2. In general, this type of oxidation process at a given temperature in steady state can be expressed as [Encinas-Oropesa et al. 2009]

$$(\Delta m)^n = k_p(T)\, t \qquad \text{(Steady state)} \tag{11.5.4}$$

and

$$h(T, t) = \Delta m = k_p(T)^{1/n}\, t^{1/n} \tag{11.5.5}$$

For example, the final weight gain for the alloy exhibits a parabolic behavior after a 200 h of exposure of AISI 301S stainless steel at 800°C [Raj 2013], given that n = 2.

Thus, the acceleration factor for predicting the increase in rust thickness in response to a given working environment at load step q can be related to that at load step time = 0 as

$$\Lambda_q = \frac{h(T_q, t_q)}{h(T_0, t_0)} = \frac{k_p(T_q)^{1/n}\, t_q^{1/n}}{k_p(T_0)^{1/n}\, t_0^{1/n}} \tag{11.5.6}$$

11.5.3. Influence of Laser Etching on Marine Grade 316 Stainless Steel

Markings, such as part number, brand name, and manufacturing date, on a stainless product by laser etching has been a standard industrial practice. However, this process may cause high surface roughness and ablation, and eventually being vulnerable to corrosion, as observed in AISI 304 [Kucera et al. 2014]. A change in the corrosion behavior of a base material caused by thermal effects (ablation) during laser processing is of concern. An experiment was set up to examine the effect of three operating parameters of laser etching process on AISI 316 stainless steel in terms of surface roughness and input energy [Korakana et al. 2020]. The operating levels of these three parameters are listed in Table 11.5.1. Design matrix 3^{3-1} is employed to deploy the tests, and the test results are exhibited in Tables 11.5.1 and 11.5.2.

TABLE 11.5.1 Design factors and levels for investigating effects of laser marking on AISI 316 stainless steel [Korakana et al. 2020].

Factors	Level 1	Level 2	Level 3
A (amp): Current	30	31.5	32
B (mm/sec): Speed	10	15	20
C (kHz): Frequency	15	20	25

TABLE 11.5.2 Design matrix and test results for investigating effects of laser marking on AISI 316 stainless steel [Korakana et al. 2020].

Run	A	B	C	R_a (μm)	H (kJ/mm)
1	30	10	15	0.409	0.660
2	30	15	20	0.242	0.250
3	30	20	25	0.563	0.010
4	31.5	10	20	0.306	0.200
5	31.5	15	25	0.547	0.300
6	31.5	20	15	0.820	1.500
7	32	10	25	0.248	0.900
8	32	15	15	0.294	0.480
9	32	20	20	0.503	0.280

The data analysis of surface roughness is done with Minitab (Stat → Regression → Regression → Fit Regression Model). The "most insignificant" factor or interaction that comes with the largest p-value is weeded out one by one until all p-values are less than 10% that is here defined as the parting significance level based on F-distribution. In light of the analysis of variances presented in Table 11.5.3, the final predictive equations for the surface roughness and input heat because of the laser marking are, respectively,

$$R_{a,p} = -249 + 16.11\ A - 0.260\ A^2 + 0.001058\ B^2$$

$$(R = 87.5\%,\ R_{adj} = 79.1\%,\ \text{and}\ R_{pred} = 47.3\%)$$

and

$$H_p = 25.83 + 7.09\ B - 8.00\ C - 0.02467\ A^2 - 0.2229\ A\ B + 0.2510\ A\ C$$

$$(R = 97.4\%,\ R_{adj} = 93.0\%,\ \text{and}\ R_{pred} = 61.5\%)$$

The correlations and adjusted correlations are good enough to justify the effectiveness of the above two equations. Low predictive correlations mean that the nuisance in the experiment is profound. One major reason is that the three levels of the applied electric current (i.e., 30, 31.5, and 32 amp) are not uniformly distributed, and therefore, it is not a balanced rotatable design matrix.

TABLE 11.5.3 ANOVA on laser marking.

(a) R_a:	Source	SS_{adj}	DOF	MS_{adj}	F-value	P-value
	Regression	0.22289	3	0.07430	5.44	0.049
	A	0.07039	1	0.07039	5.16	0.072
	A^2	0.07038	1	0.07038	5.15	0.072
	B^2	0.15249	1	0.15249	11.17	0.021
	Error	0.06827	5	0.01365		
Subtotal	0.29116	8				
Grand ave.	—	1				
Total	—					
(b) Heat:	**Source**	**SS_{adj}**	**DOF**	**MS_{adj}**	**F-value**	**P-value**
	Regression	1.58140	5	0.31628	11.18	0.037
	B	0.86451	1	0.86451	30.55	0.012
	C	1.10029	1	1.10029	38.89	0.008
	A^2	0.31950	1	0.31950	11.29	0.044
	AB	0.85529	1	0.85529	30.23	0.012
	AC	1.08541	1	1.08541	38.36	0.008
	Error	0.08489	3	0.0283		
Subtotal	1.66629		8			
Grand ave.	—		1			
Total	—		9			

(a) Surface Roughness: A plot of response surface of the surface roughness in terms of electric current (factor A) and moving speed (factor B) is shown in Figure 11.5.2. The surface roughness gets worsened with an increasing moving speed of the laser beam. Nevertheless, there is a certain setting of electric current that could lead to the minimal surface roughness as

$$\partial R_{a,p} / \partial A = 0 \rightarrow A = 31.0\,\text{amp}$$

This is similar to what was observed by the original analysis conducted by [Korakana et al. 2020].

(b) Heat Input versus Surface Roughness: Both demand that the smaller the better. A compromised solution can be decided using a Pareto frontier plot as shown in Figure 11.5.3.

FIGURE 11.5.2 Surface roughness (R_a) of laser marking as a function of electric current and moving speed of laser beam.

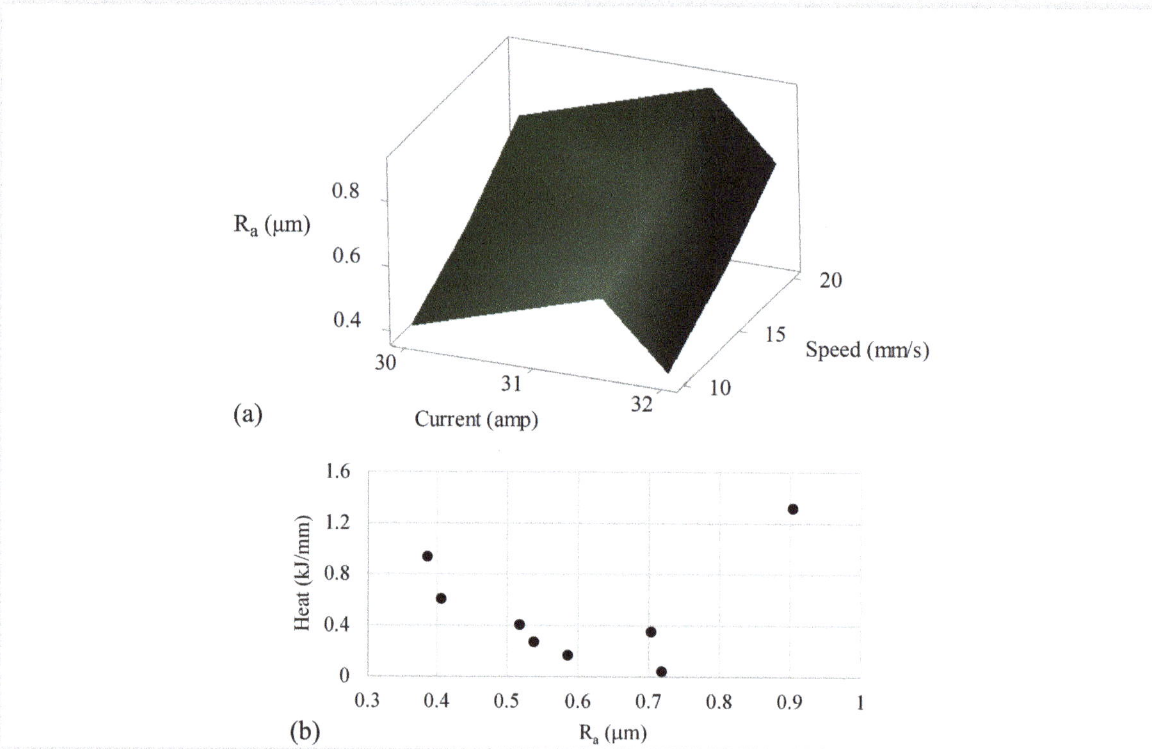

11.5.4. Erosion Wear

Erosion refers to the removal of surface material from a part. Erosion wear of tubing components, such as impellers of pumps, pipelines/elbows, valves, and nozzles, causes a lot of trouble to buildings, power plants, mining facility, and chemical factories. The synergistic effect between erosion and corrosion is generally confirmed to have a significant effect on erosion–corrosion wear properties of tubing components, especially slurry consisting of corrosive medium with solid particles.

A study on the erosion–corrosion rate of 304 stainless steel in an acidic slurry via DOEs was conducted by [Li et al. 2019]. The top surface (10 mm × 10 mm) of each rectangular steel block (10 mm × 10 mm × 20 mm) is exposed to the corrosive slurry medium for the erosion wear experiment. Slurry is made of tap water + acids + 10% sand (20–40 mesh; mainly silica). Three design factors and their related applied levels are used in the experiment as listed in Table 11.5.4. The weight loss rate is here employed to index the corrosion–erosion rate ($g/m^2/h$) (Table 11.5.5).

TABLE 11.5.4 Design factors and applied levels for investigating erosion wear [Li et al. 2019].

Factor	(−1)	(+1)	Coded variable
A (mol/L): Concentration of acid	0.25	0.50	a = (A − 0.375)/0.125
B (°C): Temperature	25 ± 3	45 ± 3	b = (B − 35)/10
C (rpm): Rotating speed of the motor	800	1200	c = (C − 1000)/200

TABLE 11.5.5 Factorial design 2^3 for investigating erosion wear rate ($g/m^2/h$).

Run	A	B	C	AB	AC	BC	ABC	Y_{ave}	Y_1, Y_2, Y_3	Y_p
1	−1	−1	−1	1	1	1	−1	28.17	28/26/30.5	31.42
2	1	−1	−1	−1	−1	1	1	39.33	37/43/38	39.42
3	−1	1	−1	−1	1	−1	1	69.50	70/73.5/65	70.50
4	1	1	−1	1	−1	−1	−1	115.7	22.6/24.6/22	115.2
5	−1	−1	1	1	−1	−1	1	34.67	113/123/111	35.33
6	1	−1	1	−1	1	−1	−1	47.33	37.5/33.5/33	47.25
7	−1	1	1	−1	−1	1	−1	79.33	15.4/15.6/17	78.33
8	1	1	1	1	1	1	1	122.7	123/128/118	123.1
Contrast	28.33	59.42	7.83	16.42	−0.33	0.58	−1.08			
Effect	14.17	29.71	3.92	8.21	−0.17	0.29	−0.54	67.083 (average)		
Error (S)	3.882	3.882	3.882	3.882	3.882	3.882	3.882			
t-ratio	3.649	7.653	1.009	2.114	0.043	0.075	0.140			
$t_{16,90\%}$	1.746	1.746	1.746	1.746	1.746	1.746	1.746			
Significant	Yes	Yes	No	Yes	No	No	No			

where
Y_{ave} is the averaged erosion rate (three tests for each treatment); obtained from [Li et al. 2019]
Y_p is the predicted erosion rate, based on the DOE model

All the contrasts and effects are calculated using the standard procedure based on Student's t distribution as illustrated in Volume 1, Chapter 2. The grand error is estimated using Equation (2.5.3) as

$$S = \left(\frac{\nu_1 S_1^2 + \nu_2 S_2^2 + \ldots + \nu_q S_q^2 + \ldots + \nu_Q S_Q^2}{\nu_1 + \nu_2 + \ldots + \nu_q + \ldots + \nu_Q}\right)^{1/2}$$

$$= \left[\frac{2\,(5.08) + 2\,(10.33) + \ldots + 2\,(3.21) + 2\,(5.01)}{2+2+2+2+2+2+2+2}\right] = 3.882$$

After comparing the ratio of each effect (both main and interactive) to the error, one has the following predictive equation of erosion wear rate,

$$Y_p = 67.083 + 14.17\,a + 29.71\,b + 8.21\,a\,b$$

where constant 67.083 is the grand average. The above equation can be rewritten as

$$Y_p = 6.793 - 116.52\,A + 0.508\,B + 6.568\,A\,B$$

A plot of response surface of erosion wear rate in terms of concentration of acid (factor A) and temperature (factor C) is shown in Figure 11.5.3. It is observed that the temperature has the greatest impact on the erosion wear rate. Note that the effect of two-factor interaction between concentration of acid and temperature is also statistically significant.

FIGURE 11.5.3 Erosion wear rate (g/m^2/h) in terms of concentration of acid (factor A) and temperature (factor B).

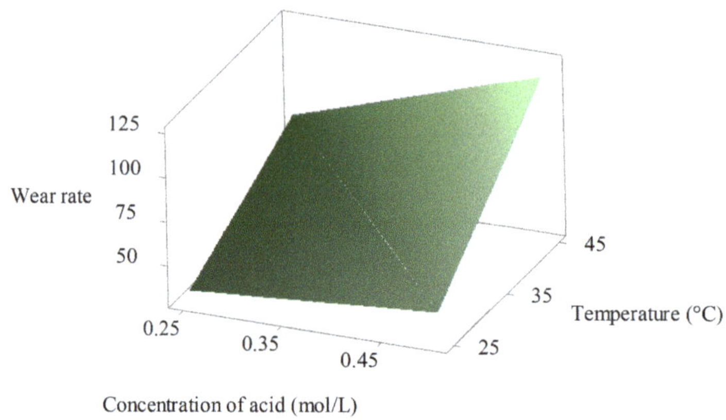

11.5.5. Corrosion Protection of Offshore Wind Power Constructions by Repair Coatings

Effects of surface preparation and coating type on the corrosion protection performance of repair coatings for offshore wind power constructions were investigated by [Momber et al. 2017]. The type of coating and surface preparation are the two factors of interest. The recipes for repair coating, surface treatments, and their design levels are listed in Table 11.5.6. Exposure conditions for the accelerated aging tests are described in detail in [ISO 20340 2009]. Test results are given in Table 11.5.7.

TABLE 11.5.6 Process parameters and their design levels of repair coatings [Momber et al. 2017].

Factor	Level 1	Level 2	Level 3	Level 4
A: Coating#:	1	2	3	4
Base	Epoxy	Polyaspartate	Polyaspartate	Epoxy/siloxane
Hardener	D*	H-D*	H-D*	—
Solid content	100%	100%	100%	100%
Dry film thickness	1000 μm	1000 μm	500 μm	1000 μm
Roughness Rz	>50 μm	>30 μm	>30 μm	>30 μm
Preparation grade	PSa 2½	PSa 2½/PMa	PSa 2½/PMa	PSa 2½/PMa
UV resistant	No	Yes	Yes	Yes
Applicability	Cartridge	Cartridge	Cartridge	Cartridge
B: Surface preparation:	1: Blast	2: Grinding	3: Impact brush	
Roughness Rz	81 ± 13 μm	35 ± 4 μm 5	3 ± 14 μm	
Preparation grade	PSa 2½	PSt 2	Equivalent PSa 2½	
Drive	Pneumatic	Electric	Electric	
Power (drive)	—	590 W	550 W	
Rotational speed	—	8800 rpm	2200 rpm	
Grit type	Garnet	Corundum	(Metal wires)	
Grit size	200–600 μm	P60 (270 μm)	0.73 mm (wire diameter)	
Pressure	7 bar	—	—	
Nozzle diameter	8 mm	—	—	

where

*: D and H-D refer to ethylenediamine and hexamethylene diisocyanate, respectively,

#: ISO 20340 (2009), *Paints and varnishes—performance requirements for protective paint systems for offshore and related structures*, ISO

TABLE 11.5.7 Design matrix 4 × 3 for investigating rust protection by coating [Momber et al. 2017].

Run	A	B	Strength (MPa)#	Creep (mm)*	Rusting*	Blistering*
1	1	1	5.2 ± 0.2	6.9 ± 1.2	1	2
2	1	2	3.9 ± 0.4	29.3 ± 2.4	0	0
3	1	3	4.8 ± 0.5	20.7 ± 0.8	0	0
4	2	1	9.9 ± 0.5	4.0 ± 0.6	0	0
5	2	2	5.2 ± 1.1	18.5 ± 3.3	1	0
6	2	3	8.6 ± 0.2	16.6 ± 2.2	1	0
7	3	1	8.0 ± 0.5	2.6 ± 0.4	0	0
8	3	2	5.9 ± 0.2	18.7 ± 1.6	1	0
9	3	3	7.6 ± 0.9	18.7 ± 2.5	1	0
10	4	1	7.1 ± 0.2	12.4 ± 3.0	0	0
11	4	2	2.8 ± 0.6	30.5 ± 3.1	1	0
12	4	3	5.7 ± 0.2	28.4 ± 2.3	1	0

#: ISO 4624 (2015), *Paints and varnishes—pull-off test for adhesion.*

*: ISO 4628, *Paints and varnishes-evaluation of degradation of coatings-designation of quantity and size of defects, and of intensity of uniform changes in appearance—Part 2: Assessment of degree of blistering, Part 3: Assessment of degree of rusting, and Part 8: Assessment of degree of delamination and corrosion around a scribe or other artificial defect.*

The data analysis of surface roughness is done with Minitab (Stat → ANOVA → General Linear Model → General Linear Model). The result reveals that both factors are statistically significant given that the related p-values are less than 10% based on F-distribution. In light of analysis of variance, the pull-off strength and creep width can be expressed, respectively, as

$$\text{Strength} = 6.225 - 1.592\ A\_1 + 1.675\ A\_2 + 0.942\ A\_3 - 1.025\ A\_4 + 1.325\ B\_1 - 1.775\ B\_2 + 0.450\ B\_3$$

$$\text{Creep} = 17.275 + 1.69\ A\_1 - 4.24\ A\_2 - 3.94\ A\_3 + 6.49\ A\_4 - 10.800\ B\_1 + 6.975\ B\_2 + 3.825\ B\_3$$

The pull-off strength is the more the better, while the creep width is the less the better. A compromised solution can be decided using a Pareto frontier plot as shown in Figure 11.5.4. Run 4 would lead to the highest pull-off strength, while Run 7 would lead to the least creep width.

FIGURE 11.5.4 Pareto plot of pull-off strength versus creep width.

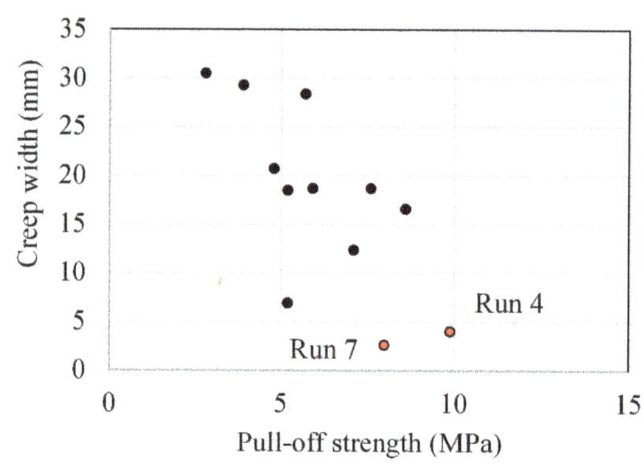

11.6. Crack Propagation

Fracture development may be divided into three stages: crack initiation, crack propagation, and then fracture failure.

(a) Crack Initiation: The initial crack occurs in this stage. The crack is generally intrinsically caused by slip bands or dislocations intersecting the surface as a result of previous cyclic loading or work hardening. It may be as well created by surface defects caused by tooling or handling. US Navy defines arbitrarily that the presence of a crack of 0.25 mm in length is the onset of crack initiation.

(b) Crack Propagation: The crack continues to grow during this stage as a result of continuously applied stresses.

(c) Fracture: Failure occurs when the material that has not been affected by the crack cannot withstand the applied stress. This stage happens rapidly.

11.6.1. Total Life Cycle

The total life cycle is a combination of fatigue cycles to the crack nucleation (from $a = 0$ to a_i) and the fracture propagation cycles (from a_i to a_c),

$$N = N_i + N_p \tag{11.6.1}$$

where

- N is the total useful product life cycles
- N_i is the fatigue life to initiating a crack
- N_p is the life cycles accounting for allowable crack propagation

Note that opening failure mode occurs more often than the other two (in-plane sliding and out-of-plane shearing).

11.6.2. Crack Propagation and Stress Intensity Factor Range

During the stage of a crack propagation, the fatigue crack growth rate is of great concern, for it is used for predicting the fatigue life of material. The fatigue crack growth rate with respect to fluctuating loading cycles, i.e., the hazard rate to the structural health of material with a crack of length a, can be related to its driving stress intensity factor range, ΔK, as

$$\frac{da}{dN_p} = f(R)\,|\Delta K - \Delta K_{th}|^m \tag{11.6.2}$$

where

- a is the crack size
- N_p is the number of cycles in the crack propagation stage
- K is the stress intensity factor, obtained by ASTM E1681 standard test method
- ΔK is the stress intensity factor range; $\Delta K = K_{max} - K_{min}$
- K_{max} and K_{min} are the maximum and minimum stress intensity factors under cyclic loading
- ΔK_{th} is the threshold stress intensity factor range, which is strongly influenced by corrosion
- m is the stress intensity factor exponent (Figure 11.6.1)
- R is the load ratio, $R = K_{min}/K_{max}$
- f(R) is the parameter as a function of load ratio R

The stress intensity factor range given above may be because of any one of the following failure mode: opening, in-plane sliding, and out-of-plane tearing. A simplified model can be derived from Equation (11.6.1), called the modified Paris equation [Sih et al. 1962] that accounts for the load ratio (R), because of [Kurihara et al. 1986], as

$$\frac{da}{dN_p} = C\left|\frac{\Delta K}{1-R}\right|^m \qquad \text{if } -5.0 \le R \le 0.5 \tag{11.6.3}$$

or

$$\frac{da}{dN_p} = C\,|\Delta K|^m \qquad \text{if } R > 0.5 \tag{11.6.4}$$

Another frequently used equation is due to [Forman et al. 1967],

$$\frac{da}{dN_p} = C \frac{|\Delta K|^m}{(1-R)\,K_c - \Delta K} \tag{11.6.5}$$

of which K_c is the stress intensity factor at critical crack length a_c. For 2024-T3 aluminum alloy, K_c = 110 MPa $m^{1/2}$ and C = 3.838 × 10^{-10} (m/cycle)/(MPa $m^{1/2})^{m-1}$, where exponent m = 3.94.

FIGURE 11.6.1 Schematic plot of fatigue crack growth as a function of stress intensity factor range.

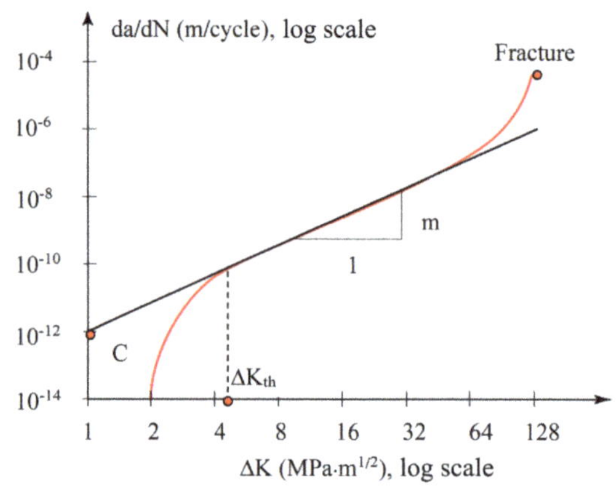

Note that opening failure mode occurs more often than the other two (in-plane sliding and out-of-plane shearing). Parameter ΔK bears the same format for the fracture toughness K [Stienstra and Anderson 1988] as

$$\Delta K = Y' \,\Delta\sigma_{max}\, (\pi\, a')^{1/2} \tag{11.6.6}$$

where
- Y' is the correction factor subject to crack tip configuration
- $\Delta\sigma_{max}$ is the maximum fluctuating stress component
- a' is the crack length in response to elastoplastic deformation

Thus, the acceleration factor for predicting a crack length in response to elastoplastic deformation with a consistent crack tip configuration can be defined by their respective converting factors as

$$\Lambda_q = \frac{h(a_0', \Delta\sigma_{max}, R)}{h(a_q', \Delta\sigma_{max}, R)} \tag{11.6.7}$$

where
- a_0' (m) is the original crack length
- a_q' (m) is the crack length of observation at the qth operating condition (treatment q)

11.6.3. Cyclic Crack Propagation

Consider the case with crack propagation that has merely minor plastic effect (i.e., more brittle fracture). Substituting Equation (11.6.5) into Equation (11.6.4) gives

$$N_p = \int_{a_i}^{a_c} C^{-1} (\Delta K)^{-m} da = \int_{a_i}^{a_c} C^{-1} [Y' \Delta\sigma_{max} (\pi a)^{1/2}]^{-m} da$$

$$= C^{-1} [Y' \Delta\sigma_{max} \pi^{1/2}]^{-m} \int_{a_i}^{a_c} a^{-m/2} da$$

$$= \frac{2 [a_c^{(1-m/2)} - a_i^{(1-m/2)}]}{(2-m) C [Y' \Delta\sigma_{max} \pi^{1/2}]^m} \quad (11.6.8)$$

The above equation is suitable for a thin wide strip. A solution with finite width correction can be found in [Bowie and Neal 1970]. Nevertheless, DOEs can be conducted to examine the influences of the finite width with orthotropic materials.

11.6.4. Crack Propagation Emanating from a Hole in a Uniformly Stressed Strip

This is a case study of low-cycle fatigue crack propagation emanating from a hole. It is intended to examine the fatigue crack propagation life of a strip (300 mm × 100 mm) of 2024-T3 aluminum alloy having a center hole (e.g., rivet hole in an aircraft body) with two existing opposite initial cracks in the radial direction as shown in Figure 11.6.2. The test specimen is subjected to uniform tension in the crack-opening mode.

FIGURE 11.6.2 Test specimen used to predict cyclic crack propagation.

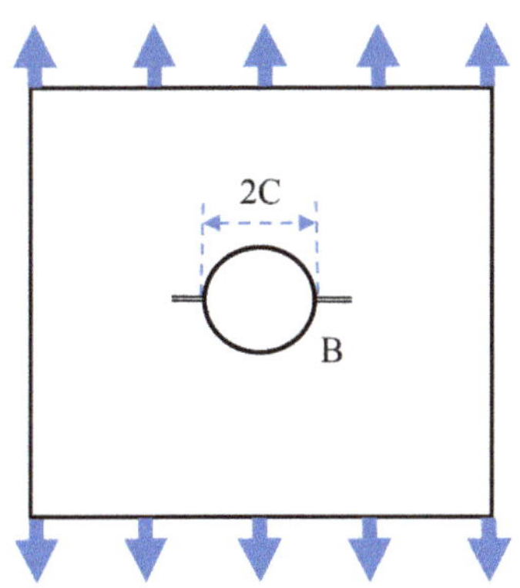

TABLE 11.6.1 Design factors for investigating fatigue crack propagation [Shabestari 2019].

Factor	Level (1)	Level (2)	Level (3)
A: Tensile stress (MPa), applied	60	105	150
B: Initial crack length (mm)	1	3	5
C: Hole radius (mm)	1	3	5

DOEs based on the design levels given in Table 11.6.1 were conducted by [Shabestari 2019]. The crack propagation cycles (N) for the 15 treatments listed in Table 11.6.2 are calculated using the extended finite element method (XFEM) that is a special FEA equipped with crack extension and remeshing algorithms [Belytschko and Black 1999]. The fatigue life is defined as the calculated fatigue cycles when the crack length reaches 30 mm. As shown in Equation (11.6.7), the number of cycles in the crack propagation stage follows the power law of both the tensile stress and the crack length. Thus, the corresponding ln(N) values for the 15 treatments are used for modeling.

TABLE 11.6.2 Design matrix for investigating fatigue crack propagation [Shabestari 2019].

Run	A	B	C	N (cycles)	ln(N)	ln(N_p)	N_p
1	60	1	1	1,172,540	13.97	13.97	1,161,903
2	60	1	5	258,723	12.46	12.42	248,364
3	60	3	3	277,468	12.53	12.56	283,830
4	60	5	5	120,307	11.70	11.66	116,407
5	60	5	1	258,723	12.46	12.53	276,026
6	105	1	3	47,868	10.78	10.81	49,721
7	105	3	1	53,714	10.89	10.96	57,307
8	105	3	3	28,417	10.25	10.27	28,722
9	105	3	5	16,964	9.74	9.75	17,206
10	105	5	3	18,780	9.84	9.72	16,591
11	150	1	1	29,313	10.29	10.22	27,515
12	150	1	5	5901	8.68	8.68	5881
13	150	3	3	6356	8.76	8.81	6721
14	150	5	1	6912	8.84	8.79	6536
15	150	5	5	2570	7.85	7.92	2757

where
A (MPa) is the uniform tensile stress
B (mm) is the initial crack length
C (mm) is the hole radius
N (Cycles) is the life cycles

The data analysis is done with Minitab (Stat ➔ Regression ➔ Regression ➔ Fit Regression Model). The "most insignificant" factor or interaction that comes with the largest p-value is weeded out one by one until all p-values are less than 10% that is here defined as the parting significance level based on F-distribution. In light of the analysis of variance presented in Table 11.6.3, the final predictive equation for the tangential force in the turning process based on the factorial design is summarized as follows:

$$\ln(Y_p) = 19.223 - 0.08506\,A - 0.4018\,B - 0.5620\,C + 0.04247\,B\,C + 0.000207\,A^2 + 0.0223\,C^2$$
$$(R = 99.95\%,\ R_{adj} = 99.91\%,\ \text{and } R_{pred} = 99.82\%) \quad (11.6.9)$$

High correlations given above confirm the effectiveness of the above predictive equation, which can be rewritten as

$$Y_p = \exp(19.223 - 0.08506\,A - 0.4018\,B - 0.5620\,C + 0.04247\,B\,C + 0.000207\,A^2 + 0.0223\,C^2) \quad (11.6.10)$$

The predicted values for $\ln(N_p)$ and N_p are listed in the last two columns of Table 11.6.2.

TABLE 11.6.3 ANOVA on fatigue crack propagation.

Source	SS_{adj}	DOF	MS_{adj}	F-value	P-value
Regression	42.6042	6	7.10070	1290.64	0.0%
A	1.8598	1	1.85977	338.04	0.0%
B	1.6941	1	1.69412	307.93	0.0%
C	0.3514	1	0.35140	63.87	0.0%
BC	0.2309	1	0.23087	41.96	0.0%
A^2	0.4920	1	0.49201	89.43	0.0%
C^2	0.0222	1	0.02219	4.03	8.0%
Error	0.0440	8	0.00550		
Subtotal	42.6482	14			
Grand ave.	—	1			
Total	—	15			

The number of cycles in the crack propagation stage varies quadratically with respect to increasing tensile stress (factor A) and hole radius (factor C), but linearly with the initial crack length (factor B). The interaction of the hole radius and the initial crack size is also statistically significant. Assume that the crack length is 1 mm. Then, Equation (11.6.9) turns into

$$Y_p = \exp(18.8212 - 0.08506\,A - 0.51953C + 0.000207\,A^2 + 0.0223\,C^2) \quad (11.6.11)$$

A plot of response surface of fatigue cycles in terms of tensile stress (factor A) and hole radius (factor C) is shown in Figure 11.6.3. It is observed that the tensile stress has the greatest impact on the reduction in the fatigue crack propagation cycles.

FIGURE 11.6.3 Crack propagation cycles versus tensile stress (factor A) and hole radius (factor C) with an initial crack length of 1 mm on each side.

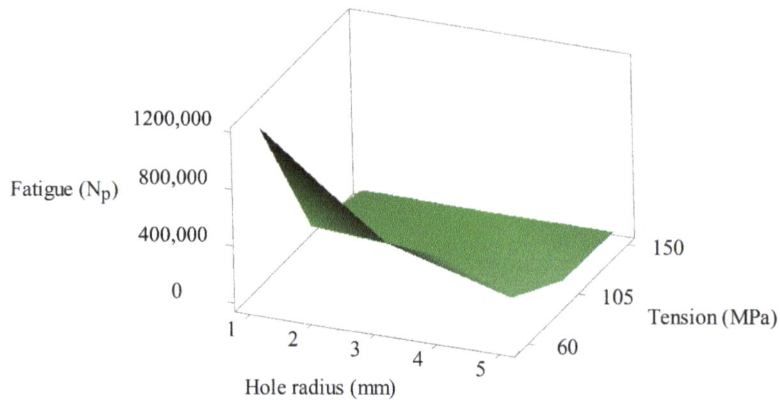

11.7. Creep of Materials

Creep is a time-dependent deformation under a certain applied load. It takes place as a result of dislocation motion in a crystalline specimen by the movement known as "slip" (glide) as a result of long-term exposure to certain levels of stresses that are still below the yield stress of the material. It occurs during a long-term exposure to heat. It occurs mostly at an elevated temperature, but can also happen at room temperature for certain materials. Schematically, the creep of materials can be described as follows:

(a) Creep: It is quantified by the increase in deformation with respect to time under constant loads. Creep in materials may result in multi-axial time-dependent plasticity nonlinearly.

(b) Relaxation: Another time-dependent behavior related to creep is relaxation, where the continuous decrease in the stress level occurs in a material subjected to constant strains at elevated temperatures.

11.7.1. Stress Relaxation

The transformation mechanism of material creep is often because of a viable load factor, by which the conversion rate (hazard rate) varies with respect to mechanical stress S (because of relaxation) inversely as

$$h(S) \propto S^{-m} \tag{11.7.1}$$

of which m is the exponent. The temperature plays a major role in influencing the creep deformation referring to the use of distinct materials as the Findley model applies. The specific load factor, i.e., mechanical stress S, may work with the temperature as

$$h(S, T, t) = A\, t^{n}\, S^{-m} \exp\left(\frac{-E_a}{K\,T}\right)] \tag{11.7.2}$$

where

A is the parameter that depends on the material and its particular creep mechanism
t (sec, min, h) are the time
n is the time exponent
S (Pa or MPa) are the mechanical stress
m is the stress exponent
E_a (J/mol/°K) is the activation energy of the creep mechanism
K is Boltzmann's constant, K ≈ 8.31696 J/mol/°K
T (°K) is the temperature

The acceleration factor for the stress level of the qth step is then

$$A_f = \frac{h_q(S, T)}{h_0(S,T)} = \left(\frac{S_q}{S_0}\right)^{-m} \exp\left[\frac{-E_a}{K}\left(\frac{1}{T_q} - \frac{1}{T_0}\right)\right] \quad (11.7.3)$$

where T_0 and S_0 are initial temperature and stress at step q = 0, respectively, for reference. For a working condition with continuous-varying temperature,

$$A_f = \frac{1}{t_{q2} - t_{q1}} \int^{t_{q2}} \left\{\left(\frac{S(t)}{S_0}\right)^{-m} \exp\left[\frac{-E_a}{K}\left(\frac{1}{T(t)} - \frac{1}{T_0}\right)\right]\right\} dt \quad (11.7.4)$$

The above equation is valid as long as the applied conditions are kept the same or approximately unchanged over the time duration ($t_{q1} \le t \le t_{q2}$). It means that the shape factor is kept constant, if working with the Weibull statistics.

For a constant stress, Equation (11.7.2) can be simplified as

$$h(S, T) = \varepsilon_i + A\, t^n \quad (11.7.5)$$

where

ε_i is the initial strain (t = 0)
A is the coefficient

11.7.2. Two-Parameter Weibull Model

The Weibull model exhibits temperature- and material-dependent behavior using a greater sensitivity range of the parameters applied than the Findley model. In light of the two-parameter Weibull distribution as given by Equation (9.2.2), the accumulated strain subject to material creep can be modeled using the following equation [Fancey 2005]:

$$\varepsilon(t) = \varepsilon_0 + \varepsilon_c \left[1 - \exp\left[-\left(\frac{t}{\eta_c}\right)^{\beta_c}\right]\right] \quad (11.7.6)$$

where

$\varepsilon(t)$ is the accumulated strain over time
ε_0 is the initial strain (t = 0)
ε_c is the creep strain, a material property subject to the material resistance to creep
η_c is the characteristic lifetime
β_c is the shape parameter

The conversion rate is the rate of creep, also called hazard rate, can be estimated using Equation (9.2.1), with $t_o = 0$, as

$$h(t) = \left(\frac{\beta_c}{\eta_c^{\beta_c}}\right) t^{(\beta_c - 1)} \qquad (11.7.7)$$

Note that the inversion of hazard rate is the MTTF. Thus, the hazard rate is representative of the severity of creep because of the effect of time, while creep strain ε_c is subject to the material.

11.7.3. Creep Models Fitting with Polymeric Composites

Creep tests were carried out using DOEs to define load levels that aim to study the creep behaviors of structural composites while keeping reliability relevance [Monticeli et al. 2020]. Test specimens are laminates made of the following four different composites:

1. PEI/CF (5-harness satin)
2. PEI/GF (5-harness satin)
3. Epoxy/CF (plain weave)
4. Epoxy-GF (plain weave)

of which GF and CF stand for glass and carbon fibers, respectively. The fiber reinforcement of each laminate falls between 53 and 54% by weight. PEI is a thermoplastic material with $T_g = 217°C$ (glass transition temperature) and density $\rho = 1.27$ g/cm^{-3} (density), while epoxy is thermoset material with $T_g = 150°C$ and $\rho = 1.33$ g/cm^{-3} °C.

Each test specimen is subject to 5 MPa in tension for 1200 sec (20 min) at the following three different temperature levels: 50, 190, and 250°C. Test results as fitted with the Findley and Weibull models and the related parameters as calculated are given in Tables 11.7.1 and 11.7.2, respectively.

TABLE 11.7.1 Creep behaviors of composites modeled with the Findley model [Monticeli et al. 2020].

Run	Matrix	Fibers	T (°C)	A	n	ε_0 (μm/m)
1. PEI-C	PEI	Cf-Satin	50	1.51×10^{-18}	0.34	30.1
2. PEI-G	PEI	Gf-Satin	50	4.83×10^{-20}	0.29	44.2
3. PEI-C	PEI	Cf-Satin	190	7.22×10^{-15}	0.32	4.49
4. PEI-G	PEI	Gf-Satin	190	3.48×10^{-15}	0.33	19.8
5. PEI-C	PEI	Cf-Satin	250	2.23×10^{-6}	0.49	1500
6. PEI-G	PEI	Gf-Satin	250	2.96×10^{-6}	0.60	2230
7. Epoxy-C	Epoxy	Cf-Plain	50	1.50×10^{-17}	0.29	307
8. Epoxy-G	Epoxy	Gf-Plain	50	3.40×10^{-16}	0.30	404
9. Epoxy-C	Epoxy	Cf-Plain	190	1.56×10^{-17}	0.25	404
10. Epoxy-G	Epoxy	Gf-Plain	190	6.21×10^{-15}	0.27	798
11. Epoxy-C	Epoxy	Cf-Plain	250	1.16×10^{-11}	0.46	992
12. Epoxy-G	Epoxy	Gf-Plain	250	1.32×10^{-10}	0.39	3220

TABLE 11.7.2 Creep behaviors of composites modeled with the Weibull model [Monticeli et al. 2020].

Run	Matrix	Fibers	T (°C)	ε_0	ε_c	η_c	β_c
1. PEI-C	PEI	Cf-Satin	50	29.9	13.2	5.38×10^2	0.45
2. PEI-G	PEI	Gf-Satin	50	33	63.9	2.47×10^4	0.16
3. PEI-C	PEI	Cf-Satin	190	6.94	348	9.41×10^2	0.46
4. PEI-G	PEI	Gf-Satin	190	22.6	253	1.54×10^3	0.43
5. PEI-C	PEI	Cf-Satin	250	1620	13,000	6.70×10^5	0.50
6. PEI-G	PEI	Gf-Satin	250	1830	29,200	6.08×10^2	1.06
7. Epoxy-C	Epoxy	Cf-Plain	50	309	101	4.29×10^2	0.44
8. Epoxy-G	Epoxy	Gf-Plain	50	467	3860	5.48×10^8	0.23
9. Epoxy-C	Epoxy	Cf-Plain	190	2150	1790	5.22×10^4	0.11
10. Epoxy-G	Epoxy	Gf-Plain	190	198	3940	4.19×10^4	0.12
11. Epoxy-C	Epoxy	Cf-Plain	250	993	7300	1.66×10^6	0.47
12. Epoxy-G	Epoxy	Gf-Plain	250	3230	2920	7.17×10^5	0.41

where
ε_0 (μm/m) is the initial strain
ε_c (μm/m) is the strain coefficient

11.8. PV

PV is the direct conversion of light into electricity at the atomic level. Some materials exhibit a property known as the photoelectric effect that causes them to absorb photons of light and release electrons—freed electrons that result in electric current, as shown in Figure 11.8.1.

FIGURE 11.8.1 Energy activation in a PV cell.

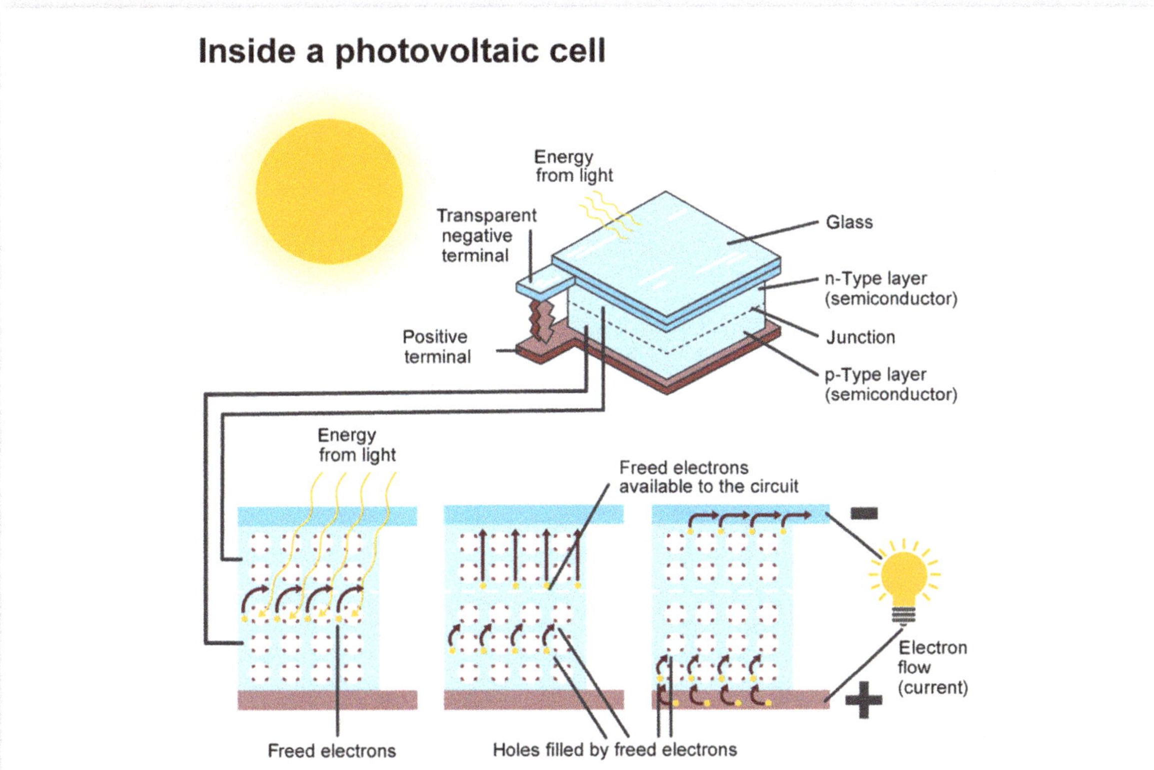

An individual PV cell is small, typically generating approximately 1–2 W of power. Each cell is made of semiconductor materials with crystalline and thin films [McEvoy et al. 2012], such as

1. Silicon: It includes mono-crystalline silicon and polycrystalline silicon.
2. Gallium Arsenide (GaAs): The GaAs solar cell shows an efficiency of 24.3%, i.e., the highest value ever reported for a single-junction solar cell.
3. Thin Films: It includes cadmium telluride (CdTe), copper indium gallium diselenide (CIGS), and a-Si/A-SiGe/a-SiGe (tandem). It is often a little bit thicker than a human hair.
4. Organic PVs (OPV): Cells are composed of carbon-rich (organic) compounds and can be tailored to enhance a specific function of the PV cell, such as bandgap, transparency, or color.

CdTe is the second-most common PV material after silicon. PV cells are electrically connected in a packaged, weather-tight PV module or panel. In order to withstand the outdoors for years, cells are sandwiched between protective materials such as plastics and glass.

11.8.1. Solar Irradiance

The sun is able to provide more than 1 kW/m^2 of free irradiance for several hours per day. When an exposed semiconductor absorbs photons effectively, it is activated to generate electricity. A schematic voltage–current curve produced in a solar cell is depicted in Figure 11.8.2. After the semiconductor material absorbs enough solar energy to overcome the bandgap, i.e., in the figure, electrons are dislodged from the semiconductor material's atoms. Each solar cell may provide 1–2 W in power.

FIGURE 11.8.2 Schematic operating curve (voltage–current curve) in a solar cell as activated.

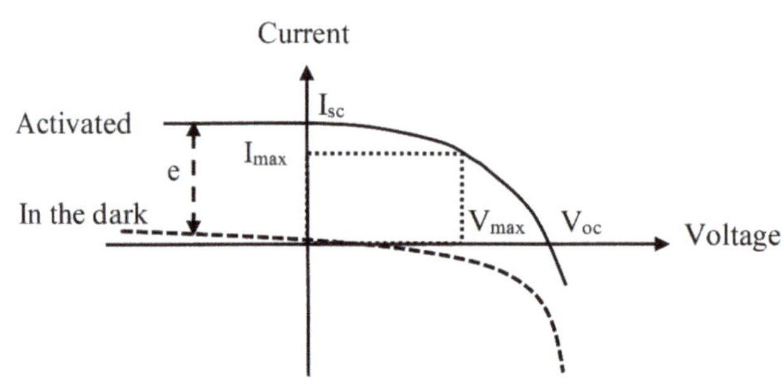

A PV device (e.g., module or panel) that converts sunlight into electrical energy generally consists of PV modules (including cells), batteries, inverter, and regulator. Based on a standard test procedure under 0.8 kW/m^2 irradiance on cell surface, the cell temperature is approximated using the following equation:

$$T_{cell} = T_{ambient} + (T_{NOC} - 20\ ^{\circ}C)\ G / 0.8$$

where

T_{cell} (°C) is the operating temperature of cells
$T_{ambient}$ (°C) is the operating temperature
T_{NOC} (°C) is the nominal operating cell temperature
G (W/m^2 or kW/m^2) is the solar insolation

The performance of a PV cell is characterized by the following five parameters:

(a) Output power P_{max}
(b) Short-circuit current I_{sc}
(c) Open-circuit voltage (OCV) V_{oc}
(d) Fill factor (FF)
(e) Efficiency

FF is the ratio of the capacity of a module (generally represented by maximum power P_{max}) to the product of V_{oc} and I_{sc}. The FF is essentially a measure of the quality of a PV module, the theoretical maximum value depending on factors such as the type of silicon used to construct the module and packaging (e.g., how close the cell is laminated on the module). Note that FF is an influential factor for calculating efficiency.

The efficiency of a PV cell is simply the amount of electrical power coming out of the cell compared to the energy from the light shining on it. It is estimated by exposing the PV cell to a constant, standard level of light while maintaining a constant cell temperature, and measuring the current and voltage that are produced for different load resistances. The theoretical energy conversion efficiency limit of a single-junction solar cell is approximately 31%, while actual efficiencies are even lower than 20%. Solar cells with multi-junctions can reach a higher efficiency because the light that does not get absorbed by the first semiconductor layer is captured by a layer beneath it. A concentration PV (CPV) makes the sunlight onto a solar cell by using a mirror or lens. By focusing sunlight onto a small area, less PV material is required for having the same efficiency.

11.8.2. Performance of PV Cell

A full factorial design 2^2 is utilized to assess the influences of two operating parameters (i.e., A for ambient temperature and B for irradiance) on the performance of a PV module, including the maximum power P_{max}, short-circuit current I_{sc}, and OCV V_{oc} [Kessaissia et al. 2020]. The term OCV refers to the voltage measured across the battery when no current is applied. The design matrix and test results are listed in Table 11.8.1.

TABLE 11.8.1 2^2 factorial design for assessing the performance of a PV panel [Kessaissia et al. 2020].

Run	A	B	P_{max}	I_{sc}	V_{oc}	FF	P_p	I_p	V_p
1	34.6	823	12.92	20.1	0.918	70.0%	12.94	20.11	0.928
2	37.0	823	12.52	19.7	0.908	70.0%	12.52	19.71	0.919
3	34.2	1317	18.12	20.5	1.263	70.0%	18.13	20.51	1.279
4	37.1	1317	18.03	20.3	1.269	70.0%	18.06	20.31	1.287

where
A (°C) is the temperature, ambient
B (W/m^2) is the irradiance, i.e., intensity of irradiation
P_{max} (W) is the maximum output power, as measured
I_{sc} (A) is the short-circuit current, as measured
V_{oc} (V) is the OCV, as measured
P_p (W) is the maximum output power, predicted by the DOE model
I_{sc} (A) is the short-circuit current, predicted by the DOE model
V_{oc} (V) is the OCV, predicted by the DOE model

The data analysis of the maximum power P_{max}, short-circuit current I_{sc}, and OCV V_{oc} is done with Minitab (Stat ➔ Regression ➔ Regression ➔ Fit Regression Model). The "most insignificant" factor or interaction that comes with the largest p-value is weeded out one by one until all p-values are less than 10% that is here defined as the parting significance level based on F-distribution. In light of the analysis of variances presented in Table 11.8.2, the final predictive equations are

$$P_{max} = 18.981 - 0.42388\ A + 0.000303\ A\ B$$

$$(R = 100\%,\ R_{adj} = 100\%,\ \text{and}\ R_{pred} = 99.99\%) \tag{11.8.1}$$

$$V_{oc} = 30.88 - 0.3294\ A - 0.006089\ B + 0.000198\ A\ B \tag{11.8.2}$$

and

$$I_{sc} = 0.8455 - 0.01456\ A + 0.000263\ B + 0.000013\ A\ B \tag{11.8.3}$$

Predicted values based on the above three equations are also listed in Table 11.8.2 in columns P_p, I_p, and V_p, respectively. Both factors and their interactions have influence on the maximum output voltage, short-circuit current, and OCV, except that factor B (irradiation intensity) has no influence on the maximum output voltage.

TABLE 11.8.2 ANOVA tables for assessing the performance of a PV panel.

(1) P_{max}	Source	SS_{adj}	DOF	MS_{adj}	$F_{u,v}$	p-value
	Regression	28.7597	2	14.3799	43,964.13	0.3%
	A	1.2611	1	1.2611	3855.58	1.0%
	AB	28.4785	1	28.478	87,068.51	0.2%
	Error	0.0003	1	0.0003		
Subtotal	28.7601	3				
Grand ave.	—	1				
Total	—	4				
(2) V_{oc}	**Source**	**SS_{adj}**	**DOF**	**MS_{adj}**	**$F_{u,v}$**	**p-Value**
	Regression	0.350000	3	0.11667	*	*
	A	0.034696	1	0.03470	*	*
	B	0.012093	1	0.01209	*	*
	AB	0.016316	1	0.01632	*	*
	Error	0.000000	0	*		
Subtotal	0.350000	3				
Grand ave.	—	1				
Total	—	4				
(3) I_{sc}	**Source**	**SS_{adj}**	**DOF**	**MS_{adj}**	**$F_{u,v}$**	**p-Value**
	Regression	0.124677	3	0.041559	*	*
	A	0.000068	1	0.000068	*	*
	B	0.000023	1	0.000023	*	*
	AB	0.000066	1	0.000066	*	*
	Error	0.000000	0	*		
Subtotal	0.1246773	3				
Grand ave.	—	1				
Total	—	4				

11.8.3. Performance of PV Panels

A PV panel consists of many cells connected in series to provide the required terminal voltage, which are connected in parallel to provide the current ratings. A comprehensive DOEs was conducted by [Almusaied 2016] as it aims to improve the performance of PV panels. The design factors and related application levels are listed in Table 11.8.3. An example test stand is shown in Figure 11.8.3.

TABLE 11.8.3 Design factors and related levels for the performance of PV panels [Almusaied 2016].

	Level (−1)	Level (0)	Level (1)
A (°): Tilt angle, from horizon	0 (Figure 11.8.1)	30	60
B (°): Azimuth angle, i.e., orientation angle	−45	0	45
C (km/h): Wind level	—	5.5	—
D (Gram of talc): Cleanness level	0	20	40
E (kW/m^2): Irradiance	Given in Table 11.8.4		
F (°C): Ambient temperature	Given in Table 11.8.4 (uncontrollable)		

FIGURE 11.8.3 Example test arrangement: tilt = 0° (factor A), mid-level cleanness (factor D), and panel toward the east (longitudinal) [Almusaied 2016].

Experimental data, as assembled in Table 11.8.4, are taken from [Almusaied 2016]. Since the wind level (factor D) is not effective statistically according to the original author, it is assumed to be constant (5.5 km/h) for advancing the further analysis proposed here. The data analysis of the maximum power P_{max} is done with Minitab (Stat ➔ Regression ➔ Regression ➔ Fit Regression Model). The "most insignificant" factor or interaction that comes with the largest p-value is weeded out one by one until all p-values are less than 10% that is here defined as the parting significance level based on F-distribution. In light of the ANOVA presented in Table 11.8.5, the final predictive equation for the maximum power is

$$P_{max,p} = -1.9 - 1.120\ A - 0.867\ B + 1.810\ D + 9.73\ F + 0.000041\ E^2 - 0.441\ F^2$$
$$+ 0.01112\ A\ D + 0.001737\ A\ E + 0.0526\ B\ F - 0.00444\ D\ E + 0.00280\ E\ F$$
$$(R - 90.1\%,\ R_{adj} - 88.6\%,\ \text{and}\ R_{pred} = 86.2\%) \quad (11.8.4)$$

Subjected to the randomness of uncontrollable ambient temperature (factor F) and noise factors (that are not quantifiable), the correlations given above are considered as a strong support of the analysis. The predicted values of P_{max} based on Equation (11.8.4) are listed as column $P_{max,p}$ in Table 11.8.4. According to the p-values listed in the ANOVA table, the two-factor interaction between irradiance (factor E) and cleanness level (factor D) has the greatest influence on the power generation.

TABLE 11.8.4 Design matrix and test results for the performance of PV panels [Almusaied 2016].

Run	A	B	C	D	E	F	P_{max} (W)	$P_{max,p}$ (W)
1	30	-45	5.5	0	904	15.6	156	131.2
1	30	-45	5.5	0	650	15.6	80.7	90.6
2	0	-45	5.5	20	720	15.3	70.6	70.9
2	0	-45	5.5	20	652	15.3	81.5	70.2
7	30	45	5.5	0	944	17.3	142	134.2
7	30	45	5.5	0	653	17.3	81	85.9
10	30	45	5.5	40	1046	17.5	72	52.7
10	30	45	5.5	40	640	17.5	77	55.7
11	30	0	5.5	20	996	18	73	94
11	30	0	5.5	20	631	18	75	64.6
12	30	0	5.5	20	981	17.73	93	93.4
12	30	0	5.5	20	625	17.73	73	65.4
14	60	45	5.5	20	990	18.5	100	121
14	60	45	5.5	20	601	18.5	68.8	69.5
16	0	45	5.5	20	553	19	27.8	58.8
16	0	45	5.5	20	58	19	65.5	64
17	60	0	5.5	40	1014	19	97	77.4
17	60	0	5.5	40	577	19	62.6	57.7
18	0	0	5.5	40	563	19.2	23.5	38
18	0	0	5.5	40	558	19.2	59.9	38.4
21	60	0	5.5	0	964	19.2	98	145.5
21	60	0	5.5	0	521	19.2	52	48.6
24	30	-45	5.5	40	539	20	25	34.6
24	30	-45	5.5	40	500	20	48.5	35.6
27	30	45	5.5	0	888	20.4	137	118.1
27	30	45	5.5	0	467	20.4	43.9	48.7
28	30	0	5.5	20	834	20.1	80	69.6
28	30	0	5.5	20	459	20.1	41.5	42.4
30	30	0	5.5	20	811	20.5	77	65.3

(Continued)

TABLE 11.8.4 (Continued) Design matrix and test results for the performance of PV panels [Almusaied 2016].

Run	A	B	C	D	E	F	P_{max} (W)	$P_{max,p}$ (W)
30	30	0	5.5	20	445	20.5	38	38.8
31	0	0	5.5	0	405	20.5	46	42.2
31	0	0	5.5	0	433	20.5	37.7	44.8
32	30	-45	5.5	40	386	21	17.2	29.8
32	30	-45	5.5	40	424	21	36.5	28.5
33	0	-45	5.5	20	323	21.4	18.6	23.9
33	0	-45	5.5	20	408	21.4	34	24.0
36	0	0	5.5	0	297	21.5	37.3	24.9
36	0	0	5.5	0	378	21.5	29.6	32.1
37	0	-45	5.5	20	228	21.3	15.5	25.6
37	0	-45	5.5	20	366	21.3	28.6	24.9
38	30	45	5.5	40	854	21.2	50.6	42.9
38	30	45	5.5	40	354	21.2	26.5	51.2
39	60	45	5.5	20	955	25	91.4	87.2
39	60	45	5.5	20	339	25	24.5	1.9
40	60	0	5.5	40	870	21.5	25	54.9
40	60	0	5.5	40	328	21.5	23.5	35.4
42	30	45	5.5	40	663	11.2	45.5	47
42	30	45	5.5	40	582	11.2	67	50.5
43	30	0	5.5	20	995	10.4	85	93.9
43	30	0	5.5	20	589	10.4	68	70.6
46	0	0	5.5	40	714	10.8	16.8	39.8
46	0	0	5.5	40	569	10.8	76	53.6
47	60	-45	5.5	20	902	11.4	103	122.1
47	60	-45	5.5	20	575	11.4	76.9	86.9
49	30	0	5.5	20	986	11.2	94.8	95.6
49	30	0	5.5	20	586	11.2	79	72
51	60	0	5.5	0	1064	11.7	187	176.5
51	60	0	5.5	0	591	11.7	80.4	79.6
52	60	-45	5.5	20	879	12.3	132	118.8
52	60	-45	5.5	20	594	12.3	81	87.4
53	0	45	5.5	20	707	12.7	50	60.6
53	0	45	5.5	20	597	12.7	81	60.6
54	30	0	5.5	20	1005	12.4	100	99.6
54	30	0	5.5	20	601	12.4	23.3	73.8
59	30	-45	5.5	0	848	14	142	127.6
59	30	-45	5.5	0	610	14	83.5	91.1
60	30	-45	5.5	40	843	13.9	55	62.6
60	30	-45	5.5	40	609	13.9	83.9	68.9
62	0	0	5.5	0	641	14.5	84.4	89.3
62	0	0	5.5	0	608	14.5	83.3	86.3
65	60	0	5.5	40	1059	14.4	101	89.6
65	60	0	5.5	40	602	14.4	81	73.6
68	0	0	5.5	40	621	15.3	29.5	48.3

(Continued)

TABLE 11.8.4 (Continued) Design matrix and test results for the performance of PV panels [Almusaied 2016].

Run	A	B	C	D	E	F	P_{max} (W)	$P_{max,p}$ (W)
68	0	0	5.5	40	595	15.3	80.4	50.3
71	60	0	5.5	0	1047	15.3	188.8	175.5
71	60	0	5.5	0	586	15.3	78	76.8
73	30	−45	5.5	0	671	15.5	116.5	94.2
73	30	−45	5.5	0	578	15.5	76.5	80.6
74	30	0	5.5	20	961	16	95	95.8
74	30	0	5.5	20	574	16	76	68.3
75	60	45	5.5	20	982	16.4	119	121
75	60	45	5.5	20	568	16.4	73.6	69.3
76	30	0	5.5	20	1000	16.7	62	97.9
76	30	0	5.5	20	562	16.7	72	65.5
77	60	−45	5.5	20	477	16.5	55	59.6
77	60	−45	5.5	20	553	16.5	71.2	67.5
78	0	45	5.5	20	782	17	24.7	66.3
78	0	45	5.5	20	546	17	69	63.2
80	30	45	5.5	0	954	17.5	163	135.9
80	30	45	5.5	0	534	17.5	67	67.8

TABLE 11.8.5 ANOVA table for the performance of PV panels [Almusaied 2016].

Source	SS_{adj}	DOF	MS_{adj}	$F_{u,v}$	p-Value
Regression	99,135	11	9012.2	30.67	0.0%
A	3023	1	3023.1	10.29	0.2%
B	2234	1	2234.5	7.60	0.7%
D	5237	1	5237.4	17.83	0.0%
F	1070	1	1070.2	3.64	6.0%
E^2	820	1	819.8	2.79	9.9%
F^2	2771	1	2771.0	9.43	0.3%
AD	821	1	821.5	2.80	9.9%
AE	3846	1	3846.2	13.09	0.1%
BF	2434	1	2434.0	8.28	0.5%
DE	12,835	1	12,835.2	43.68	0.0%
EF	903	1	903.0	3.07	8.4%
Error	22,918	78	293.8		
Subtotal	122,052	89			
Grand ave.	—	1			
Total	—	90			

Once the solar panel is installed, there are two controllable factors that can be adjusted to improve the power generation. They are tilt angle (factor A) and cleanness level (factor D). The response surface plot of power generated versus these two factors is depicted in Figure 11.8.4, with the following fixed conditions: B = 0° (Azimuth angle), C = 5.5 km/h (wind speed), E = 1000 W/m^2 (irradiance), and F = 16.46°C (ambient temperature). It can be seen that good cleanness (0 g of talc) improves the power generation, so does an increasing tilt angle. Their two-factor interaction is also statistically significant.

FIGURE 11.8.4 Improvement of the maximum power generated by adjusting two operating factors (tilt angle and cleanness).

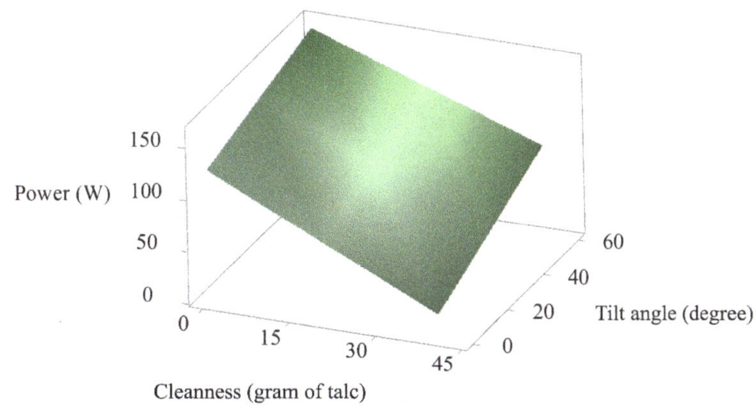

11.9. Automobile Ride

The vibration and noise transmitted to passengers by tactile and visual paths are the major ride characteristics, by which a vehicle occupant is used to tell the ride quality of a vehicle. As being integrated from the roof to the ground, a simplified model of a vehicle dynamic system may consist of the following:

(a) Sprung Mass: Vehicle components, including the body, engine, battery, seating system, and cross-members, being attached to the suspensions and moving as an integral unit are regarded as the sprung mass, denoted by M_S.

(b) Suspension: The suspension is regarded as a combination of links, springs, and dampers (shock absorbers). As to the formulation and calculation for a closed-form solution to vehicle bounce, there is no rule defined for dividing the suspension. In practice, half the suspension mass, including control arms, struts, tie rods, and springs, is added to the sprung mass; the other half to the unsprung mass.

(c) Unsprung Mass: The axles and tire/wheel systems form the unsprung mass, which is regarded as a rigid body and transmits excitation forces from the ground to the sprung mass.

(d) A tire is represented as a spring, because the amount of damping inherent to the viscoelastic tire is relatively small when compared with the shock absorber.

The ratio of the output amplitude to the input amplitude is defined as the gain of a dynamic system. The transmissibility ratio is a gain of a dynamic system and is the non-dimensional ratio of the vehicle response amplitude to the excitation amplitude. An excitation source can be an individual or a combination of road roughness, rotations of tire wheels, rotation of the driveline, cranking of the piston engine, and others. A quarter-car test rig is given in Figure 11.9.1.

FIGURE 11.9.1 Quarter-car test rig [Mitra et al. 2016].

11.9.1. Modal Analysis of Bounce

There are two DOFs (sprung mass and unsprung mass) in the vehicle dynamic system associated with a quarter-vehicle model, as shown in Figure 11.9.2. The balance of forces in the z-direction for the sprung mass results in

$$M_S \left(\frac{d^2 z_S}{dt^2}\right) + C_S \left(\frac{dz_S}{dt} - \frac{dz_U}{dt}\right) + K_S (z_S - z_U) = F_S \tag{11.9.1}$$

and the balance of forces in the z-direction for the unsprung mass results in

$$M_U \left(\frac{d^2 z_S}{dt^2}\right) + C_S \left(\frac{dz_U}{dt} - \frac{dz_S}{dt}\right) + C_U \left(\frac{dz_U}{dt} - \frac{dz_0}{dt}\right) + K_S (z_U - z_S) + K_U (z_U - z_0) = F_U \tag{11.9.2}$$

where z_S, z_U, and z_0 are z-displacements of sprung mass, unsprung mass, and road roughness. M_S, C_S, and K_S are the mass, damping, and spring constant associated with the sprung mass; M_U, C_U, and K_U are the mass, damping, and spring constant associated with the unsprung mass. F_S accounts for the forces acting on the sprung mass, including the aerodynamic drag force, crosswind effect, and engine vibration. F_U stands for the forces acting on the unsprung mass, e.g., undulation of a road surface and unbalanced rotations of tire wheels.

FIGURE 11.9.2 A quarter-car model depicting vehicle vibrations in vertical direction.

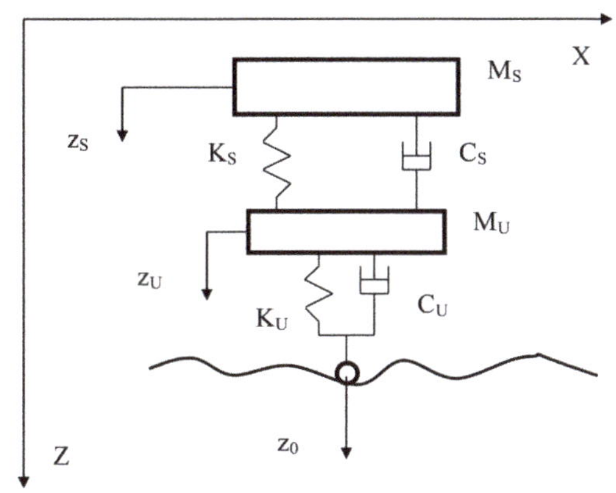

The first step in the identification of a mechanical system under vibration is to extract the undamped natural frequencies. By eliminating the damping effect ($C_S = 0$ and $C_U = 0$) and zeroing the forcing functions ($F_S = 0$ and $F_U = 0$), one can simplify Equations (11.9.1) and (11.9.2) to

$$M_S \left(\frac{d^2 z_S}{dt^2}\right) + K_S (z_S - z_U) = 0 \tag{11.9.3}$$

and

$$M_U \left(\frac{d^2 z_U}{dt^2}\right) + K_S (z_U - z_S) + K_U z_U = 0 \tag{11.9.4}$$

Assuming a solution consisting of harmonic functions for the above two equations,

$$z_S = Z_S \cos(\omega t) \tag{11.9.5}$$

and

$$z_U = Z_U \cos(\omega t) \tag{11.9.6}$$

Substituting the above two equations into Equations (11.9.3) and (11.9.4), one has two simultaneous equations for Z_S and Z_U.

$$(-M_S \omega^2 + K_S) Z_S - K_S Z_U = 0 \tag{11.9.7}$$

and

$$-K_S Z_S + (-M_U \omega^2 + K_S + K_U) Z_U = 0 \tag{11.9.8}$$

A zero determinant for the above two equations guarantees a nontrivial solution, and it leads to the equation for natural frequencies,

$$(\omega^2)^2 - \left(\frac{K_S}{M_U} + \frac{K_U}{M_U} + \frac{K_S}{M_S}\right) \omega^2 + \frac{K_S K_U}{M_S M_U} = 0 \tag{11.9.9}$$

The above equation is solved for two natural frequencies associated with the two DOFs as

$$\omega^2 = \frac{1}{2}\left\{\left(\frac{K_S}{M_U} + \frac{K_U}{M_U} + \frac{K_S}{M_S}\right) \pm \left[\left(\frac{K_S}{M_U} + \frac{K_U}{M_U} + \frac{K_S}{M_S}\right)^2 - \frac{4\,K_S\,K_U}{M_S\,M_U}\right]^{1/2}\right\} \tag{11.9.10}$$

The two positive values calculated using the above equation give the natural frequencies of the quarter-car system. For a typical quarter passenger car, $M_U = 45.45$ kg, $M_S = 454.5$ kg, $K_S = 22$ kN/m, and $K_U = 176$ kN/m. The calculated undamped natural frequencies are $\omega_S = 1.04$ Hz for the sprung mass and $\omega_U = 10.5$ Hz for unsprung mass, respectively. The wide separation of ω_S and ω_U provides vibration isolation from each other. Thus, the transmissibility in vibration from the unsprung mass to the sprung mass is low and vice versa. For example, after a wheel hits a bump, unsprung mass oscillates at its own frequency while the sprung mass does not act in accordance.

Usually, damping of a shock absorber of a passenger vehicle is a variable curve, ranging from 0.8 to 2.5 kN/(m/sec) in normal operation. Damping also varies from bump to rebound. For example, 1.2 kN/(m/sec) in bump and 1.8 kN/(m/sec) in rebound are nominal design values of damping of a front shock absorber for a full-size sedan. The power consumption of a damper is

$$W_D = C_S\left(\frac{dz_S}{dt} - \frac{dz_U}{dt}\right)$$

11.9.2. Steady-State Response: Exact Solution

Assume that there is no external force exerted on the sprung mass ($F_S = 0$) and unsprung mass ($F_U = 0$). The tire is subjected to an excitation in the longitudinal direction from the harmonic ground surface with an amplitude of Z_0 and a frequency of ω_0; i.e., $z_0 = Z_0 \sin(\omega_0 t)$. Because damping of a shock absorber is much higher than that of the tire, C_U is neglected, and Equations (11.9.1) and (11.9.2) are reduced to

$$M_S\left(\frac{d^2 z_S}{dt^2}\right) + C_S\left(\frac{dz_S}{dt} - \frac{dz_U}{dt}\right) + K_S(z_S - z_U) = 0 \tag{11.9.11}$$

and

$$M_u\left(\frac{d^2 z_u}{dt^2}\right) + C_S\left(\frac{dz_U}{dt} - \frac{dz_S}{dt}\right) + K_S(z_u - z_S) + K_U z_U = K_U Z_0 \sin(\omega_0 t) \tag{11.9.12}$$

The steady-state response of the system to the forced vibration can be obtained by solving the above equation for a particular solution. Let the responses be harmonic.

$$z_U = a_U \cos(\omega_0 t) + b_U \sin(\omega_0 t) \tag{11.9.13}$$

and

$$z_S = a_S \cos(\omega_0 t) + b_S \sin(\omega_0 t) \tag{11.9.14}$$

After substituting the above two equations into Equations (11.9.11) and (11.9.12) and collecting the terms associated with $\cos(\omega_0 t)$ and $\sin(\omega_0 t)$ together for each equation, one can obtain the following four simultaneous equations,

$$\begin{bmatrix} K_S - M_S\omega_0^2 & C_S\omega_0 & -K_S & -C_S\omega_0 \\ -C_S & K_S - M_S\omega_0^2 & C_S & -K_S \\ -K_S & -C_S\omega_0 & K_S - M_S\omega_0^2 & C_S\omega_0 \\ C_S\omega_0 & -K_S & -C_S\omega_0 & K_S - M_S\omega_0^2 \end{bmatrix} \begin{Bmatrix} a_S \\ b_S \\ a_U \\ b_U \end{Bmatrix} = \begin{Bmatrix} 0 \\ 0 \\ 0 \\ K_U Z_0 \end{Bmatrix} \tag{11.9.15}$$

By solving these four simultaneous equations for a_S, b_S, a_U, and b_U, one obtains the amplitudes of vibration for both the sprung mass, Z_S, and the unsprung mass, Z_U, respectively, as

$$Z_S = (a_S^2 + b_S^2)^{1/2}$$

$$= \frac{Z_0\,[(K_U\,K_S)^2 + (K_U\,C_S\,\omega_0)^2]^{1/2}}{\{[(K_S - M_S\,\omega_0^2)\,(K_U - M_U\,\omega_0^2) - M_S\,K_S\,\omega_0^2]^2 + (C_S\,\omega_0)^2\,(M_S\,\omega_0^2 + M_U\,\omega_0^2 - K_U)^2\}^{1/2}} \quad (11.9.16)$$

$$Z_U = (a_U^2 + b_U^2)^{1/2}$$

$$= \frac{Z_0\,\{[K_U\,(K_S - M_S\,\omega_0^2)]^2 + (K_U\,C_S\,\omega_0)^2\}^{1/2}}{\{[(K_S - M_S\,\omega_0^2)\,(K_U - M_U\,\omega_0^2) - M_S\,K_S\,\omega_0^2]^2 + (C_S\,\omega_0)^2\,(M_S\,\omega_0^2 + M_U\,\omega_0^2 - K_U)^2\}^{1/2}} \quad (11.9.17)$$

Quality of ride with regard to bouncing can be decided using the following three criteria:

1. Transmissibility ratio of sprung mass to road waviness is Z_S/Z_0. A high value means more vibration is imparted to the occupants.
2. Suspension travel ratio is $(Z_S - Z_U)/Z_0$. It determines the space required for suspension spring movement allowed between bump and rebound stops. A high value means more suspension spring deflection to be experienced as the vehicle travels. A typical suspension travel is between ±90 mm for a passenger vehicle and ±150 mm for an off-road vehicle.
3. Dynamic tire deflection ratio is $(Z_U - Z_0)/Z_0$. It determines the road-holding capability. A high value means a larger separation between the wheel spindle and ground, resulting in less normal load between the tire and ground.

EXAMPLE 11.9.1

Nominal mechanical properties of a quarter-car model are listed as follows: M_S = 400 kg, M_U = 55 kg, K_S = 18 kN/m, K_U = 225 kN/m, C_S = 1.5 kN/(m/sec), and C_U = 0 kN/(m/sec). A statistical design is to be conducted to study the impact of M_S, C_S, and K_S on the transmissibility ratios based on Equations (11.9.16) and (11.9.17). There are two design levels associated with each design factors (independent variables) as follows:

	Low	High
M_S (kg)	300	500
C_S [kN/(m/sec)]	1	2
K_S (kN/m)	12	24

Determine the impacts of M_S, C_S, K_S, and their interactions on the transmissibility ratio of sprung mass and the dynamic tire deflection ratio because of road waviness, given a road surface full of sinusoidal waviness in the longitudinal direction.

Solution:

Because interactions between factors are of great concern in addition to the effects of individual factors, a full factorial design matrix is specified and given in **Table 11.9.1**.

Peak values of the transmissibility-ratio curves are calculated using Equations (11.9.16) and (11.9.17) and reported in **Table 11.9.1**. Values of ω_{SD} and ω_{UD}, damped natural frequencies for sprung mass and unsprung mass, respectively, are the frequencies at which peak values occur. Transmissibility ratio curves of sprung mass and unsprung mass for run #0, based on the nominal design values, are given in **Figures 11.9.3** and **11.9.4**, and the peak values are also reported in **Table 11.9.1**. Using data in the statistical design matrix for runs #1-#8, one can obtain the predictive equation for the transmissibility ratio of sprung mass to road waviness as

$$Z_S / Z_0 = 1.076 + 0.03125\, m_S - 0.0356\, c_S - 0.0008\, k_S$$
$$- 0.0165\, m_S c_S + 0.00094\, m_S k_S - 0.00294\, c_S k_S \qquad \text{(a)}$$

Because the effects of k_S, $m_S k_S$, and $c_S k_S$ are small, these effects are truncated and not shown in the above equation. Thus,

$$Z_S / Z_0 \approx 1.076 + 0.03125\, m_S - 0.0356\, c_S - 0.0165\, m_S c_S \qquad \text{(b)}$$

The lower-case variables m_S, c_S, and k_S are dimensionless variables ranging from −1 to 1, and they are defined as

$$m_S = \frac{M_S - \text{Mean}}{(\text{High} - \text{Low})/2} = \frac{M_S - 400}{(500 - 300)/2}$$

$$c_S = \frac{C_S - \text{Mean}}{(\text{High} - \text{Low})/2} = \frac{C_S - 1.5}{(2 - 1)/2}$$

and $$k_S = \frac{K_S - \text{Mean}}{(\text{High} - \text{Low})/2} = \frac{K_S - 18}{(24 - 12)/2}$$

The constant 1.076 in Equation (a) is the mean of all the eight data (#1-#8) collected for Z_S/Z_0. Based on Equations (11.9.16) and (11.9.17), run #0 yields 1.0454, which is close to the mean value of 1.076 (3% in error). The effect of m_S is calculated using the (+, −) signs of m_S column in **Table 11.9.1**, as (−1.064 + 1.115 − 1.029 + 1.059 − 1.064 + 1.164 − 1.022 + 1.051)/4 = 0.0625. The denominator, 4, means there are four contrasts, i.e., from −1 to +1, in the (+, −) signs of m_S column. The coefficient for m_S in Equation (a) is then calculated as 0.0625/2 = 0.03125. The denominator, 2, is the range of variation of m_S, which can go from −1 to +1. Similarly, coefficients (effects) for other factors and interactions between factors can be calculated as follows:

$$cs:\ (-1.064 - 1.115 + 1.029 + 1.059 - 1.064 - 1.164 + 1.022 + 1.051)/4/2 = -0.0356$$

$$ks:\ (-1.064 - 1.115 - 1.029 - 1.059 + 1.064 + 1.164 + 1.022 + 1.051)/4/2 = -0.0008$$

$$m_s c_s:\ (1.064 - 1.115 - 1.029 + 1.059 + 1.064 - 1.164 - 1.022 + 1.051)/4/2 = -0.0165$$

$$m_s k_s:\ (1.064 - 1.115 + 1.029 - 1.059 - 1.064 + 1.164 - 1.022 + 1.051)/4/2 = 0.00094$$

$$c_s k_s:\ (1.064 + 1.115 - 1.029 - 1.059 - 1.064 - 1.164 + 1.022 + 1.051)/4/2 = -0.00294$$

By the same token, the predictive equation for the transmissibility ratio of unsprung mass to road waviness is

$$Z_U / Z_0 = 4.761 - 1.264\ m_S + 1.497\ c_S - 0.4399\ m_S\ c_S \quad \text{(c)}$$

Run #0 yields 4.389, which is close to the mean value of 4.761 given in the above equation (with 8.5% in error), as to Z_S/Z_0. Thus, the dynamic tire deflection is

$$(Z_U - Z_0) / Z_0 = 3.761 - 1.264\ m_S + 1.497\ c_S - 0.4399\ m_S\ c_S \quad \text{(d)}$$

As a case study, the dynamics tire deflection ratio is 3.389 for run case #0. Assume that the road waviness is $Z_0 = 10$ mm and then $(Z_U - Z_0)/Z_0 = 33.89$ mm. The static deflection of a tire falls usually in the range of 28 and 32 mm subjected to the vehicle weight, when the vehicle rests on the level ground. That $(Z_U - Z_0)/Z_0 = 33.89$ mm means the tire has been deflected upward and left the ground. In other words, there will be no normal force between the tire and the ground to generate cornering force, tractive force, or brake force, when the vehicle travels at a speed corresponding to 3.55 Hz of the road waviness. The vehicle loses the ground traction and is endangered.

Another concern is the mixed impacts of M_S and C_S on the transmissibility ratio and dynamic tire deflection ratio. As shown in Equations (b) and (c), increasing M_S will increase Z_S/Z_0, but reduce $(Z_U - Z_0)/Z_0$. It means that the ride comfort deteriorates, but the road-holding capability improves. On the other hand, increasing C_S will decrease Z_S/Z_0, but increase $(Z_U - Z_0)/Z_0$. It means that the ride is more comfortable, but the road-holding capability is worsened.

This conflicting phenomenon calls the need of an active or semi-active suspension system, which produces variable rates of C_S for different road conditions. A proper compromise must be struck between vibration isolation and road-holding capability.

TABLE 11.9.1 Calculated transient responses of an example quarter-car model. ω_{SD} and ω_{UD} are frequencies (in Hz) at which peak transmissibility ratios occur.

Run #	M_S	C_S	K_S	$M_S\ C_S$	$M_S\ K_S$	$C_S\ K_S$	$Zs/Z_0\ (\omega_{SD})$	$Zu/Z_0\ (\omega_{UD})$
0	0	0	0	0	0	0	1.045 (0.31)	4.389 (3.55)
1	−	−	−	+	+	+	1.064 (0.32)	4.099 (4.02)
2	+	−	−	−	−	+	1.155 (0.37)	2.454 (3.23)
3	−	+	−	−	+	−	1.029 (0.19)	7.967 (4.01)
4	+	+	−	+	−	−	1.059 (0.22)	4.561 (3.21)
5	−	−	+	+	−	−	1.064 (0.45)	4.078 (4.01)
6	+	−	+	−	+	−	1.164 (0.53)	2.426 (3.21)
7	−	+	+	−	−	+	1.022 (0.25)	7.956 (4.01)
8	+	+	+	+	+	+	1.051 (0.30)	4.546 (3.20)

FIGURE 11.9.3 Transmissibility ratio of sprung mass to road waviness.

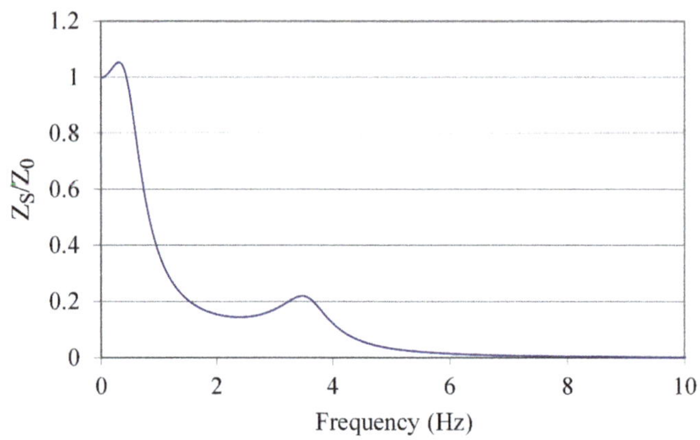

FIGURE 11.9.4 Transmissibility ratio of unsprung mass to road waviness (vehicle data given in Example 11.9.2).

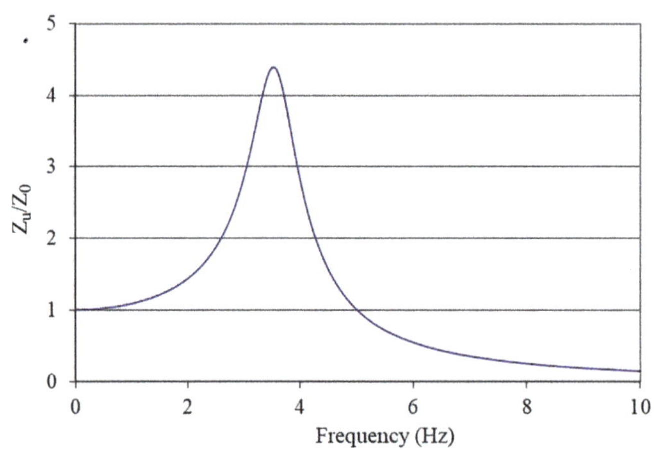

The natural frequencies of a free vibration can be also obtained by setting $C_S = 0$ for Equations (11.9.16) and (11.9.17), and they are

$$\frac{Z_S}{Z_0} = \frac{K_U K_S}{M_S M_U (\omega_{n1}^2 - \omega_0^2)(\omega_{n2}^2 - \omega_0^2)} \tag{11.9.18}$$

and

$$\frac{Z_U}{Z_0} = \frac{K_S (K_S - M_S \omega^2)}{M_S M_U (\omega_{n1}^2 - \omega_0^2)(\omega_{n2}^2 - \omega_0^2)} \tag{11.9.19}$$

where ω_{n1} and ω_{n2} are the two natural frequencies associated with the quarter-car vibration system.

$$\omega_{n1} = (K_U / M_U)^{1/2} \tag{11.9.20}$$

and

$$\omega_{n2} = (K_S / M_S)^{1/2} \tag{11.9.21}$$

11.9.3. Bounce Analysis via Random Vibrations

The traditional procedure of fatigue analysis, as described in Section 11.9, is available only if the loading sequence is a function of time. However, there are cases, in which the occurrence of fatigue is because of random vibrations such as an occupant in an on-ground passenger vehicle, a boat on the water and aircraft wings in the air. Then, the problem has to be resolved in the frequency domain. In light of spectral analysis, the frequency distribution of a response signal can be used to check whether a vibration system is experiencing resonances, which would yield an unaccepted structural unreliability. The automobile ride comfort is given in [ISO 2631-1 1997] that

1. A ride with $Z > 0.315\ m/sec^2$ would make the occupant uncomfortable
2. A ride with $Z > 0.5\ m/sec^2$ would make the occupant very uncomfortable, even fatigued

where Z is the vertical acceleration.

A physical quarter-car model, as shown in Figure 11.9.1, is employed to test the vibration response in the vertical direction in order to simulate the ride comfort of an occupant sitting in an on-ground light passenger vehicle [Mitra et al. 2016]. The excitation source is a sinusoidal bump profile, and the follower is considered as the wheel of the quarter-car model. The cam profile is actuated with a motor so as to simulate the excitation of the wheel vibration in the vertical direction subject to the road waviness. Four design factors and their related design levels of concern are listed in Table 11.9.2.

TABLE 11.9.2 Design factors and levels for examining ride comfort.

Factor	(−1)	(+1)	Coded variable
A (kN/mm): Spring stiffness, K_S	18	26	a = (A − 22)/4
B (N·s/m): Damping coefficient, C_S	418	673	b = (B − 545.5)/127.5
C (kg): Added sprung mass, M_S	41	81	c = (C − 61)/20
D (rpm): Wheel speed	150	250	d = (D − 200)/50
E (kPa): Tire pressure	Fixed		
F (degree): Camber angle	Fixed		
G (mm): Toe	Fixed		

The objective is to study how the vertical acceleration (m/sec^2) of the sprung mass responds to the variations of the four design variables. The analysis is done following the regular procedure of DOE-t (Volume 1, Chapter 2). Since replication = 1, the underlined small effects (six of them) are used to compute the error using Equation (2.5.7) as

$$S = \left[\frac{(-0.040)^2 + (0.004)^2 + (-0.001)^2 + (0.009)^2 + (0.026)^2 + (0.002)^2}{6}\right]^{1/2} = 0.031$$

After the insignificant terms are excluded, the predictive equation for the vertical acceleration of the sprung mass is

$$Z_p = 0.67438 + 0.064\,a + 0.069\,b - 0.052\,c + 0.206\,d - 0.058\,a\,d - 0.076\,b\,c + 0.073\,b\,d - 0.069\,a\,b\,d - 0.088\,b\,c\,d \qquad (11.9.22)$$

The predicted values of vertical vibration of sprung mass are listed in column Z_p (Table 11.9.2). It can be seen that all the main effects are significant. Assume that the part of spring stiffness and damping coefficient is set at 22 kN/mm (A = 22 kN/mm ➔ a = 0) and 545.5 N·s/m (B = 545.5 N·s/m ➔ b = 0), and Equation (11.9.22) reduces to

$$Z_p = 0.67438 - 0.052\ c + 0.206\ d$$

$$\text{i.e.,} \quad Z_p = 0.00898 - 0.0026\ C + 0.004126\ D \tag{11.9.23}$$

The response surface of vertical acceleration (Z_p) in terms of factor C (added sprung mass, e.g., passenger) and factor D (wheel speed) is plotted in Figure 11.9.5. It can be seen that heavier added mass (e.g., passenger) and lower speed would reduce the vertical acceleration (Table 11.9.3).

TABLE 11.9.3 2^4 factorial design for examining ride comfort.

(a) Main effects

Run	a	b	c	d	Z	Z_p
1	−1	−1	−1	−1	0.41	0.483
2	1	−1	−1	−1	0.66	0.589
3	−1	1	−1	−1	0.34	0.313
4	1	1	−1	−1	0.665	0.695
5	−1	−1	1	−1	0.35	0.355
6	1	−1	1	−1	0.47	0.461
7	−1	1	1	−1	0.285	0.233
8	1	1	1	−1	0.565	0.615
9	−1	−1	−1	1	0.535	0.551
10	1	−1	−1	1	0.72	0.701
11	−1	1	−1	1	1.38	1.301
12	1	1	−1	1	1.1	1.175
13	−1	−1	1	1	0.71	0.775
14	1	−1	1	1	0.99	0.925
15	−1	1	1	1	0.87	0.869
16	1	1	1	1	0.74	0.743
Contrast	0.129	0.138	−0.104	0.413		
Effect	0.064	0.069	−0.052	0.206	0.674 (average)	
Error	0.699	0.699	0.699	0.699		
t-ratio	2.106	2.249	1.697	6.747		
$t_{6,1\text{-}10\%}$	1.440	1.440	1.440	1.440		
Significant?	Yes	Yes	Yes	Yes		

(b) Interactions

Run	ab	ac	ad	bc	bd	cd	abc	abd	acd	bcd	abcd
1	1	1	1	1	1	1	−1	−1	−1	−1	1
2	−1	−1	−1	1	1	1	1	1	1	−1	−1
3	−1	1	1	−1	−1	1	1	1	−1	1	−1
4	1	−1	−1	−1	−1	1	−1	−1	1	1	1

(Continued)

TABLE 11.9.3 (Continued) 2^4 factorial design for examining ride comfort.

(b) Interactions											
Run	**ab**	**ac**	**ad**	**bc**	**bd**	**cd**	**abc**	**abd**	**acd**	**bcd**	**abcd**
5	1	−1	1	−1	1	−1	1	−1	1	1	−1
6	−1	1	−1	−1	1	−1	−1	1	−1	1	1
7	−1	−1	1	1	−1	−1	−1	1	1	−1	1
8	1	1	−1	1	−1	−1	1	−1	−1	−1	−1
9	1	1	−1	1	−1	−1	−1	1	1	1	−1
10	−1	−1	1	1	−1	−1	1	−1	−1	1	1
11	−1	1	−1	−1	1	−1	1	−1	1	−1	1
12	1	−1	1	−1	1	−1	−1	1	−1	−1	−1
13	1	−1	−1	−1	−1	1	1	1	−1	−1	1
14	−1	1	1	−1	−1	1	−1	−1	1	−1	−1
15	−1	−1	−1	1	1	1	−1	−1	−1	1	−1
16	1	1	1	1	1	1	1	1	1	1	1
Contrast	−0.080	0.009	−0.115	−0.153	0.146	−0.002	0.018	−0.139	0.053	−0.176	−0.004
Effect	*−0.040*	*0.004*	−0.058	−0.076	0.073	*−0.001*	*0.009*	−0.069	*0.026*	−0.088	*−0.002*
Error	0.031	0.031	0.031	0.031	0.031	0.031	0.031	0.031	0.031	0.031	0.031
t-ratio	1.308	0.143	1.881	2.494	2.392	0.041	0.286	2.269	0.859	2.883	0.061
$t_{6,1-\alpha}$	1.440	1.440	1.440	1.440	1.440	1.440	1.440	1.440	1.440	1.440	1.440
Significant?	No	No	Yes	Yes	Yes	No	No	Yes	No	Yes	No

Notes:
Z is the data obtained from physical rests
Z_p is the predicted value of Z, based on the criterion that significance level α = 10%

FIGURE 11.9.5 Response surface of vertical acceleration (Z_p) in terms of factor C (added sprung mass, e.g., passenger) and factor D (wheel speed).

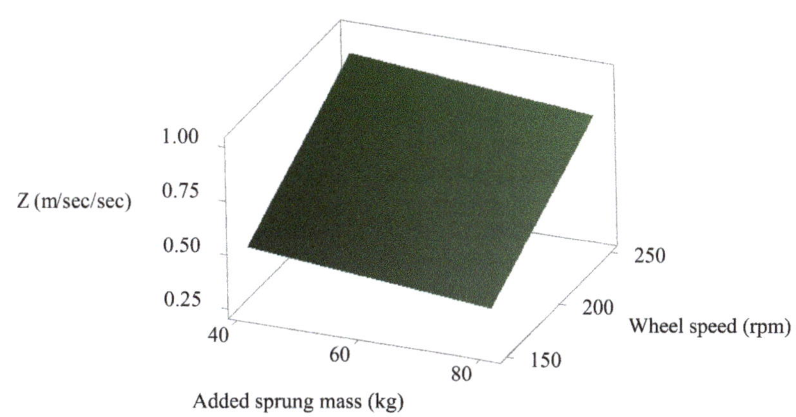

11.10. Machine Learning-Assisted DOEs (DOE+ML)

DOE+ML refers to the joint application of DOEs and ML. For solving domain-specific problems, e.g., product innovation (PI), the combined methodologies would delineate the state of the art according to the available big data and identify the feasible trends. Here, ML works as an inferential mechanism. As the results obtained from ML help identify the next set of experimental conditions (treatments), the framework becomes an active learning (AL) process. The goal is to build an effective and adequate DOE+ML model that would outperform an artificial neural network (ANN) that often appears in ML models. This can be done by assuring that information randomly collected from representatively distributed spots in the study domain is well present in the dataset. The use of balanced orthogonal design matrices (Volume 1, Chapters 2–5) or optimal designs (Volume 1, Chapter 6) fortifies the information learned from each previous experiment and thus explores the design space more efficiently in comparison with grid or random search methods [Munger and Desa 2021].

11.10.1. ML

ML, as one of the artificial intelligence (AI) techniques, is of model-based reasoning, object-oriented frame representation, hypothetical reasoning, role-based reasoning, and automated state-space search. It refers to pieces of codes that enable computer programs to figure out specific patterns or trends from historical information (data) to perform complex tasks. ML algorithms can be utilized to solve domain-specific problems or develop generic techniques that can be used to answer data-related questions by means of the following learning processes:

(a) Unsupervised Learning: The system related to the given data points is unknown and the coded algorithms label them by organizing the data or describing its structure, such as system identification and fitting statistical distributions. For example, K-means algorithm uses the proximity of an output to a cluster of data points for identification, classification, and predictive modeling.

(b) Supervised Learning: The coded algorithms make predictions based on a set of labeled examples in the system. A model may be determined to help quickly explore the unknown experimental space in order to reduce the number of required trials. Linear regression, logistic regression, naïve Bayes, decision tree, random forest algorithm, and K-nearest neighbor (KNN) algorithm are supervised learning processes.

(c) Reinforcement Learning: The coded algorithms, learned from outcomes, will decide which action to take next. For example, an autonomous vehicle learned to make decisions based on a history of reinforcement, such as keeping itself in the lane and controlling the speed limit.

A typical optimization algorithm for ML involves two primary components: a method for statistical inference (e.g., Gaussian process) and an acquisition function that decides where to sample. These are the two functional needs for DOEs when apparent-designed test data are not available.

A schematic drawing utilizes the data generated by ML as an invisible inferential mechanism to generate data for individual treatments (runs) in the DOEs, as shown in Figure 11.10.1. While ML lends researchers to speed up product development progress, such as reduction in time needed to determine battery lifetimes by as much as 98% [Chui 2021], it did not reveal the underlying physics or chemistry that made some batteries last longer than others.

FIGURE 11.10.1 ML as an inferential mechanism [Arboretti et al. 2022].

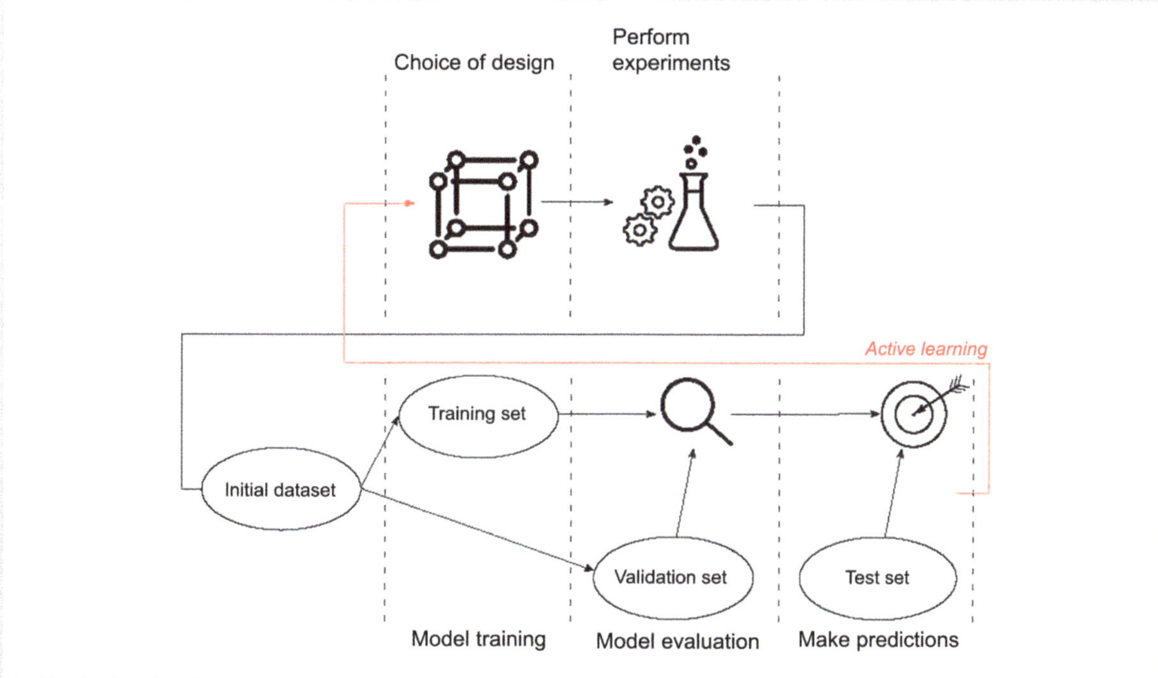

11.10.2. Example of Applying DOE+ML to Exploring OPVs

One of the important features of machine-assisted DOEs is the ability to predict applicable working ranges of the factors of interest, when a fundamental understanding of the inferential mechanism is not unknown. OPV technologies are of great interest because of the potential to make "plastic" solar cells through ambient processing, such as roll-to-roll printing, inkjet printing, and spray coating.

Four process parameters, as given in Table 11.10.1, were selected to study their influences on the power conversion efficiency (%) of an OPV device by [Cao et al. 2018]. An experimental design, as listed in Table 11.10.2, was outlined on the basis of these four process factors to improve the power conversion efficiency (P) of a bulk heterojunction (BHJ), which is the core of the OPV device that comprises a low band gap polymer—poly[N-9′-heptadecanyl-2,7-carbazole-alt-5,5-(4′,7′-di-2-thienyl-2′,1′,3′-benzothiadiazole)] (PCDTBT). Due to the complexity and not-so-well-known facts, applicable ranges of these four parameters are selected using ML techniques. The test results are posted as Column P (%) in Table 11.10.2. There are only 15 data available, because the 16th treatment is not physically feasible [Cao et al. 2018].

$$\begin{aligned}\ln(P_p) = {} & 2.14 + 2.081\,A - 51.20\,C - 0.4410\,D - 0.014278\,A^2 + 6.531\,C^2 \\ & - 0.05087\,A\,B + 0.07238\,A\,C - 0.04404\,A\,D + 1.2228\,B\,C \\ & + 0.02695\,B\,D + 0.9272\,C\,D \\ & (R = 99.97\%,\ R_{adj} = 99.76\%,\ \text{and } R_{pred} = 98.17\%) \qquad (11.10.1)\end{aligned}$$

TABLE 11.10.1 Process parameters and their design levels for exploring power conversion efficiency of OPVs [Cao et al. 2018].

Factor	Level 1	Level 2	Level 3	Level 4
A: Donor (weight %)	10	25	40	55
B: Concentration (mg/mL)	10	15	20	25
C: Spin speed (10^3 rpm)	0.6	1	2	3
D: Additive (volume %)	0	2	8	12

TABLE 11.10.2 Fractional factorial design 4^{4-2} for exploring power conversion efficiency, P(%), of OPVs [Cao et al. 2018].

Run	A	B	C	D	P (%)	P_p (%)
1	10	20	3	2	0.05	0.0471
2	10	25	1	8	3.24	3.2736
3	10	10	0.6	0	0.016	0.0156
4	10	15	2	12	0.0004	0.0004
5	25	20	0.6	12	7.14	6.2749
6	25	15	1	2	3.22	3.4089
7	25	10	3	8	0.00033	0.0003
8	25	25	2	0	7.21	6.8930
9	40	10	1	12	1.85	1.7693
10	40	20	2	8	6.16	6.0014
11	40	25	0.6	2	3.9	4.1805
12	40	15	3	0	2.27	2.4690
13	55	10	2	2	1.16	1.0928
14	55	15	0.6	8	3.18	3.4562
15	55	20	1	0	3.89	3.3701
16	55	25	3	12	NA	NA

The above predictive equation is effective with high correlations, including unadjusted model correlation, adjusted model correlation, and predictive correlation. Equation (11.10.1) can be rewritten as

$$\text{i.e.,}\quad P_p = \mathrm{Exp}(2.14 + 2.081\,A - 51.20\,C - 0.4410\,D - 0.014278\,A^2 + 6.531\,C^2 - 0.05087\,A\,B + 0.07238\,A\,C - 0.04404\,A\,D + 1.2228\,B\,C + 0.02695\,B\,D + 0.9272\,C\,D) \tag{11.10.2}$$

The power conversion efficiencies of the top 15 treatments (Table 11.10.2) predicted using Equation (11.10.2) are also listed in Table 11.10.2 for the purpose of comparison with the test data (Table 11.10.3).

TABLE 11.10.3 ANOVA for exploring power conversion efficiency of OPVs.

Variance	SS_{adj}	DOF	MS_{adj}	$F_{u,v}$	p-Value
Regression	165.496	11	15.0451	521.3	0.0%
A	7.530	1	7.5296	260.9	0.1%
C	4.023	1	4.0233	139.4	0.1%
D	1.186	1	1.1855	41.08	0.8%
A^2	9.722	1	9.7223	336.9	0.0%
C^2	3.214	1	3.2136	111.4	0.2%
AB	5.353	1	5.3532	185.5	0.1%
AC	1.600	1	1.6001	55.45	0.5%
AD	3.861	1	3.8607	133.8	0.1%
BC	5.393	1	5.3934	186.9	0.1%
BD	2.192	1	2.1922	75.96	0.3%
CD	2.568	1	2.5676	88.97	0.3%
Error	0.087	3	0.0289		
Subtotal	165.58	14			
Grand ave.	—	1			
Total	—	15			

11.11. Monte Carlo Simulations as an Inferential Mechanism

After participating factors and their feasible variations are identified, Monte Carlo simulations can be employed to generate the data contributed by each individual factor in the design matrix of the DOEs. Because of a large amount of potential factorial variations in the real world, Monte Carlo simulations would fortify a statistical distribution for a factor of concern as long as the number of data generated for each factor is large enough. It means that Monte Carlo simulations serve as an inferential mechanism. A typical application is the statistical tolerance design as delineated in Chapter 7, Section 7.5.

11.12. Digital Twin (DT)-Assisted DOEs

DTs refer to the digital counterparts of a physical system (product of process) of interest. DT technology is to have a virtual model or representation of the system and is intended for helping designers fine-tune their decision-making before finalizing the system development. For example, a vehicular system is equipped with sensors that generate data critical to its performance just like an extension of design for testability, by which the data are then relayed to a computer processor and applied to its virtual copy, i.e., called a DT. The exchange of relevant data and information between the physical and virtual allows early hardware and software integration as well as frontload testing [Grieves and Vickers 2016, Serughetti 2022]. Employing fuzzy logic in ADAS DT systems is considered crucial for facilitating the transition from manned to autonomous vehicles. A concise graphical presentation of DT is exhibited in Figure 11.12.1.

DOEs sometimes require extensive expert knowledge about the investigated factors and their applicable ranges, and it may consequently lead to a good number of rounds of time-consuming and costly experiments. The DOEs assisted by a virtual representation of the system (product of process) of interest based on DT technology, including Internet of things, cloud computing, AI, and simulation/emulation, can be used as an option to reduce the number of experiments [Kuchemüller et al. 2021]. The DT-aided DOE approach (DTADOE) is useful when the product or process of interest is an unfamiliar system to the experimenter.

FIGURE 11.12.1 Concept of DT [Piromalis and Kantaros 2022].

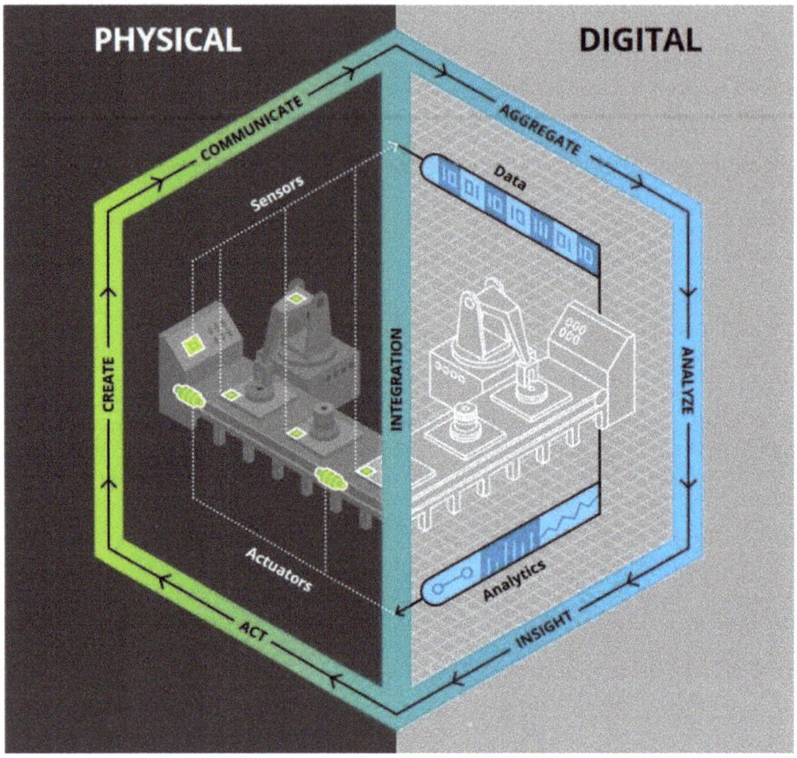

References

Almusaied, Z. I. (2016), "Optimization of Solar Energy Harvesting: Building Infrastructure and Statistical Optimization," MS thesis, Texas State University.

Arboretti, R. et al. (2022), "Design of Experiments and Machine Learning for Product Innovation: A Systematic Literature Review," *Quality and Reliability Engineering International*, 38, pp. 1131-1156.

Arrhenius, S. (1895), "On the Influence of Carbonic Acid in the Air upon the Temperature of the Ground," Paper Presented to the *Stockholm Physical Society.*

Balasubramanian, J., Lalwani, R., and Kumar, V. (2019), "Mitigation for Edge Corrosion Protection for Automotive Bodies," *SAE Technical Paper 2019-28-2561*, https://doi.org/10.4271/2019-28-2561.

Bannantine, J., Comer, J., and Handrock, L. (1990), *Fundamentals of Metal Fatigue Analysis*, Prentice-Hall, Englewood Cliffs.

Belytschko, T. and Black T. (1999), "Elastic Crack Growth in Finite Elements with Minimal Remeshing," *Int'l Journal for Numerical Methods in Engineering*, 45, pp. 601-620.

Benavides, E. M. (2011), "Reliability Model for Step-Stress and Variable-Stress Situations," *IEEE Transactions on Reliability*, 60(1), pp. 219-233.

Bouarroudj, M., Khatir, Z., and Ousten, J. P. (2007), "Degradation Behavior of 600V-200A IGBT Modules under Power Cycling and High Temperature Environment Conditions," *Microelectronics Reliability*, 47, pp. 1719-1724.

Bowie, O. L. and Neal, D. M. (1970), "A Note on the Central Crack in a Uniformly Stressed Strip," *Engineering Fracture Mechanics*, 2(2), pp. 181-182.

Cao, B. et al. (2018), "How to Optimize Materials and Devices via Design of Experiments and Machine Learning: Demonstration Using Organic Photovoltaics," *ACS Nano*, 12(8), pp. 7434-7444.

Chen, G. and Li, J. L. (2011), "Accelerated Test Design of Thermal-Oxidative Aging of Nylon 66 Airbag Material," *Advanced Materials Research*, 332/334, pp. 1202-1205.

Chiao, C. and Hamada, M. (1996), "Robust Reliability for Light Emitting Diodes Using Degradation Measurements," *Quality and Reliability Engineering International*, 12, pp. 89-94.

Chui, G. (2021), "Battery Research Gets Machine Learning Boost," Battery Informatics Lab at SLAC, Stanford University, https://hai.stanford.edu/news/battery-research-gets-machine-learning-boost.

DeSourza, D. I. and Rocha, R. (2012), "Applying Eyring's Model to 'Times to Breakdown' of Insulating Fluid," *Int'l Journal of Performability Engineering*, 8(3), pp. 279-288.

Do, K. H. and Roy, S. K. (1994), "Corrosion of Steel in Tropical Sea Water," *British Corrosion Journal*, 29(3), pp. 233-236.

Elsayed, E. A. and Liao, H. T. (2003), "A Geometric Brownian Motion Model for Field Degradation Data," *Int'l Journal of Material and Product Technology*, 19(6), pp. 51-72.

Elsayed, E. A. and Liao, H. T. (2004), "Reliability Prediction and Testing Plan Based on an Accelerated Degradation Rate Model," *Int'l Journal of Material and Product Technology*, 21(5), pp. 402-422.

Encinas-Oropesa, A. et al. (2009), "Evaluation of Oxidation Related Damage Caused to a Gas Turbine Disc Alloy between 700 and 800°C," *Materials at High Temperatures*, 26(3), pp. 241-249.

Eyring, H., Gladstone, S., and Laidler, K. J. (1941), *The Theory of Rate Processes*, McGraw Hill, New York.

Fancey, K. S. (2005), "A Mechanical Model for Creep, Recovery and Stress Relaxation in Polymeric Materials," *Journal of Materials Science*, 40, pp. 4827-4831.

Forman, R.G., Kearney, V. E., and Engle, R. M. (1967), "Numerical Analysis of Crack Propagation in Cyclic-Loaded Structures," *Journal of Basic Engineering*, 89(3), pp. 459-463.

Fukuda, M., Wakita, K., and Iwane, G. (1983), "Dark Defects in InGaAsP/InP Double Heterostructure Phasers under Accelerated Aging," *Journal of Applied Physics*, 54, pp. 1246-1250.

Grieves, M. and Vickers, J. (2016), "Origins of the Digital Twin Concept," Working Paper, Florida Institute of Technology/NASA, accessed April 5, 2022, https://www.researchgate.net/publication/307509727_Origins_of_the_Digital_Twin_Concept?channel=doi&linkId=57c6f44008ae9d64047e92b4&showFulltext=true.

Hallberg, O. and Peck, D. S. (1991), "Recent Humidity Accelerations, a Base for Testing Conditions," *Quality and Reliability Engineering International*, 7, pp. 169-180.

Hardikar, K. et al. (2013), "A Methodology for Testing, Characterization and Prediction of Edge Seal Performance in PV Modules," *Photovoltaics International*, 22.

Held, P., Jacob, G., Nicoletti, P., Scacco, P., and Pooch, M. (1997), "Fast Power Cycling Test for IGBT Modules in Traction Application," *Int'l Journal of Electronics*, 86(10), pp. 1193-1204.

ISO 20340 (2009), "Paints and Varnishes—Performance Requirements for Protective Paint Systems for Offshore and Related Structures," International Standard Organization.

ISO 2631-1 (1997), "Mechanical Vibration and Shock—Evaluation of Human Exposure to Whole-Body Vibration—Part 1: General Requirements," Amendment 1: 2010-07-01, May 1, 1997.

Jiang, R. et al. (2017), "Effects of Oxygen-Related Damage on Dwell-Fatigue Crack Propagation in a P/M Ni-Based Superalloy: from 2D-3D Assessment," *Int'l Journal of Fatigue*, 99, pp. 175-186.

Karabela, A. et al. (2011), "Effects of Cyclic Stress and Temperature on Oxidation Damage of a Nickel-Based Superalloy," *Materials Science and Engineering, A*, 528(19-20), pp. 6194-6202.

Kessaissia, F. Z. et al. (2020), "Factorial Design and Response Surface Optimization for Modeling Photovoltaic Module Parameters," *Energy Reports*, 6, pp. 299-309; Tmrees, *EURACA*, Athens, Greece, September 4–6, 2019.

Konishi, H., Yamashita, M., Uchida, H., and Mizuki, J. (2005), "Characterization of Rust Layer Formed on Fe, Fe-Ni and Fe-Cr Alloys Exposed to Cl-Rich Environment by Cl and FeK-EdgeXANES Measurements," *Materials Transactions*, 46(2), pp. 329-336.

Korakana, A., Korakana, S., and Ulmek, N. (2020), "Analyzing the Effect of the Parameters of Laser Etching Process Influencing the Corrosion Resistance and Surface Roughness of Marine Grade 316 Stainless Steel," *Materials Today: Proceedings*, 32(3), pp. 452-462.

Kucera, M. et al. (2014), "Influence on Laser Marking on Stainless Steel Surface and Corrosion Resistance," *Metal 2014*, Brno, Czech Republic, May 21–23, 2014, EU.

Kuchemüller, K. B. et al. (2021), "Digital Twins and Their Role in Model-Assisted Design of Experiments," *Advances in Biochemical Engineering/Biotechnology*, 177, pp. 29-61.

Kurihara, M. et al. (1986), "Analysis on Fatigue Crack Growth Rates under a Wide Range of Stress Ratio," *Journal of Pressure Vessel Technology*, 108(2), pp. 209-213.

Lakshminarayanan, V. (2001), "Environmental-Stress Screening Improves Electronic-Design Reliability," *EDN/Design/Design How-To*, pp. 73-84.

Lee, J. and Pan, R. (2010), "Analyzing Step-Stress Accelerated Life Testing Data Using Generalized Linear Models," *IIE Transactions*, 42, pp. 589-598.

Li, P., Zhao, Y., and Wan, L. (2019), "Research on Erosion-Corrosion Rate of 304 Stainless Steel in Acidic Slurry via Experimental Design Method," *Materials*, 12(14), p. 2330.

McEvoy, A., Markvart, T., and Castaner, L. (2012), *Practical Handbook of Photovoltaics, Fundamentals and Applications*, 2nd Edition, Elsevier, Amsterdam, the Netherlands; ISBN: 978-0-12-385934-1.

Mitra, A. C. et al. (2016), "Optimization of Automotive Suspension System by Design of Experiments: A Nonderivative Method," *Advances in Acoustics and Vibration*, 2016, Article ID 3259026, 10 pages.

Momber, A. W. et al. (2018), "Statistical Effects of Surface Preparation and Coating Type on the Corrosion Protection Performance of Repair Coatings for Offshore Wind Power Constructions," *Materials and Corrosion*, 69, pp. 460-471.

Monticeli, F. M. et al. (2020), "Statistical Analysis of Creep Behavior in Thermoset and Thermoplastic Composites Reinforced with Carbon and Glass Fibers," *Journal of Strain Analysis for Engineering Design*, 56(7), pp. 452-461.

Munger, T. and Desa, S. (2021), "A Statistical Design of Experiments Approach to Machine Learning Model Selection in Engineering Applications," *Journal of Computing and Information Science in Engineering*, 21(1), p. 011008.

Peck, D. C. (1986), "Comprehensive Model for Humidity Testing Correlation," *The 24th IEEE International Reliability Physics Symposium*, New York, April 1, 1986, pp. 44-50.

Piromalis, D. and Kantaros, A. (2022), "Digital Twins in the Automotive Industry: The Road toward Physical-Digital Convergence," *Applied System Innovation*, 5, p. 65.

Raj, A. K. (2013), "On High-Temperature Materials: A Case on Creep and Oxidation of a Fully Austenitic Heat-Resistant Superalloy Stainless Steel Sheet," *Journal of Materials*, 2013, Article ID 124649, 6 pages.

Roy, S. and Sahoo, P. (2012), "Corrosion Study of Electroless Ni-P-W Coatings Using Electrochemical Impedance Spectroscopy," *Portugaliae Electrochimica Acta*, 30(3), pp. 203-220.

Seo, K. (2017), "Data Analysis and Experimental Design for Accelerated Life Testing with Heterogeneous Group Effects," Doctoral dissertation, Arizona State University.

Serughetti, M. (August 17, 2022), "Driving the Future of Innovation in Automotive with Digital Twins for Electronic Systems Development," New Horizons for Chip Design, Synpsys.

Shabestari, S. (2019), "Development of a Regression Model for the Fatigue Life Assessment of Open-Hole Specimens with Double through the Thickness Cracks," MS thesis, Middle East Technical University.

Sih, G. C., Paris, P. C., and Erdogan, F. (1962), "Crack-Tip Stress Intensity Factors for the Plane Extension and Plate Bending Problem," *Journal of Applied Mechanics*, 29, pp. 306-312.

Smet, V., Forest, F., and Huselstein, J. J. (2011), "Ageing and Failure Modes of IGBT Modules in High Temperature Power Cycling," *IEEE Transactions on Industrial Electronics*, 58(10), pp. 4931-4941.

Stienstra, D. and Anderson, T. L. (1988), "Statistical Interferences on Cleverage Fracture Toughness Data," *Journal of Engineering Materials and Technology*, 112(1), pp. 31-37.

US Energy Information Administration (2022), "Solar Explained Photovoltaics and Electricity," Retrieved April 27, 2022, https://www.eia.gov/energyexplained/solar/photovoltaics-and-electricity.php.

Zelen, M. (1959), "Factorial Experiments in Life Testing," *Technometrics*, 1(3), pp. 269-288.

Zhao, W. and Elsayed, E. A. (2005), "A General Acceleration Life Model for Step-Stress Testing," *IIE Transactions*, 37(11), pp. 1059-1069.

Problems

P11.1: On Example 11.3.1, if the damage model is $C(T) = 1 + 2T + 3T^2$, assume that the normal operating temperature is 90°C. Please identify the past hours and failed hours by cumulative exposure model.

P11.2: On Example 11.3.1, if the damage model is $C(T) = 1 + 2T + 3T^2$, assume that the normal operating temperature is 140°C. Please identify the past hours and failed hours by cumulative exposure model.

P11.3: An electronic component was tested to failure at three different temperatures: 133, 143, and 153°C. Six units were tested at each temperature level. The time-to-failure data (hours) are given as follows:

1. 406°K (133°C): 1 (because of test equipment failure), 248, 456, 528, 731, and 813
2. 416°K (143°C): 164, 176, 289, 319, 340, and 543
3. 426°K (153°C): 92, 105, 155, 184, 219, and 235

How many hours can the electronic component survive at the anticipated operating temperature, i.e., 356°K (83°C)? Does the product meet the allowed one-year warranty time, with an expected return of 10% of the population?

P11.4: On the lifetime testing of glass capacitors, assume that the test stand is a block factor. Please formulate the predictive equation using the maximum-likelihood method for the data given in Table 11.4.1, of which the data are obtained from [Zelen 1959].

P11.5: Factorial experiments were conducted to study the effects of the following four factors: (A) carbon content of steel (%), (B) surface preparation (wet- or dry-blasted), (C) temperature (°C), and (D) agitation (yes or no), respectively, on the corrosion rate (mm/year) of blasted steel by seawater, with two replicates for each treatment (Y_1 and Y_2). Data collected from the laboratory tests with 16 treatments (factorial design 2^4) are listed as follows [Do and Roy 1994].

Run	A	B	C	D	Y_1	Y_2
1	0.06	Dry	20	No	0.131	0.120
2	0.43	Dry	20	No	0.095	0.098
3	0.06	Wet	20	No	0.126	0.111
4	0.43	Wet	20	No	0.127	0.092
5	0.06	Dry	32	No	0.255	0.261
6	0.43	Dry	32	No	0.181	0.196
7	0.06	Wet	32	No	0.225	0.231
8	0.43	Wet	32	No	0.170	0.174
9	0.06	Dry	20	Yes	0.217	0.212
10	0.43	Dry	20	Yes	0.225	0.198
11	0.06	Wet	20	Yes	0.137	0.164
12	0.43	Wet	20	Yes	0.200	0.177
13	0.06	Dry	32	Yes	0.378	0.524
14	0.43	Dry	32	Yes	0.290	0.291
15	0.06	Wet	32	Yes	0.283	0.243
16	0.43	Wet	32	Yes	0.197	0.181

Please identify the influential main effects and interactions that have significant influences on the corrosion rate.

P11.6: Corrosion protection of mild steel by Ni-P-W coatings. Electroless Ni-P-W coatings can be used for slowing down the corrosion of mild steel (e.g., SAE 1020 steel). The extent of rusting can be characterized by the following two parameters:

(a) R ($\Omega \cdot cm^2 \times 10^5$): Charge transfer resistance; the larger the better

(b) Y (Faraday $\times 10^{-5}$): Double layer capacitance; the smaller the better

which can be measured using an electrochemical impedance spectroscopy.

Fractional factorial analysis based on 3^{4-1} (D = ABC) was applied to exploring the effects of four design factors [Table P11.6(a)] on the level corrosion [Table P11.6(b)]. Please make a plot of Pareto frontier in terms of R and Y.

TABLE P11.6(a) Process parameters and their design levels applied for investigating rust protection by coating [Roy and Sahoo 2012].

Process factor	(0)	(1)	(2)
A (g/L): Concentration of source of nickel (nickel sulfate)	20	25	30
B (g/L): Concentration of reducing agent (sodium hypophosphite)	14	17	20
C (g/L): Concentration of source of tungsten (sodium tungstate)	15	20	25
D (°): Annealing temperature of SAE1040 steel	400	500	600

Courtsey of Prasanta Sahoo.

TABLE P11.6(b) Design matrix 3^{4-1} (D = ABC) for investigating rust protection by coating [Roy and Sahoo 2012].

Run	a	b	c	d	A	B	C	D	R	Y
1	0	0	0	0	20	14	15	400	4.277	1.128
2	0	0	1	1	20	14	20	500	9.955	0.6007
3	0	0	2	2	20	14	25	600	8.023	0.4938
4	0	1	0	1	20	17	15	500	0.9784	0.7661
5	0	1	1	2	20	17	20	600	8.948	0.8043
6	0	1	2	0	20	17	25	400	8.321	0.7096
7	0	2	0	2	20	20	15	600	5.101	0.5066
8	0	2	1	0	20	20	20	400	10.44	0.6805
9	0	2	2	1	20	20	25	500	11.08	1.593
10	1	0	0	1	25	14	15	500	0.3652	0.5353
11	1	0	1	2	25	14	20	600	5.501	0.437
12	1	0	2	0	25	14	25	400	6.512	0.7335
13	1	1	0	2	25	17	15	600	3.633	0.5992
14	1	1	1	0	25	17	20	400	5.377	0.8081
15	1	1	2	1	25	17	25	500	19.33	0.4801
16	1	2	0	0	25	20	15	400	3.227	0.7853
17	1	2	1	1	25	20	20	500	5.781	1.612
18	1	2	2	2	25	20	25	600	8.129	0.5208
19	2	0	0	2	30	14	15	600	9.891	0.9541
20	2	0	1	0	30	14	20	400	3.597	0.563
21	2	0	2	1	30	14	25	500	2.5405	0.632
22	2	1	0	0	30	17	15	400	2.384	0.615
23	2	1	1	1	30	17	20	500	3.86	0.6616
24	2	1	2	2	30	17	25	600	13.54	0.4931
25	2	2	0	1	30	20	15	500	5.23	0.5265
26	2	2	1	2	30	20	20	600	7.283	0.5368
27	2	2	2	0	30	20	25	400	4.728	0.3827

Bibliography: Books

Akao, Y. (Editor; 1990), *Quality Function Deployment: Integrating Customer Requirements into Product Design*, Productivity Press, Cambridge, MA, 392 pages; ISBN: 9781563273131.

Allen, P. (2020), *Design of Experiments for 21st Century Engineers*, Lulu.com, 214 pages; ISBN: 978-0244584504.

Altshuller, G. (1984), *Creativity as an Exact Science*, Gordon & Breach, New York.

American Supplier Institute (1993), *Quality Function Deployment*, American Supplier Institute, Dearborn, MI.

Anderson, M. J. and Whitcomb, P. J. (2015), *DOE Simplified: Practical Tools for Effective Experimentation*, 3rd Edition, Productivity Press, London, UK, 268 pages; ISBN: 978-0429258022.

Anderson, M. J. and Whitcomb, P. J. (2016), *RSM Simplified: Optimizing Processes Using Response Surface Methods for Design of Experiments*, Productivity Press, New York, 311 pages; ISBN: 978-1315382326.

Anderson, V. L. and McLean, R. A. (1974), *Design of Experiments: A Realistic Approach*, Marcel Dekker, New York.

Antony, J. (2023), *Design of Experiments for Engineers and Scientists*, 3rd Edition, Elsevier, San Diego; ISBN: 978-0-44-315173-6.

Armold, S. F. (1990), *Mathematical Statistics*, Prentice-Hall International, Englewood Cliffs.

Atkinson, A. C., Donev, A. N., and Tobias, R. D. (2007), *Optimum Experimental Designs, with SAS*, Oxford University Press, Oxford, UK; ISBN: 978-0-19-929660-6.

Automotive Industry Action Group (2010), *Measurement System Analysis Manual*, 2nd Edition, AIAG, Southfield, MI.

Bailey, R. A. (2008), *Design of Comparative Experiments*, Cambridge University Press, Cambridge, UK.

Beckwith, T. G. (1990), *Mechanical Measurements*, Addison-Wesley Publishing Company, Reading, MA.

Bertsche, B. (2008), *Reliability in Automotive and Mechanical Engineering*, Springer, Berlin, 511 pages; ISBN: 978-3540681892.

Bhote, K. (1996), *Going beyond Customer Satisfaction to Customer Loyalty*, American Management Association, New York, 140 pages; ISBN: 978-0814423622.

Bhote, K. R. and Bhote, A. K. (1999), *World Class Quality Using Design of Experiments to Make It Happen*, 2nd Edition, American Management Association, New York, 487 pages; ISBN: 978-0814426425.

Bicheno, J. (2004), *New Lean Toolbox: Towards Fast Flexible Flow*, PICSIE Books, Buckingham, UK.

Bjorke, O. (1989), *Computer-Aided Tolerancing*, 2nd Edition, ASME Press, New York.

Blischke, W. R., Karim, M. R., and Prabhakar Murthy, D. N. (2011), *Warranty Data Collection and Analysis*, Springer, London, UK; ISBN: 978-0-85729-646-7.

Boothroyd, G., Dewhurst, P., and Knight, W. (2010), *Product Design for Manufacture and Assembly*, 3rd Edition, CRC Press, Hoboken.

Box, G. E. P. and Draper, N. R. (1998), *Evolutionary Operation: A Statistical Method for Process Improvement*, First Printing Edition, John Wiley & Sons, New York, 237 pages; ISBN: 978-0471255512.

Box, G. E. P. and Draper, N. R. (2007), *Response Surfaces, Mixtures, and Ridge Analyses*, 2nd Edition, John Wiley & Sons, Hoboken, 857 pages; ISBN: 978-0470053577.

Box, G. E. P., Hunter, J. S., and Hunter, W. G. (2005), *Statistics for Experimenters: Design, Innovation, and Discovery*, 2nd Edition, John Wiley & Sons, London, UK, 672 pages; ISBN: 978-0-471-71813-0.

Brook, R. J. and Arnold, G. C. (1985), *Applied Regression Analysis and Experimental Design*, Taylor & Francis Group/CRC Press, Boca Raton, 256 pages; ISBN: 0-8247-7252-0.

Brue, G. and Launsby, R. G. (2003), *Design for Six Sigma*, McGraw-Hill, New York.

Bryson, A. E. and Ho, Y. C. (1975), *Applied Optimal Control*, John Wiley & Sons, New York.

Chiang, Y. J. (2023), *Automotive Engineering Materials—Thermomechanical Properties*, Chongqing University Press, Chongqing, China; ISBN: 978-7-5689-3293-6.

Cochran, W. G. and Cox, G. M. (1957), *Experimental Design*, John Wiley & Sons, New York.

Cogorno, G. R. (2020), *Geometric Dimensioning and Tolerancing for Mechanical Design*, 3rd Edition, McGraw-Hill, New York; ISBN: 978-1260453782.

Coles, S. (2001), *An Introduction to Statistical Modeling of Extreme Values*, Springer-Verlag, London, UK.

Condra, L. (2019), *Reliability Improvement with Design of Experiment*, 2nd Edition, CRC Press, Boca Raton; ISBN: 978-0824705275.

Cooper, W., Lawrence, M., and Joe, Z., (2011), *Handbook on Data Envelopment Analysis*, Springer, New York; ISBN: 978-1-4419-6150-1.

Cox, N. D. (2006), *How to Perform Statistical Tolerance Analysis*, ASQC, Milwaukee, WI; ISBN: 978-0-87389-010-6.

Czitrom, V. and Spagon, P. D. (1997), *Statistical Case Studies for Industrial Process Improvement*, ASA-SIAM Series on Statistics and Applied Probability, SIAM, Philadelphia.

Das, R. N. (2014), *Robust Response Surfaces, Regression, and Positive Data Analyses*, Chapman and Hall/CRC Press, Boca Raton, 336 pages; ISBN: 978-1466506770.

Davenport, W. B. and Root, F. S. (1980), *An Introduction to the Theory of Random Signals and Noise*, McGraw-Hill, New York.

Davies, O. L. (Editor; 1971), *The Design and Analysis of Industrial Experiments*, McMillan Co., New York.

Davim, P. J. (Editor; 2016), *Design of Experiments in Production Engineering*, Springer, Cham; ISBN: 978-3319238371.

Dean, A., Voss, D., and Draguljic, D. (2017), *Design and Analysis of Experiments*, 2nd Edition, Springer, Cham; ISBN: 978-3319522487.

Deming, W. E. (1986), *Out of the Crisis*, Massachusetts Institute of Technology, Cambridge, MA.

Diamond, W. J. (1981), *Fractional Experimental Designs*, Wadsworth, Belmont, CA.

Dodson, B. (2006), *The Weibull Analysis Handbook*, 2nd Edition, American Society for Quality, Quality Press, Milwaukee, WI.

Doebelin, E. O. (1990), *Measurement Systems, Applications and Design*, 4th Edition, McGraw-Hill, New York.

Drake, P. Jr. (1999), *Dimensioning and Tolerancing Handbook*, McGraw-Hill, New York; ISBN: 0-07018131-4.

Eriksson, L. et al. (2008), *Design of Experiments Principles and Applications*, 3rd Edition, Umetrics Academy, Umeå, Sweden; ISBN: 91-973730-0-1.

Fang, K., Li, R., and Sudjianto, A. (2006), *Design and Modeling for Computer Experiments*, Chapman & Hall/CRC Press, London, UK; ISBN: 978-1584885467.

Fienberg, S. E. (1987), *The Analysis of Cross-Classified Categorical Data*, The MIT Press, Cambridge, MA.

Fisher, R. A. (1966), *The Design of Experiments*, 8th Edition, Hafner Publishing Company, New York, 236 pages.

Fleiss, J. L. (1981), *Statistical Methods for Rates and Proportions*, 2nd Edition, John Wiley & Sons, New York.

Foster, T. (2004), *Managing Quality: An Integrative Approach*, 2nd Edition, Prentice Hall, Upper Saddle River, 544 pages; ISBN: 978-0138759643.

Fowlkes, W. Y. and Creveling, C. M. (1995), *Engineering Methods for Robust Product Design: Using Taguchi Methods in Technology and Product Development*, Addison-Wesley Publishing Company, Reading, MA; ISBN: 978-0133007039.

Funkenbusch, P. D. (2004), *Practical Guide to Designed Experiments: A Unified Modular Approach*, CRC Press, Boca Raton; ISBN: 978-0824753887.

Good, P. I. and Hardin, J. W. (2009), *Common Errors in Statistics (And How to Avoid Them)*, 3rd Edition, John Wiley & Sons, Hoboken.

Goos, P. and Jones, B. (2011), *Optimal Design of Experiments: A Case-Study Approach*, Wiley, New York, 459 pages; ISBN: 978-0470744611.

Grant, E. L. and Leavenworth, R. S. (2004), *Statistical Quality Control*, Indian Edition, McGraw-Hill, Noida, Uttar Pradesh; ISBN 13: 9780070435551.

Grunditz, E. A. (2016), "Design and Assessment of Battery Electric Vehicle Powertrain, with Respect to Performance, Energy Consumption and Electric Motor Thermal Capability," PhD thesis, Department of Energy and Environment, Chalmers University of Technology, Göteborg, Sweden.

Gunter, B. and Coleman, D. (2014), *A DOE Handbook: A Simple Approach to Basic Statistical Design of Experiments*, Createspace Independent Publishing Platform, 118 pages; ISBN: 978-1497511903.

Gwet, K. L. (2014), *Handbook of Inter-Rater Reliability*, 4th Edition, Advanced Analytics, LLC, Gaithersburg; ISBN: 978-0970806284.

Harrington, M. (2011), *The Design of Experiments in Neuroscience*, 3rd Edition, SAGE, Thousand Oaks, CA, 261 pages; ISBN: 978-1412974325.

Hicks, C. R. and Turner K. V. (1999), *Fundamental Concepts in the Design of Experiments*, 5th Edition, Oxford University Press, New York, 565 pages; ISBN: 978-0195122732.

Hines, W. W. and Montgomery, D. G. (1990), *Probability and Statistics in Engineering and Management Science*, John Wiley & Sons, New York.

Hinkelmann, K. and Kempthorne, O. (2008), *Design and Analysis of Experiments Set*, Wiley, Hoboken, 1411 pages; ISBN: 978-0-470-38551-7.

Hite, J. A., Schmidt, J. W., and Bennett, G. H. (1975), *Analysis of Queuing Systems*, Elsevier, New York.

Huhn, S. and Drechsler, R. (2021), *Design for Testability, Debug and Reliability: Next Generation Measures Using Formal Techniques*, Springer, Cham; ISBN: 978-3030692087.

Huitema, B. E. (2011), *The Analysis of Covariance and Alternatives: Statistical Methods for Experiments, Quasi-Experiments, and Single-Case Studies*, 2nd Edition, Wiley, Hoboken, 688 pages.

International Bureau of Weights and Measures (2006), *The International System of Units*, 8th Edition; ISBN: 92-822-2213-6.

Joens, B. and Montgomery, D. C. (2019), *Design of Experiments: A Modern Approach*, 1st Edition, Wiley, Hoboken, 272 pages; ISBN: 978-1-119-61119-6.

Juran, J. M. and Gryna, F. M. (1993), *Quality Planning and Analysis from Product Development through Use*, 3rd Edition, McGraw-Hill, New York, p. 256.

Kalpakjian, S. (Editor; 1992), *Manufacturing, Engineering and Technology*, 2nd Edition, Addison Wesley, Reading, MA.

Kapur, K. C. and Lamberson, L. R. (1991), *Reliability in Engineering Design*, Wiley, Hoboken, 608 pages; ISBN: 978-0-471-51191-5.

Keyte, B. and Locher, D. (2004), *The Complete Lean Enterprise: Value Stream Mapping for Administrative and Office Processes*, Productivity Press, New York.

Khuri, A. I. and Cornell, J. A. (2019), *Response Surfaces Design and Analyses*, Taylor & Francis Group, Boca Raton, 536 pages; ISBN: 978-0367401252.

King, J. P. and Jewett, W. S. (2010), *Robustness Development and Reliability Growth: Value-Adding Strategies for New Products and Processes*, Prentice Hall, Upper Saddle River; ISBN: 978-0-13-222551-9.

Kuehl, R. O. (2000), *Design of Experiments: Statistical Principles of Research Design and Analysis*, 2nd Edition, Duxbury Press, North Scituate, MA; ISBN: 978-0534368340.

Kverneland, K. O. (1996), *Metric Standards for Worldwide Manufacturing Engineering*, ASME, New York.

Lawless, J. F. (2003), *Statistical Models and Methods for Lifetime Data*, 3rd Edition, John Wiley & Sons, New York.

Lawson, J. (2014), *Design and Analysis of Experiments with R*, Chapman & Hall/CRC Press, Boca Raton; ISBN: 978-1439868133.

Liker, J. Z. (2000), *The Toyota Way: 14 Management Principles from the World's Greatest Manufacturer*, 2nd Edition, McGraw-Hill, New York, 448 pages; ISBN: 0071392319.

Liston, C. and Sheth, N. J. (1973), *Statistical Design and Analysis of Engineering Experiments*, McGraw-Hill, New York.

Long, J. S. (1997), *Regression Models for Categorical and Limited Dependent Variables*, SAGE Publications, Thousand Oaks, CA.

Lorenzen, T. J. and Anderson, V. L. (1993), *Design of Experiments: A No-Name Approach*, Marcel Dekker, Inc., New York.

Mabie, H. H. and Reinholtz, C. F. (1987), *Mechanisms and Dynamics of Machinery*, 4th Edition, John Wiley & Sons, New York.

Mantz, H. F. and Waller, R. A. (1977), *Bayesian Reliability Analysis*, John Wiley & Sons, New York.

Mascitelli, R. (2011), *Mastering Lean Product Development: A Practical, Event-Driven Process for Maximizing Speed*, Profits and Quality, Northridge, CA; ISBN: 978-0966269741.

Mason, R. L., Gunst, R. F., and Hess, J. L. (2003), *Statistical Design & Analysis of Experiments with Applications to Engineering and Science*, 2nd Edition, John Wiley & Sons, Inc., New York, 760 pages; ISBN: 978-0-471-37216-5.

Mathews, P. G. (2004), *Design of Experiments with MINITAB: Homework Problems*, ASQ Quality Press, Milwaukee, WI.

Meeker, W. Q., Escobar, L. A., and Pascual, F. G. (1998), *Statistical Methods for Reliability Data*, 2nd Edition, John Wiley & Sons, New York, 704 pages; ISBN: 978-1-118-11545-9.

Military and Government Specs & Standards (Naval Publications and Form Center) (NPFC) (1980), "Reliability Program for Systems and Equipment Development and Production," MIL-STD-785, DOD, Washington, DC.

Milliken, G. A. and Johnson, D. E. (1984), *Analysis of Messy Data*, Vol. 1: Designed Experiments, Van Nostrand Reinhold, New York.

Modarres, M., Amiri, M., and Jackson, C. (2017), *Probabilistic Physics of Failure Approach to Reliability: Modeling, Accelerated Testing, Prognosis and Reliability Assessment*, Wiley-Scrivener, Hoboken, 288 pages; ISBN: 978-1119388630.

Moen, R. et al. (2012), *Quality Improvement through Planned Experimentation*, McGraw-Hill, New York; ISBN: 978-0071759663.

Montgomery, D., Rigdon, S., Pan, R., and Freeman, L. (2022), *Design of Experiments for Reliability Achievement*, Wiley, Hoboken; ISBN: 978-11-19237693.

Montgomery, D. C. (2019), *Design and Analysis of Experiments*, 10th Edition, John Wiley & Sons, Inc., Hoboken, 688 pages; ISBN: 978-1-119-49244-3.

Morris, M. D. (2017), *Design of Experiments: An Introduction Based on Linear Models*, CRC Press, Boca Raton; ISBN: 978-1138628021.

Murthy, D. P., Xie, M., and Jiang, R. (2004), *Weibull Models*, John Wiley & Sons, Inc., Hoboken.

NASA (2010), NASA System Safety Handbook, Volume 1, System Safety Framework and Concepts for Implementation, Version 1, SP-2010-580, Washington, DC.

Nelson, B. L. (2013), *Foundations and Methods of Stochastic Simulation*, Springer, Boston; ISBN: 978-1-4614-6160-9.

Nelson, W. B. (2004), *Accelerated Testing-Statistical Models, Test Plans, and Data Analysis*, John Wiley & Sons, New York; ISBN: 978-0471697367.

Nevins, J. L. and Whitney, D. E. (1990), *Concurrent Design of Products and Processes*, McGraw-Hill, New York.

NIST (2003), *Engineering Statistics Handbook*, National Institute of Standards and Technology, Washington, DC.

NIST/SEMATECH (2012), *e-Handbook of Statistical Methods*, Springer, Berlin, http://www.itl.nist.gov/div898/handbook.

O'Connor, P. and Kleyner, A. (2011), *Practical Reliability Engineering*, John Wiley & Sons, New York.

Oehlert, G. W. (2010), *A First Course in Design and Analysis of Experiments*, University of Minnesota Digital Conservancy, Minneapolis, MN, 683 pages; ISBN: 9780716735106, Retrieved from the University of Minnesota Digital Conservancy, https://hdl.handle.net/11299/168002.

Ohring, M. and Kasprzak, L. (2014), *Reliability and Failure of Electronic Materials and Devices*, 2nd Edition, Academic Press, Cambridge, MA, 734 pages; ISBN: 978-0120885749.

Onyiah, L. C. (2008), *Design and Analysis of Experiments: Classical and Regression Approaches with SAS*, Chapman and Hall/CRC Press, London, UK; ISBN (eBook): 9780429140273.

Peace, G. S. (1993), *Taguchi Methods: A Hands-On Approach*, Addison-Wesley Publishing Company, Reading, MA.

Petroski, H. (1996), *Invention by Design How Engineers Get from Thought to Thing*, Harvard University Press, Cambridge, MA.

Petrov, V. (2019), *Laws of System Evolution*, TRIZ, 57 pages; ISBN: 978-1696068833.

Phadke, M. S. (1989), *Quality Engineering Using Robust Design*, Prentice Hall, Piscataway, NJ.

Proust, M. (2010), *Design of Experiments Guide, JMP*, A Business Unit of SAS, Cary, NC.

Pyzdek, T. and Keller, P. (2019), *The Six Sigma Handbook*, 5th Edition, McGraw-Hill, New York, 720 pages; ISBN: 9781260121827.

Rhinehart, R. R. (2016), *Nonlinear Regression Modeling for Engineering Applications: Modeling, Model Validation, and Enabling Design of Experiments*, Wiley, Chichester, UK, 400 pages; ISBN: 978-1-118-59796-5.

Rigdon, S. E., Pan, R., Montgomery, D. C., and Borror, C. M. (2022), *Experiments for Reliability Achievement*, Wiley, Boca Raton, 416 pages; ISBN: 978-1119237693.

Rodrigues, M. and Iemma, A. (2014), *Experimental Design and Process Optimization*, CRC Press, Boca Raton, 336 pages; ISBN: 978-0429161865.

Ross, P. J. (1988), *Taguchi's Techniques for Quality Engineering*, McGraw-Hill, New York.

Ross, S. M. (1987), *Introduction to Probability and Statistics for Engineers and Scientists*, John Wiley & Sons, New York.

Rössler, A. (2014), *Design of Experiment for Coatings*, Vincentz Network, Hanover, Germany; ISBN: 978-3-86630-885-5.

Rother, M. (2009), *Toyota Kata*, McGraw-Hill, New York.

Rother, M. and Shook, J. (2009), *Learning to See*, Lean Enterprise Institute, Boston.

Roy, R. K. (2001), *Design of Experiments Using the Taguchi Approach: 16 Steps to Product and Process Improvement*, Wiley, New York, 560 pages; ISBN: 978-0-471-36101-5.

Russell, K. G. (2021), *Design of Experiments for Generalized Linear Models*, Chapman and Hall/CRC Press, Boca Raton, 240 pages; ISBN: 978-1032094052.

Sahai, H. and Ageel, M. I. (2000), *The Analysis of Variance*, Birkhauser, Boston.

Samuels, M. L., Witmer, J. A., and Schaffner, A. A. (2016), *Statistics for the Life Sciences*, 5th Edition, Pearson Education, Harlow, UK; ISBN: 978-1-292-10181-1.

Santner, T. J., Williamns, B. J., and Notz, W. I. (2018), *The Design and Analysis of Computer Experiments*, 2nd Edition, Springer, New York; ISBN: 978-1493988457.

Schleich, B. (2017), *Skin Model Shapes: A New Paradigm for the Tolerance Analysis and the Geometrical Variations Modelling in Mechanical Engineering*, VDI Verlag, Düsseldorf.

Schmidt, S. R. and Launsby, R. G. (1994), *Understanding Industrial Designed Experiments*, 4th Edition, Air Academy Press, Colorado Springs, CO, 768 pages.

Shigley, J. E. and Mischke, C. R. (1989), *Fundamentals of Machine Component Design*, 5th Edition, McGraw-Hill, New York; ISBN: 978-1118012895.

Shingo, S. (1985), *A Revolution in Manufacturing: The SMED System*, Productivity Press, Stamford, CT; ISBN: 978-0915299034.

Silva, V. (2018), *Statistical Approaches with Emphasis on Design of Experiments Applied to Chemical Processes*, Intechopen, London, UK, 180 pages; ISBN: 978-953-51-3878-5.

Taguchi, G. (1987), *System of Experimental Design*, UNIPUB, Kraus International Publications, New York.

Taguchi, G., Chowdhur, S., and Wu, Y. (2005), *Taguchi's Quality Engineering Handbook*, John Wiley & Sons, Inc., Hoboken.

Taguchi, G., Chowdhury, S., and Taguchi, S. (2000), *Robust Engineering*, McGraw Hill, New York.

Taguchi, G., Elsayed, E, and Hsiang, T. (1989), *Quality Engineering in Production Systems*, McGraw-Hill Book Company, New York.

Taguchi, G. and Konishi, S. (1987), *Taguchi Methods, Orthogonal Arrays and Linear Graphs: Tools for Quality Engineering*, American Supplier Institute, Dearborn, MI.

Tamhane, A. C. (2009), *Statistical Analysis of Designed Experiments Theory and Application*, Wiley, New York, 720 pages; ISBN: 978-0-471-75043-7.

Taylor, W. (1991), *Optimization and Variation Reduction in Quality*, McGraw Hill, New York; ISBN: 0-07-063255-3.

Thomke, S. H. (2003), *Experimentation Matters: Unlocking the Potential of New Technologies for Innovation*, Harvard Business School Press, Boston.

Toutenburg, H. and Shalabh (2009), *Statistical Analysis of Designed Experiments*, 3rd Edition, Springer, New York, 633 pages; ISBN: 978-1441911476.

Vuchkov, I. N. and Boyadjieva, N. L. (2001), *Quality Improvement with Design of Experiments: A Response Surface Approach*, Springer, Dordrecht, 508 pages; ISBN: 978-1402003929.

Watson, G. H. (1993), *Strategic Benchmarking: How to Rate Your Company's Performance against the World's Best*, ASQC Quality Press, Milwaukee, WI.

Weber, D. C. and Skillings, J. H. (1999), *A First Course in the Design of Experiments: A Linear Models Approach*, CRC Press, New York.

Winer, B. J. (1971), *Statistical Principles in Experimental Design*, 2nd Edition, McGraw-Hill, New York.

Wu, C. F. J. and Hamada, M. S. (2009), *Experiments: Planning, Analysis and Parameter Design Optimization*, 2nd Edition, Wiley, New York; ISBN: 978-0471699460.

Yates, F. (1937), *The Design and Analysis of Factorial Experiments*, Imperial Bureau of Social Science, Harpenden, England.

Index

About the Authors

Young J. Chiang, PhD The author earned his Bachelor of Science (BS) degree from the Department of Power Mechanical Engineering, Tsing Hua University, Taiwan, in 1976. When studying for his PhD degree (Department of Engineering Mechanics, University of Wisconsin, Madison), the author was honored to take two graduate courses from Professor George Box in the early 1980s as part of his minor program (Department of Statistics, University of Wisconsin, Madison). He obtained his PhD degree in 1983. Ever since, Dr. Chiang has had opportunities to apply the design of experiments in combination with finite element methods (computer simulation) and experimental mechanics (experimental life tests) to improve product performance and solve complex reliability problems in the automotive industry. He had 30 years of engineering and management experience with product research and development (R&D), computer-aided engineering (CAE), design, reliability, quality, manufacturing, and program management in the US automotive industry, including Uniroyal (Michelin), Cummins, Magna, Edscha, Danaher, CTS, Coda, and Chrysler. Dr. Chiang was a member of American Society of Mechanical Engineers (ASME) and Society of Automotive Engineers (SAE). After retiring from the US automotive industry, he worked as the vice director of Weichai Power Research Center (Shandong, China) and founded the Weichai Institute of Reliability Research (>100 research engineers). He was also named the vice director of the China Key State Laboratory of Engine Reliability and the executive commissioner on the Reliability Committee of the Chinese Mechanical Engineering Society.

Besides technical papers and reports, Dr. Chiang has put forth a great effort into a nonprofit business engaged in the development of potential automotive engineering textbooks for graduate students and practitioners in the automotive industry, ever since the early 1990s. He has published two books on how to improve product performance and grow product reliability based on the physics of failure and material constituents: *Mechanics and Design for Product Life Prediction* (ISBN 978-7-5689-1917-6) and *Automotive Engineering Materials-Thermomechanical Properties* (ISBN 978-7-5689-3293-6) with Chongqing University Press (Chongqing, China). This publication (*Product Design and Testing for Automotive Engineering*) is the second volume of the book series on *Design of Experiments for Product Reliability Growth* brought forth by the author. The corresponding manuscript was even used as a textbook-to-be by the author, in the capacity of full professor, for teaching graduate students in the Department of Automotive Engineering, Chongqing University.

Amy L. Chiang, MS After the author obtained her BS degree in electrical engineering from the University of Michigan, Ann Arbor, in 2012, she worked for Viasat Inc. (Carlsbad, CA), with much of her focus on product development and testing. While working for the company, she completed the Master of Science (MS) degree of "Masters of Advanced Study Degree (MAS) in Wireless Embedded Systems," offered by the Departments of Electrical and Computer Engineering and Computer Science and Engineering at the University of California, San Diego.

The author is an active participant in businesses engaged in the development of new technologies and multidisciplined products. With insights into viable paradigms and advancements in practice, she assists product teams in developing wireless embedded software, program strategy, and methodology of statistical thinking.

www.ingramcontent.com/pod-product-compliance
Lightning Source LLC
LaVergne TN
LVHW071342210926
848486LV00003B/209
9781468607703